PRION DISEASES OF HUMANS AND ANIMALS

PRION DISEASES OF HUMANS AND ANIMALS

Editors:
STANLEY PRUSINER
Departments of Neurobiology, Biochemistry and Biophysics,
University of California
JOHN COLLINGE
Department of Biochemistry and Molecular Genetics,
St Mary's Hospital Medical School, London
JOHN POWELL and BRIAN ANDERTON
Institute of Psychiatry,
De Crespigny Park, London

ELLIS HORWOOD
NEW YORK   LONDON   TORONTO   SYDNEY   TOKYO   SINGAPORE

First published in 1992 by
ELLIS HORWOOD LIMITED
Market Cross House, Cooper Street,
Chichester, West Sussex, PO19 1EB, England

A division of
Simon & Schuster International Group
A Paramount Communications Company

Printed and bound in Great Britain
by Bookcraft, Midsomer Norton

British Library Cataloguing in Publication Data

A Catalogue Record for this book is available from the British Library

ISBN 0–13–720327–6

Library of Congress Cataloging-in-Publication Data

Available from the publisher

At every crossway on the road that leads to
the future, tradition has placed against each
of us ten thousand men to guard the past

Maurice Maeterlinck, 1907

when you have eliminated the impossible
whatever remains, however improbable, must
be the truth

Arthur Conan Doyle, 1929

List of contributors

Judd M. Aiken Department of Veterinary Science, University of Wisconsin-Madison, Madison, WI 53706, USA.

Tikvah Alper Birkholt, Sarisbury Green, Hampshire SO3 6AL, UK.

Michael Alpers Papua New Guinea Institute of Medical Research, PO Box 60, Goroka, EHP, Papua New Guinea.

Brian Anderton Department of Neuroscience, Institute of Psychiatry, De Crespigny Park, Denmark Hill, London SE5 8AF, UK.

Lucila Autilio-Gambetti Division of Neuropathology, Institute of Pathology, Case Western Reserve University, Cleveland, OH 44106, USA.

Pastrizia Avoni Neurological Institute, University of Bologna Medical School, Bologna, Italy.

H.F. Baker Division of Psychiatry, Clinical Research Centre, Watford Road, Harrow, Middlesex HA1 3UJ, UK.

Michael A. Baldwin Department of Neurology, University of California, San Francisco, CA 94143, USA.

Ron Beavis Rockefeller University, New York, NY 10021, USA.

Elizabeth Beck 34 Downs Road, Epsom, Surrey, KT18 5LD, UK.

C. Bennett Division of Medical Genetics, Guy's Hospital, St Thomas Street, London SE1 9RT, UK.

J.W. Boellaard Institute für Hirnforschung der Universität, Tübingen, Germany.

David R. Borchelt Department of Neurology, University of California, San Francisco, CA 94143, USA.

R. Bradley Central Veterinary Laboratory, New Haw, Weybridge, Surrey KT15 3NB, UK.

G. Brown Department of Biochemistry and Molecular Medicine, St Mary's Hospital, London W2 1NY, UK.

Paul Brown Laboratory of Central Nervous System Studies, NINDS, National Institutes of Health, Bethesda, MD 20892, USA.

M.E. Bruce AFRC and MRC Neuropathogenesis Unit, Institute for Animal Health, Ogston Building, West Mains Road, Edinburgh EH9 3JF, UK.

J. Brugere-Picoux Ecole Nationale Vétérinaire, Maisons-Alfort, France.

O. Bugiani Instituto Neurologico, Carlo Besta, Via Celona 11, 20133 Milano, Italy

Alma L. Burlingame Department of Pharmaceutical Chemistry, University of California, San Francisco, CA 94143, USA.

George A. Carlson McLaughlin Research Institute, 1625 Third Avenue North, Great Falls, MT 59401, USA.

B. Caughey NIH–NIAID, Laboratory of Persistent Viral Diseases, Rocky Mountain Laboratories, Hamilton, MT 59840, USA.

Brian Chait Rockefeller University, New York, NY 10021, USA.

J. Chatelain FRA C. Bernard 'Neurochimie des Communications Cellulaires' and Service de Biochimie, Hôpital Saint-Louis, 1 Avéneu C. Vellefaux, 75010 Paris, France.

B. Chesebro NIH–NIAID, Laboratory of Persistent Viral Diseases, Rocky Mountain Laboratories, Hamilton, MT 59840, USA.

Joanne Clinton Serious Mental Afflications Research Team, Department of Anatomy and Cell Biology, St Mary's Hospital Medical School, Imperial College, Norfolk Place, London W2 1PG, UK.

J. Collinge Department of Biochemistry and Molecular Medicine, St Mary's Hospital Medical School, Norfolk Place, London W2 1PG, UK.

M.P. Conneally Department of Medical & Molecular Genetics, Indiana University Medical Center, Indianapolis, IN 46202, USA.

Pietro Cortelli Neurological Institute, University of Bologna Medical School, Bologna, Italy.

T.J. Crow Division of Psychiatry, Clinical Research Centre, Watford Road, Harrow, Middlesex HA1 3UJ, UK.

P.M. Daniel Department of Applied Physiology, The Royal College of Surgeons of England, 35–43 Lincoln's Inn Fields, London WC2A 3PN England.

D. Davies AFRC and MRC Neuropathogenesis Unit, Institute for Animal Health, Ogston Building, West Mains Road, Edinburgh EH9 3JF, UK.

Stephen J. DeArmond Departments of Pathology and of Neurology, University of California, San Francisco, CA 94143, USA.

Linda Detwiler United States Department of Agriculture, Hyattsville, MD 10782, USA.

A.G. Dickinson Institute for Animal Health, AFRC and MRC Neuropathogenesis Unit, Ogston Building, West Mains Road, Edinburgh EH9 3JF, UK.

S.R. Dlouhy Department of Medical & Molecular Genetics, Indiana University Medical Center, Indianapolis IN 46202, USA.

K. Doh-ura Department of Neuropathology, Neurological Institute, Faculty of Medicine, Kyushu University, Fukuoka, Japan.

R. Doshi Department of Pathology, Brook General Hospital, Shooters Hill Road, London SE18 4LW, UK.

M. Dussaucy   FRA C. Bernard 'Neurochimie des Communications Cellulaires' and Service de Biochimie, Hôpital Saint-Louis, 1 Avénue C. Vellefaux, 75010 Paris, France.

T.F.G. Esmonde   CJD Surveillance Unit, Department of Clinical Neurosciences, Western General Hospital, Edinburgh, UK.

M.R. Farlow   Department of Neurology, Indiana University Medical Center, Indianapolis IN 46202, USA.

C.F. Farquhar   AFRC and MRC Neuropathogenesis Unit, Institute for Animal Health, Ogston Building, West Mains Road, Edinburgh EH9 3JF, UK.

Dallas Foster   Department of Neurology, University of California, San Francisco, CA 94143, USA.

B. Frangione   Department of Pathology, New York University Medical Center, New York, NY 10016, USA.

H. Fraser   AFRC and MRC Neuropathogenesis Unit, Institute for Animal Health, Ogston Building, West Main Road, Edinburgh EH9 3JF, UK.

C.D. Frith   Division of Psychiatry, Clinical Research Centre, Watford Road, Harrow, Middlesex HA1 3UJ, UK.

Ruth Gabizon   Department of Neurology, Hadassah University Hospital, Jerusalem, Israel.

Jean-Marc Gabriel   Department of Neurology, University of California, San Francisco, CA 94143, USA.

D. Carleton Gajdusek   Laboratory of Central Nervous System Studies, NINDS, National Institutes of Health, Bethesda, MD 20892, USA.

Pierluigi Gambetti   Division of Neuropathology, Institute of Pathology, Case Western Reserve University, Cleveland, OH 44106, USA.

B. Ghetti   Department of Pathology, Indiana University School of Medicine, 635 Barnhill Drive, Indianapolis, IN 46202-5120, USA.

G. Giaccone   Instituto Neurologico, Carlo Besta, Via Celona 11, 20133 Milano, Italy

Clarence J. Gibbs, Jr   Laboratory of Central Nervous System Studies, National Institute of Neurological Disorders and Stroke, National Institutes of Health, Bethesda, MD 20892, USA.

Bradford W. Gibson   Department of Pharmaceutical Chemistry, University of California, San Francisco, CA 94143, USA.

Robert Glasse   Department of Anthropology, Graduate Center, City University of New York, 35 West 42nd Street, New York, NY 10026, USA.

Lev G. Goldfarb   Laboratory of Central Nervous System Studies, NINDS, National Institutes of Health, Bethesda, MD 20892, USA.

Wilfred Goldmann   AFRC and MRC Neuropathogenesis Unit, Institute for Animal Health, Ogston Building, West Mains Road, Edinburgh EH9 3JF, UK.

D. Groth   Department of Neurology, University of California, San Francisco, CA 94143, USA.

Tibor Gyuris   Department of Neurology, University of California, San Francisco, CA 94143, USA.

W.J. Hadlow   908 South Third Street, Hamilton, MT 59840-2924, USA.

A.E. Harding   The National Hospital, Queen Square, London WC1N 3BG, UK.

J. Hardy   Department of Biochemistry and Molecular Medicine, St Mary's Hospital Medical School, London W2 IPG, UK.

Stephen A.C. Hawkins   Department of Pathology, Central Veterinary Laboratory, New Haw, Weybridge, Surrey KT15 3NB, UK.

Rolf Hecker   Department of Neurology, University of California, San Francisco, CA 94143, USA.

M.E. Hodes   Dept of Medical & Molecular Genetics, Indiana University Medical Center, Indianopolis, IN 46202, USA.

Leroy Hood   California Institute of Technology, Pasadena, CA 91125, USA.

K.K. Hsiao   Department of Neurology, University of California, San Francisco, CA 94143, USA.

Gordon Hunter   21 Cleveland Grove, Newbury, Berks, UK.

Nora Hunter   AFRC and MRC Neuropathogenesis Unit, Institute for Animal Health, Ogston Building, West Mains Road, Edinburgh EH9 3JF, UK.

Klaus Jendroska   Department of Neurology, Universitäts Klinikum Rudolf Virchow, Berlin, Germany.

Esther Kahana   Neurological Unit, Barzilai Medical Centre, Ashkelon, Israel.

Klaus Kellings   Institut für Physikalische Biologie, Heinrich-Heine-Universität Düsseldorf, Universitätsstrasse 1, D-400 Düsseldorf, Germany.

T. Kitamoto   Department of Neuropathology, Neurological Institute, Faculty of Medicine, Kyushu University, Fukuoka, Japan.

Hans Kretzschmar   Neuropathology Institute, University of Munich, Munich, Germany.

J.L. Laplanche   FRA C. Bernard 'Neurochimie des Communications Cellulaires' and Service de Biochimie, Hôpital Saint-Louis, 1 Avénue C. Vellefaux, 75010 Paris, France.

J.M. Launay   FRA C. Bernard 'Neurochimie des Communications Cellulaires' and Service de Biochimie, Hôpital Saint-Louis, 1 Avénue C. Vellefaux, 75010 Paris, France.

M. Leach   Division of Psychiatry, Clinical Research Centre, Watford Road, Harrow, Middlesex HA1 3UJ, UK.

Andréa C. LeBlanc   Division of Neuropathology, Institute of Pathology, Case Western Reserve University, Cleveland, OH 44106, USA.

Shirley Lindenbaum   Department of Anthropology, Graduate Center, City University of New York, 35 West 42nd Street, New York, NY 10026.

R. Lofthouse   Division of Psychiatry, Clinical Research Centre, Watford Road, Harrow, Middlesex HA1 3UJ, UK.

R.C. Lowson   Animal Health Division, Ministry of Agriculture, Fisheries and Food, Hook Rise South, Tolworth, Surbiton KT6 7NF, UK.

Elio Lugaresi   Neurological Institute, University of Bologna Medical School, Bologna, Italy.

Valeria Manetto   Division of Neuropathology, Institute of Pathology, Case Western Reserve University, Cleveland, OH 44106, USA.

R.F. Marsh   Department of Veterinary Science, University of Wisconsin-Madison, Madison, WI 53706, USA.

W.B. Matthews University Department of Neurology, Radcliffe Infirmary, Oxford, UK.

P.A. McBride AFRC and MRC Neuropathogenesis Unit, Institute for Animal Health, Ogston Building, West Mains Road, Edinburgh EH9 3JF, UK.

Michael P. McKinley Department of Neurology, University of California, San Francisco, CA 94143, USA.

Rossella Medori Neurological Institute, University of Bologna Medical School, Bologna, Italy.

Zeev Meiner Department of Neurology, Hadassah University Hospital, Jerusalem, Israel.

Norbert Meyer Institut für Physikalische Biologie, Heinrich-Heine-Universität Düsseldorf, Universitätsstrasse 1, D-400 Düsseldorf, Germany.

Carol Mirenda Department of Neurology, University of California, San Francisco, CA 94143, USA.

Mirella Mochi Neurological Institute, University of Bologna Medical School, Bologna, Italy.

Pasquale Montagna Neurological Institute, University of Bologna Medical School, Bologna, Italy.

M.J. Mullan Department of Biochemistry and Molecular Medicine, St Mary's Hospital Medical School, London W2 1PG, UK.

Sara Neuman Department of Neurology, University of California, San Francisco, CA 94143, USA.

Bruno Oesch Brain Research Institute, University of Zurich, 8029 Zurich, Switzerland.

F. Owen Department of Physiological Sciences, Stopford Building, Manchester University, Oxford Road, Manchester M13 9PT, UK.

Mark S. Palmer Department of Biochemistry and Molecular Genetics, St Mary's Hospital Medical School, London W2 1PG, UK.

Keh-Ming Pan Department of Neurology, University of California, San Francisco, CA 94143, USA.

I.H.Pattison Deceased.

M. Poulter Division of Psychiatry, Clinical Research Centre, Watford Road, Harrow, Middlesex HA1 3UJ, UK.

John Powell Department of Neuroscience, Institute of Psychiatry, De Crespigny Park, Denmark Hill, London SE5 8AF, UK.

Stanley B. Prusiner Departments of Neurology and of Biochemistry and Biophysics, University of California, San Francisco CA 94143-0518, USA.

R. Race NIH–NIAID, Laboratory of Persistent Viral Diseases, Rocky Mountain Laboratories, Hamilton, MT 59840, USA.

Alex Raeber Department of Neurology, University of California, San Francisco, CA 94143, USA.

D Rapp Department of Neurology, University of California, San Francisco, CA 94143, USA.

R.M. Ridley Division of Psychiatry, Clinical Research Centre, Watford Road, Harrow, Middlesex HA1 3UJ, UK.

Detlev Riesner Institut für Physikalische Biologie, Heinrich-Heine-Universität Düsseldorf, Universitätsstrasse 1, D-400 Düsseldorf, Germany.

Gareth W. Roberts Serious Mental Afflictions Research Team, Department of Anatomy and Cell Biology, St Mary's Hospital Medical School, Imperial College, Norfolk Place, London W2 1PG, UK.

Mark Rogers Department of Neurology, University of California, San Francisco, CA 94143, USA.

J.R. Scott Institute for Animal Health, AFRC and MRC Neuropathogenesis Unit, Ogston Building, West Mains Road, Edinburgh EH9 3JF, UK.

Michael Scott Department of Neurology, University of California, San Francisco, CA 94143, USA.

A. Serban Department of Neurology, University of California, San Francisco, CA 94143, USA.

Dan Serban Department of Neurology, University of California, San Francisco, CA, 94143, USA.

T. Shah Division of Psychiatry, Clinical Research Centre, Watford Road, Harrow, Middlesex HA1 3UJ, UK.

Yvonne I. Spencer Department of Pathology, Central Veterinary Laboratory, New Haw, Weybridge, Surrey KT15 3NB, UK.

Neil Stahl Department of Neurology, University of California, San Francisco, CA 94143, USA.

G. Steinmetz Clinique Neurologique, Hôpital Civil, Strasbourg, France.

Linda Stowring Department of Pathology (Neuropathology Unit), University of California, San Francisco, CA 94143, USA.

F. Tagliavini Istituto Neurologico C. Besta, Divisione di Neuropatologia, Via Celonia II, 20133 Milano, Italy.

Albert Taraboulos Departments of Neurology and of Biochemistry and Biophysics, University of California, San Francisco, CA 94143, USA.

J. Tateishi Department of Neuropathology, Neurological Institute, Faculty of Medicine, Kyishu University, Fukuoka, Japan.

David Teplow Calfornia Institute of Technology, Pasadena, CA 91125, USA.

S. Thomas FRA C. Bernard 'Neurochimie des Communications Cellulaires' and Service de Biochimie, Hôpital Saint-Louis, 1 Avénue C. Vellefaux, 75010 Paris, France.

M. Torchia Department of Neurology, University of California, San Francisco, CA 94143, USA.

C. Tranchant Clinique Neurologique, Hôpital Civil, Strasbourg, France.

Hans-Juergen Tritschler Division of Neuropathology, Institute of Pathology, Case Western Reserve University, Cleveland, OH 44106, USA.

Federico Villare Division of Neuropathology, Institute of Pathology, Case Western Reserve University, Cleveland, OH 44106, USA.

J.M. Warter Clinique Neurologique, Hôpital Civil, Strasbourg, France.

Charles Weissmann Institut für Molekularbiologie 1, Universität Zürich, 8093 Zürich, Switzerland.

Gerald A.H. Wells   Department of Pathology, Central Veterinary Laboratory, New Haw, Weybridge, Surrey KT15 3NB, UK.

David Westaway   Department of Neurology, University of California, San Francisco, CA 94143, USA.

J.W. Wilesmith   Epidemiology Department, Central Veterinary Laboratory, New Haw, Weybridge, Surrey KT15 3NB, UK.

R.G. Will   CJD Surveillance Unit, Department of Clinical Neurosciences, Western General Hospital, Edinburgh, UK.

Shu-Lian Yang   Department of Neurology, University of California, San Francisco, CA 94143, USA.

R.D. Yee   Dept. of Neurology, Indiana University Medical Center, Indianapolis, IN 46202, USA.

Vincent Zuliani   Department of Neurology, University of California, San Francisco, CA 94143, USA.

Some of the speakers and participants at the international conference 'Prion Diseases in Humans and Animals', held at the Royal Institute of British Architects, London, 2–4th September 1991.

(a)   (b)

(a) Jim Hope. (b) Richard Kimberlin.

(a) Moira Bruce. (b) More than 250 participants. (c) Neil Stahl. (d) Hugh Frazer. (e) Tikvar Alper. (f) Dmitry Goldgaber, Karen Hsiao, Mark Palmer. (g) Jeff Almond.

(a) Charles Weissmann. (b) Dave Westaway. (c) Anita Harding. (d) Richard Kimberlin and Charles Weissmann. (e) Nora Hunter. (f) Pierluigi Gambetti and Bernadino Ghetti. (g) Dmitry Goldgaber and Richard Kascsak.

(a) John Collinge. (b) Jun Tateishi and Hans Kretzschmar. (c) Detlev Reisner. (d) The organizers (Brian Anderton, John Collinge, John Powell and Stan Prusiner). (e) Ray Bradley. (f) Dick Marsh. (g) Delegates in informal discussion.

(a) Ray Bradley. (b) Tim Crow and Harry Baker. (c) Gareth Roberts, Brian Anderton and Stan Prusiner. (d) Alison Goate, Karen Hsiao and Gareth Roberts. (e) Hank Baron. (f) Gerald Wells. (g) Bob Will.

Table of contents

Preface

By any measure, a book on prions causing neurodegenerative diseases in humans and animals is timely. The study of prions has advanced so rapidly over the past few years that gathering the contributions made by a large number of investigators seems a most worthy goal. We hope that this book will serve to stimulate further research activities in the study of prion diseases.

While much more remains to be learned about the prion diseases, many advances in our knowledge of this area of biology and medicine seem to be unprecedented. We hope that the information summarized in this book will serve to provoke some young investigators to enter the field and to become involved in a very exciting and new area of biomedical research. While the prion diseases are relatively rare, there are quite common neurodegenerative disorders which may have important parallels. Certainly Alzheimer's disease, Parkinson's disease and amyotrophic lateral sclerosis (ALS), all of which are neurodegenerative diseases, present important challenges to physicians and biomedical investigators. It is significant that neurodegenerative diseases increase with age, and as the longevity of populations of developed nations increases, problems of neurodegenerative disease are becoming of greater and greater magnitude. Equally important are the prion diseases of animals. The epidemic of bovine spongiform encephalopathy in Great Britain has raised enormous concern, and clearly our knowledge of the prion diseases is inadequate to answer all of the questions which have been posed.

Many of the manuscripts included in this book are based on presentations made at a meeting in London at the Royal Institute of British Architects on September 2–4, 1991, entitled 'Prion Diseases in Humans and Animals' organized by Brian Anderton, John Collinge, John Powell and Stanley Prusiner.

The aim of the editors was to organize a meeting which would bring together all of the leading workers from many disciplines involved in prion research for a comprehensive discussion of the state of the field. The work presented at this meeting was then to form the nucleus of a book which gathered the historical reviews and classic papers marking milestones in the field, as well as the most up to date coverage of current research. While many questions remain in the study of prion diseases, many have been answered and a rather clear picture of the nature of the prion particle is beginning to emerge. It is for the reader to judge how far the field has progressed and where it is likely to move next. It is our hope that this collection of papers will provide a convenient point of departure for anyone wishing to learn much more about this fascinating and provocative area of biomedical research.

We are deeply indebted to many sources of support, the largest of which were provided by the John Douglas French Foundation for Alzheimer's Disease Research in Los Angeles, California and The Commission of the European Communities. All of these sources are listed in an acknowledgement. Many people contributed to the organization of the meeting, and to all of them we wish to express our most sincere thanks.

December 7, 1991

Stanley B. Prusiner
John Collinge
John Powell
Brian Anderton

Acknowledgement

The organizations listed below made the assembly and publication of this book possible through their sponsorship of an international conference on Prion Diseases in Humans and Animals.

We wish to express our extreme gratitude and appreciation.

John Douglas French Foundation for Alzheimer's Disease Research, Los Angeles, California, USA
The Commission of the European Communities, Brussels, Belgium
Alzheimer's Disease and Related Disorders Association, Inc., Chicago, Illinois, USA
American Health Assistance Foundation, Rockville, Maryland, USA
Abbott Laboratories
Agricultural and Food Research Council
British Bio-Technology Limited
The British Society for Immunology
Burroughs Wellcome Company
Foundation Ipsen pour la Recherche Thérapeutique, Paris, France
Genentech Inc., San Francisco, California, USA
Hollywood Park Racing Charities, Inc.
ICI Americas Inc., Delaware, USA
Ministry of Agriculture, Fisheries and Food, London, UK
Pasture Mérieux, Lyon, France
Rhône-Poulenc Rorer
Roche Products Limited, Welwyn Garden City, Hertfordshire, UK
Ross Laboratories Division of Abbott Laboratories
The Royal Society, London, UK
Sandoz Pharmaceuticals

St. Ivel, London, UK
Sigma-Tau Pharmaceuticals, Inc.
SmithKline Beecham
The Wellcome Foundation Ltd., Beckenham, Kent, UK
The Wellcome Trust, London, UK

Part I Introduction

1

An introduction to prion research

Brian Anderton, John Collinge, John Powell and Stanley B. Prusiner

Prion diseases of humans are unique in that they are simultaneously inherited and yet often transmissible experimentally to laboratory animals. After kuru of New Guinea natives and Creutzfeldt–Jakob disease found world wide were shown to be transmissible to laboratory animals, the observation that about 10% of cases of Creutzfeldt–Jakob disease were familial posed an interesting conundrum for the field. While it is doubtful that the significance of this observation was appreciated when the studies were initially reported, recent research has focused heavily on the genetic aspects of prion diseases of humans. This has been possible for two reasons primarily: (1) the discovery of the prion protein and the chromosomal gene encoding it and (2) the explosive development of modern molecular biological and genetic techniques.

The second area in which research on prion diseases has scored impressive advances is the use of transgenic mice. The results of these studies have established firmly many aspects of our current understanding of the infectious prion particle and the diseases that it causes. We have learned that the product of the prion protein gene can control the incubation time, dictate the neuropathology, and direct the synthesis of new prions in conjunction with the scrapie isoform of the prion protein in the inoculum. While many studies have been carried out to identify the molecular components of the infectious particle in experimental scrapie of rodents, there is an increasing perception, in large part because of the results of the experiments with transgenic mice, that the infectious prion particle is composed largely, if not entirely, of the scrapie isoform of the prion protein. This protein, designated PrP^{Sc}, is formed from the cellular prion protein (PrP^C) or a precursor by a post-translational process, the nature of which is still not understood.

Of equal significance to the advances made in research on prion diseases is the emergence of a new prion disease of epidemic proportions, that being bovine spongiform encephalopathy. At the time of publication, over 60 000 cows will have died of bovine spongiform encephalopathy. Presumably the cause of bovine spongiform encephalopathy is the contamination of meat and bonemeal supplements derived from the offal of sheep carrying scrapie prions. Changes in rendering procedures used to process sheep offal in the late 1970s appear to be responsible for the production of scrapie-contaminated meat and bonemeal which was fed as a dietary supplement to large numbers of cattle in Great Britain. This contaminated meat and bonemeal may have also produced disease in a variety of other ungulates, greatly expanding the list of animals to which the disease can be transmitted. Presumably, any animal with a prion protein gene can develop a prion disease.

That prion-contaminated meat and bonemeal was fed to domestic animals is a testimony to how primitive our knowledge of prions is, how poor the methods for detection are, and how great the need for expanded research in this area is.

There seems to be little doubt that the study of prions is a new and emerging field which seems to lie at the intersection of genetics, cell biology, and virology. Although scrapie was long the province of virologists, molecular genetics has transformed our understanding of the disease process, and we are beginning to re-evaluate where these diseases fit best among the more classical disciplines in biology. Certainly, investigators will continue to use virological techniques for measuring incubation periods, titres of the infectious pathogen, and learning about the mode of replication of the transmissible particle. On the other hand, molecular genetics has clearly defined the prion diseases as unique disorders whose pathogenesis is not that of a virus infection. Furthermore, molecular biological and biochemical studies have effectively eliminated the possibility that the prion particle is a virus. What remains is to determine whether PrP^{Sc} alone can transmit the disease or a cofactor is necessary. To date, there is no firm evidence for a cofactor.

The history of scrapie and the human prion diseases is rich. To provide a background for what has been done in recent years, and to allow the reader a wonderful glimpse of some of the early work, we asked some of the truly outstanding investigators who have made superb contributions to the field to prepare historical reviews of their work. These unusual documents can be found in Part II. What makes this section unique is that the people who made the history are the ones who have written it.

2

Terminology of prion diseases

John Collinge and Stanley B. Prusiner

Many terms have been used over the past 40 years to encompass the group of neurodegenerative diseases that are reviewed in this book under the title *Prion Diseases of Humans and Animals* (Table 1). Initial terminology was based on descriptive terms that had arisen in relation to common clinical features of these disorders. The naturally occurring disease of sheep was described in England as 'rubbers' or the 'goggles', in France as 'la tremblante' (the tremble), in Germany as 'Gnubberkrankeheit' (itching disease) and 'traberkrankheit' (trotting disease). Scrapie was a Scottish term which unlike the alternatives of 'cuddy trot' and 'yeukie pine' has stuck (Greig, 1940). The human disease, kuru, derives from the Fore word meaning 'tremor' (Gajdusek and Zigas, 1957).

Table 1. Prion diseases of animals and humans

Disease	Species
Scrapie	Sheep and goat
Transmissible mink encephalopathy	Mink
Chronic wasting disease	Mule, deer and elk
Bovine spongiform encephalopathy	Cattle
Kuru	Human
Creutzfeldt–Jakob disease (CJD)	Human
Gerstmann–Sträussler-Scheinker syndrome (GSS)	Human
Fatal familial insomnia	Human

Creutzfeldt–Jakob disease (CJD) was a term coined by Spielmeyer drawing from the case reports of Creutzfeldt (Creutzfeldt, 1920) and Jakob (Jakob, 1921a, b, c). In subsequent years numerous additional reports appeared introducing a large number of other names for the condition including disseminated encephalopathy, spastic pseudosclerosis, cortico-pallido-spinal degeneration, cortico-striato-spinal degeneration, Jakob's syndrome, presenile dementia with cortical blindness, Heidenhain's syndrome, subacute vascular encephalopathy with mental disorder, subacute presenile spongiosis atrophy, Nevin–Jones disease and Brownell–Oppenheimer syndrome (Gibbs and Gajdusek, 1978). Gerstmann–Sträussler syndrome (GSS) was first described by Gerstmann in a single case (Gerstmann, 1928) with a more detailed report of this and seven other cases from the same Austrian family by Gerstmann, Sträussler and Scheinker in 1936. This condition is also referred to as Gerstmann–Sträussler–Scheinker disease.

The term 'status spongiosis' was introduced by Spielmeyer in 1922 to describe the gross vacuolation seen histologically in CJD and these diseases have become known as 'spongiform encephalopathies' because of this characteristic histological feature. Spongiform change may affect any part of the cerebral grey matter and is usually accompanied by neuronal loss and astrocytic proliferation. In addition, amyloid plaques were frequently reported. Several spongiform encephalopathies of captive animals have subsequently been described: transmissible mink encephalopathy, a disease of ranched mink (Hartsough and Burger, 1965); chronic wasting disease of mule deer and elk (Williams and Young, 1980); bovine spongiform encephalopathy (Wells *et al.*, 1987).

In 1936 Cuillé and Chelle demonstrated the transmissibility of scrapie by experimental inoculation, following a long incubation period. Some type of virus was assumed to be the causative agent and Sigurdsson coined the term 'slow virus infection' in 1954. In 1959 Hadlow drew attention to the similarities between kuru and scrapie at the neuropathological, clinical and epidemiological levels. A landmark in the field was the demonstration of transmission (by intracerebral inoculation) of kuru to the chimpanzee by Gajdusek *et al.* (1966). CJD was transmitted to the chimpanzee in 1968 by Gibbs *et al.* (1968). This work led to the concept of the 'transmissible dementias'. The term 'transmissible spongiform encephalopathy' was introduced by Gadjusek and Gibbs in 1969 to encompass scrapie, transmissible mink encephalopathy, kuru and CJD. Transmissibility to experimental animals then became the hallmark against which diagnostic criteria for CJD were assessed and refined. Masters *et al.* (1979) defined 'probably CJD' as a case presenting with progressive dementia with at least one of the following features: myoclonus, pyramidal signs, a characteristic electroencephalographic appearance, cerebellar signs, or extrapyramidal signs. The diagnosis became 'definite CJD' if neuropathological confirmation of spongiform encephalopathy was also available. Atypical CJD cases were subsequently identified on the basis of their transmissibility. Successful transmission of GSS to experimental animals was first reported in 1981 (Masters *et al.*, 1981). GSS is generally regarded as a rare chronic cerebellar disorder. The neuropathologic hallmark of GSS is multicentric amyloid plaques.

The nature of the transmissible agent in the spongiform encephalopathies has been a subject of heated debate for many years. The natural assumption that the agent must be some sort of virus was challenged both by the failure to demonstrate directly such a virus (or any immunological response to it) and by the large body of evidence indicating that the agent, whatever it was, had unusual properties that were not in keeping with any known type of virus. In particular, the infectious pathogen showed remarkable resistance to treatments that would normally be expected to inactivate the nucleic acid genome of a virus such as ultraviolet irradiation or treatment with nucleases (Alper *et al.*, 1966, 1967; Hunter, 1972; Prusiner, 1982). The term 'prion' was proposed to distinguish the infectious pathogen from viruses or viroids (Prusiner, 1982). Prions were defined as 'small proteinaceous infectious particles that resist inactivation by procedures which modify nucleic acids'.

Enriching brain fractions for scrapie infectivity led to discovery of a protease-resistant sialoglycoprotein, designated prion protein (PrP) of 27–30 kDa denoted PrP 27–30 (Bolton *et al.*, 1982; McKinley *et al.*, 1983; Prusiner, *et al.*, 1982, 1983). This protein was found to accumulate in the brains of animals or humans with spongiform encephalopathies and in some cases to form amyloid deposits (Bendheim *et al.*, 1984; Bockman *et al.*, 1985; DeArmond *et al.*, 1985; Kitamoto *et al.*, 1986; Prusiner *et al.*, 1983; Roberts *et al.*, 1986). While this protein was first thought to be encoded by a nucleic acid carried within the infectious pathogen, amino acid sequencing of the *N*-terminus of PrP 27–30 (Prusiner *et al.*, 1984) led to the recovery of cognate cDNA clones using an isocoding mixture of oligonucleotides (Chesebro *et al.*, 1985; Oesch *et al.*, 1985). PrP 27–30 was found to be encoded by a single-copy chromosomal gene rather than by a putative nucleic acid in fractions enriched for scrapie infectivity (Basler *et al.*, 1986; Oesch *et al.*, 1985). PrP 27–30 was found to be derived from a larger molecule of 33–35 kDa designated PrP^{Sc}. The normal or cellular product of the PrP gene is a protease-sensitive sialoglycoprotein of 33–35 kDa designated PrP^C (Oesch *et al.*, 1985). PrP genes have been cloned from hamsters, mouse, rat, sheep, cow, mink, chicken and humans (Basler *et al.*, 1986; Goldman *et al.*, 1990, 1991; Harris *et al.*, 1991; Kretzschmar *et al.*, 1986; Locht *et al.*, 1986; Westaway *et al.*, 1987; H. Kretzschmar, in preparation). The PrP gene is highly conserved across mammalian species and hybridization studies with hamster PrP cDNA indicated hybridizing sequences in *Drosophila* and yeast (Westaway and Prusiner, 1986). The human prion protein gene designated PRNP was mapped to the short arm of chromosome 20 and the mouse PrP gene designated *Prn-p* was mapped to the homologous chromosome (Sparkes *et al.*, 1986). Suggested nomenclatures for prion proteins and their genes are given in Table 2.

Prion protein extracted from the brains of scrapie-affected animals (designated PrP^{Sc}) differs from PrP extracted from normal brain (designated PrP^C) in that it is partially resistant to proteolysis (Barry and Prusiner, 1986; Meyer *et al.*, 1986; Oesch *et al.*, 1985). No difference in primary structure has been demonstrated between PrP^C and PrP^{Sc} (Stahl and Prusiner, 1991) and PrP^{Sc} is known to be derived from a protease-sensitive precursor by a post-translational process (Borchelt *et al.*, 1990). The nature of this post-translational modification is unknown.

The demonstration of genetic linkage between GSS and a missense mutation at codon 102 of the PrP gene established GSS to be an autosomal dominant genetic

Table 2. Prion protein terminology

(A) The normal, cellular isoform and the disease-related, partially protease-resistant isoform are both encoded by the same chromosomal prion protein gene designated PRNP in humans on chromosome 20 and *Prn-p* in mice on chromosome 2.

PrP	prion protein
PrP^C	cellular isoform of PrP
PrP^{Sc}	disease-related PrP isoform
PrP 27-30	protease-resistant core of PrP^{Sc}

It remains possible that only a subset of PrP^{Sc} molecules constitute or contribute to the infectious prion particle. PrP^{Sc} cannot be more precisely defined at present and simply designates the disease-related isoform.

(B) It may be desirable to designate the disease-related isoform of PrP in all prion diseases as PrP^{Sc} rather than use notations such as PrP^{CJD}, PrP^{GSS} or PrP^{BSE}. Such notation can become complex especially when describing various PrP isoforms in transgenic animals. Thus, we suggest that the superscript 'Sc' be used to designate the disease-related PrP isoform for all species.

(C) A prefix of two or three letters preceding PrP should be used to denote the species.

HuPrP	human prion protein
MoPrP	mouse prion protein
SHaPrP	Syrian hamster prion protein
BovPrP	bovine prion protein

---

disorder (Hsiao *et al.*, 1989). This mutation in a proline-to-leucine substitution in the prion protein. Now many families have been identified with this and other mutations in the PrP gene at codons 117, 178, 198, 200 and 217 as well as insertions of varying numbers of octarepeats. These mutations co-segregate with familial CJD or GSS and have been identified only in affected or at-risk individuals from families with a history of neurodegenerative disease and not in normal controls. A single family is described with a LOD score in excess of 11 (Poulter *et al.*, 1992) and the combined LOD score for all the families probably exceeds 20 at zero recombination. This work argues persuasively that prion protein mutations are the cause of the inherited prion diseases of humans. Direct demonstration of the pathogenicity of the codon 102 mutation has been provided by experiments with transgenic mice. Multiple copies of a mouse prion protein transgene encoding the proline→leucine substitution found in GSS produced spontaneous spongiform neurodegeneration in mice (Hsiao *et al.*, 1990).

The availability of DNA markers for the inherited forms of the human prion diseases has enabled a fresh assessment of the phenotype of these conditions. Demonstration of the 144 bp insertion mutation in a family not suspected to have a spongiform encephalopathy enabled a diagnosis of familial CJD–GSS for the first time to be made by a DNA test (Collinge *et al.*, 1989). Previously, the clinical diagnosis in this family was familial Alzheimer's disease. Although it is not surprising that such clinical overlap exists, further investigation identified an individual with not only an atypical clinical picture but absence of histological features of spongiform encephalopathy (Collinge *et al.*, 1989). While it was accepted that spongiform change is a variable feature of these disorders (Prusiner, 1982), this case also lacked histological evidence of neuronal loss, astrocytic proliferation and amyloid deposition. Indeed, the histological appearance of brain sections was essentially normal. Neuropathological examination therefore cannot exclude these CNS degenerative diseases. This study led to the realization that the clinical syndromes of CJD and GSS are part of a wider spectrum of CNS degenerative disorders and the full extent of this spectrum has yet to be defined. For this reason the term 'inherited prion disease' seems to be a more appropriate diagnostic description of these illnesses. Subclassification can be designated by the specific mutation; a preliminary list is given in Table 3. This classification, unlike previous descriptive terms, relies only on aetiological mutations.

Table 3.   Inherited prion disease of humans

Notation 'Inherited prion disease (PrP mutation)'

(1)   Inherited prion disease (PrP 144 bp insertion)
(2)   Inherited prion disease (PrP leucine 102)
(3)   Inherited prion disease (PrP valine 117)
(4)   Inherited prion disease (PrP asparagine 178)
(5)   Inherited prion disease (PrP serine 198)
(6)   Inherited prion disease (PrP lysine 200)
(7)   Inherited prion disease (PrP arginine 217)

Since it is now clear that the prion protein and its gene are central to the aetiology of these disorders, whether sporadic, inherited or 'infectious', we suggest that all of them be designated as prion diseases. It could be argued that prion protein disease is a more accurate term and leaves open the question of the possibility that non-transmissible forms of these diseases will be demonstrated since the term 'prion' implies transmissibility. However, we favour the term 'prion disease' on grounds of simplicity and since this term is already in wide usage.

REFERENCES

Alper, T., Haig, D.A. and Clarke, M.C. (1966) The exceptionally small size of the scrapie agent. *Biochem. Biophys. Res. Commun.* **22** 278–284.
Alper, T., Cramp, W.A., Haig, D.A. and Clarke, M.C. (1967) Does the agent of scrapie replicate without nucleic acid? *Nature (London)* **214** 764–766.
Barry, R.A. and Prusiner, S.B. (1986) Monoclonal antibodies to the cellular and scrapie prion proteins. *J. Infect. Dis.* **154** 518–521.

Basler, K., Oesch, B., Scott, M., Westaway, D., Wälchli, M., Groth, D.F., McKinley, M.P., Prusiner, S.B. and Weissmann, C. (1986) Scrapie and cellular PrP isoforms are encoded by the same chromosomal gene. *Cell* **46** 417–428.

Bendheim, P.E., Barry, R.A., DeArmond, S.J., Stites, D.P. and Prusiner, S.B. (1984) Antibodies to a scrapie prion protein. *Nature (London)* **310** 418–421.

Bockman, J.M., Kingsbury, D.T., McKinley, M.P., Bendheim, P.E. and Prusiner, S.B. (1985) Creutzfeldt–Jakob disease prion proteins in human brains. *N. Engl. J. Med.* **312** 73–78.

Bolton, D.C., McKinley, M.P. and Prusiner, S.B. (1982) Identification of a protein that purifies with the scrapie prion. *Science* **218** 1309–1311.

Borchelt, D.R., Scott, M., Taraboulos, A., Stahl, N. and Prusiner, S.B. (1990) Scrapie and cellular prion proteins differ in their kinetics of synthesis and topology in cultured cells. *J. Cell Biol.* **110** 743–752.

Chesebro, B., Race, R., Wehrly, K., Nishio, J., Bloom, M., Lechner, D., Bergstrom, S., Robbins, K., Mayer, L., Keith, J.M., Garon, C. and Haase, A. (1985) Identification of scrapie prion protein-specific mRNA in scrapie-infected and uninfected brain. *Nature (London)* **315** 331–333.

Collinge, J., Harding, A. E., Owen, F., Poulter, M., Lofthouse, R., Boughey, A. M., Shah, T. and Crow, T.J. (1989) Diagnosis of Gerstmann–Sträussler syndrome in familial dementia with prion protein gene analysis. *Lancet* **2** 15–17.

Creutzfeldt, H.G. (1920) Über eine eigenartige herdförmige Erkrankung des Zentralnervensystems. *Z. Gesamte Neurol. Psychiat.* **57** 1–18.

DeArmond, S.J., McKinley, M.P., Barry, R.A., Braunfeld, M.B., McColloch, J.R. and Prusiner, S.B. (1985) Identification of prion amyloid filaments in scrapie-infected brain. *Cell* **41** 221–235.

Gajdusek, D.C. and Zigas, V. (1975) Degenerative disease of the central nervous system in New Guinea—the endemic occurrence of "kuru" in the native population. *N. Engl. J. Med.* **257** 974–978.

Gajdusek, D.C., Gibbs, C.J. Jr, and Alpers, M. (1966) Experimental transmission of a kuru-like syndrome to chimpanzees. *Nature (London)* **209** 794–796.

Gerstmann, J. (1928) Über ein noch nicht beschriebenes Reflex-Phanomen bei einer Erkrankung des zerebellaren System. *Wien. Med. Wochenschr.* **78** 906–908.

Gibbs. C.J., Jr, and Gajdusek, D.C. (1978) Subacute spongiform virus encephalopathies: the transmissible virus dementias. In: Katzman, R., Terry, R.D. and Bick, K.L. (eds) *Alzheimer's Disease: Senile Dementia and Related Disorders, Aging*, Vol. 7. Raven Press, New York, pp. 559–577.

Gibbs, C.J., Jr, Gajdusek, D.C., Asher, D.M., Alpers, M.P., Beck, E., Daniel, P.M. and Matthews, W.B. (1968) Creutzfeldt–Jakob disease (spongiform encephalopathy): transmission to the chimpanzee. *Science* **161** 388–389.

Goldman, W., Hunter, N., Foster, J.D., Salbaum, J. M., Beyreuther, K. and Hope, J. (1990) Two alleles of a neural protein gene linked to scrapie in sheep. *Proc. Natl. Acad. Sci. USA* **87** 2476–2480.

Goldman, W., Hunter, N., Martin, T., Dawson, M. and Hope, J. (1991) Different forms of the bovine PrP gene have five or six copies of a short, G–C-rich element within the protein-coding exon. *J. Gen. Virol.* **72** 201–204.

Greig, J.R. (1940) Scrapie. *Trans. Highlands Agric. Soc.* **52** 71–90.

Harris, D.A., Falls, D.L., Johnson, F.A. and Fishbach, G.S. (1991) A prion-like protein from chicken brain copurifies with an acetylcholine receptor-inducing activity. *Proc. Natl. Acad. Sci. USA* **88** 7664–7668.

Hartsough, G.R. and Burger, D. (1965) Encephalopathy of mink. I. Epizootiologic and clinical observations. *J. Infect. Dis.* **115** 387–392.

Hsiao, K., Baker, H.F., Crow, T.J., Poulter, M., Owen, F., Terwilliger, J.D., Westaway, D., Ott, J. and Prusiner, S.B. (1989) Linkage of a prion protein missense variant to Gerstmann–Sträussler syndrome. *Nature (London)* **338** 342–345.

Hsiao, K.K., Scott, M., Foster, D., Groth, D.F., DeArmond, S.J. and Prusiner, S.B. (1990) Spontaneous neurodegeneration in transgenic mice with mutant prion protein of Gerstmann–Sträussler syndrome. *Science* **250** 1587–1590.

Hunter, G.D. (1972) Scrapie: a prototype slow infection. *J. Infect. Dis.* **125** 427–440.

Jakob, A. (1921a) Über eigenartige Erkrankungen des Zentralnervensystems mit bemerkenswertem anatomischen Befunde (spastische Pseudosclerose-Encephalomyelopathie mit disseminierten Degenerationsherden). Preliminary communication. *Dtsch. Z. Nervenheilkd.* **70** 132–146.

Jakob, A. (1921b) Über eigenartige Erkrankungen des Zentralnervensystems mit bemerkenswertem anatomischen Befunde (spastische Pseudosklerose-Encephalomyelopathie mit disseminierten Degenerationsherden). *Z. Gesamte Neurol. Psychiat.* **64** 147–228.

Jakob, A. (1921c) Über eine der multiplen Sklerose klinisch nahestehende Erkrankung des Centralnervensystems (spastische Pseudosclerose) mit bemerkenswertem anatomischem Befunde. *Med. Klin.* **17** 372–376.

Kitamoto, T., Tateishi, J., Tashima, I., Takeshita, I., Barry, R.A., DeArmond, S.J. and Prusiner, S.B. (1986) Amyloid plaques in Creutzfeldt–Jakob disease stain with prion protein antibodies. *Ann. Neurol.* **20** 204–208.

Kretzschmar, H.A., Stowring, L.E., Westaway, D., Stubblebine, W.H., Prusiner, S. B. and DeArmond, S.J. (1986) Molecular cloning of a human prion protein cDNA. *DNA* **5** 315–324.

Locht, C., Chesebro, B., Race, R. and Keith, J.M. (1986) Molecular cloning and complete sequence of prion protein cDNA from mouse brain infected with the scrapie agent. *Proc. Natl. Acad. Sci. USA* **83** 6372–6376.

Masters, C.L., Gajdusek, D.C., Gibbs, C.J. Jr, Bernouilli, C. and Asher, D.M. (1979) Familial Creutzfeldt–Jacob disease and other familial dementias: an inquiry into possible models of virus-induced familial diseases. In: Prusiner, S.B. and Hadlow, W.J. (eds) *Slow Transmissible Diseases of the Nervous System*, Vol. 1. Academic Press, New York, pp. 143–194.

Masters, C.L., Gajdusek, D.C. and Gibbs, C.J. Jr (1981) Creutzfeldt–Jakob disease virus isolations from the Gerstmann–Sträussler syndrome. *Brain* **104** 559–588.

McKinley, M.P., Bolton, D.C. and Prusiner, S.B. (1983) A protease-resistant protein is a structural component of the scrapie prion. *Cell* **35** 57–62.

Meyer, R.K., McKinley, M.P., Bowman, K.A., Braunfeld, M.B., Barry, R.A. and Prusiner, S.B. (1986) Separation and properties of cellular and scrapie prion proteins. *Proc. Natl. Acad. Sci. USA* **83** 2310–2314.

Oesch, B., Westaway, D., Wälchli, M., McKinley, M.P., Kent, S.B.H., Aebersold, R., Barry, R.A., Tempst, P., Teplow, D.B., Hood, L.E., Prusiner, S.B. and Weissmann, C. (1985) A cellular gene encodes scrapie PrP 27–30 protein. *Cell* **40** 735-746.

Poulter, M., Baker, H.F., Frith, C.D., Leach, M., Lofthouse, R., Ridley, R.M., Shah, T., Owen, F., Collinge, J., Brown, G., Hardy, J., Mullan, M.J., Harding, A.E., Bennett, C., Doshi, R. and Crow, T.J. (1992) Inherited prion disease with 144 base pair gene insertion. 1. Genealogical and molecular studies. *Brain* **115**, in press.

Prusiner, S. B. (1982) Novel proteinaceous infectious particles cause scrapie. *Science* **216** 136–144.

Prusiner, S.B., Bolton, D.C., Groth, D.F., Bowman, K.A., Cochran, S.P. and McKinley, M.P. (1982) Further purification and characterization of scrapie prions. *Biochemistry* **21** 6942–6950.

Prusiner, S.B., McKinley, M.P., Bowman, K.A., Bolton, D.C., Bendheim, P.E., Groth, D.F. and Glenner, G.G. (1983) Scrapie prions aggregate to form amyloid-like birefringent rods. *Cell* **35** 349–358.

Prusiner, S.B., Groth, D.F., Bolton, D.C., Kent, S.B. and Hood, L.E. (1984) Purification and structural studies of a major scrapie prion protein. *Cell* **38** 127–134.

Roberts, G.W., Lofthouse, R., Brown, R., Crow, T.J., Barry, R.A. and Prusiner, S.B. (1986) Prion-protein immunoreactivity in human transmissible dementias. *N. Engl. J. Med.* **315** 1231–1233.

Sparkes, R.S., Simon, M., Cohn, V.H., Fournier, R.E.K., Lem, J., Klisak, I., Heinzmann, C., Blatt, C., Lucero, M., Mohandas, T., DeArmond, S.J., Westaway, D., Prusiner, S. B. and Weiner, L. P. (1986) Assignment of the human and mouse prion protein genes to homologous chromosomes. *Proc. Natl. Acad. Sci. USA* **83** 7358–7362.

Stahl, N. and Prusiner, S.B. (1991) Prions and prion proteins. *FASEB J.* **5** 2799–2807.

Wells, G.A.H., Scott, A.C., Johnson, C.T., Gunning, R.F., Hancock, R.D., Jeffrey, M., Dawson, M. and Bradley, R. (1987) A novel progressive spongiform encephalopathy in cattle. *Vet. Rec.* **121** 419–420.

Westaway, D. and Prusiner, S.B. (1986) Conservation of the cellular gene encoding the scrapie prion protein. *Nucleic Acids Res.* **14** 2035–2044.

Westaway, D., Goodman, P.A., Mirenda, C.A., McKinley, M.P., Carlson, G.A. and Prusiner, S.B. (1987) Distinct prion proteins in short and long scrapie incubation period mice. *Cell* **51** 651–662.

Williams, E.S. and Young, S. (1980) Chronic wasting disease of captive mule deer: a spongiform encephalopathy. *J. Wildl. Dis.* **16** 89–98.

Part II Historical reflections

3

A sideways look at the scrapie saga: 1732–1991

I.H. Pattison†

I first met scrapie disease of sheep at the Moredun Institute, Edinburgh, in 1939, three years after the claim by Cuillé and Chelle from France that they had reproduced the disease by intraocular injection of healthy sheep with spinal cord from an affected sheep. When first made, this claim had been viewed with the scepticism that had been a cross to be borne by scrapie research workers since the days of Roche-Lubin (1848), who suggested that the cause was sexual excess or lightning. However, scepticism in 1936 was not unreasonable because Cuillé and Chelle reported that the disease had not appeared until between 14 and 22 months after injection. In due course it was established that earlier investigators had not waited long enough for clinical disease to follow attempted experimental transmission.

The history of the investigation of scrapie, notably in France and Germany, back in 1732 has been recorded by M'Gowan (1914), and the significance of the disease in Britain can be gauged from the fact that it was included (with tuberculosis, Johne's disease, and zinc poisoning) in 1910 in the first-ever Government grant for research at the Royal Veterinary College, London.

There was excitement at Moredun Institute in 1939 because the 1936 report from France had recently been confirmed on a massive scale by the Institute's deputy director W.S. Gordon. In 1938 Gordon had taken over the entire sheep stock of a hill farm in Eskdalemuir, Selkirkshire, and had set up a scrapie transmission experiment involving 697 animals. The experiment was to run for four years, after

†Deceased September 17, 1991.

which surviving sheep would be killed and the land planted with trees by the Forestry Commission. The experimental inocula used were various dilutions of brain, spinal cord, or spleen from clinical scrapie, plus appropriate control materials from normal sheep. In the event, there were 198 cases of scrapie. My appearance at Moredun Institute coincided with the earliest flush of cases from this experiment, and I became deeply involved in histological diagnosis of the bilaterally symmetrical spongiform encephalopathy characteristic of the disease. Regrettably, that remarkable experiment has not been recorded in the literature; a summary was given by Gordon in 1966.

In 1939, then, experimental transmission of scrapie, sheep to sheep, had been confirmed beyond doubt. The disease was limited to sheep, and in Britain it was under investigation only at Moredun Institute. The consensus was that a virus must be involved.

War restricted work at the Institute, and scattered the staff. Only D.R.Wilson continued to struggle with scrapie. In 1942 Gordon was appointed director of the Agricultural Research Council's Field Station at Compton in Berkshire (renamed in 1963 the Institute for Research on Animal Diseases), and I joined the staff in 1946.

Scrapie was not on the research programme at Compton until its appearance in the 1940s and early 1950s in sheep from Britain exported to Canada, the USA, Australia, and New Zealand. Those countries placed an embargo on British sheep unless they could be guaranteed free from scrapie. Further research was essential. It was agreed that Wilson's programme would be extended, and that work would start at Compton.

Wilson had been searching for a virus since 1939. His achievements were remarkable (see Pattison, 1988) because his only method of detecting the transmissible agent was by inoculation of sheep of about 25% susceptibility after up to a year's incubation. He had not found a virus, but he had established that the agent was highly resistant to heat (100°C for 30 min), formalin, phenol, and chloroform, that it would pass filters of APD 650 nm and 410 nm, was not completely sedimented by centrifugation at 40000 rev min^{-1} for 2 h, remained viable in dried brain for at least two years, and resisted a considerable dose of ultraviolet light. It could be transmitted sheep to sheep intradermally, intravenously, intracerebrally, and subcutaneously. There was no growth in tissue culture, and no antibodies were detected. Histopathology showed fewer vacuolated cells in the brain stem of experimental than of natural cases. In summary, by 1953 Wilson had produced evidence that this was a very eccentric transmissible agent.

Control of scrapie by Government in Britain was considered during the 19th century, but variable incidence, and concealment of outbreaks for economic reasons, posed difficulties in compulsory control, and the matter was not pursued. In the 1950s, however, closure of important export markets created renewed calls for control by Government. Late in 1955 the Minister of Agriculture, Fisheries and Food was asked in Parliament to consider control on behalf of 'sheep breeders who through this terrible disease of scrapie have lost the whole of their exports'. He declined to make scrapie notifiable because it was 'very difficult to diagnose'. What he really meant was that diagnosis of sheep scrapie can be costly in time and money if histopathology is involved. This was not the only occasion on which the Animal

Health Division has sidestepped the responsibility of really researching the epidemiology and control of naturally occurring scrapie in sheep.

When research on scrapie was expanded in the mid-1950s, it was agreed that because Moredun Institute had departments of biochemistry and virology, whereas Compton at that time had neither but had extensive farm animal accommodation, experiments would be planned to match the facilities at each institute, with liason between the two. I assumed day-by-day responsibility for experiments at Compton, that were essentially in the field of experimental pathology.

An experiment set up at this time to examine the susceptibility of different breeds of sheep to scrapie had results of great significance. The experiment, planned by Gordon in his uniquely expansive manner, involved assembling at Compton between 30 and 57 sheep of 24 different breeds (1027 in all), and injecting 10 of each breed with a suspension of scrapie sheep brain; control animals were similarly injected with normal brain. The remaining animals of each breed were injected subcutaneously with the same scrapie material. Injection day for that experiment was the most concentrated of my professional life. The effort was infinitely worthwile, however, because over the next two years differences in clinical response to the same inoculum were highlighted between breeds, and between families within breeds, in a manner that established once and for all the importance of genetic susceptibility in the expression of scrapie in sheep. The results of this experiment were summarized by Gordon (1959), recording an incidence of scrapie ranging from 78% in Herdwicks to nil in Dorset Downs. On the basis of these findings, and using 43 rams and 946 ewes of the Herdwick breed, Gordon initiated at Compton in 1961 a programme to create by breeding two distinct flocks of sheep, one highly susceptible and the other highly resistant to scrapie.

Shortly after scrapie research was started at Compton, I examined a little known report from France in 1939 that scrapie had developed in both of two goats 26 months after intraocular injection of lumbar cord from a scrapie-affected sheep (Cuillé and Chelle, 1939). There had been no other mention in the literature of experimental exposure of goats to scrapie. I injected 12 goats intracerebrally with scrapie sheep brain, and was astonished when all developed scrapie between 15 and 22 months later. I had become so used to partial susceptibility (usually about 25%) in sheep that this 100% response in goats was wholly unexpected. The validity of the observation was confirmed by further passage in goats, and the fact that the disease was indeed scrapie was confirmed by back-pass to sheep (Pattison et al., 1959).

This 100% susceptibility of goats simplified and extended research, and a large number of experiments examined the nature of the disease and of the transmissible agent. Most are described in a paper I published in 1988. I mention here five of special significance.

(1) *The clinical picture in goats:* During passage of scrapie through goats two distinct clinical forms emerged that I called 'drowsy' and 'scratching' (Pattison and Millson, 1961a). This was not surprising, because similar variations in clinical signs have long been recognized in sheep. Indeed, the name 'scrapie' suggests a scratching syndrome, whereas the French 'la tremblante' indicates that shaking

is the predominant sign. The importance of recognition of these different signs in goats was that it stimulated my colleague R.L. Chandler at Compton to suggest to me that he inject brain from a 'drowsy' goat and brain from a 'scratching' goat into each of three strains of mice in which he had been testing the pathogenicity of Mycobacterium johnei. No suggestion could have been more prescient. Seven months later, one strain of mice injected with 'drowsy' goat scrapie developed an encephalopathy that was subsequently shown to be scrapie. Other strains of mice were found to be susceptible (Chandler, 1961, 1962, 1963). In 1963 transmission of the disease to rats was described by Chandler and Fisher (1963) at Compton, and to hamsters by Zlotnik (1963) at Moredun Institute. In this manner scrapie was created in small laboratory animals, and interest in the disease spread across the world.

(2) *Experimental transmission to goats and sheep by the oral route:* It has been recorded that before research was started at Compton in the mid-1950s, scrapie had been produced experimentally sheep to sheep by various routes of injection. However, the possibility of transmission by mouth had not been examined. In May 1958 I dosed two goats orally with 100 ml of scrapie brain emulsion; both developed scrapie, 15 and 21 months respectively later. In view of this result, 50 sheep of various breeds were drenched with the same volume of scrapie sheep brain emulsion, and 46 of the same breeds received the same volume of normal sheep brain. Seven cases of scrapie developed in five different breeds between 6 and 11 months after dosing with scrapie brain, but none in the controls. The importance of this observation was that it offered a possible explanation for the generally recognized but very rare occurrence of transmission of scrapie between sheep at pasture. Over a decade later this observation of Pattison and Millson (1961b) became more plausible when Pattison et al. (1972) reported the presence of the scrapie agent in foetal membranes voided at lambing. Sheep will eat each other's foetal membranes, and cannibalistic 'contact' was a possible explanation for spread of scrapie at pasture.

(3) *Recognition of a 'species barrier' in experimental production of scrapie*: During passage of scrapie between animals of different species (e.g. sheep–goat, goat–mouse, mouse–rat) in many experiments, I had noted that there was a characteristic response to the scrapie agent typified by a longer incubation period than between animals of the same species, as well as by atypical clinical signs, and atypical histopathology. Further, in the first passage between animals of the same species, once the barrier had been bridged, the incubation period shortened and was fixed, and the histopathology characteristic for that species was also fixed (Pattison, 1965a). I interpreted these observations as an expression of difference either between animal species per se or between scrapie agents in different species—or possibly both. To the best of my knowledge, the phenomenon has not been explained.

(4) *Relationship between size of experimental dose of the scrapie agent and incubation period for histologial and clinical response*: At an early stage in the study of experimental scrapie in goats, there were indications that the occurrence of clinical signs in inoculated animals was related in time to the amount of scrapie agent

in the inoculum. Thus, for example, the incubation period in animals given scrapie brain (where the highest titre might be expected) was much shorter than in animals given scrapie muscle. Such observations were, of course subjective, and could not be examined quantitatively until serial dilutions could be tested—an experiment that would require an unrealistically large number of goats. When scrapie was passed to mice, however, such an experiment became logistically possible. A large number of mice were inoculated intracerebrally with dilutions to 10^{-6} of scrapie or normal mouse brain. Earliest histological lesions were seen at 56 days, and earliest clinical signs at 112 days in scrapie-inoculated mice, and the observations of a 'rather precise step-by-step progression of the disease in relation to time after inoculation and to dilution of inoculum', is now accepted as a guide to variations in titre of the agent (Pattison and Smith, 1963).

(5) Resistance of the scrapie agent to formalin (100% formalin is 40% formaldehyde in water): It has been mentioned that Wilson of Moredun Institute recorded considerable resistance of the scrapie agent to formalin. It may seem odd that the action of formalin on the scrapie agent had been examined, but the formalin at 0.35% which Wilson realized had been resisted by the scrapie agent was in a batch of louping-ill vaccine, made at Moredun Institute in 1935 and later found to be contaminated with scrapie. Louping-ill virus had been killed by the formalin, but the scrapie agent survived (Gordon, 1966). To extend this observation, Wilson exposed the agent in scrapie sheep brain to higher concentrations of formalin, and found that it was still active after treatment with 3.0% acting at 37°C for 13 days. To examine further this unexpected resistance, I injected goats intracerebrally, in the early 1960s, with scrapie goat brain treated for 18 h at 37°C with concentrations of formalin ranging from 0.25% to 20.0%. To my astonishment, all goats developed scrapie. Disbelieving the observation, I obtained from the histology laboratory small pieces of scrapie goat brain that had been held in 10% formol–saline or 12% neutral buffered formalin (as used for histological fixation), for periods ranging from 6 to 28 months, and pieces of scrapie mouse brain that had been in 10% formal–saline for 4 months. Goat tissue was injected intracerebrally into goats, and mouse tissue into mice. To my amazement, all these formalin-fixed tissues produced scrapie (Pattison, 1965b). I ended that paper with the comment 'if the agent is a living virus it is likely to be a virus of a kind as yet unrecognized'.

Recognition in 1965 of the fantastic resistance to formalin of the scrapie agent was, for me, a watershed in investigation of the disease. My own researches, and investigations elsewhere, in particular extensive efforts at Moredun Institute to grow the agent or to demonstrate antibodies, had failed to identify a virus. It seemed that a non-viral agent must exist. In this frame of mind, I was impressed by descriptions in the literature of the physiochemical properties of the encephalitogenic factor, a basic protein, responsible for experimental allergic encephalomyelitis—notably resistance to heat, formalin, autolysis, ultrasonication, freezing, drying, and some organic solvents (Paterson, 1965). Therefore I applied to the scrapie agent techniques used to purify encephalitogenic factor. In retrospect—especially from a world of molecular biological skills unknown two decades ago—the resultant paper (Pattison and Jones, 1967) contains more than a whiff of sealing-wax and string. However, it provides in detail reasons for the conclusion that 'the scrapie transmissible agent may be, or may be associated with, a small basic protein'.

At the time of my examination of encephalitogenic factor vis à vis scrapie, Tikvah Alper of the Hammersmith Hospital, London, and my colleagues D.A. Haig and M.C. Clarke of the virology department at Compton, published in 1966 the first in a series of exceptionally important papers on the effect of ionizing and ultraviolet irradiation on the scrapie agent (Alper et al. 1966). They concluded that 'the evidence that no inactivation results from exposure to a huge dose of ultraviolet light, of wavelength specifically absorbed by nucleic acids, suggests that the agent may be able to increase in quantity without itself containing nucleic acid'. I find it well-nigh unbelievable that this first-class work by first-class people was not immediately accepted, and the virus theory abandoned forthwith, but that is how it was. My own non-viral suggestion of a small protein was dismissed as buffoonery.

The sticking point, of course, was increase in amount of the scrapie agent when passed through animals. No one doubted this increase; most called it multiplication, and multiplication implied involvement of nucleic acid. I postulated that another mechanism might be involved, because in a variety of experiments I had observed scrapie in control animals injected with normal tissue (Pattison and Jones, 1968). The occurrence of these cases of scrapie had been patchy and unpredictable, but there had been too many to be written off as experimental errors. I suspected that the small protein—the scrapie agent-might be present in normal tissues. In the 1968 paper I wrote 'If the possibility is considered that the scrapie agent is present in an inhibited form in normal tissue and in a released form in scrapie tissue, such unmasking would provide an alternative explanation to self-replication'. This suggestion was impartially reviewed in the *Nature-Times News Service* of 8 April 1968, but was disbelieved by almost all workers in the scrapie field, the consensus being that the results could be explained by laboratory contamination.

At this time (1967), also, my colleague D.A. Haig at Compton and I discovered a marked tendency for pieces of scrapie-affected brain to grow in artificial culture, with mitotic figures, by contrast with very slight, or nil, growth of normal brain (Haig and Pattison, 1967). This finding reinforced my belief that the bilateral symmetry of scrapie histopathology is more like neoplasia than inflammation.

What I consider to be my most significant observation on scrapie was published in 1974. I have mentioned that in 1961 Gordon set up at Compton a large-scale breeding experiment with Herdwick sheep, designed to create for experimental purposes two flocks, one highly resistant and the other highly susceptible to scrapie. My contribution to this long-term endeavour was to identify clinically and his-topathologically the hundreds of scrapie-affected animals involved. Gordon died in 1967, but the programme continued to completion in 1973. In the following year I reported an unexpected consequence (Pattison, 1974). A ram and a ewe, aged 40 and 43 months respectively, in the flock bred for high susceptibility developed scrapie, histopathologically typical of the naturally occurring disease. This was experimental confirmation of the long-held conviction of many shepherds that scrapie in sheep can have a spontaneous, exclusively genetic origin, especially through in-breeding. This disease is not identical with the experimental disease in mice, rats, and hamsters. In-breeding of these species never results in scrapie. To the best of my knowledge, no attempt has been made to confirm this observation. I was too close to retirement to provide the investment in time required.

I retired in 1976, with the nature of the transmissible agent of scrapie still obscure, and virologists as adamant as ever that theirs was the only possible point of view.

Years passed without progress. Then, out of the fog, a new name emerged. In 1982 S.B. Prusiner put forward the prion hypothesis, postulating that 'novel proteinaceous particles cause scrapie' (Prusiner, 1982).

After a long time in the dog-house, I was sufficiently reassured by this proposition to send a letter to *Nature* summarizing my own concept of the scrapie agent as a protein. This was published under the editor's heading 'Scrapie a "gene"?' (Pattison, 1982). My conclusions could not have been more aptly summarized. In 1984 I suggested that experimental transmission of scrapie may involve an oncogene (Pattison, 1984).

In conclusion, I express admiration for the tenacity, diligence, and ingenuity with which Prusiner and his colleagues have addressed the spongiform encephalopathies. With deference, I add a single caveat. Naturally occurring scrapie in sheep is not the same as experimental scrapie in any species. It should be studied, in depth, as a unique disease.

REFERENCES

Alper, T., Haig, D.A. and Clarke, M.C.(1966) The exceptionally small size of the scrapie agent. *Biochem. Biophys. Res. Commun.* **22** 278–284.

Chandler, R.L. (1961) Encephalopathy in mice produced by inoculation with scrapie brain material. *Lancet* **1** 1378–1379.

Chandler, R.L. (1962) Encephalopathy in mice. *Lancet* **1** 107–108.

Chandler, R.L. (1963) Experimental scrapie in the mouse. *Res. Vet. Sci.* **4** 276–285.

Chandler, R.L. and Fisher, J. (1963) Experimental transmission of scrapie to rats. *Lancet* **2** 1165.

Cuillé, J. and Chelle, P.L. (1939) Experimental transmission of trembling to the goat. *C. R. Seances Acad. Sci.* **208** 1058–1060.

Gordon, W.S. (1959) Scrapie panel. In: *Proc. 63rd annu. meeting US Livestock Sanitary Association,* pp. 286–294.

Gordon, W. S. (1966) Report of scrapie seminar held at Washington DC, January 1964. In: *ARS 91-53.* Department of Agriculture, Washington DC, US p.8.

Haig, D.A. and Pattison, I.H. (1967) In-vitro growth of pieces of brain from scrapie-affected mice. *J. Pathol. Bacteriol.* **93** 724–727.

M'Gowan, J.P. (1914) *Investigation into the Disease of Sheep Called "Scrapie".* Blackwood, Edinburgh.

Paterson, P.Y. (1965) Scrapie and allergic encephalomyelitis. In: *Slow, Latent and Temperate Virus Infections.* NINDB Monograph 2, Gajdusek, D.C., Gibbs, C.J., Jr., and Alpers, M.P. (eds) US Government Printing Office, Washington, DC pp. 169–175.

Pattison, I.H. (1965a) Experiments with scrapie and with special refernce to the nature of the agent and the pathology of the disease. In: *Slow,Latent and Temperate Virus Infection.* NINDB Monograph 2, Gajdusek, D.C., Gibbs, C.J., Jr., and Alpers, M.P. (eds) US Government Printing Office, Washington DC, pp. 249–257.

Pattison, I.H. (1965b) Resistance of the scrapie agent to formalin. *J. Comp. Pathol.* **75** 159–164.

Pattison, I.H. (1974) Scrapie in sheep selectively bred for high susceptibility.*Nature, (London)* **248** 594–595.

Pattison, I.H. (1982) Scrapie a 'gene'? *Nature, (London)* **299** 200.

Pattison, I.H.(1984) Oncogenes—implications for human cancer. *J. R. Soc. Med.* **77** 805.

Pattison, I.H. and Jones, K.M. (1967) The possible nature of the transmissible agent of scrapie. *Vet.Rec.* **80** 1–8.

Pattison, I.H. and Jones, K.M. (1968) Detection of the scrapie agent in tissues of normal mice and in tumours of tumour-bearing but otherwise normal mice. *Nature (London)* **218** 102–104.

Pattison, I.H. and Millson, G.C. (1961a) Scrapie produced experimentally in goats with special reference to the clinical syndrome. *J. Comp. Pathol.* **71** 101–108.

Pattison, I.H. and Millson, G.C. (1961b) Experimental transmission of scrapie to goats and sheep by the oral route. *J. Comp. Pathol.* **71** 171–176.

Pattison, I.H. and Smith, K. (1963) Histological observations on experimental scrapie in the mouse, *Res. Vet. Sci.* **4** 269–275.

Pattison, I.H., Gordon, W.S. and Millson, G.C. (1959) Experimental production of scrapie in goats. *J.Comp.Pathol. Ther.* **69** 300–312.

Pattison, I.H., Hoare, M.N., Jebbett, J.N. and Watson, W.A. (1972) Spread of scrapie to sheep and goats by oral dosing with foetal membranes from scrapie-affected sheep. *Vet. Rec.* **90** 465–468.

Prusiner, S.B. (1982) Novel proteinaceous infectious particles cause scrapie. *Science* **216** 136–144.

Roche-Lubin (1848) *Rec. Med. Vet. Ser. 3.* **5** 698

Zlotnik, I. (1963) Experimental transmission of scrapie to golden hamsters. *Lancet* **2** 1072.

4

The search for the scrapie agent: 1961–1981

Gordon Hunter

INTRODUCTION

I have been invited to recall those two exciting decades when the foundations of modern scrapie research were laid. They commenced in 1961 which was something of an *annus mirabilis* as far as scrapie research in Britain was concerned. Firstly, Bill Hadlow's seminal letter was published in the *Lancet*: he pointed out the close similarity between scrapie and kuru and suggested that the human nervous disease might like scrapie be transmissible. Secondly, Dick Chandler showed that scrapie could be established in the mouse, and the uniform susceptibility and incubation period displayed in this species made it possible for the first time to carry out accurate quantitative experiments on the disease. Thirdly, at a more mundane level, the generous allocation of US Public Law 480 funds made it possible for the Institute at Compton in England and at Edinburgh (Moredun) in Scotland to mount realistically supported scrapie research programmes for the first time since the end of the Second World War. The subsequent history can conveniently be subdivided into three phases.

(1) The experimental phase (1961–1967): The combined virological and biochemical attack established with quantitative supporting data that scrapie disease was indeed caused by a most unusual agent, thus confirming the suspicions of earlier workers such as John Stamp and Bill Gordon.
(2) The hypothetical phase (1967–1973): Scrapie research workers paused from their experimental labours and spent a lot of time in speculation. Was the scrapie agent a virus? If so, why did it behave so unlike most known viruses? If not, what was the chemical basis for this new form of life?

(3) The period of false trails (1973–1981): Impatient with the slow rate of progress, several scientists overinterpreted preliminary results during the long waits between successive titration experiments in scrapie mice. The identification of the prion protein by Stan Prusiner and his colleagues gradually brought this period to an end: scientists now had something solid and concrete to work with.

I shall now deal sequentially with each of these three historical phases of scrapie research. However, it must be stressed that this form of treatment is for clarity of exposition: in reality there was of course considerable overlap between the three periods.

THE EXPERIMENTAL PHASE (1961–1967)

Dick Chandler's mouse model proved a gold mine for researchers like me entering scrapie research in the early 1960s. It was soon established that it was possible to obtain accurate reproducible quantitative results using tissue extracts and fractions, both by applying the classical techniques of virus-type titration and by measurements of length of incubation period. The only snag was the length of the incubation period (four months or more in the early mouse models), which meant that titration experiments took at least seven months. The hamster model with its shorter incubation period was thus a considerable advance on this and was extensively used in the 1970s by Prusiner and his colleagues. For the early workers, however, the long waits for the results of experiments became almost intolerable at times, especially as one realized on reflection that the parameters one had chosen to examine were perhaps not the ideal ones. It was also essential, in order to achieve a reasonable rate of progress, to set up new experiments before the experiments in the previous round were completed and analysed. Frustration was inevitable when, as was often the case, the results of the first round showed that the procedures chosen for the subsequent round of experiments were less than ideal.

Nevertheless, the situation was vastly better than it had been for earlier workers struggling with the inadequate sheep and goat models of the disease. Thus, at Compton (and subsequently confirmed at the Moredun Institute and elsewhere) we were able to make a quantitative study of the effect of heat applied to suspensions containing the scrapie agent and subsequently inoculated into mice. There was a steady drop in the amount of surviving biological activity as the temperature was increased, but some activity did survive boiling (as earlier reported for the large animal models), and prolonged autoclaving above 120°C was necessary to eliminate the scrapie agent completely. It is even more resistant to dry heat and will survive temperatures well above 120°C for considerable periods when water is excluded.

However, at that time most scrapie workers were sure that the agent must be a virus and the main thrust of research was directed towards the isolation of this putative virus. Cellular fractionation of scrapie brain homogenates was an obvious approach, and it soon became clear that scrapie activity was not associated primarily with the nuclear or soluble fractions of brain cells. Brain homogenates are highly complex mixtures of fractions from several cell types, so that the information that can be derived directly from their examination is limited, but it was clear that scrapie activity was always concentrated in fractions rich in microsomal and plasma membranes. Of course, it was impossible to determine whether the scrapie association was limited to one type of membrane and, if so, from what cell type it was derived. However, the virus hypothesis seemed still valid, and ultrasonic dispersion of cell fractions was used in size determinations. Scrapie activity would not pass through filters with pores less than 30 nm in diameter, but it would pass through filters with pores of larger size. This behaviour was consistent, for instance, with that of the picorna group of small RNA viruses.

So to the next stage! Scrapie workers in Britain and America tried to release the 'scrapie virus' from the scrapie-enriched membrane fractions, but now a prolonged period of intense frustration ensued. Try how we may, using solvents, enzymes, detergents and other chaotropic agents, it proved impossible to separate scrapie activity from membranous components. True to its previous reputation, the scrapie agent also proved to be exceptionally hardy and most treatments failed to reduce its biological titre very much. Powerful chaotropic agents such as phenol were, of course, capable of the complete dispersion of membrane fractions, but when such treatments were applied scrapie activity dramatically vanished: there seemed to be no halfway-house, the scrapie agent being very 'sticky', i.e. requiring a hydrophobic environment to survive. Even today, we would only modify these conclusions to a limited extent.

The failure of the biochemical approach to deliver the strapie agent was paralleled by similar negative results in the virological and other fields. Thus, extensive searches using the electron microscope failed to reveal the presence of a virus. Immunological studies likewise failed to turn up a scrapie antibody, while immune-suppressed or interferon-treated animals developed scrapie in much the same way as normal mice. David Haig and Michael Clark, working at Compton in association with Tikvah Alper at the Hammersmith Hospital, did make spectacular progress in one direction, however. They showed that the scrapie agent was remarkably resistant to ionizing radiation, with a target possessing the apparent size of about 100 000 molecular weight. This was, of course, well below the size of a typical virus, more in line with the response of the naked plant viroids or of large proteins. The latter possibility was enhanced by the discovery that the inactivation spectrum in the ultraviolet region was more typical of a protein than of a nucleic acid. It suggested at least that a protein component was essential for scrapie activity. The size estimate of 100 000 was, of course, at odds with the filtration estimate of size. Both results were valid and repeatable, so perhaps the radiation experiments were telling us that, in the preparations used in the filtration tests, the scrapie agent was bound to material not essential for biological activity.

Haig and Clarke also succeeded partially where others had failed and, using explants from scrapie brain, they obtained a cell line where scrapie activity could be maintained, albeit at a relatively low level, for many generations, perhaps indefinitely like a transformed line of cells. They were unable to improve on this situation, nor, I believe, has anyone since; and again it proved impossible to identify in any way a particulate agent in their system. They were, however, able to carry out some interesting experiments with their cell line, and it was, for instance, possible to use enzyme markers to show that, in their cell line at least, scrapie activity was predominantly associated with the plasma membrane of the cell rather than with microsomal membranes.

So, in the mid to late 1960s, progress in scrapie research ground to a halt. Where to go from here? Researchers turned to the development of hypotheses that might be tested and show the way forward.

THE HYPOTHETICAL PHASE (1967–1973)

As early as 1959, John Stamp had claimed that it was 'unlikely that the (scrapie) factor would be nucleoprotein in nature'. Iain Pattison, the Compton pathologist, was first in the field following the mouse work to support Stamp's earlier claim. He was particularly impressed with the resistance of the scrapie agents to reagents such as formalin, and in a heretical paper he emphasized his view that scrapie could not be classed with conventional viruses. Like so many others, however, he overstated his case. On the basis of experiments with an electro-osmometer, where cross-contamination was obviously quite inadequately controlled, he claimed that the scrapie agent was a basic protein. This was easily disproved in general terms, since basic proteins isolated from scrapie brain under fairly mild conditions contained no biological activity whatsoever. Of course, more complex systems containing scrapie activity might have components that were basic proteins.

However, Pattison's claim was the forerunner of many unconventional suggestions about the scrapie agent that continue to this day. Thus, J.S. Griffith proposed in *Nature* a more general proposition that the scrapie agent was a protein. Griffith made some interesting suggestions about the possible self-replication of proteins, but there were at that time no ways of testing for the transmission of information back from protein into nucleic acid to reverse the functional direction of the genetic code. It was really pure speculation rather than hypothesis.

Alan Dickinson produced an interesting hypothesis, somewhat more on conventional lines. He had been struck by the very great variation in the incubation periods produced by different strains of scrapie mice. Thus the ME7 strain of scrapie induced a short incubation period of around 180 days in one strain of mice and longer incubation period of about 350 days in another mouse strain. However, the 22A scrapie agent showed precisely opposite characteristics, 480 days of incubation in the first strain of mouse and only 200 days in the second strain. The mice appeared to differ only in one single operational gene, the *Sinc* gene, and Dickinson suggested that the alleles of the *Sinc* gene produced products that interacted with various forms of the scrapie agent (perhaps naked nucleic acid) such that complex variations in incubation period and other unusual properties of the scrapie agent could be explained. This is one of the two hypotheses produced at that time in which there is still some interest today. It does seem probable that prion protein was the product of the *Sinc* gene described by Dickinson; but the rest remains speculation.

The other hypothesis to retain interest is the membrane hypothesis of Gibbons and myself. Gibbons, a brilliant physical chemist, was fascinated by the strong association between the scrapie agent and cell membrane and felt sure that oligosaccharides must be involved in the structure of the scrapie agent. We proposed, first of all, that the scrapie agent might multiply or reproduce as a component of a large membrane structure. I suppose this may be true, but it did not really have much to say about structure or mechanisms. What was probably more interesting in our hypothesis was the suggestion that the scrapie agent might be a glycoprotein or glycolipid with an oligosaccharide chain or chain differing from normal. We pointed out that oligosaccharides are in a formal sense genetic materials. Cell enzymes are used to synthesize them, but can only do so if presented with a short chain of sugar molecules to build on. It might be possible that the cell enzymes could reproduce the scrapie chains if the differences from normal were small. There is a real problem here in view of the extent of individual variation within a species in the amount and nature of glycosylation processes. How do you define the norm when it comes to a pattern of oligosaccharides decorating the surface of a glycoprotein? Certainly, the hypothesis is relevant today when one considers the problems of people researching the structure of the prion protein and variations in the attached oligosaccharides.

Way back in the early 1970s, and it does now seem a long way back, we found a lot of evidence consistent with our membrane hypothesis, but we were never able to challenge it adequately. Suckling in my laboratory did show that glycosyltransferase activity increased in scrapie brain quite early in the incubation period, which did indeed suggest that there were changes in protein glycosylation in scrapie brain. Unfortunately we failed to follow this work up, being diverted into a false trail that at first seemed to be leading to the Holy Grail of a scrapie nucleic acid.

Adams and Field promulgated a modified version of the membrane hypothesis which they termed the 'linkage–substance' hypothesis, but it was no easier to test than the original hypothesis. Workers in the scrapie field instead were occupied in analysing a number of claims for spectacular advances, none of which stood the test of time, usually a relatively short space of time too.

THE PERIOD OF FALSE TRAILS (1973–1981)

In the early 1970s two groups of workers produced evidence for the existence of scrapie-related factors that affected the behaviour of blood cells. At the Mental Retardation Laboratory on Staten Island, New York, Richard Carp and his colleagues thought that the level of the white blood cells known as polymorphs was lowered in scrapie. They further believed that a factor in the blood supposed to be responsible for the change could also affect the growth of cells in tissue culture. Hopes that this factor was or would soon lead to the scrapie agent were, however, rapidly dashed when the potential importance of the finding led other workers to repeat the experiments. It became obvious that the results were not reproducible, and, what was to become a feature of research of scrapie, too much reliance had been placed on the results obtained by an inexperienced junior worker.

Back in England, a somewhat similar claim had been made by E.J. Field and his colleagues, working for the MRC in Newcastle. They were interested in another type of white cell, the lymphocyte, and studied in the cytopherometer its interaction with the macrophage. They came to believe that lymphocytes could be stimulated by scrapie-modified tissues to emit a factor or factors that altered the mobility of macrophages in the cytopherometer. However, the changes were small and the system complicated; other workers did not succeed in detecting anything reliably repeatable. Perhaps there was some substance in the observations, but interest petered out when it became clear that the system was too unreliable to be useful.

Even more publicity was attached to reports of the 'Cho particles', about which most people came upon first in the Canadian National Press. Cho was fresh from his outstanding success in isolating the virus that caused Aleutian mink disease, and perhaps understandably he underestimated the difficulties inherent in the scrapie situation. At Compton, we subsequently received details of Cho's work and pictures of particles alleged to be the scrapie virus photographed in the electron microscope. In view of the very extensive programme of research that had been carried out at Compton and elsewhere on the electron microscopy of scrapie, we were very sceptical. Sure enough we, like others, found the same particles in normal brain. Cho had probably been somewhat unlucky in that some preparations from scrapie brain did seem to contain a lot more of the particles than did normal brain. However, more extensive studies using large numbers of controls showed clearly that the particles were definitely not specific to scrapie brain; they were probably mostly particles containing ferritin.

A somewhat more solid observation was made by Narang, who observed unusual particles in scrapie brain. He was not the first to do so, but his experiments were more precise than anything previously carried out. The Narang particles, characterized by a circular or rectilinear profile in enlarged nerve cell processes, were certainly not found in normal brain. However, as Chandler and others showed they were also very rare or absent in some scrapie systems. It became clear that they were not the scrapie agent but produced secondarily in some of the disease situations that have been examined.

Most interest of all, however, was excited by claims to have obtained a scrapie-specific nucleic acid. Malone, a graduate student working with Semancik at Riverside Laboratory, California, was experimenting with scrapie-rich particulate material. After preliminary treatments involving ammonium sulphate precipitation and enzymic digestion, fractionation by gel electrophoresis was carried out; Malone was convinced that one of the fractions she isolated contained a small DNA molecule which was specific for the scrapie material. She was probably unaware that several people had previously carried out similar experiments without success, and indeed nobody was able to repeat the work. Probably there had been some cross-contamination in the original electrophoretic experiments, and the enzyme assay for DNA carried out on crude materials was also of doubtful validity. However, it took a period of years before the claims were finally withdrawn.

It took even longer to establish that some more extensive claims made at Compton were also invalid. Corp, a graduate student of Richard Kimberlin, working in my department, made essentially two claims. One very interesting piece of work involved a study of the levels of polyadenylated RNA in scrapie brain. Corp claimed that the amounts of this type of RNA were greatly reduced in scrapie brain from an early stage in the incubation period. This work was developed in considerable detail, and it was only when other scrapie workers had difficulty in repeating published work that suspicions were aroused. Lax and Manning in my own laboratory showed conclusively that once again experiments had been inadequately controlled, and in fact a whole edifice had been built on false foundations.

Corp had also claimed to have isolated a DNA molecule from scrapie brain. The work here was not taken so far, but as with Malone, the claim could not be substantiated. This debacle unfortunately coincided with a period of contracting budgets, and the Agricultural Research Council closed down the Compton scrapie programme. It is a pleasure to note that, after a ten-year gap, scrapie research has been returned there with the expansion of the BSE programme.

CONCLUSIONS

History is history: some people enjoy it, others find it boring; and we all know about someone who considered it to be 'bunk'. The only real value of an article like this is derived from any lessons that can be learnt that are still applicable today: perhaps there are a few.

Firstly, I would suggest that research on a long time scale still requires the same statistical evaluation as short-term experiments. Many of the early mistakes arose from jumping to conclusions before there had been adequate experimentation to justify them. This is understandable, and of course even way back in 1965, ludicrous though it may seem today, researchers were rushing to publish early results because they feared that someone would discover the scrapie agent before them. The discipline required in long-term research is very severe, but 'if you don't like the heat'.

Secondly, scrapie is not normally for inexperienced workers or graduate students. The pressures on new entrants to research to achieve quick results is such that they will be hard pressed not to overinterpret their preliminary findings. If graduate students are used, there is a heavy responsibility on supervisors to apply close scrutiny to their work.

Finally, there is perhaps a lesson in not having preconceived ideas. Most research workers entering the scrapie field have wasted a lot of time because they have not believed earlier results, even when well substantiated. Early in the 1970s, Millson and I stared at the prion protein as gels, but because we were expecting to find a new protein, we did not appreciate its significance and thought it was just the secondary overproduction of a normal brain protein. A full understanding of the scrapie problem will probably only come when somebody looks at the problem.

I have not burdened the text with references: all the work mentioned has been extensively reviewed. Thus, for instance, the 1960s are covered by Hunter, G.D. (1972) *J.Infect. Dis.* **125** 427–440, while for the 1970s and just beyond read S.B. Prusiner's 'Prions—novel proteinaceous particles' in the 1984 volume of *Advances in Virus Research*.

5

Photo- and radiobiology of the scrapie agent

Tikvah Alper

ABSTRACT

From work done on scrapie by D. A. Haig, M. C. Clarke and me some years ago, inferences may be drawn which strongly support the membrane hypothesis for prion replication but argue against 'protein only'. In our outline action spectrum no peaks were observed at wavelengths specifically absorbed by nucleic acid or protein. Exposure of dilute suspensions of the agent to ionizing radiation, with oxygen present or absent, showed that its response was the opposite of that of nucleic acid and protein macromolecules, and was matched only by that of membrane systems testable for function, demonstrating that lipid peroxidation is lethal to the agent. If infection occurs through integration of a 'diseased' membrane moiety into the plasma membrane of a new cell, the membrane hypothesis can accommodate also the evidence of genetic factors in prion diseases.

INTRODUCTION

All the inferences drawn from our results, and from those with which comparisons are made, depend on measurement of loss of biological or biochemical function: with scrapie, only infectivity can be measured. The meticulous end-point dilution technique used by my colleagues, and our method of analysis, enabled us to state 95% confidence intervals for our datum points, something that seems to be exceptional in this field.

The first problem we addressed was the size of the agent, for which we used the method based on 'target theory', namely exposure of dry preparations to ionizing radiation and determination of the dose needed to give an average of one inactivating event per macromolecule. We concluded that the agent was smaller, perhaps by a factor of 10, than any known virus (Alper *et al*, 1966). Simultaneously with our initial electron irradiations of dry preparations from scrapie-affected mouse brain, we exposed aqueous suspensions to the light from a low pressure mercury lamp, which emits most of its energy at 254 nm. Since this is near enough to the peak absorption band of nucleic acids, such lamps are in general use as 'germicidal'. We could discern no inactivating effect of doses that were enormous compared with those required to destroy the function of any known nucleic acid entity that had been tested up to that time. It was these observations, considered together, that led us to moot the possibility that the agent has a mode of replication independent of the integrity of a nucleic acid moiety (Alper *et al.*, 1967).

INFORMATION FROM ACTION SPECTRA

If that were true, some other biological macromolecule must obviously be essential for replication. Evidently the chromophores which result in absorption by nucleic acids of energy at wavelengths near 260 nm are absent from the agent. It is, in fact, the chromophores present in biological macromolecules that make the use of spectrophotometers so important a tool of biochemists: absorption peaks occur where chromophores absorb photons of appropriate energy. With unpurified suspensions of entities the composition of which is not known, absorption spectra cannot help in identification. However, measurement of effects of UV at different wavelengths on the entity of interest will show which of these are most effectively absorbed. The resulting 'action spectrum' may then be compared with known absorption spectra as a guide to the critical constituent(s).

Our own early attempts to derive an action spectrum for the agent were unsuccessful (Alper *et al.*, 1967); the most powerful monochromator available to us did not deliver enough energy at the wavelengths that later proved to be the most effective in inactivating the agent. Fortunately, colleagues of Dr. Latarjet at the Institut du Radium had developed chemical filters allowing them to obtain much more powerful near-monochromatic beams at a few wavelengths from a 500 W high pressure mercury lamp (Muel and Malpièce, 1969). In Paris, therefore, we could investigate the effects of those wavelengths; the results were illuminating. Fig. 1 shows dose–effect curves for wavelengths of 237, 254, 267 and 280 nm; Results for the three latter were identical. Later, irradiations were performed at 225 and 210 nm, but the results were less reliable (Alper *et al.*, 1978). Fig. 2 shows the data points for scrapie for all six wavelengths, and, for comparison, the response of bacteriophage T_2 to exposure at the four shortest wavelengths, as well as an absorption spectrum for purified bacterial endotoxin.

Action spectra have been undervalued as an aid to solving problems regarding the nature of biological systems. When the information they convey has seemed to be in conflict with the conventional wisdom of the time, the reaction has been to attempt to explain away the discrepancy, rather than to think that the conventional wisdom

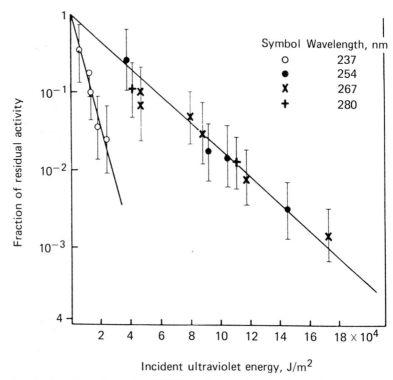

Fig. 1. Inactivation of scrapie agent by UV light at four wavelengths. Vertical bars show 95% confidence intervals for all observations.

might be challenged. An example is the interpretation of action spectra for mutation induction and killing in a fungus, shown as a Cold Spring Harbor Symposium (Hollaender and Emmons, 1941). At that time, of course, genes were '*known*' to be proteins, but Dr. Hollaender remarked on the 'outstanding fact' that the most effective region for producing toxic and genetic effects was the one most highly absorbed by nucleic acid. However, luminaries present, including in particular a Nobel prize winner, were at pains to point out how fortunate it was that these results could be accounted for without any necessity to challenge the current belief in the proteinaceous nature of genes!

If our outline action spectrum for the scrapie agent is regarded as genuine information, and not something to be explained away, some deductions therefrom have been overlooked. The absence of a peak at 260 nm plus the marked effectiveness of 237 nm, where action spectra for all nucleic acid entities show a trough (Figs 2 and 3), should, in my opinion, long ago have served to spare all the labour that has gone into the search for 'the' nucleic acid component of the agent. As shown in Fig. 2, there is some similarity with an absorption spectrum taken on bacterial endotoxin, thought by the authors to be characteristic of a lipo-polysaccharide–protein complex (Marsh and Crutchley, 1967). In the light of the current consensus on the importance

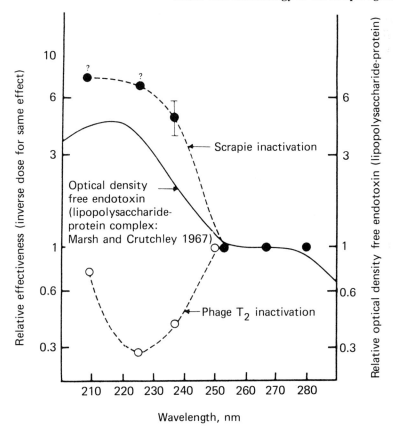

Fig. 2. Relative effectiveness of UV at six wavelengths in inactivating scrapie agent (●). Parameters at the two shortest wavelengths are less reliable than the other four. For comparison, relative effectiveness of the four shortest wavelengths in inactivating bacteriophage T2 (○) and optical density of a lipo-polysaccharide–protein complex are also shown.

of PrP in prion infectivity, it is puzzling that there is no sign of special lethal effectiveness of UV at 280 nm.

THE GIBBONS AND HUNTER MEMBRANE HYPOTHESIS

Our inference that the agent does not depend for replication on coding by nucleic acid met with considerable scepticism, but was accepted by some, including Gibbons and Hunter (1967), who consequently put forward their 'membrane hypothesis'. Some of the supporting evidence was cited by Hunter et al. (1973) and Hunter (1974). An essential feature is the plasma membrane location of the agent (Clarke and Millson, 1976), since this is crucial to the precedent for their proposal cited by Gibbons and Hunter.

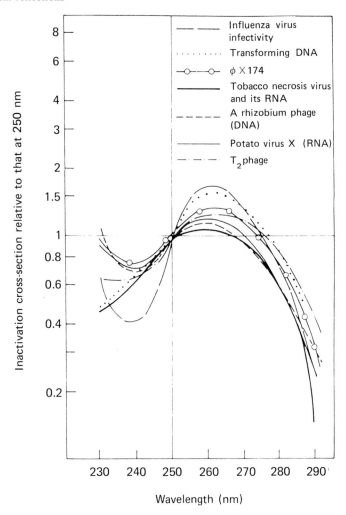

Fig. 3.   Action spectra for a number of viruses and virus core nucleic acids (by courtesy of Academic Press).

SUPPORT FROM 'INDIRECT ACTION' OF IONIZING RADIATION

Further inferential evidence in strong support comes from our use of ionizing radiation for a purpose quite different from that of the target theory investigations (Alper *et al.*, 1978). We wanted to see the effect of so-called 'indirect action', i.e. reactions of the agent with the active radicals formed from radiolysis products of water when dilute aqueous suspensions are irradiated. Both oxidizing and reducing species are engendered. If dissolved oxygen is present it diminishes the potential for reductive action by reacting with the reducing radicals at a high rate, whereas the product of the reaction [radical + O_2] will greatly increase the probability of oxidative reactions.

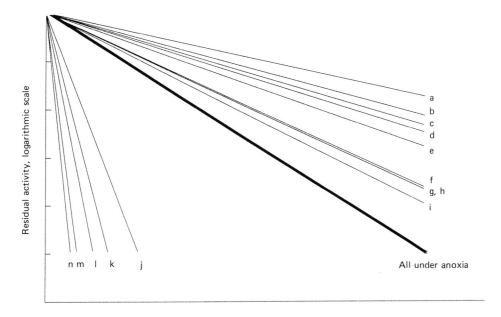

Relative dose of ionizing radiation

Fig. 4. Residual activity of biological test systems after exposure to radiolysis products of water. Normalized 'anoxic' inactivation curve for all systems shown by thick line. With oxygen present, radiation effects are reduced for (a) loss of infective titre, phage T1, (b) loss of esterase activity, α-chymotrypsin, (c) loss of transfer activity, tRNA of *E. coli*, (d) loss of infective titre, phage S13, (e) loss of infective titre, DNA of ϕ X174, (f) loss by lima bean protease of inhibitory action towards β-trypsin, (g) loss by polyuracil of ability to synthesize phenylalanine, (h) loss of enzyme activity, ribonuclease, (i) loss of enzyme activity, lysozyme. With oxygen present, radiation effects are enhanced for (j) inactivation of GAPDH in erythrocyte membrane ghosts, (k) sensitization to penicillin, membrane component of *E. coli* cell wall, (l), (n) release of lysosomal enzymes, (m) loss of infective titre, scrapie agent. Reference for (j), Kong *et al.* (1981). References for all others in Alper *et al.* (1978).

When nucleic acid or protein model systems have been irradiated in these conditions, oxygen has always protected them against loss of biological or biochemical function; with the scrapie agent, however, the presence of oxygen in the irradiated suspensions enhanced the inactivating effect, to such a great and unexpected extent that in our first pilot experiment we lost measurable titre after the higher doses. Fig. 4 is an attempt to convey the enormous difference between the responses of scrapie and of nucleic acid and protein macromolecules. Only membrane systems respond in the same way as the scrapie agent to the action of oxidizing radicals in destroying function. From known facts of radiation chemistry it is clear that this must be due to lipid peroxidation. In this regard, it is remarkable that the lethal action of 0.01 M periodate (Hunter *et al.*, 1973; Hunter, 1974) has hardly been noticed.

It would be informative if radiation chemistry type experiments could be carried out on purified preparations of PrP but, before this could be done, some means of testing for function would have to be devised.

PRECEDENT FOR MEMBRANE HYPOTHESIS, AND PREDICTION THEREFROM

As a precedent for their proposal, Gibbons and Hunter (1967) cited observations of Beisson and Sonneborn (1965), who had found that foreign pieces of *Paramecium* cortex could be grafted into a whole cell, yielding a new pattern which was maintained through both sexual and asexual reproduction. The authors attributed the replication of the new material with each cell division to the 'informational potential of existing structures'. The demonstration that lipid peroxidation is lethal to the scrapie agent suggests that infectivity depends on a membrane moiety. The established association of infectivity also with PrP (Prusiner, 1989) suggests that, by analogy with the observations on *Paramecium*, 'infection' of a cell might result from integration into its plasma membrane of a PrP–membrane complex. It could be predicted that each cell division would then give rise to one other infected cell. That is precisely what was found by Clarke and Haig (1970), who titrated scrapie as it developed in tissue cultures from scrapie-affected mouse brain. The titre of scrapie infectivity exactly matched the increase in numbers of cells in the cultures, whether titrations were on whole or disrupted cells from dispersed monolayers (Fig. 5).

GENETIC CONSIDERATIONS

If the burden of my argument in favour of the membrane hypothesis rested mainly on evidence of the importance of lipid peroxidation, it could be argued that the 'second component' suggested by Prusiner (1989), or the 'other unidentified host components' suggested by Bolton and Benheim (1988), might be one or more lipid molecules, without the involvement of any structural elements. But account must be taken of observations such as the hereditary nature of scrapie in sheep (Parry, 1962), evidence of the different incubation times for different 'strains' of scrapie inoculated into the same strain of inbred mice (Dickinson and Meikle, 1969), the preservation of genotypic identity of some scrapie strains passed between different host species (Kimberlin *et al.*, 1989), and the evidence of familial incidence of the human diseases CJD and GSS, summarized by Prusiner (1989). Clearly both hereditary and infective forms of prion diseases must be accounted for and the answer cannot be that infectivity depends simply on attachment of lipid or other host molecules to PrP. The evidence recapitulated here compels the conclusion that the answer lies in the self-replicating properties of membrane, the evidence for which was reviewed by Palade (1983). If PrP is involved in infectivity, this must come from its association with a membrane fragment originating in the first instance in a cell in the nervous system in which a gene mutation is expressed. It seems obvious that PrP must have an important role in membrane function and it is plausible that a mutation in the 'PrP gene' may be responsible for a structural change in the membrane at the site of its attachment. Thereafter, this changed, 'wrong' element would be replicated in daughter cells without reference back to the genetic code. The potentiality would then exist not only for spread of disease to the whole nervous system, but also of infectivity arising from breakdown of any cell affected in this way, and the infectivity titre would increase exactly in the manner observed by Clarke and Haig (1970).

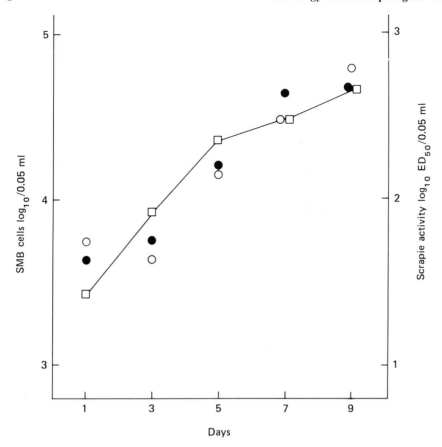

Fig. 5. Increase in cell number in culture from scrapie mouse brain (×) and increase in scrapie activity in suspensions from whole (●) and disrupted (○) cells. (By courtesy of Mr M.C. Clarke and Research in Veterinary Science.)

As an example, consider how different 'strains' of scrapie might have different incubation times. Together, the characteristics of both the infective membrane fragment and the plasma membrane of the potentially recipient cell would determine the probability of integration, and incubation time would be a function of that probability.

CONCLUSION

If this simple-minded interpretation of the membrane hypothesis should be confirmed, the implications lead to some new aspects of biology, as Diener (1987) suggested. Integration of a prion unit of infectivity into the plasma membrane of another cell leads ultimately to disaster for the host animal. But suppose that a gene mutation occurred in a somatic cell, resulting in a subtle change in the plasma membrane that conferred a selective advantage. The same mechanism that operates to integrate 'bad' membrane moieties into other cells might result in the integration also of the advantageous one, so that there would be a contribution to the evolutionary process additional to that depending entirely on the occurrence of gene mutations in germ cells.

ACKNOWLEDGMENTS

I am greatly indebted to Mr Michael Clarke for permission to use Fig. 5 and for the trouble he has taken since my retirement to keep me in touch with the scrapie literature. I am grateful to Drs W. A. Cramp, T. O. Diener, J. Hope and P. Wardman for useful discussions and to Dr. Hope also for sending me his absorption spectrum for PrP purified from scrapie mouse brain. I thank Dr J. A. C. Sterne for preparing Fig. 4.

REFERENCES

Alper, T., Haig, D.A. and Clarke, M.C. (1966) The exceptionally small size of the scrapie agent. *Biochem. Biophys. Res. Commun.* **22** 278–284.

Alper, T., Cramp, W.A., Haig, D.A. and Clarke, M.C. (1967) Does the agent of scrapie replicate without nucleic acid? *Nature (London)* **214** 764–766.

Alper, T., Haig, D.A. and Clarke, M.C. (1978) The scrapie agent: evidence against its dependence for replication on intrinsic nucleic acid. *J. Gen. Virol.* **41** 503–516.

Beisson, J. and Sonneborn, T.M. (1965) Cytoplasmic inheritance of the organization of the cell cortex in *Paramecium aurelia. Proc. Natl. Acad. Sci. USA* **53** 275–282.

Bolton, D.E. and Bendheim, R.E. (1988) In: *Novel Infectious Agents of the Nervous System.* Ciba Foundation Symposia, Vol. 135, pp. 164–181.

Clarke, M.C. and Haig, D.A. (1970) Multiplication of scrapie agent in cell culture. *Res. Vet Sci.* **11** 500–501.

Clarke, M.C. and Millson, G.C. (1976) The membrane location of the scrapie agent. *J. Gen. Virol.* **31** 441–445.

Dickinson, A.G. and Meikle, V.M.H. (1969) A comparison of some biological characteristics of the mouse-passaged scrapie agents 2A and ME7. *Genet. Res. Camb.* **13** 213–225.

Diener, T.O. (1987) PrP and the nature of the scrapie agent. *Cell* **49** 719–721.

Gibbons, R.A. and Hunter, G.D. (1967) Nature of the scrapie agent. *Nature (London)* **215** 1041–1043.

Hollaender, A. and Emmons, C.W. (1941) Wavelength dependence of mutation production in the ultraviolet with special emphasis on fungi. *Cold Spring Harbor Symp. Quant. Biol.* **9** 179–185.

Hunter, G.D. (1974) Scrapie. *Prog. Med. Virol.* **18** 289–306.

Hunter, G.D., Kimberlin, R.H., Collis, S. and Millson, G.C. (1973) Viral and non-viral properties of the scrapie agent. *Ann. Clin. Res.* **5** 262–272.

Kimberlin, R.H., Walker, C.A. and Fraser, H. (1989) The genomic identity of different strains of mouse scrapie is expressed in hamsters and preserved on re-isolation in mice. *J. Gen. Virol.* **70** 2017–2025.

Kong, S., Davison, A.J. and Bland, J. (1981) Actions of gamma-radiation on resealed erythrocyte ghosts. A comparison with intact erythrocytes and a study of the effects of oxygen. *Int. J. Radiat. Biol.* **40** 19–29.

Marsh, D.G. and Crutchley, M.J. (1967) Purification and physico-chemical analysis of fractions from the culture supernatant of *Escherichia coli* 078K80: free endotoxin and a non-toxic fraction. *J. Gen. Microbiol.* **47** 405–420.

Muel, B. and Malpièce, O. (1969) Chemical filters for narrow band UV irradiation between 235 and 300 nm. *Photochem. Photobiol.* **10** 283–291.

Palade, G.E. (1983) Membrane biogenesis: an overview. *Methods Enzymol.* **96** XXIX–LV.

Parry, H.B. (1962) Scrapie: a transmissible and hereditary disease of sheep. *Heredity* **17** 75–105.

Prusiner, S.B. (1989) Scrapie prions. *Annu. Rev. Microbiol.* **43** 345–374.

6

The scrapie–kuru connection: recollections of how it came about

W. J. Hadlow

During the late 1950s the United States embargo on British sheep, the source of scrapie in North American sheep, set off renewed interest in the disease in the United Kingdom. Two existing groups were enlivened, often seemingly as rivals in the game—the Moredun Institute in Edinburgh led by John Stamp and the Agricultural Research Council Field Station in Compton, England, led by William Gordon. Outside these groups was H. B. (James) Parry at Oxford who was carrying on his own study of sheep diseases. His 1956 leading article in *The Lancet* (Bosanquet *et al.*, 1956) claiming that scrapie is a primary myopathy likened to human hereditary muscular dystrophy had set him apart from those who all along had considered scrapie a neurologic disease, however uncertain its true nature might be. The big issue then was, is scrapie a transmissible neurologic disease or a hereditary myopathic one? Many long-held notions were brought into question, at least for a while.

The embargo and Parry's heretical view of scrapie prompted both groups to extend the solid early work of D. R. Wilson, who did much but wrote little (Wilson *et al.*, 1950). The group at Moredun Institute sought to strengthen Wilson's conclusion that scrapie is indeed a transmissible disease—one caused by an infectious agent with some properties of a virus. Their findings confirmed the transmissibility of the disease, provided more evidence of the causative agent's extreme resistance to heat, and gave some indication of its extraneural distribution (Stamp *et al.*, 1959). One of the group, Israel Zlotnik, long a sceptic about the meaning of vacuolated neurons in the diagnosis of scrapie, was engaged in a histologic marathon that convinced him that scrapie-affected sheep have more of them than do normal sheep (Zlotnik, 1957). In a meticulous study of skeletal muscle, Tom Hulland, a visiting veterinary pathologist from Canada, had collected evidence eliminating forever the contention that scrapie is a primary myopathy (Hulland, 1958). Less well known, John MacKay was trying to grow the transmissible scrapie agent in cell culture. He may have been the first to do so.

At Compton, the great variation in susceptibility to scrapie of different breeds of sheep when exposed parenterally was coming to light in William Gordon's impressive 24-breed experiment (Gordon, 1959). Well suited to the extensive animal holding facilities at Compton, it was a prelude to later efforts to breed highly susceptible and highly resistant lines of sheep (Nussbaum *et al.*, 1975). Beyond that it drew attention more generally to the role of the genetic make-up of the sheep and the occurrence of the disease. Iain Pattison's study of the susceptibility of goats to experimental scrapie was well underway and was providing clues about the extraneural distribution of the scrapie agent (Pattison *et al.*, 1959). His study was especially welcomed, because goats proved to be much more uniformly susceptible than sheep. For instance, the morbidity in animals inoculated intracerebrally was 100% in goats but usually only 25% to 35% in sheep. With such morbidity in goats, various things could be done in them that could not be done in sheep. For one, the causative agent could be assayed (titrated), which was so desperately needed for studies on pathogenesis. (Dick Chandler's report on the susceptibility of the mouse, which then became the standard bioassay for the scrapie agent, was still three years away (Chandler, 1961).)

And, at the University of Cambridge, Tony Palmer was winding up his own study of scrapie made while he was a research fellow at National Hospital, Queen Square, London. Among other contributions, he offered fresh thoughts on the clinical signs of the disease and on the importance of vacuolated neurons in explaining its pathology (Palmer, 1957). At the time, most persons in Britain making a postmortem diagnosis of scrapie still counted on seeing vacuolated neurons to do so. Indeed, it was the fashion.

Back then, the human health hazard of scrapie was not talked about, except that I was assured none existed. No special precautions were taken in handling the scrapie agent or in dealing with tissues from infected animals. In fact, carcasses of infected sheep and goats freely entered the meat trade whether from farm or research laboratory.

However commonplace all this may seem today, it made up the fascinating milieu in Britain when I went there in the spring of 1958 to help with studies on scrapie at Compton. After six years as pathologist at Rocky Mountain Laboratory, Hamilton, Montana, I had become restive and readily accepted an offer from the United States Department of Agriculture (USDA) to join one of the British groups. I came to Compton knowing little about the disease; to me it was nothing more than a strange malady notably of British sheep that was allotted little space in veterinary textbooks of the time. So, I arrived with no fixed notions about the disease, which I found advantageous in view of the uncertainties and controversies that prevailed then about scrapie.

As a general veterinary pathologist brought up in the late 1940s on morbid anatomy and little else, I set about looking for changes in brains of animals, mostly goats, gleaned from various experiments intended for other ends. I tried to put to good use what formal training I had gotten in neuropathology as a graduate student at the University of Minnesota Medical School. There I came under the tutelage of Fae Tichy, a junior associate of A. B. Baker in the Department of Neurology and Psychiatry. (In those days of 'heads without bodies and bodies without heads', brains did not come to the Department of Pathology, where I spent most of my time.) Using Baker's syllabus, *An Outline of Neuropathology* (Baker, 1949), and many profitable sessions at the microscope, Dr Tichy gave me the background I needed to make some sense of what I saw in animal brains then and later. In the mid-1950s, my experience was extended by what I got from a brief stay in Webb Haymaker's department at the Armed Forces Institute of Pathology and from examining brains and spinal cords of monkeys used to figure out a better safety test for the Salk polio vaccine.

Having acquired a slight bent toward neuropathology, I thought I could best contribute to the British effort by looking at brains, which otherwise were usually discarded. My doing this disappointed the Director at Compton, who had hoped a virologist would show up instead of me. Apparently, the last thing he needed at Compton was another pathologist! However, because of his good nature, I was left to work full time on the pathology of scrapie in any way I chose, and so I did.

One expression of the generous support I received at Compton was having Peter Dennis, the senior histotechnologist there, assigned to me. The beautiful sections he turned out made my job most satisfying, even fun. Right off, I saw more in them than holes in nerve cells—the well-entrenched diagnostic hallmark of scrapie. Neurons were changed in other ways as well. Many were shrunken and deeply basophilic. Others had fallen away. Also, a characteristic sponginess often permeated the grey matter neuropil. More impressive, though, was the astrocytic response, hypertrophy and hyperplasia of fibrous astrocytes as shown most dramatically by Cajal's gold sublimate technique. Moreover, the distribution of the degenerative lesion had a regular topographic pattern that set scrapie apart as a neuropathologic entity. However, not all of what I saw was new. Ivan Bertrand, the renowned French neuropathologist, described some of the same findings when he wrote about the histopathology of scrapie in 1937 (Bertrand *et al.*, 1937). His contribution was ignored or overlooked by persons preoccupied with vacuolated neurons and by those who dismissed outright any changes in the brain in their wanting to call scrapie a myopathy.

Despite all the doubts and uncertainties then about scrapie, one could not avoid becoming excited about it, perhaps in large part because of them. Every new finding, great or small, was welcomed, even though it might be looked upon suspiciously by the other group. After a year of watching the disease evolve clinically in many animals, of looking at many of their brains, and of absorbing scrapie lore from sundry sources, I thought I had a good idea of what scrapie is like: a protracted degenerative disease of the brain, not an inflammatory one, caused by an infectious agent best thought of then as a virus. I did not know of another disease like it in man or animal.

That was the way things were when a friend and colleague from Rocky Mountain Laboratory, William Jellison, a parasitologist, showed up at Compton the afternoon of June 28, 1959. He was returning from meetings in eastern Europe and stopped by as a social call. At dinner that evening he casually mentioned I might be interested in an exhibit he saw the previous day at the Wellcome Medical Museum in London.

It had to do with a strange brain disease of a primitive people in New Guinea. My curiosity was aroused. So five days later I took the train to London and sought out the exhibit at the Wellcome Medical Museum.

It was on the first floor just inside the main door and comprised several panels much like those of today's poster sessions. Large photographs, mostly if not solely in colour as I recall, were used to convey the story of kuru in New Guinea. I am not certain who put the exhibit together, but presumably the photographs of the people and their environs were from Carleton Gajdusek's collection at the National Institute of Neurological Disease and Blindness. The revealing photomicrographs of the characteristic neurohistologic changes, which caught my eye, no doubt were taken from Igor Klatzo's preparations at the same institute (Klatzo *et al.*, 1959). From the start, I was drawn to the neurohistologic changes, especially the vacuolated neurons, unusual in human brains, that were so much like those of scrapie.

All this information was new to me, though I remembered weeks later that a colleague at Rocky Mountain Laboratory, Carl Eklund, who was familiar with my work on scrapie, had written to me in April 1959 about an article on kuru that had just appeared in the *American Journal of Medicine* (Gajdusek and Zigas, 1959). Regrettably, I did not follow up the suggestion, implicit in his letter, that I should read it.

Before leaving London that day, I went to the Royal Society of Medicine Library, where I was an overseas associate, to look up several references on kuru cited in the exhibit, including the one in the *American Journal of Medicine*. I returned to Compton laden with information to mull over in the days ahead. In doing so, I found the overall resemblance of kuru and scrapie to be uncanny. The similarities in epidemiologic features, general clinical pattern, and neurohistologic changes could not be put aside. From these similarities I realized that scrapie might not be unique after all. Still to be described were other examples of this encephalopathy in animals, yet here was a human disease that fit the scrapie mould. But, was it transmissible?

The significance of this likeness seemed too important to keep from others. So, the letter to the editor of *The Lancet* (Hadlow, 1959), put together in about a week of struggling with the idea, was sent off July 18, 1959. Because I thought a printers' strike at the time might delay its publication, a few days later I sent a copy to Carleton Gajdusek—the one person who would be interested in it. The suggestion to test the transmissibility of kuru in laboratory primates seemed essential, even though much uncertainty remained then about what actually took place when scrapie was transmitted. Also, despite its transmission, some persons were still reluctant to look upon it as an infectious disease.

I have no record of any comments the editor of *The Lancet* may have made about the letter. The galley proof arrived August 8. Except for deletion of the word 'recent' in the parenthetical comment on the top of page 290 and for several changes to British spelling, the letter was as I had written it, and that is the way it appeared in the September 5 issue of the journal. Notes I made shortly thereafter indicate I received favourable comments from several medical people, including Harry Lander, an Australian physician who had studied kuru in New Guinea. He gave an illustrated lecture on the disease, which I attended, at St Thomas' Hospital Medical School in

London on September 14. After talking with him, the first person I met who had original knowledge of kuru, I felt the conclusions in my letter were not all that far fetched.

The one laudatory comment I remember well was from my mentor, James R. M. Innes, a British veterinary pathologist living in the United States. Long before the idea of comparative medicine took hold and studies on animal models of human disease became fashionable (and fundable), he had been extolling the merits of the comparative approach in pathology. So he was especially delighted with my comparison of scrapie and kuru; it was the kind of connection he had always sought in his studies on animal diseases.

During November and December 1959, John Stamp, William Gordon, and I along with James Hourrigan and Maurice Shahan of the USDA toured the United States talking about scrapie, a highly emotional topic then among sheepmen. The tour, intended to appease them with facts about the disease, began on November 23 in Washington, DC. That was when I first met Carleton Gajdusek, the young man with a crew cut who stood silently in the back of the room while I gave my talk. After the meeting, he came up to me and introduced himself. I am sure we talked about my letter, but I have no recollections of what was said. We parted.

If we remained in contact my mail, phone, or telegraph during the months that followed my return to England, I have no record of it until March 28, 1961. Then, as I was about to leave England after my three-year stint there, I received a letter, followed by a telegram a few days later, from Carleton inviting me to join him in a study of slow infections of the nervous system, including inoculating chimpanzees with kuru brain tissue. I was only slightly interested, for by then I was halfway committed to stay with the USDA. Nevertheless, late in July 1961, after my unintended return to Rocky Mountain Laboratory, I met with Carleton and Joe Smadel, Associate Director of the National Institutes of Health, to discuss their proposition. I declined their offer, concluding, unfairly as it turned out, that anyone who took the job would become little more than an exalted handler of apes. Besides, by then Carl Eklund and I were ready to tackle scrapie. So I lost out in helping to do what I had suggested should be done with kuru—inoculate laboratory primates—but I have no regrets.

Before long Joe Gibbs, a former associate of Joe Smadel, joined Carleton. Thus began the oft-recounted study of kuru and other human neurologic diseases in laboratory primates. The study first came in the fore in 1966 when Carleton's group reported the transmission of kuru to chimpanzees after an incubation period of 18 to 21 months (Gajdusek et al., 1966). Of course I was pleased to hear this exciting news. It bore out what I had only surmised seven years earlier. Obviously, scrapie and kuru were far more alike than was apparent then. Perhaps here was a whole new approach to understanding the causes of other subacute neurologic diseases of man. Beyond these thoughts I do not remember having any special sentiments about this historic event. Anyway, by then Carl Eklund and I were too busy coping with our own sources of excitement, even if less momentous than transmitting kuru to apes.

Over the years the administration at the National Institutes of Health had tried repeatedly to promote collaboration between Carleton's laboratory and us at Rocky Mountain Laboratory. Fortunately, the administration then had the good sense not to use coercion to bring this about when it did not happen on its own. So the idea was dropped, and we continued to go our separate ways. Carleton and Joe Gibbs worked on kuru and other human diseases, and Carl Eklund and I worked on scrapie and other animal diseases. In the end this was probably best for both groups; we complemented one another's studies, kept up an enduring professional relationship, and remained good friends. Each of us has been rewarded for his contribution, only in different ways.

That is how I recall the scrapie–kuru connection came about and some of its consequences. Over the years much has been made of my letter pointing out the likeness of the two diseases and its implications. That, of course, pleases me. Yet, as the events bear out, the observation was largely fortuitous. Certainly knowing something about scrapie, especially its pathology, helped me to see the resemblance at first glance. Nevertheless, the likeness of the two diseases is such that no doubt sooner or later someone would have become aware of it. Apart from anything else that might be said about the scrapie–kuru connection, I think it fits nicely the idea of one medicine as put forth long ago by Rudolph Virchow and others (Klauder, 1958). At least I like to remember it that way.

REFERENCES

Baker, A.B. (1949) *An Outline of Neuropathology*, 4th edn. W. C. Brown Co., Dubuque, IA, 172 pp.

Bertrand, I., Carré, H. and Lucam, F. (1937) La "tremblante" du mouton. (Recherches histo-pathogiques). *Ann. Anat. Pathol.* **14** 565–586.

Bosanquet, F.D., Daniel, P.M. and Parry, H.B. (1956) Myopathy in sheep: its relationship to scrapie and to dermatomyositis and muscular dystrophy. *Lancet* **2** 737–746.

Chandler, R.L. (1961) Encephalopathy in mice produced with scrapie brain material. *Lancet* **1** 1378–1379.

Gajdusek, D.C. and Zigas, V. (1959) Kuru. Clinical, pathological and epidemiological study of an acute progressive degenerative disease of the central nervous system among natives of the Eastern Highlands of New Guinea. *Am. J. Med.* **26** 442–469.

Gajdusek, D.C., Gibbs, C.J., Jr, and Alpers, M. (1966) Experimental transmission of a kuru-like syndrome to chimpanzees. *Nature (London)* **209** 794–796.

Gordon, W.S. (1959) Scrapie panel. *Proc. 63rd Annu. Meet. US Livestock Sanitary Association, December 15–18, 1959*, pp. 286–294.

Hadlow, W.J. (1959) Scrapie and kuru. *Lancet* **2** 289–290.

Hulland, T.J. (1958) The skeletal muscle of sheep affected with scrapie. *J. Comp. Pathol.* **68** 264–274.

Klatzo, I., Gajdusek, D.C. and Zigas, V. (1959) Pathology of kuru. *Lab. Invest.* **8** 799–847.

Klauder, J.V. (1958) Interrelations of human and veterinary medicine: discussion of some aspects of comparative dermatology. *N. Engl. J. Med.* **258** 170–177.

Nussbaum, R.E., Henderson, W.M., Pattison, I.H., Elcock, N.V. and Davies, D.C. (1975) The establishment of sheep flocks of predictable susceptibility to experimental scrapie. *Res. Vet. Sci.* **18** 49–58.

Palmer, A.C. (1957) Studies in scrapie. *Vet. Rec.* **69** 1318–1328.

Pattison, I.H., Gordon, W.S. and Millson, G.C. (1959) Experimental production of scrapie in goats. *J. Comp. Pathol.* **69** 300–312.

Stamp, J.T., Brotherston, J.G., Zlotnik, I., MacKay, J.M.K. and Smith, W. (1959) Further studies on scrapie. *J. Comp. Pathol.* **69** 268–280.

Wilson, D.R., Anderson, R.D. and Smith, W. (1950) Studies on scrapie. *J. Comp. Pathol.* **60** 267–282.

Zlotnik, I. (1957) Significance of vacuolated neurons in the medulla of sheep affected with scrapie. *Nature (London)* **180** 393–394.

7

Kuru and scrapie

D. Carleton Gajdusek

In view of the widespread interest in early investigation on kuru:

Agricultural Research Council
Field Station
Compton
Nr. Newbury
Berkshire.

21st July, 1959.

WJH/AAW

Dr. D. Carleton Gajdusek
N.I.N.D. & B.
National Institutes of Health
Bethesda 14, MARYLAND, U.S.A.

Dear Doctor Gajdusek:
   Many thanks for the several reprints on Kuru.
   I've been impressed with the overall resemblance of Kuru and an obscure degenerative neurologic disorder of sheep called Scrapie.

Since leaving the Public Health Service (Rocky Mountain Laboratory) about 18 months ago, I've been representing the U.S.D.A. at this station to further studies on the neuropathology of this fascinating disease. I've been concerned primarily with the syndrome induced experimentally in the goat by intracerebral or subcutaneous inoculation of brain tissue from scrapie-affected sheep (and goats in passage). The lesions in the goat seem to be remarkably like those described for Kuru. However, aside from this aspect of the diseases, other features appear to have much in common. All this has suggested to me that an experimental approach similar to that adopted for scrapie might prove to be extremely fruitful in the case of Kuru. It may seem far fetched in view of the present evidence, but the same could be said about scrapie in the absence of knowledge about the successful experimental induction of the disorder in the sheep and the goat, a species not naturally affected.

Because I've been greatly impressed by the intriguing implication, I've submitted a letter to The Lancet. The current printing strike may delay publication so I've enclosed a copy of the short note for your information.

I would be grateful for your comments on the suggestion, which may have broader biologic implications than are now apparent, especially with regard to human neurologic disease.

Sincerely yours,

W. J. Hadlow, D.V.M.

Scrapie and kuru

Sir,

While attempts to draw too close an analogy between diseases of man and lower animals are attended by numerous pitfalls, many valuable clues contributing to the understanding of the fundamental nature of a disease can be gained from a broad comparative viewpoint.

With this in mind, I should like to present a few brief comments on the similarity of two progressive degenerative disorders of the central nervous system, namely scrapie[1] affecting sheep and Kuru[2] affecting the Fore natives in the Eastern Highlands of New Guinea. I do not suggest that these diseases are identical or even counterparts, but in my opinion their overall resemblance is too impressive to be ignored.

When viewed collectively, the general epizootiologic, etiologic, clinical, and pathologic features of scrapie closely parallel those that characterize Kuru, as recently described by American[2] and Australian[3-5] investigators. Thus, each disease is endemic in certain confined populations, whether this be flock or tribe, in which the usual incidence is low, about one to two per cent. Sheep, or persons, may become clinically affected months after they have been moved from flocks, or communities, where the respective disease is endemic. Moreover, scrapie may appear in a previously non-affected flock following introduction of a ram or ewe from a flock in which the disease is known to exist. Likewise, Kuru may be introduced through marriage to populations previously free of the disease. Yet, despite these well-established observations, the exact mechanism underlying the "spread" of scrapie or Kuru is obscure; certainly evidence that contagion is responsible is either equivocal or non-existent.

The cause of each disease is obscure. No microbiological agent has been isolated, nor is there convincing evidence that either scrapie or Kuru is an infectious disease in the generally recognized sense. Nutritional and toxic factors have not been incriminated as having etiologic significance in either disorder. However, a genetic predisposition is strongly suggested by the accumulated epidemiologic and epizootiologic data.

The natural history and general clinical aspects of the two diseases are strikingly similar. Onset of each disorder is insidious and occurs in the absence of signs of antecedent illness. The course is afebrile and almost invariably is relentlessly progressive. Both diseases usually end fatally within three to six months after onset; only rarely have remissions and recoveries been observed. Detailed consideration of individual clinical signs and physical findings seems less important in illustrating the broad similarity, for these features are largely reflections of variations in the characteristic distribution of the lesions and in the functional development and integration of the nervous systems of the two species. Nevertheless, such signs as ataxia, which becomes progressively more severe, tremors, and changes in behaviour are features of both diseases. In neither case have laboratory studies revealed regular or significant abnormalities in the blood or cerebrospinal fluid.

Lastly, the neuropathologic changes, though essentially non-specific, are remarkably similar in the two diseases. Widespread neuronal degeneration largely characterized by shrinkage and hyperchromia of the cell body and by various degrees of vacuolation of the cytoplasm is typical of both disorders. Astrocytic gliosis, often of pronounced intensity, is another common feature, as is minimal degeneration of myelin in some instances. Inflammatory changes of the extent and severity usually associated with an active encephalite process are not observed. Large single or multilocular "soap bubble" vacuoles in the cytoplasm of nerve cells have long been regarded as a characteristic finding in scrapie; this extremely unusual change, apparently seldom seen in human neuropathologic material, also occurs in Kuru,[5] and first aroused my curiosity about the possible similarity of the two diseases.

This brief comparison of scrapie and Kuru is presented not merely to enumerate interesting similarities, but, more significantly, to indicate an intriguing implication when the analogy is viewed against the background of accumulated data on experimental investigations of scrapie. Despite the lack of indications suggesting that scrapie is an infectious disease, the disorder can be induced experimentally in the sheep[1,6] and in the closely related goat[7] (but not in other species so far tested) inoculated intracerebrally or subcutaneously with suspensions of brain tissue, as well as certain other tissues, from scrapie-affected sheep. Strangely, clinical disease never appears in inoculated animals in less than three-and-a-half months and may not become evident in some animals until 30 months after inoculation. The nature of the pathogenetic factor in the inoculum and the fundamental biochemical alteration that gives rise to the disorder in the inoculated animal remain to be elucidated. Nevertheless, such experimental induction of a progressive degenerative disorder of the central nervous system and its successful "transmission" in series would seem to provide a valuable clue to the eventual understanding of the broadly similar and equally perplexing human neurologic disorder represented by Kuru. Thus, it might be profitable, in view of veterinary experience with scrapie, to examine the possibility of the experimental induction of Kuru in a laboratory primate, for one might surmise that the pathogenetic mechanisms involved in scrapie—however unusual they may be—are unlikely to be unique in the province of animal pathology.

Yours faithfully,

W. J. Hadlow D.V.M.
Agricultural Research Council,
Field Station, Compton, Nr. Newbury, Berks.

REFERENCES

*(1) Stamp, J. T., *Vet. Rec.*, 1958, **70**, 50.
(2) Gajdusek, D. C., Zigas, V., *Amer. Jour. Med.*, 1959, **26**, 442.
(3) Bennett, J. H., Rhodes, Y. R., Robson, H. N., *Aust. Ann. Med.*, 1958, **7**, 269.
(4) Simpson, D. A., Lander, H., Robson, H. N., *Aust. Ann. Med.*, 1959, **8**, 8.
(5) Fowler, N., Robertson, E. G., *Aust. Ann. Med.*, 1959, **8**, 16.
(6) Gordon, W. S., *Vet. Rec.*, 1957, **69**, 1324.
† (7) Gordon, W. S., Pattison, L. H., *Vet. Rec.*, 1957, **69**, 1444.

*—Good presentation of current knowledge about scrapie, and many references to earlier work.

†—A more detailed paper on the goat studies will appear in the next (July) issue of the *Journal of Comparative Pathology*—Pattison, Gordon, & Millson of this station.

July 28, 1959

Refer to: NBI-COLR

Dr. W. J. Hadlow
U.S. Department of Agriculture
Agricultural Research Council
Field Station, Compton, Nr. Newbury
Berkshire, England

Dear Dr. Hadlow:

I write to inform you that I have dispatched a copy of your interesting letter to Dr. Gajdusek and to the LANCET with reference to SCRAPIE AND KURU. Dr. Gajdusek is now in New Guinea for further investigations of the disease Kuru and is due to return to the Institute in September.

I am certain the Doctor would appreciate any of your other publications which may be available. An extensive report on the neuropathology of Kuru has just appeared in LABORATORY INVESTIGATION, but reprints are not yet available. Several other reports on the disease, as well as a report on a newly discovered Kuru-like syndrome, prevalent in the Western Highlands of New Guinea, are to appear before long. I shall be happy to send you reprints as they become available.

Dr. Gajdusek will undoubtedly contact you from the Territory upon receipt of your letter.

Sincerely yours,

(Mrs.) Marion Poms
Secretary to
D. Carleton Gajdusek, M.D.
National Institute of Neurological
Diseases and Blindness

KURU RESEARCH CENTER
KAINANTU, E. H.

W. J. Hadlow, D. V. M.   Territory of Papua and New Guinea
U.S. Dept. of Agriculture   August 6, 1959
Animal Disease and Parasite Research Division
Field Station
Compton, Nr. Newbury, Berkshire

Dear. Dr. Hadlow,

I am deeply grateful to you for your letter of 21 July which was quite properly addressed but which has only now reached me since I have been off on Kuru patrol work. I retain my main office at the NIH in Bethesda, Maryland but our field base is here and at Okapa, to the south.

As you may have been able to gather from our articles on Kuru, we are pursuing the matter of possible infectious etiology extensively—I am, in fact, a virologist by training. However we have thus far had poor luck with inoculation experiments and the possibility of doing more extensive inoculation work has until now, been small. We are, however, proceeding accordingly at the present time and frozen and fresh materials are being injected into a number of animal hosts during this years work on Kuru. In your note to LANCET, which I am deeply grateful to you for bringing to my attention, I note that you have probably not seen our extensive pathological descriptions of KURU which include some features which were little stressed in the report you have quoted. Therefore I take this opportunity to list for you the appropriate references:

Klatzo, I., Gajdusek, D. C., and Zigas:
1959 Journal of Neuropathology and Experimental Neurology. **18**:2, 335–336, 1959.

Klatzo, I., Gajdusek, D. C., and Zigas, V. Pathology of Kuru. Laboratory Investigation, **8**:4, 799–847, 1959.

Furthermore, at the International Congress on the Actual Encephalitities which was held in May of 1959 in Antwerp a large section of the discussion was centered on Kuru and Kuru pathology and the results of this Congress will soon be published. Until August 15 or so I believe the NIH Exhibit on Kuru will still be on display at the Welcome Museum of Medical Science, 183 Euston Road, London, N.W.1, England. This exhibit has included a very large display of the neuropathology of the disease with enlarged color transparancies of photomicrographs of the peculiar plaque-like body of Kuru.

I thank you for your interest and your interesting suggestions. We too believe that toxic and/or infectious factors will still prove to be important in the disease but as yet we are operating largely on suspicion. I hope you will be willing to send me at my NIH address any publications and references you can on Scrapie.

Sincerely,

D. Carleton Gajdusek

8

Spongiform encephalopathies — slow, latent, and temperate virus infections — in retrospect

Clarence J. Gibbs Jr

It is extremely difficult to write about the past when the present and the future present scientific challenges that mandate forward motion of intellect and creativity. Thus it was with reluctance that I accepted the invitation to record some of my observations made during the more than 30 years of work in the field of slow infections of the nervous system.

In 1959 I had left the Walter Reed Army Medical Service Graduate School Department of Hazardous Operations where I had successfully developed an effective inactivated Rift Valley Fever (RVF) vaccine for immunoprophylatic use in man and animals at risk to this highly infectious virus and often fatal disease. I left WRAIR to accept a position in the Arthropod-Borne Virus Section, Laboratory of Tropical Virology (LTV), of the National Institute of Allergy and Infectious Diseases (NIAID) with the intent of ultimately going to work in LTV's laboratory in Panama. However, after only one year it was patently obvious that LTV was on a collision course with destiny and my scientific future lay not in shoring up a sinking ship; rather, it lay in new challenges—as yet unidentified.

It was in search of these new challenges that 1961 brought me once again into the office of Dr Joseph E. Smadel, under whom I had worked at Walter Reed and who moulded and guided much of my career in virology. Having informed the Chief of LTV, Dr Alex Shelokov, that I planned to accept Dr Wilber Down's offer of a Rockefeller Foundation Fellowship to work in San Paulo, Brazil, on arthropod-borne viruses, I sought the advice of Dr Smadel.

Smadel's reaction was immediate and violent and in his inimitable fashion he pointed his finger in my face and, said 'Goddam it Gibbs you're not going to Brazil!'. Having recovered his composure he went on to say that Dr Marston, head of the Rockefeller Foundation, was pulling out of their overseas laboratories since they were no longer interested in isolating and characterizing viruses from mosquitoes. To which I replied that if I wasn't going to Brazil just where was I to be banished. His reply was that I was going to the Patuxent Wildlife Research Center, Laurel, Maryland, where I would study scrapie disease of sheep and attempt to transmit to chimpanzees, smaller non-human primates and laboratory rodents kuru and other subacute progressive degenerative diseases of the nervous system of man and animals under the direction of Dr D. Carleton Gajdusek.

I was already vaguely familiar with scrapie disease of sheep since my father had been a federal government and state of Maryland field veterinarian and I frequently participated in his investigations of infectious diseases among domestic animals. At the same time I was familiar with Carleton Gajdusek from Walter Reed and of his and Vincent Zigas' work on the discovery and investigations of kuru among the Fore people of the then Eastern Highlands of New Guinea. More importantly was the fact that I had observed mice that Carleton Gajdusek and J. Anthony Morris, another of the Smadel team from WRAIR working at NIH, had inoculated with scrapie Carleton had brought back from Compton, England. Knowing of my desire to leave the Laboratory of Tropical Virology Tony Morris had recommended me to Drs Smadel and Gajdusek for inclusion in the program on 'slow infections' of the nervous system. I must admit that I was somewhat reluctant to leave completely the field of arbovirology for the unknowns of kuru and related diseases. I told this to Carleton during a luncheon meeting we had concerning my joining his programme at which time he agreed that I could continue to pursue my interest in this field and any other conventional virus I might wish to work on. This concession led to my isolating, for the first time, Chickungunya virus from patients with haemorrhagic fever in India. This work was done in collaboration with Keerti Shah of the School of Public Health, Johns Hopkins, who also had a small laboratory at the Patuxent Research Center. It is important to note, however, that Carleton Gajdusek did not have to 'recruit' me for the Patuxent part of his laboratory; his invitation to join his 'very nebulous and unstabilized section' where 'we can pick and choose what we want but we *do not* want to do what Rockefeller, CDC, WHO and others are doing. Virusologically speaking we have immediately on our hands a vast wealth of possible studies among the best "sentinels" in the world—the primitive communities still existent—some even being shifted around by social forces in almost "animal sentinel", fashion.' What a challenge! My decision was a simple one when Dr Smadel told me in very fartherly terms 'Joe, I want you to undertake this work and I guarantee that within five years you will either have golden positive or golden negative results and I am sure they will be golden positive.' He was right but he never lived long enough to see the golden positive results he predicted for the transmissibility of kuru and the resultant emergence of an entirely new and challenging field in neurology, neuropathology, neuroimmunology, neurovirology, microbiology, molecular biology and more recently molecular genetics.

Thus, 'slow virus' investigations at NIH, under the direction of D. Carleton Gajdusek, evolved in an unforeseeable and circuitous manner. It was established within the Division of Collaborative and Field Research, National Institute of Neurological Diseases and Blindness (NINDB), as part of Gajdusek's laboratory of 'Child Growth and Development and Disease Patterns in Primitive Cultures' and was designated as the 'study of slow, latent and temperate virus infections'. It became the first such laboratory in the world, and its work in the 1960s phrased the problem of slow virus infections. In 1963 the Neurological Institute (NINDB) sponsored the first meeting on slow virus infections and two years later a monograph entitled *Slow, Latent, and Temperate Virus Infections* was published. These events and the subsequent transmission of kuru and Creutzfeldt–Jakob disease (CJD) from man to primates, demonstrating infection as the aetiology of both human diseases,launched worldwide awareness of slow virus infections in man.

This work, along with Dr John Sever's laboratory also in NINDB, was largely responsible for the demonstration of slow and delayed measles infection as the cause of Dawson's type-A inclusion encephalopathy, which, we, along with John Sever, and Wolfgang Zeman, renamed subacute sclerosing panencephalitis (SSPE) during the first international meeting on this disease, held at NIH in 1967. It was the demonstration that chronic non inflammatory human disease was caused by long-term incubating viruses that led to the demonstration of a papovavirus aetiology for progressive multifocal leukoencephalopathy and to tick-borne (Russian spring–summer) encephalitis virus in Soviet Siberia and Europe as a cause of chronic infection in man and primates. Subsequently SSPE-like syndromes caused by the rubella virus were established, and other chronic and persistent central nervous system virus infections caused by adenoviruses, echoviruses, herpesviruses and retroviruses were identified. It is an interesting fact that to date all recognized virus-induced slow infections result almost exclusively in subacute progressive degenerative diseases of the nervous system.

In 1961, before I joined Gajdusek, he and Dr J. Anthony Morris had obtained permission from Dr Carlton Herman and his superiors at the Patuxent Wildlife Research Center, Laurel, Maryland, the Bureau of Sports Fisheries and Wildlife, United States Department of Interior, to utilize very modest and minimally adequate laboratory facilities augmented by NINDB's construction of a cinder block non-insulated building for housing chimpanzees and smaller non human primates inoculated with brain materials from patients that had died with kuru, amyotropic lateral sclerosis, and parkinsonism–dementia of Guam, multiple sclerosis and a wide variety of other neurological diseases. Gajdusek and Morris had in August 1961 inoculated several hundred mice with homogenates of brains from scrapie-infected sheep in the United States and these animals were housed in laboratories on the NIH campus. It would be two years before an NIH–Department of Interior Interagency Agreement authorizing the construction of facilities and allocation of space would be signed.

In spite of construction delays we moved rapidly to establish and activate the Laboratory of Slow, Latent and Temperate Viruses experimental studies by the inoculation of literally thousands of mice with scrapie and many species of non human primates with homogenates of brain from patients that had died with kuru, amyotropic lateral sclerosis–parkinsonism dementia of the Chamorro people on Guam, multiple sclerosis, Pick's and Schilder's diseases and Alzheimer's disease just to name a few. On February 17, 1963, Gajdusek, Tony Morris and I inoculated the first chimpanzee (which we affectionately named Daisey (A-1)) intracerebrally with 0.2 ml of a 10% homogenate of brain from kuru patient Kigea. On August 21, 1963, we inoculated a newborn rhesus monkey, which we allowed Tony Morris to keep in his laboratory, and a second chimpanzee (which we named George (A-4) until we subsequently learned she should be called Georgette) with a 10% homogenate of brain from kuru patient Enage. These early inoculations took place in Gajdusek's laboratory on the NIH campus. Animals were anaesthetized with a light exposure to ether, and injected into left frontal cortex. Within minutes following inoculation the chimpanzees were once again roaming the laboratory and sitting on the secretarys' desks. It is of interest to note that this first rhesus—which we identified as 'Morris monkey' remains alive and well up to the present time—$20\frac{1}{2}$ years after inoculation.

These early inoculations were quickly followed by the acquisition and inoculation of more than 150 non human primates, primarily rhesus and African green monkeys with kuru, ALS, PD and other neurological diseases and the vast majority of these animals were housed in contracted 'isolated' facilities of a private biomedical research facility in Falls Church, Virginia. Six months into this study disaster struck in the form of virulent tuberculosis involving all animals in the contractor's facilities necessitating the killing of these animals and the incineration of their bodies. The TB had been accidently introduced by the contractor.

Meanwhile, in April 1963, scrapie-infected mice were transferred from the NIH campus to the less than adequate facilities at the Patuxent Wildlife Research Center where they were housed in vacant laboratories. As we set up a plethora of experiments on the physical, chemical and biological properties of scrapie as determined *in vivo* in mice and as we continued to isolate additional strains of scrapie from the brains of sheep naturally affected with scrapie, brains provided by Drs James Hourrigan and Albert Klingsporn of the US Department of Agriculture, we also re-initiated and expanded our studies on the human neurological diseases in additional chimpanzees and more than a dozen species of smaller non-human primates. All in all by the end of 1963 I had inoculated about 10 000 mice, 7 chimpanzees, and 75 smaller non-human primates and along with one technician (Mike Sulima who is still working with me) we not only conducted experiments but we did all of the animal care work alone. The actual physical support provided to our work at Patuxent was minimal and my working philosophy was 'get results and we won't have to ask or beg'.

In addition to the early problem caused by the outbreak of TB in monkeys, in September 1963, in response to our request to import scrapie, visna and maedi from Iceland, we were subjected to enormous scrutiny by Dr Robert C. Reisinger, Agricultural Research Service, Animal Inspection and Quarantine Division, USDA, who inspected our Patuxent facilities and concluded (1) the facilities provided inadequate containment and, (2) inadequate airflow, and (3) were not mosquito proof and thus did not provide for acceptable standards for work on scrapie. In his report he paid particular attention to 'our irregular method of acquiring scrapie virus'. He continued his harassment, almost succeeding in blocking our work, until he left the USDA to join the NCI at my strongest recommendation to my sister institute.

While all of this was occurring Gajdusek was in New Guinea and other areas of the South Pacific conducting extended field studies on kuru and other medical problems for almost one full year. In a letter dated January 9, 1964, Carleton writes to me as follows:

> In view of Sir Mac's [MacFarlane Burnett] ⟨speculative⟩ paper on kuru, which is devoid of data and yet so filled with wild guesses without any further quantitation or any really new data, that we shall have a barrage of requests as to whether we have ever thought of this or that which he suggests—it is all there in my earlier papers in more direct, less speculative version—I write to suggest that you get me the data together for a note by yourself and myself to NATURE which must not be over one–two typewritten pages long. I know what I want to say in it, but it might contain a very short table, and that is what I want. Its purpose is to officially document what primate and animal inoculates of frozen kuru inocula have been done and are in progress. In early days we had some ice-refrigerated material reach Melbourne, it was frozen there, and inoculated into animals and tissue cultures in our lab, in Tony's [Morris] lab, and in Nancy Roger's [in Dr Smadel's laboratory] hands. There was even one early specimen passed into chick embryos, mice, tissue culture at Hall Institute in 1957.

Thus, during the early kuru investigations the possibility that the disease was a viral infection or a post-infectious process was seriously investigated even in the face of neuropathological studies which failed to reveal the type of CNS pathology commonly associated with virus infections and clinical-pathology laboratory studies which also failed to reveal the findings common to all known CNS viral infections. As noted in the letter from Gajdusek quoted above, a few brains from kuru patients were preserved in buffered glycerine and, later, even in ice refrigeration within a few hours of death and shipped to Melbourne where they were inoculated in one case into small laboratory animals and chick embryos, and several other cases frozen ($-20°C$) and shipped frozen to the NIH where they were inoculated into mice, chick embryos and tissue cultures with no resulting evidence of infection after three months of observation. Again quoting from a letter by Gajdusek to me in 1964,

> ... after Hadlow pointed out the similarity of kuru and scrapie, we were forced to consider strange slow virus infections, perhaps even with genetic dependence and even—at a widely speculative level—acting somewhat like lysogenic phage and because of unusual heat stability, perhaps masked nucleic acids originating from the host's genetic apparatus itself.

In the October 17, 1964, issue of *nature* (*Nature* **204** (4955) 257–259) our paper entitled 'Attempts to demonstrate a transmissible agent in kuru, amyotropic lateral sclerosis and other subacute and chronic nervous system degenerations of man' gave the details of our work and established our position in the field of slow infections.   In preparing this report I reviewed the original agreement between NINDB and the Bureau of Sports Fisheries and Wildlife, USDA, dated July 27, 1962. In that document Gajdusek had provided a list of diseases that he felt should be included in our inoculation protocols. These were as follows:

In man
Multiple sclerosis
Dementia parkinsonism complex and ALS (Guam)
Kuru
Schilder's disease
Infectious hepatitis
Jacob, Creutzfeldt syndrome
Kozhenikov's epilepsy in the USSR, and other acute epilepsies of childhood
Chronic epidemic encephalitis
Leukoencephalitis (Van Bogaert's)
Inclusion body encephalitis

In primates
Confluent leukoencephalomyelosis
Van Bogaert's disease
Acute amaurotic epilepsy

Wild animals
Staggering gait of deer and moose
Demyelinating processes of raccoons
Lipid dystrophy of coyotes
Infectious hepatitis of ducks

Domestic animals
Scrapie (rida)
Visna in sheep
Lipid dystrophy of dogs

In 1963 Gajdusek and I called the first meeting on slow virus infections. Two years later the proceedings were published as the first monograph in the field. The NIH team was joined by close collaborators from Papua New Guinea (Vin Zigas, Mike Alpers, Frank Schofield), England (Elisabeth Beck) and others from many countries. In addition to being the very first publication to present in detail the current knowledge of slow infections this monograph contained the first reports on transmissible mink encephalopathy and the putative role of 'downer' cattle as the cause of this disease. It also contained an addendum to my paper describing for the first time the development of 'progressive incapacitating cerebellar signs with ataxia, tremors and some wasting lassitude' in the first two chimpanzees (A-1, Daisey and A-4, Georgette) inoculated 20 months and 21 months previously with homogenates of brain from kuru patients Kigea and Enage respectively.

I had never seen a patient with kuru nor had I reviewed the many hundreds of feet of cinema film Gajdusek and Dick Sorgenson had taken of kuru patients and thus I was not at all sure that both chimpanzees were responding to their inoculations or whether they had merely developed intercurrent infections. However, it soon became evident that the symptoms were slowly progressive and they were not associated with any haematological or clinical chemistry abnormalities associated with acute infections. I had not seen or heard directly from Gajdusek for several months while he was in New Guinea so I cabled him describing the conditions of the two chimpanzees. His reply was 'say nothing of this development; we may well have contaminated the animals with scrapie'. I doubted this very much since I had been very careful to keep the scrapie animals and their inocula totally separate from the human disease program. Moreover, I doubted then, and I still doubt, the reports describing the spread of scrapie by contact, or horizontal routes of infection.

As the disease progressed in both chimpanzees we documented their symptoms by cinematography, still photographs and daily written observations. I was convinced that kuru had been transmitted. We established an elaborate nursing schedule during the daylight hours seven days a week and as the animals became progressively worse we extended the nursing hours to 24 h per day. Not being a neurologist I decided to invite some of my colleagues to conduct detailed neurological examinations of the animals. No human patient would have received the medical attention these animals received, being examined by Doctors E.A. Carmicheal, Karin Nelson, Terry Elizan, Richard T. Johnson, Fred Bang, J.R.M. Innes, M.P. Alpers, Morris Victor and a plethora of nationally and internationally recognized neurologists.

The superb nursing care provided by my technical associates, Mr Sulima, Mr Bacote and Mr VanSteinberg, along with my newly arrived staff physician, Dr David M. Asher, allowed us to study the progressive course of the disease over several months with no apparent discomfort to the animals. I really believe the care given was to a large degree provided out of affection that we all had developed for these very remarkable animals and their loss, even in the establishment of a remarkable scientific event, was felt by the staff.

Great plans were made in preparation for the autopsy of the animals. At each step of our planning we notified Gajdusek and requested his input from New Guinea. It was decided that we would fly Mrs Beck, our consultant neuropathologist, from London in order that tissues would be handled correctly and properly fixed for the histopathological studies she would be conducting. I was the prosector assisted by Mike Alpers, Dave Asher and the technical staff. All procedures, conducted under absolute aseptic conditions, went as planned and as Elisabeth boarded the aircraft with the tissues to return to London she commented that 'a Nobel Prize' will come as a result of this breakthrough.

Thus we had established 'slow virus infections' as a cause of subacute progressive degenerative disease of the central nervous system of man. This was followed by the development of experimentally transmitted kuru in a number of additional chimpanzees in rapid succession along with the demonstration that the disease could be serially passaged in chimpanzees. Later the disease was extended to Old World and New World monkeys but not to ordinary laboratory rodents.

The interest and excitement surrounding the demonstration that infection ws the aetiology of kuru was significantly magnified by my paper reporting on the transmission of subacute spongiform encephalopathy of the Creutzfeldt–Jakob type to chimpanzees employing the same techniques employed in the study of kuru. In a way this overshadowed the kuru transmissions since now we were reporting on a disease that was admittedly rare but not unknown to Western medicine. At the time of my report there were approximately 100 cases of CJD reported in the literature. Within 2–3 years following my article we had received case summaries and brain tissues obtained by surgical biopsy or early autopsy on several hundred patients. Our original transmissions to chimpanzees were quickly followed by our transmission of CJD to Old World and New World monkeys and domestic cats and subsequently Jun Tateishi, in Fukuoka, successfully transmitted the disease to mice and Manuelidis transmitted the disease to guinea pigs. The experimental host range for CJD has been expanded by ourselves and others over the years. This provided a much better model disease to study human disease and provided the opportunity to study the transmissible agent's biological, physical and chemical properties as compared to those of scrapie and we found no appreciable differences suggesting that these were separate and distinctly different agents.

It is of interest to note that Igor Klatzo of NINDB, and Meta Newman of St Elizabeth's Hospital, Washington, DC, after studying kuru patient brains histologically, independently suggested that the pathological lesions observed in the brains of kuru patients were strikingly similar to those seen in the brains of patients that had died with Jakob–Creutzfeldt disease. As we searched for cases of CJD we found that this disease had been reported in the literature under at least 25 different synonyms. Thus, the first case brought to our attention by Elisabeth Beck was a brain biopsy specimen exhibiting marked status spongiosis—however, when the patient came to autopsy several months later severe atrophy had resulted in collapse of the cytoarchitecture of the brain and the spongiosis was no longer apparent. When I brought this to the attention of Ken Earle, then Chief of Neuropathology at the Armed Forces Institute of Pathology, he merely commented that European pathologists were splitters whereas American pathologists were groupers. Whatever the case we soon learned that there are wide variations in the clinical picture and in the neuropathologic lesions in what we now call Creuzfeldt–Jakob disease. As an interesting side note, prior to my report on the transmissibility of this 'presenile' dementia the disease was referred to as Jakob's disease or Creutzfeldt's disease or finally Jakob–Creutzfeldt syndrome. Because my initials are CJ, I referred to the disease in my paper as CJD.

Subsequently, along with Colin Master, we identified the transmissibility of the Gerstmann–Straussler–Schenker syndrome as a subtype of CJD and were the first to report an autosomal dominant form of brain disease that had infection as its etiology. These findings led ultimately to the work of other investigators, describing genetic mutations in familial forms of CJD and GSS.

Following our initial transmission of kuru to chimpanzees at the Patuxent Wildlife Center where we were also working on scrapie we were the first to 'cast a stone' at our successful transmissions. I petitioned the Director of our Institute for permission to confirm and extend these transmissions in a 'scrapie-free' environment. After several months searching we located a newly established primate facility in New Iberia,

Louisiana, under the direction of Dr William E. Greer, one of the country's leading veterinary primatologists. We very quickly initiated a greatly expanded inoculation program in chimpanzees, smaller non human Old World and New World primates, domestic dogs, cats, and a variety of other animals. As would be expected we confirmed and extended our findings on kuru, CJD and GSS, and added new species of animals to the host range of these diseases. We have maintained our human disease programme at the New Iberia Research Center for more than 20 years and have valued very highly the friendship and contributions of Bill Greer and his staff.

It is thus a strangely interesting fact that from the investigation of the exotic disease called kuru in a cannibal tribal culture in the interior highlands of Eastern New Guinea that a virus-like organism was first implicated in chronic, progressive, degenerative, non-inflammatory, and heredofamilial diseases of man. These findings have also led to the establishment of a new field of infectious diseases, new to microbiologists, to specialists in neurology, neuropathology and virology. They have made possible the identification of two forms of cerebral amyloidosis—the transmissible dementias and the non transmissible dementias—and have led to significant leads into the basic mechanisms of aging. The studies of myself and Carleton Gejdusek initiated and inspired investigations of many other chronic degenerative diseases of the central nervous system as possible slow virus infections.

Although Gajdusek, Zigas and their coworkers in the early studies of kuru strongly considered infection as the aetiology of kuru it was William J. Hadlow, working with W. S. Gordon on scrapie disease of sheep and goats at the Agricultural Research Council's laboratory in Compton, England, who pointed out similarities between this strange disease in animals and kuru in man. Hadlow's report in a letter to the editor of *Nature* prompted Gajdusek to visit the group at Compton, the Scottish investigators under J. T. Stamp at the Moredun Institute in Edinburgh and Pall Palsson and his veterinary virologists in the laboratory of Bjorn Sigurdsson in Iceland, where slow virus infections of animals were first defined and the criteria for slow infections were first established in the 1950s. This trip led to the establishment of our laboratory of Slow, Latent, and Temperate Viruses. It also led to the formation, development and enduring friendships with colleagues around the world—especially Bill Hadlow and Carl Ecklund for whom I have had the greatest respect. I gained much from my association with Dr Dick Masland, Eckert Wipf, Jenny Arliss and Marion Poms, Carleton's long-term secretary, who made the impossible possible. Never breaking the rules they taught me how to bend them legally in order to get things done in spite of the obstacles. Also, I owe a debt of gratitude to Tony Morris for introducing me to scrapie, to the late Joe Smadel for his counsel, guidance and encouragements and to Jim Hourrigan and Albert Klingsporn who taught me much about scrapie. It goes without saying that I have developed a friendship with Carleton Gajdusek that, despite the differences in our approach to problems, has provided for a unique form of scientific accomplishments—a relationship that has been great to experience and to treasure but one which is not definable.

Finally, although trained as a medical microbiologist with speciality in virology in acute infections, unlike many of my closest colleagues, I approached my involvement in the study of kuru and scrapie in a very positive manner and with enthusiasm assured that we would establish infection as the aetiology of degenerative diseases of the central nervous system. There are those who have stated that what most militated against starting such a project was their own lack of conviction that a transmissible agent would ever be isolated from the diseases in question where in all cases complex autoimmune mechanisms, genetic determination, metabolic derangements and extrinsic toxins or nutrional deficiencies had not been excluded as aetiological possibilities, or those who felt negative about the study for fear that they might become 'zoo keepers' over a large number of non human primates. Their doubts and their fears prevented them from experiencing the thrill of discovery and the establishment of new frontiers in medicine.

REFERENCES

Gajdusek, D.C. (1962) Letter to C.J. Gibbs, Jr, from Moraei-Teiwan ground near Morandugai, Southwestern Section of Kuk-type People, January 18, 1962.

Gajdusek, D.C. (1964a) Letter to C.J. Gibbs, Jr, from Amusa, Gimi Ling Group, January 9, 1964.

Gajdusek, D.C. (1964b) Letter to C.J. Gibbs, Jr, from Wanitabi, South Fore, Okapa, E.H., January 24, 1964.

Gajdusek, D.C., Gibbs, C.J., Jr, and Alpers, M. (eds) (1965) *Slow, Latent, and Temperate Virus Infections.* NINDB Monograph No. 2, Public Health Service Publication No. 1378, US Government Printing Office, Washington, DC.

Gajdusek, D.C., Gibbs, C.J., Jr, and Alpers, M. (1967) Transmission and passage of experimental — kuru — to chimpanzees. *Science* **155** (3759) 212–214.

Gibbs, C.J., Jr, and Gajdusek, D.C. (1965) Attempts to demonstrate a transmissible agent in kuru, amyotrophic lateral sclerosis, and other subacute and chronic progressive nervous system degenerations of man. In: Gajdusek, D.C., Gibbs, C.J., Jr., and Alpers, M. (eds) *Slow, Latent, and Temperate Virus Infections.* US Government Printing Office, Washington, DC, pp. 39–48.

Gibbs, C.J., Jr, and Gajdusek, D.C. (1969) Infection as the etiology of spongiform encephalopathy (Creutzfeldt–Jakob disease). *Science* **165** (September 5) 1023–1025.

Hadlow, W.J. (1959) Scrapie and kuru. *Lancet* **2** (7097) 289–290.

9

Prion diseases from a neuropathologist's perspective

Elisabeth Beck and P. M. Daniel

Our interest in the transmissible spongiform encephalopathies (prion diseases) dates back to 1962. At that time we were engaged, together with H. B. Parry, in the systematic investigation of the brains from a series of 34 sheep with natural scrapie. Our studies revealed, apart from generalized lesions, a degeneration of the olivo-ponto-cerebellar and the hypothalamo-neurohypophysial systems (degeneration in the latter system was often as severe as in animals, previously studied, in whom the pituitary stalk had been severed (Beck *et al.*, 1969b)). With degeneration in these two systems a pathological basis was provided for some of the clinical signs such as ataxia, tremor and diabetis insipidus from which many of the animals suffered (Beck *et al.*, 1964). In sheep and goats with experimental scrapie an identical picture was seen.

In the early 1960s Dr Gajdusek visited England and came to see us at the Institute of Psychiatry in London. He enquired whether we would be interested to study some kuru material, an offer which we readily accepted. By 1965, when the workshop was organized, we had received the brains from eight cases dying of kuru. These formed the basis for our paper which was read at the meeting (Beck and Daniel, 1965). In the early stages of kuru research status spongiosus of the gray matter, if present at all, was only slight. In fact, it was not noted by Klatzo *et al.* (1959) in the 12 cases which they examined, nor was it an outstanding feature in our own material (as years went by it became more marked). On the other hand we were impressed by the severe degeneration of the olivo-ponto-cerebellar system in every case with a marked atropy of the cerebellar vermis, particularly of its vestibular component. Such pathological features correlated well with the clinical symptoms. We were struck by the overall similarity of the neuropathological changes to those found in scrapie and in certain human system degenerations. We even contemplated '...whether some "agent" may be implicated in the aetiology of such conditions'.

Since the 1965 symposium we have closely collaborated with the NIH team and were involved in many of their exciting discoveries.

The next step was the successful transmission of kuru to the chimpanzee in 1966 (Gajdusek *et al.*, 1966) when I was summoned to Bethesda to collect the brain of this animal. When the microscopic sections came through they presented the most amazing picture. The entire gray matter, but especially the cortical mantle, was affected by the severest grade of status spongiosus we had ever come across with no obvious atrophy. Degeneration of the olivo-ponto-cerebellar system, which we had expected to see, was rather moderate (Beck *et al.*, 1966, 1973). Thus, the neuropathological appearance of the brain was much more akin to what was then known by British neurologists as 'subacute spongiform encephalopathy' than it was to kuru. (Terminology initially led to some confusion since a distinction between subacute spongiform encephalopathy and Creutzfeldt–Jakob disease with myoclonus, as common in Britain, is not recognized in the United States. Here both conditions are designated Creutzfeldt–Jakob disease.) It therefore occurred to us that in 'subacute spongiform encephalopathy' we may have another neurological disease which could be transmitted to animals. In due course a biopsy specimen was obtained from a typical case and sent to Bethesda. When a chimpanzee was eventually inoculated with this material, the animal developed neurological disease after an incubation time of 13 months. Thus, in 1968 Creutzfeldt-Jakob disease was shown to be a second of the obscure neurological conditions transmissible to animals (Gibbs *et al.*, 1968; Beck *et al.*, 1969a). It is noteworthy that the success of this transmission was solely due to morphological observations.

One further point to which neuropathology could provide an answer wanted elucidation, i.e. what are the first lesions to be recognized in the brain during early stages of incubation, before the severe pathological picture has obliterated every clue and in animals which do not happen to die of intercurrent disease? To answer this question, ten spider monkeys were inoculated with kuru material. They were killed, under anaesthesia, by perfusion of the brain with fixative at given intervals, while healthy and free from neurological signs. These brains were studied at the Max-Planck-Institut für Hirnforschung in Frankfurt in the Departments of Professors Hassler and Krücke, where one of us (E.B.) received a grant for 2 years. The essential results pointed to a re-activation of embryonic growth mechanisms (Beck *et al.*, 1975, 1982).

REFERENCES

Beck, E. and Daniel, P.M. (1965) Kuru and scrapie compared: are they examples of system degeneration? In: Gajdusek, D.C., Gibbs, C.J. and Alpers, M. (eds) *Slow, Latent and Temperate Virus Infections*. Monograph No. 2, National Institute of Neurological Diseases and Blindness, US Government Printing Office, Washington, DC, pp. 85–93.

Beck, E., Daniel, P.M. and Parry, H.B. (1964) Degeneration of the cerebellar and hypothalamo-neurohypophysial systems in sheep with scrapie; and its relationship to human system degenerations. *Brain* **87** 153–176.

Beck, E., Daniel, P.M., Gajdusek, D.C. and Gibbs, C.J., Jr (1966) Experimental "kuru" in chimpanzees. A pathological report. *Lancet* **ii** 1056–1059.

Beck, E., Daniel, P.M., Matthews, W.B., Stevens, D.L., Alpers, M.P., Asher, D.M., Gajdusek, D.C. and Gibbs, C.J., Jr (1969a) Creutzfeldt–Jakob disease. The neuropathology of a transmission experiment. *Brain* **92** 699–716.

Beck, E., Daniel, P.M. and Pritchard, M.L. (1969b) Regeneration of hypothalmic nerve fibres in the goat. *Neuroendocrinology* **5** 161–182.

Beck, E., Daniel, P.M., Asher, D.M., Gajdusek, D.C. and Gibbs, C.J., Jr (1973) Experimental kuru in the chimpanzee: a neuropathological study. *Brain* **96** 441–462.

Beck, E., Bak, I.J., Christ, J.F., Gajdusek, D.C., Gibbs, C.J., Jr, and Hassler, R. (1975) Experimental kuru in the spider monkey. Histopathological and ultrastructural studies of the brain during early stages of incubation. *Brain* **98** 595–612.

Beck, E., Daniel, P.M., Davey, A.J., Gajdusek, D.C. and Gibbs, C.J., Jr (1982) The pathogenesis of transmissible spongiform encephalopathy. An ultrastructural study. *Brain* **105** 755–786.

Gajdusek, D.C., Gibbs, C.J., Jr, and Alpers, M. (1966) Experimental transmission of a kuru-like syndrome to chimpanzees. *Nature (London)* **209** 794–796.

Gibbs, C.J., Jr, Gajdusek, D.C., Asher, D.M., Alpers, M.P., Beck, E., Daniel, P.M. and Matthews, W.B. (1968) Creutzfeldt–Jakob disease (spongiform encephalopathy): transmission to the chimpanzee. *Science* **161** 388–389.

Klatzo, I., Gajdusek, D.C. and Zigas, V. (1959) Pathology of kuru. *Lab. Invest.* **8** 799–847.

10

Reflections and highlights: a life with kuru

Michael Alpers

For the whole of my professional life I have been involved with kuru. I first read about it in the local newspaper when I was a medical student. I thought at once that here was the opportunity I was looking for. I am still seeing patients with kuru and, though I had for a long time anticipated documenting the last case of the disease, it seems now that the disease will outlast my professional life.

I remember the excitement of reading about the disease. Being in medical school was the most stultifying experience of my life. I was interested in neurology, genetics, paediatrics, and—especially—anthropology. Kuru had it all. Moreover, I had to get away. It was a perfect solution. Norrie Robson, the Professor of Medicine in Adelaide, was very sympathetic, and the matter was arranged. Nevertheless, it was two years or so before I made it to Okapa. First I graduated, then worked in the Adelaide Children's Hospital and finally had special training in neurology. Even after I reached Papua New Guinea I had 6 weeks in Goroka studying linguistics before I ever saw Okapa. However, from the moment that my aspirations to study kuru had been given Norrie Robson's full support I was a person with a purpose, and inwardly quite transformed.

This was 1959. I had never had cause to use the medical library before. Now I was there devouring all the early papers on kuru (beginning with Gajdusek and Zigas (1957) and Zigas and Gajdusek (1957)). There were not many but I kept busy since I was swept away by every end reference. Then Bill Hadlow's letter came out in the *Lancet* (Hadlow, 1959). I read everything that I could find about scrapie—which was not a great deal. I could not believe that I had stepped into this magic world. My grades in medical school became even worse until I realized that somehow I had to pass in order to be able to realize my new-found opportunities. However, my intellectual life was reserved for the accumulating pile of notes I was taking in the library. I knew all those in Adelaide who had seen kuru patients or had worked on kuru material: apart from Norrie Robson there was Henry Bennett, Professor of Genetics and an old friend of mine from Cambridge (Bennett *et al.*, 1958), Harry Lander, Donald Simpson (Simpson *et al.*,1959), Clive Auricht, Bronte Gabb, Donald Perriam and the pathologist Malcolm Fowler from the Children's Hospital. Malcolm used to demonstrate to medical students in pathology; I have a strong memory of ending a practical class with a long argument with him over the merits of Dr Zhivago. Malcolm wrote an excellent paper on the pathology of kuru (Fowler and Robertson, 1959) with Graeme Robertson, the doyen of Australian neuropathologists, who worked in Melbourne and whom I did not meet for a number of years. Malcolm is also immortalized in *Naegleria fowleri*, but that is another story.

I did make it to Okapa, with a young family (a son was born later in Papua New Guinea—at Yagaum, where I go now to work on malaria, but that too is another story). We were met at the Tarabo Mission airstrip by the Assistant District Officer (ADO) and driven the 15 kilometres to Okapa station. The car stopped at various points along our progress so that we could be given a good demonstration of 'berating the natives'. The ADO soon learned that in this particular school I was an even worse pupil than I had been in medical school.

I was introduced to kuru by Andrew Gray, the doctor in Okapa. I previously had met Andrew in Adelaide, when he was undertaking a short period of special training in obstretics at the Queen Victoria Hospital. Why he was in Okapa and what his principal brief was from the government is a long and interesting story, not for this occasion. When Andrew left Okapa he gave me all his books, which I still have and which gave me new insights into English and French literature. One of the questions that had emerged from my reading of the kuru literature was whether it was clinically a cerebellar or an extrapyramidal disease. One of my aims was to follow a number of patients throughout the course of their disease to determine the clinical evolution of kuru in detail and to try and resolve this question. This I was eventually able to do and showed that kuru while predominantly a cerebellar disease did show extrapyramidal features, particularly at certain stages, often dramatic in their intensity but quite fleeting. The first few cases I saw were purely cerebellar and since kuru has a remarkably similar course in all patients I thought this had settled the issue; however, as I learned more I began to realize that the truth did not simply reside in one-half of the dichotomy but was more complex. The clinical descriptions that were painstakingly gathered from patients seen in their home setting over the next two years were written up when I returned to Adelaide and later published in mimeographed form by the National Institutes of Health (Alpers, 1964). Subsequent observations with neurologist colleagues have confirmed this clinical view of kuru (Alpers and Rail, 1971; Prusiner *et al.*, 1982).

The very first patient that I saw with kuru did not in fact have kuru. There were no patients in the hospital in Okapa with the disease, despite the fact that it had been built by the Lutheran Mission for the express purpose of looking after kuru patients; the patients did not want to come and Andrew Gray refused to admit them unless they had some other concomitant clinical condition. Kuru was considered by the local people to be caused by sorcery—on that there are many stories, best recorded by the two anthropologists Bob Glasse and Shirley Glasse (later Lindenbaum) who were working in the South Fore at this time, also under the auspices of the University of Adelaide (Lindenbaum, 1979). Sorcery was always performed by a sorcerer; often in retaliation for a death from kuru an alleged sorcerer would be murdered. One such murderer was in jail at Okapa when I arrived and it was said that he had come down himself with kuru. Andrew and I examined him and I followed him carefully. It became apparent that he had many bizarre signs and clear features of hysterical disease. We were convinced that he did not have kuru but everybody else was certain that he did and that it was advancing at a galloping rate. The ADO decided that he should be released from jail and sent home immediately. I subsequently examined and followed him up in his home village; he then moved to a village called Weya which was very remote, a full day's walk from my own base in Waisa in the South Fore. I took movies of his clinical condition, which was unchanged from what it had been in jail. However, on a subsequent visit I learned that he was recovering from kuru and indeed this was quite clear from the clinical signs. He eventually made a full recovery. We have never seen any patient with unequivocal signs of kuru recover and so of all the many 'recoveries' from kuru recorded in our files there are none that we believe are recoveries from true kuru. In most cases the disease is another one which temporarily gives shaking and weakness—in recent years this has most commonly been malaria. This causes the patient and her (or his) relatives great concern and when recovery eventually ensues it is believed that the patient has recovered from kuru, particularly if traditional measures have been undertaken to heal the disease. Occasionally, as in the case I first encountered, recovery occurs from hysterical kuru; this particular case is the best documented that we have of this phenomenon. The experience had a secondary benefit for my work in the area since it was believed that the movie camera of the new doctor who had started to wander around looking at kuru patients had been instrumental in curing a kuru patient from Weya: consequently requests came in from all the communities around me asking that I come and see their patients and offering me every inducement to do so. I was gratified by the support and assistance that I was receiving from the communities in which I worked but I could not initially understand this extraordinary new wave of enthusiasm until it finally dawned on me—probably after the third or fourth explanation of the matter by one of my assistants—what the basis for it was.

I remember my first encounter with Carleton Gajdusek in Okapa. Carleton was a mythological figure. About him I knew many complex and contradictory myths so that meeting this extraordinary man and finding that indeed he was a man was rather an anticlimactic experience. More importantly, he was a person whom I could relate to (I am a good listener) and with whom I was immediately on sympathetic terms. The meeting began a life-long warm and collegial friendship. Not long after meeting Carleton I set off with him on a walk between Okapa and Menyamya, during which, *inter alia*, we encountered a patrol officer making the first government patrol to the area across which we were walking, and we discovered that Mount Yelia, one of the major peaks on our route, was a by no means dormant volcano. It was a marvellous journey.

Frank Schofield, the Assistant Director for Research in the Department of Public Health, was my boss. He was based in Port Moresby. There were not many members of the research division: myself and the small group working in Maprik on Frank's project on neonatal tetanus. How I got to be in the group in the first place is a rather complex story. In brief, when Norrie Robson first agreed to send me to work on kuru he had a Rockefeller Fellowship in mind, of the same kind which supported Bob and Shirley Glasse in the field. In the meantime, however, the government of the then Australian Territory of Papua and New Guinea had decided that since the current dominant hypothesis was that kuru was a genetic disease the whole of the kuru region should be quarantined in order to prevent the spread of this fatal gene (see Dobzhansky, 1960). To compensate for this restriction on human movement the government decided to appoint a second doctor for the area who would be engaged full-time on research into the disease. Robson was approached to see whether he had anyone who might be prepared to come and fill this role and of course he already had somebody ready and waiting. So I came as a government medical officer not as a poor research fellow on a grant. This was, incidentally, the only effect of the government's quarantine policy since they wisely never attempted to carry it out. Nobody had a clue what I should be doing but that was good for me since I knew what I wanted to do and, with all the arrogance of youth, believed that somehow it would be useful. After my first meeting with Carleton I turned in my lot with him—and we have collaborated ever since. When I set out from Okapa to walk to Menyamya I was not going officially beyond Agakamatasa, the village overlooking the Lamari River on the edge of Fore territory where Carleton had a house. I was already living in the village of Waisa and my house was being built there, but I also had a base in Okapa. While in Agakamatasa, I received a message that Frank Schofield had arrived in Okapa to see me and so we arranged to meet half-way between Agakamatasa and Purosa, where the road from Okapa ended. We sat on the edge of the path in the middle of the forest and discussed work. Frank seemed to be happy about my general plans for research work and was kind enough to give me a free hand. After our meeting he returned to Purosa and then to Okapa and I walked back to Agakamatasa. When I subsequently disappeared to cross the Lamari with Carleton and to walk through Anga territory to Menyamya I was, as Frank subsequently explained to me, strictly AWOL. My family and the authorities in Okapa felt the same. However, I was by then hard back at work in the bush in ways more directly related to kuru and my misdemeanour was forgiven.

I remember my first autopsy. Of course I remember them all but the first one stands out particularly, mainly because of the uncertainty as to whether I could successfully carry it out. Everything had been prepared and all I could do was wait for the patient to die, be with him and his family as much as was seemly and live quietly nearby until the shouted word came that he was dead. In the house where I had seen him many times alive his father and I performed the autopsy, with my two assistants standing by, and to give me sufficient room in the confined space of the small house as few other observers as we could achieve. I made everybody wear masks. It was a limited autopsy of the brain only. Small pieces were taken in sterile fashion and placed into pre-labelled sterile bottles. The whole brain was removed and placed in a bucket of formalin. I then replaced the skull cap and sewed up the scalp. My assistants and I carried the covered bucket of formalin, the sterile samples, which were protected in an insulated box, and my instruments to Okapa. The samples were placed in the station freezer. The next morning I arranged a charter from Goroka that would pick me up at the airstrip at Tarabo and then fly me and the samples to Lae. There I placed them in the deep freezer at the hospital and obtained dry ice. The following morning the shipment went off in dry ice to Roy Simmons at the Commonwealth Serum Laboratories in Melbourne. I spoke to Roy from Lae giving him the details of the shipment. Roy held the samples at $-70°C$ until he was able conveniently to despatch them to Carleton at the National Institutes of Health. I returned by charter to Tarabo and then to Okapa. Then I went back to the patient's village and took part in the funeral ceremonies, contributing something to the mortuary feast and mourning as earnestly as the other participants. When the brain had been well fixed I sent it by air freight to the National Institutes of Health. This was the pattern which was repeated on subsequent occasions. However, once it had been successfully achieved the uncertainty and concern that the whole operation might end up as a total and embarrassing disaster was dispelled and the procedure, though equally emotionally draining, was less of a strain—and less of a triumph when successfully accomplished. However, one other particular autopsy was especially memorable since it was that of a young girl from my own village of Waisa whom I knew particularly well and whose father was a good friend. She took a long time in her terminal moribund state to die: so long indeed that her father left her after more than a week of incessant watching and care, telling her that she had better hurry up and die since he could take it no longer. It was three more days before he and I, living nearby, were told that she had just died and we both met in the house again to conduct the autopsy. Since until not many years before it had been the practice to cut up and consume their dead at each mortuary feast Fore men were accustomed to cutting into dead relatives: it was part of the ritual of death and they were much more experienced at it than I was. However, I did have scalpel, forceps, scissors and a saw—which helped us all to get the matter over with quickly and efficiently.

The other studies I carried out were epidemiological. This led to work at the National Institutes of Health rechecking the whole kuru file held there, the early records of Carleton, Vin Zigas and Jack Baker (Gajdusek *et al.*, 1961), the work of the many patrol officers who succeeded Jack at Okapa and my own more recent records, and putting the epidemiological information on computer. All my initial analyses, however, were done by hand. It was an exciting moment when Pat Hunt (now Kelly) and I added up all the figures by age group and divided them into the early period from 1957 to 1959 and the more recent one from 1961 to 1963. There was an epidemiological impression from the field that the disease was becoming less common in children and this impression was sustained as one worked systematically through the whole file. However, until the data had been cleaned I made no analyses. Suddenly there were the figures in front of us: kuru had declined indeed in children but in the younger age group it had essentially disappeared. This was an exciting new fact about kuru that cried out for an explanation. For the moment,however, all I could do was wonder at this revelation that shone through our dull charts. Further analyses revealed other changing patterns in the epidemiology of kuru (Alpers and Gajdusek, 1965) but nothing with the force of this first discovery.

At the same time I was working as a chimpanzee neurologist following the animals that had been inoculated with the kuru brain material so carefully obtained from my patients in Papua New Guinea. Material from other neurological diseases had also been inoculated to test their transmissibility and, most importantly as it turned out, to act as controls for the test of the transmissibility of kuru. I spent many hours patiently and regularly documenting the clinical state of these animals. They were held in a small facility in the US Wildlife Service establishment in the woods of Maryland at Patuxent. Joe Gibbs was in charge out there assisted by Mike Sulima and Al Bacote. About two years after inoculation one of the animals was noted by her attendants to be behaving strangely. This progressed. We had inoculated these animals with the intention of testing the transmissibility of kuru. However, the neurological and movie documentation of their clinical state became a regular chore, a routine procedure that was conducted as objectively, efficiently and mindlessly as possible. I remember one day going into the anteroom where I wrote my clinical report and after dashing it off in the usual fashion I made my clinical summary at the end, almost mechanically: 'clinical impression—kuru'. The word leaped from the page. I discussed the matter with Joe and we decided that Carleton should be brought back especially from the field to observe this phenomenon. He was somewhat reluctant to interrupt his field trip but such a request could hardly be ignored. On his return I was a little anxious. However, if the diagnosis of kuru had to be first discerned through the dry sequence of clinical notes it was magnificently displayed in the performance of the chimpanzee when Carleton first came to see her. For on this particular morning the resemblance to the human disease was uncanny. Then an enormous amount of work that had nothing directly to do with kuru began and our animals were tested for every trace element deficiency or poisoning imaginable and every micronutrient deficiency that could be tested for. We found nothing. Eventually the progress of the disease was such that we needed to kill the first of our diseased animals: an option not available for the human patients whom we could only helplessly watch die. The neuropathologist in the team, Elisabeth Beck, came over from England and she, Joe and I planned and performed the autopsy. Elisabeth took the brain back with her to London and we waited for good fixation (Elisabeth would not be hurried on that point). Then one day, unexpectedly, a telegram was received from Elisabeth saying that the neuropathology of the chimpanzee brain was indistinguishable from human kuru. The next day the report on the transmissibility of kuru was written and submitted to *Nature* (Gajdusek *et al.*, 1966).

We had changing patterns in the epidemiology of kuru and now the demonstration of transmissibility. How did they all fit together? Somehow they seemed to support the old idea of cannibalism being involved in the pathogenesis of the disease. Furthermore, though an old idea it kept recurring whenever kuru was discussed (see the discussion to Alpers, 1965). Bob and Shirley Glasse had made a special study of the details of endocannibalism among the Fore (Glasse, 1967; Lindenbaum, 1979). Yet the accepted mechanism by which cannibalism could lead to kuru was an autoimmune one, and we now had a virus as the cause of the disease. By conventional arguments of hypothesis and counterhypothesis cannibalism was clearly rejected as the cause. Yet some change that had taken place in the kuru region had to explain the cohort of children—now only just apparent but very real—growing up free of kuru. We listed everything we could think of (Alpers, 1965). There were many possibilities and we tossed them all around—but none made sense. Then one morning, working at home in my makeshift office in the porch of our house in Bethesda preparing a paper for a meeting of the International Academy of Pathology I was waving all these possibilities around in my mind when suddenly the pieces of the puzzle clicked into place. Cannibalism, though it seemed to be a plausible explanation, did not on close analysis explain the phenomena as an independent mechanism. However, cannibalism as the single mode of transmission of the transmissible virus of kuru did make sense: it was suddenly all too painfully obvious. It was obvious because nothing further needed to be explained, and painfully so because of the agonies of uncertainty that existed when the explanations seemed so close at hand and yet not quite there: until that moment when it all clicked into place. But what a moment! Endocannibalism as the mode of transmission of the transmissible agent of kuru explained the sex and age distribution of the disease, since it was women and children who ate the infective brain and not the men; and the cohort of children growing up who were free of kuru had also grown up in a community now free of endocannibalism; since the practice was endocannibalism and only relatives were consumed the disease was familial in its distribution. Fortunately the disease was not transmitted vertically from infected mother to child; and that too was now clear. These explanations formed the basis of my presentation at the meeting (Alpers, 1968).

We now had an explanation for the transmission of kuru in its epidemic form in the kuru region, but how did the disease get there in the first place? Also, how was the sporadic Creutzfeldt–Jakob disease—which was the world-wide equivalent of kuru which we subsequently showed also to be transmissible experimentally to chimpanzees—transmitted naturally, at an incidence of about 1 per million annually. The first answer was that whatever the mechanism for these phenomena it was likely to be the same for both. This insight came as I put together a paper with Leonard Rail for presentation at the annual meeting of the Australian Association of Neurologists (Alpers and Rail, 1971). By then I had moved to Perth in Australia. All my plans for working on kuru and Creutzfeldt–Jakob disease in Perth came to nought—with one exception. Why my work proved to be such a failure there is not a matter to go into now. The one exception was my annual visit to the field to continue work on the epidemiology of kuru and my regular visits to the National Institutes of Health to update the epidemiology and clinical files there. However, further insights did come in Perth. There had long been speculations as to the nature of the scrapie agent; now that Hadlow's hunch had proven to be true the agent of kuru was clearly the same or at least of the same kind as that of scrapie. Thinking and puzzling over the nature of these agents occupied many of us over many years. The first breakthrough in my own thinking had arisen—once again—when putting together a paper for presentation at a meeting. This was the essential nature of viruses as genomic parasites (Alpers, 1969). This concept informed my teaching of virology in Perth and my own attempts to clarify the nature of the infectious agent of kuru, but never matched with the concept of most virologists, who continued to be more comfortable with the view of viruses as virions, the smallest of the microorganisms.

The discussions over the nature of the scrapie agent had focused on the question posed by Alper *et al.* (1967) in the title of their paper in *Nature*—does the agent of scrapie replicate without nucleic acid? The question has been confused with the cognate but different one, does the agent of scrapie contain nucleic acid? Anybody working on scrapie—or kuru, or Creutzfeldt–Jakob disease—had to be thinking and speculating about this problem, whether or not one was able to do anything about it. My next insight came when preparing an abstract for a meeting on microbial ecology in Dunedin. I remember discussing this idea tentatively with Neville Stanley, friend and colleague from the University of Western Australia, and becoming certain of it by the end of our conversation. The substance of the idea was that these agents were host coded and their 'transmission' in the sporadic form was through the host genome (Alpers, 1977; in more accessible form Alpers, 1979). The first case of kuru in the region had been a sporadic one, a rare form of a rare but cosmopolitan human disease; it unfortunately became the source of epidemic kuru, which was spread by endocannibalism. However, horizontal transmission, as in epidemic kuru, still required an independent entity, an agent, and the nature of that agent continued to elude us.

It was Stan Prusiner who addressed that problem directly, boldly taking up the old hypothesis that the agent was a pure protein and trying to disprove it. He brought a new vitality and a new mix of intellectual and technological resources to the scrapie field. I have many happy memories of discussions and collaborative interactions with Stan, in his department in San Francisco and in the field in Papua New Guinea—by this time I had moved from Perth to the Papua New Guinea Institute of Medical Research in Goroka. I remember particularly a long lunch over seafood and chardonnay in which Stan propounded to me all his latest results on the restaurant tablecloth; and a very long day's walk in the bush to see a kuru patient; and the vicissitudes but ultimately successful outcome of an experimental study of cannibalism (Prusiner *et al.*, 1985).

Stan's work has flourished and it now seems clear that the hypothesis he made his own is not going to be disproved—by him or anyone else. The good science, the rigour, the prodigious team-work have paid off. He and his colleagues have found the answer to the question, does the agent of scrapie contain nucleic acid? It does not. Unfortunately, in my view, he invented and promoted a new terminology before he had solved the problem (Prusiner, 1989). Carleton's group has also gone into the molecular genetics of these agents, with considerable success, and Carleton has developed innovative ideas on their nature (Gajdusek, 1990). The ideas converge, but no-one would get that impression from reading papers from the two groups: it is as if they were operating in two independent worlds.

In writing a paper on the human biology of kuru (Alpers, 1992) I decided to add a note which would try to reconcile these viewpoints (even if the principals could not be brought together). In doing so I wrestled with the last conceptual problem I had with the idea of replicable conformational change in proteins being the basis of the amyloid precursor agents (prions, if you like) of scrapie–kuru. The rhetorical impression had been created that this phenomenon, since it did not involve nucleic acid, was non-biological. This had to be nonsense, but exactly how was one to make sense of the phenomenon? We were being asked to accept the notion that scrapie replicates without nucleic acid, based on a concept of conformational replication, and that the answer to Tikvah Alper's question was yes. My last memory in this series of reminiscences is the relatively recent one of waking up on a Saturday morning in Goroka with a clear realization that the answer to Tikvah Alper's question was a resounding no: conformational change was a fundamental biological property (well exemplified by much recent work in immunology) and the biological opportunities for—and constraints to—such conformational replication in proteins were set by the primary amino acid sequence coded in nucleic acid. Of course the agent of scrapie does not replicate without nucleic acid: even if it contains no nucleic acid and induces its own form on congeneric proteins, and in this limited sense does not require nucleic acid to replicate itself, if the precursor proteins were not genomically coded for there would be no replication. We were back to the key concept of host coding of these agents—except that now, with all this recent work, it was no longer necessary to postulate nucleic acid coming in and out of the genome (which can and does happen with viruses) to explain the transmissibility of these agents: in genetic terms the host genome did it all. This seems all too obvious—as always—but it had not been made explicit but rather obfuscated in the literature. In view of the membrane location of these 'amyloid precursor' proteins in their natural function in the cell it is likely that this replicative mechanism will not be found restricted to scrapie–kuru, or degenerative neurological diseases, or amyloidology, or infectious disease processes, but rather to be a widespread biological phenomenon. For me, at least, the conceptual journey had reached its destination. The old hypothesis, now substantiated (an appropriate word), made biological sense and was in fact in the mainstream of modern biological concepts emerging in a number of fields. That Saturday morning was a very gratifying one for me: even if the friends and colleagues to whom I sent copies of the paper were not all as excited about it as I was!

REFERENCES

Alper, T., Cramp, W.A., Haig, D.A. and Clarke, M.C. (1967) Does the agent of scrapie replicate without nucleic acid? *Nature (London)* **214** 764–766.

Alpers, M.P. (1964) *Kuru:a Clinical Study.* US Department of Health, Education and Welfare, Public Health Service, Washington, DC, 38 pp.

Alpers, M.P. (1965) Epidemiological changes in kuru, 1957 to 1963. In: Gajdusek, D.C., Gibbs, C.J., Jr, and Alpers, M.P. (eds) *Slow, Latent and Temperate Virus Infections.* National Institute of Neurological Diseases and Blindness, National Institutes of Health, Public Health Service, US Department of Health, Education and Welfare, Washington DC, pp. 65–82.

Alpers, M.P. (1968) Kuru: implications of its transmissibility for the interpretation of its changing epidemiological pattern. In: Bailey, O.T. and Smith, D.E. (eds) *The Central Nervous System, Some Experimental Models of Neurological Diseases.* Williams and Wilkins, Baltimore, MD, pp. 234–251.

Alpers, M.P. (1969) Kuru: clinical and aetiological aspects. In: Whitty, C.W.M., Hughes, J.T. and MacCallum, F.O. (eds) *Virus Diseases and the Nervous System.* Blackwell Scientific Publications, Oxford, pp. 83–97.

Alpers, M.P. (1977) The ecology of kuru and other subacute spongiform encephalopathies. *Programme and Abstracts of the International Symposium on Microbial Ecology, Dunedin, August 22–26, 1977,* Abstract C13, paper mimeographed, 5 pp.

Alpers, M.P. (1979) Epidemiology and ecology of kuru. In: Prusiner, S.B. and Hadlow, W.J. (eds) *Slow Transmissible Diseases of the Nervous System,* Vol.1. Academic Press, New York, pp. 67–90.

Alpers, M.P. (1992) Kuru. In: Attenborough, R.D. and Alpers, M.P. (eds) *Human Biology in Papua New Guinea: the Small Cosmos.* Oxford University Press, Oxford, pp. 313–334.

Alpers, M.P. and Gajdusek, D.C. (1965) Changing patterns of kuru: epidemiological changes in the period of increasing contact of the Fore people with western civilization. *Am. J. Trop. Med. Hyg.* **14** 852–879.

Alpers, M.P. and Rail, L. (1971) Kuru and Creutzfeldt–Jakob disease: clinical and aetiological aspects. *Proc. Aust. Assoc. Neurol.* **8** 7–15.

Bennett, J.H., Rhodes, F.A. and Robson, H.N. (1958) Observations on kuru. I. A possible genetic basis. *Australas. Ann. Med.* **7** 269–275.

Dobzhansky, T. (1960) Eugenics in New Guinea. *Science* **132** 77.

Fowler, M. and Robertson, E.G. (1959) Observations on kuru. III. Pathological features in five cases. *Australas. Ann. Med.* **8** 16–26.

Gajdusek, D.C. (1990) Subacute spongiform encephalopathies: transmissible cerebral amyloidoses caused by unconventional viruses. In: Fields, B.N., Knipe, D.M., Chanock, R.M., Hirsch, M.S., Melnick, J.L., Monath, T.P. and Roizman, B. (eds) *Fields Virology,* 2nd edn. Raven Press, New York, pp. 2289–2324.

Gajdusek, D.C. and Zigas, V. (1957) Degenerative disease of the central nervous system in New Guinea. The endemic occurrence of 'kuru' in the native population. *N. Engl. J. Med.* **257** 974–978.

Gajdusek, D.C., Zigas, V. and Baker, J. (1961) Studies on kuru. III. Patterns of kuru incidence: demographic and geographical epidemiological analysis. *Am. J. Trop. Med. Hyg.* **10** 599–627.

Gajdusek, D.C., Gibbs, C.J., Jr, and Alpers, M.P. (1966) Experimental transmission of a kuru-like syndrome to chimpanzees. *Nature (London)* **209** 794–796.

Glasse, R.M. (1967) Cannibalism in the kuru region of New Guinea. *Trans. N Y Acad. Sci.* **29** 748–754.

Hadlow, W.J. (1959) Scrapie and kuru. *Lancet* **2** 289–290.

Lindenbaum, S. (1979) *Kuru Sorcery: Disease and Danger in the New Guinea Highlands.* Mayfield Publishing Company, Palo Alto, CA.

Prusiner, S.B. (1989) Scrapie prions. *Annu. Rev. Microbiol.* **43** 345–374.

Prusiner, S.B., Gajdusek, D.C. and Alpers, M.P. (1982) Kuru with incubation periods exceeding two decades. *Ann. Neurol.* **12** 1–9.

Prusiner, S.B., Cochran, S.P. and Alpers, M.P. (1985) Transmission of scrapie in hamsters. *J. Infect. Dis.* **152** 971–978.

Simpson, D.A. Lander, H. and Robson, H.N. (1959) Observations on kuru. II. Clinical features. *Australas. Ann. Med.* **8** 8–15.

Zigas, V. and Gajdusek, D.C. (1957) Kuru: clinical study of a new syndrome resembling paralysis agitans in natives of the Eastern Highlands of Australian New Guinea. *Med. J. Aust.* **2** 745–754.

11

Fieldwork in the South Fore: the process of ethnographic inquiry

Robert Glasse and Shirley Lindenbaum

Contemporary interest in the history and philosophy of science, as well as in some branches of anthropology, has turned increasingly to the way in which medical knowledge is produced. The full story of kuru research, situated in the differing practices of social and medical science, remains to be told. As a contribution to an understanding of the early days of kuru investigation, however, we outline here the way in which we conducted our fieldwork among the Fore of Papua New Guinea. We will limit our discussion to a few key issues (the history of kuru, the practice of cannibalism, and the different disciplinary practices in medicine and anthropology) rather than attempt a broader consideration of the sociology of knowledge in the two disciplines.

Our choice of kuru research resulted from an invitation by Dr J. H. Bennett, the head of the Department of Genetics at Adelaide University, to undertake fieldwork with the Fore. He was particularly interested in the genealogical information we could gather that would confirm the genetic hypothesis. We were aware that the genealogies we might collect would not provide the kind of 'pedigrees' that geneticists would expect, but Fore notions of kinship proved to be even more flexible than we had anticipated.

Our general plan was to carry out an in-depth study of the social life and culture of several adjacent residential groups (or parishes, in anthropological terminology) and also to test our findings by ranging widely through Fore territory, as well as that of neighbouring language groups to the north and west, where kuru was also present.

We hoped to settle in a place (1) having a high incidence of kuru (as established by earlier epidemiological patrols), (2) with convenient access to the jeep road stretching from Okapa to Purosa, (3) in a community where people would welcome our presence, and (4) where no other Europeans were within close range. As a research strategy we decided to avoid direct inquiries on the subject of kuru until we and the Fore had reached a certain level of mutual trust and confidence. It is perhaps worth noting at this point that one of us (R.G.) had already spent nearly two years among the Huli of the Southern Highlands and several months travelling in Melanesia collecting artifacts for the American Museum of Natural History. This meant that little time would be lost in learning 'neo-Melanesian' pidgin and recruiting assistants.

At the 'half-way' point in our field study, which began in July 1961 and ended in May 1963, we took a break in Melbourne and then in Adelaide from April to June 1962. At the invitation of Professor Bennett, who had sponsored the first part of our field trip, we participated in workshops held at the Department of Genetics in Adelaide to talk about kuru research. Between April and June, we wrote a series of papers in response to questions posed by Dr. Bennett and others concerning the nature of the kuru epidemic.[†]

Leaving Adelaide, we returned to our fieldhouse in the South Fore, where we continued with our earlier investigations, wrote further reports[‡], and completed a total of 20 months in the field. Our second period of fieldwork, from July 1962 to May 1963, was supported by the Papua New Guinea Department of Public Health.

These early papers, especially those drafted in Adelaide, were circulated in mimeographed format and later duplicated by the National Institutes of Health, and focused more on our findings than on the methods we employed in reaching our conclusions. This essay provides a fuller picture of the context of the period, of the people we met and worked with (both Fore and non-Fore), and of the general process of day to day anthropological enquiry.

CHOICE OF A FIELD SITE AT AKERAKAMUTI

Leaving the denuded grasslands of Kainantu (in a vehicle provided by the Papua New Guinea administration), we journeyed through an Eastern Highlands region that was increasingly covered in a dense mix of tropical forest, intermittently slashed with orderly gardens of sweet potato, yam, taro, sugar and banana, until we reached

---

[†]In writing the papers, the division of labour resulted in the following titles: 'South Fore society, a preliminary report', and 'The spread of kuru among the Fore' written by R. M. Glasse, and 'The social effects of kuru', a 'Note on women in the South Fore' and a 'Note on Fore medicine' written by Shirley Glasse (later Lindenbaum).

[‡]In 1963 R.G. wrote a fuller version of 'Cannibalism in the kuru region', a manuscript drafted during the previous year, and S.G. expanded 'The social life of women', and a 'Note on Fore medicine and sorcery—with an ethnobotanical list'. This latter manuscript was inspired by the visit of the botanist John Womersley, who stayed with us for a number of days in April 1963, and the subsequent collection by S.G. of plant material, which was sent for identification to the Herbarium at Lae.

the government post of Okapa late in the afternoon on Thursday, June 29, 1961. Dr Andrew Gray, the station Medical Officer, was our generous host on this first evening in Fore territory. We talked long into the night about the current status of kuru, and left the following morning for the South Fore in a vehicle provided by the Assistant District Officer (ADO), Mert Brightwell. We had decided not to stop in the North Fore villages near Okapa where there was a lower incidence of kuru, and which we judged were too close to the government station. Moreover, Ronald and Catherine Berndt had included the hamlets around Okapa station in their ethnographic studies of this northern region between 1951 and 1953 (see Berndt, 1962). Instead, we crossed the ridge separating the North Fore (Ibusa) and South Fore (Atigina) dialect groups, pausing briefly at the divide to view the valleys and settlements to the south, and then continuing on for another three or four miles until we came to a place called Ivingoi. Leaving the Land Rover on the road, we descended into the nearest village on foot. There we were met by a group of villagers. We explained (to the few young men who spoke pidgin at that time) that we were looking for a place to settle for a year or more, and that we wished to learn the Fore language and to talk to them about the past and the present. We did not know at the time that the village, Akerakamuti, was the site of a failed cargo cult centred in this region, and that our talk of building a house and settling in the community was viewed as the delayed arrival of the 'European' goods they had expected a decade earlier. The cult leaders, now government-appointed Luluais and Tultuls (native officials), saw our presence as a vindication of their own failed promises (as they later confided to us). At this point, we were all elated by the prospect of living together, although our expectations and understandings of what this might entail must have been very different. After an hour or so, we continued our journey to Purosa, six miles further south, where the road ended. This required crossing another great mountain ridge separating the Atigina region from the populations living at these lower altitudes. Although the people we spoke to here were also welcoming, an American Protestant missionary (John James) had already built a large house and compound at Purosa, and we decided not to settle nearby. By the time we passed Ivingoi again on our return trip to Okapa, a group of local men (including the old cult leaders) had assembled on the road, and they indicated that they wanted to talk to us. If we came to live with them, they said, they would supply the building materials for a house (timber, bamboo and grass thatch). In the meantime, we could stay in their *haus lotu*, a small 'church' building constructed in response to the evangelism of Mr James, which at that time stood unused. Encouraged by this turn of events, we drove back to Okapa, and, with the help once again of the ADO, we returned next morning with our camp gear and supplies.

As soon as we moved into the church, we planned the layout of our own fieldhouse. With the consent of the villagers, we chose a site in the hamlet of Akerakamuti. The local leaders then sent out word that men and women should bring timber, cane and kunai grass, which we would purchase with money and trade goods. (At that time, the Fore were eager for salt, tobacco, newspaper, cloth, beads, tomahawks and bushknives. We subsequently also made a cash payment, which was given to the local leaders in the form of shillings for ready distribution to all who had contributed labour.) Several days later, the Adelaide University Land Rover became available for our use, and we used it to shuttle people from Ivingoi to the forest areas they owned and from which they had the right to remove timber, vines, and other forest products. During the next week, many people brought us gifts of food—bananas, sweet potato and sugarcane. They wanted no payment, they said, and we understood that we had already entered into important exchange relationships entailing delayed reciprocity, the key to our acceptance as social beings.

The construction of a sturdy fieldhouse took three weeks. Again, in this late colonial era, the ADO sent his station carpenter (a Papuan craftsman) to help with the framing of the house, doors and windows. During the third week we made a brief trip to Goroka to purchase further food supplies (canned meat, margarine, flour, powdered milk, sugar and so on), and found that, on our return, the house was almost ready. We had designed it with one door, to which everyone would have access, placing a pot-bellied stove in this half of the house as an invitation to passers-by to warm themselves on cold mornings or wet afternoons, and where a sweet potato might roast slowly as people smoked and talked, and as time passed, gathered to look at the gallery of photographs of themselves mounted on the bamboo walls. The house was divided into two parts, a second door giving access to our sleeping quarters, shower recess and work table, where we could more readily control the flow of day and night time traffic.

During the period of house construction, we began more or less systematic fieldwork. We compiled lists in Fore of various semantic domains: body parts, animals, insects, plants, soil types, personal names and so forth, using these early language-learning days to get to know the Fore as they were getting to know us. (Some preliminary studies of the Fore language by the Summer Institute of Linguistics were also available at this time.) In these 'low-tech' fieldwork days, we decided not to use our bulky tape recorder in personal interactions. Rather, when possible, we would keep more or less verbatim accounts of interviews and of certain spontaneous conversations. These were documented in shorthand by S.L., and then typed up each evening. Thus, our fieldnotes show what questions we asked and the often detailed responses we received. We reserved the tape recorder for occasional language-learning sessions, and especially for recording Fore songs, which we often replayed when people spent evenings around the pot-bellied stove.

Direct observation and inquiry was a key methodology, although, as indicated, we did not initially press queries about kuru. After several weeks, however, we realized that we were seeing too few cases, and it soon became apparent that a great exodus of patients had occurred. By July 1961, kuru victims from throughout the Fore, and some from the Keiagana, had made the arduous journey across a mountain range to Uvai in the Gimi, where a Gimi curer was said to be treating victims of the disease. In early September, we followed the route of the pilgrims, spoke to those still in residence at Uvai, and visited the hamlets of those who had taken the cure and returned. (See Lindenbaum (1979) for a fuller account.) On return to the Fore, we continued to join in day-to-day events, and attended and participated in many public ceremonies—death payments, marriage negotiations, reciprocal exchanges of food and valuables—keeping a record of who participated and why. This involved keeping a detailed account also of our own contributions and receipts. We thus compiled a dense set of observational and interview data, often going over the same events and questions with numbers of men and women in different villages. Our interviews were conducted in pidgin and some Fore, always with several interpreters, to provide different translations of the same speech or event. When we later travelled through Keiagana-, Kamano-, and Gimi-speaking areas, additional translators joined us. Sometimes we found repetition of the same questions boring, as did the Fore, who must have considered us slow learners, but every so often the same question elicited a new or unexpected response, thus changing the nature of the question. For even with an understanding of the literature on Melanesia available before fieldwork began, we arrived with a set of assumptions and ideas (perhaps not consciously recognized) that were continually revised and 'peeled away', allowing us to comprehend gradually different Fore assumptions and views of the world. In time, people became accustomed to our constant questioning and interest in extensive accounts of past and present events. While the villagers of Akerakamuti called us 'Bapu' and 'Shorlay', we were described to visitors as 'Story Masta' and 'Story Missus'.

As the months passed and as we began to feel we understood a particular social institution, such as the rules governing preferential marriage (for example, in this area a man prefers to marry a woman in the category of mother's brother's daughter), we searched for exceptions that were inconsistent with these stated principles (see R. M. Glasse, 1969). This strategy helped to identify contradictions between what people said they did and what they actually did.

THE ORIGIN AND HISTORY OF KURU

One of our major briefs for Professor Bennett, as noted earlier, was to gather data that would corroborate or invalidate the genetic hypothesis as a cause of kuru (Bennett *et al.*,1958, 1959). While we agreed with Bennett that the collection of genealogies was essential, we also began to perceive (as a result of the Fore's insistence on the recent origin of the disease) that the history of kuru was of equal moment. In establishing the antiquity of kuru, we relied on two kinds of evidence: statements made by informants spontaneously or in directed interviews, and the absence of information about kuru in myth and ritual. Let us consider the 'negative' evidence first.

In the early 1960s, it could be said that the Fore possessed an oral, pre-literate culture. (A very small number of young men spoke and could write simple notes in pidgin as a result of contact with the Seventh Day Mission or their training as Aid Post Orderlies). The lack of written records, however, does not mean the dearth of a sense of history or disinterest in the past. Rather, information may be stored in the form of narratives, ritual (prescribed behaviours invested with cultural meanings), and sometimes in material artefacts. Events remote in time may be symbolically coded in various forms, sometimes to be read directly, at other times decoded through the structural analysis of myth or ritual.

The origin story for kuru, however, was not embedded in myth or ritual, but was told in documentary fashion by those who first encountered the disease some few decades before our arrival. The symbolic manifestations of kuru also reflected the recent acceptance by the Fore of a sorcery rationale for the disease. Some people noted that, at their first encounter with kuru, they thought the tremors resembled the swaying of the casuarina tree. They assumed that the patients had a shaking disorder, which they called 'cassowary disease', by the further analogy that cassowary quills resemble waving casuarina fronds. Other people, noting the patient's emotional lability, first called the disease *negi nagi*, meaning a silly or foolish person. When it became apparent that the victims were uniformly dying, however, they changed their views, recognizing it as the product of sorcery. Sorcerers were now thought to be stealing from the victim some food remnants, hair, nail clippings, or excrement to incorporate into a leafy bundle along with a sorcerer's stone. The victim was named, the bundle was buried in muddy ground, and, as the bundle disintegrated, the victim experienced the initial symptoms of head pain and an unsteady gait. The victim's kin then attempted to identify the sorcerer to persuade him (sorcerers were always said to be men) to undo his wretched work so that the patient might revive. The Fore theory concerning the cause of kuru was not unique. Resembling other theories of disease causation, it was merely a latecomer to an inventory that accounted for the fatal illnesses afflicting the population in the 1960s.

The second line of evidence, perhaps more persuasive, concerns the Fore assertion that the disease was both new and recent. Without written records, how might we evaluate Fore accounts of the appearance of kuru?

Not unexpectedly, the Fore have no difficulty in describing the birth order of their children, living and dead (although stillborn or neonatal deaths may be occasionally neglected). Knowledge of birth order provides a grid of relative chronology. For example, Fore men and women who are born at the same time (perhaps within a few weeks of each other), and in the same or even in adjacent, friendly communities, share an enduring, life-long relationship of great importance. They address each other by a special term, and men may inherit each other's wives, the survivor sharing in the other's death payment, whether or not they also happen to be kin. When one adds to this structural feature a hamlet-wide knowledge of birth order, this sense of history begins to encompass families and small communities. Death order is a somewhat less reliable marker, but there is usually a consensus about whether or not a certain person died before or after the birth (or death) of another individual belonging to the same group. Knowledge of birth and death order gleaned from genealogies thus provides an intricate grid of the sequence of past events. If we add to this the somewhat less reliable data about the timing of male initiation and of marriages, we expand the sequence of recalled events to which other occurrences can then be related. This is only relative chronology to be sure, for specific dates are lacking.

Every so often, however, a certain event can be dated more or less precisely. For example, a severe earthquake, or a volcanic eruption with ashfall, preserved in oral narratives, may be confirmed in geological studies, thus providing absolute dating (see Blong, 1982). In the case of the Fore region, the amazing and frightening apparition of planes flying overhead (thought to be huge birds sent by the ancestors) and the subsequent downing of a Japanese plane during World War II provide fixed points for local history. Most people who witnessed these momentous events can recall whether a particular individual was alive or dead at the time, thus hitching the grid of relative chronology to better-documented occasions.

Our pursuit of the history of kuru was slow and detailed. Initially, as indicated, we often waited until people began to speak of it without direct questioning. This proved to be important, for we found that as we moved toward the area where kuru was said to have begun, many people were apprehensive about being identified as the 'originators'. The small-scale warring groups in this region had a history of armed conflict, as individuals accused each other of causing kuru deaths in their own camp. Some people also feared that the government would punish them for having set the tragedy in motion. Diary entries for this first period of fieldwork indicate that we followed the 'origin of kuru' story for some months, travelling throughout the North and South Fore, visiting almost every hamlet. We also journeyed further north into the Kamano and Keiagana region (where the disease was said to have originated), and west into the Gimi, where kuru was also present. We camped overnight, stopping for perhaps a day or two in some areas, a week or more in others.

As our early reports indicate, we believe that kuru first emerged in local consciousness in the first two decades of this century. This is not the same as saying that kuru began between 1900 and 1920. We do not know when or precisely where the first case occurred, and perhaps this may never be known. Also, the mutual accusations and suspicions among neighbouring groups make it unlikely that we could decide whether the first cases arose in the North Fore area of Awande or the nearby Keiagana villages further north. However, there seemed no reason to doubt the accounts that many people gave in the early 1960s of the first kuru cases they encountered some 30 or 40 years earlier, their initial wonder at its nature, and the names of the first individuals in various locations who were victims of the disease (see Lindenbaum, 1979, pp. 16–19).

Our conclusion that kuru was of recent origin was not welcome news for advocates of the genetic hypothesis. (Norma McArthur's re-analysis of the early census material raised further doubt. McArthur visited us in the field from May 11 to May 17, 1963 working with us in her revision of the age estimates of cases recorded by patrol officers during their first patrols. Her re-analysis showed that the apparent bimodality in the distribution of kuru deaths, upon which part of the genetic hypothesis rested, was a product of the unconscious bias of these early estimates (McArthur, 1964, 1976).)

While we now felt that we had placed the occurrence of kuru in time and space, a pattern confirmed in interviews with hundreds of informants, this information told us little about how kuru was spread. Also, the epidemiological picture available to us at the time (Gajdusek and Zigas, 1957) did not suggest any obvious mechanism.

CANNIBALISM AND KURU

Early in 1962, midway through our first field trip, we began to suspect that cannibal consumption of the kuru dead might prove to be important. In kuru research, the matter of who said what when has become as much a part of the mystery as the ambiguity surrounding Fore chronologies. For this reason, we include the following information from the documents in our files.

When anthropologists go to the field, they leave family, friends and entertainment, but often take pieces of their own culture with them (an inventory of these memorabilia and divertissements would make an interesting study). In our case, before we left Australia in June 1961, we arranged to have a wide range of serious and not-serious reading matter sent to us. In addition to a subscription to the English anthropological journal *Man* (which, as a result of an error by a clerk in the bookstore, turned out to be a trashy, erotic, Australian magazine of the same name), we sent ourselves the complete works of Zola, and a subscription to the American magazine *Time*. The May 18 1962, issue of *Time*, which probably arrived at Akerakamuti in June or July, contained an account of RNA research concerning memory transfer through cannibalism in planaria (McConnell, 1962). The article noted that when certain 'educated' flatworms were cut into pieces and fed to other flatworms, they passed their learning to the uneducated worm (McConnell, 1962, p. 48). Although the results of this study could not be confirmed (Hartry, *et al.*, 1964), the idea of the transfer through cannibalism of some key agent provided us with a 'model' for considering the parallels in our own data on Fore cannibalism.

Without pressing the analogy further, however, we continued to gather data on cannibalism and kuru in 1962 and again in 1963, since it was apparent that a case could be made that Fore cannibalism and kuru were both of recent and perhaps related origin. Cannibalism was not an ancient custom among the Fore, having been adopted (according to the Fore) in imitation of their northern neighbours a short time before the appearance of kuru in these same areas. Moreover, in the South Fore, the area with the highest incidence of kuru, cannibalism had continued later than in the north. In addition, cannibalism was largely limited to adult women, children of both sexes, and the occasional elderly man, thereby matching the sex and age incidence of kuru in the early 1960s.

Our report to John Gunther, the Director of Public Health, written in Okapa on April 10, 1963, and covering the period of fieldwork from November 1962 to 1963, indicates that, in this second phase of research, we were continuing to gather information on seven topics begun earlier: the origin and spread of kuru; cannibalism and kuru; the social effects of kuru; women's life and child-rearing practices; basic kinship studies; myths, folklore and traditional history; concepts of disease treatment. About cannibalism and kuru we wrote:

> Extensive data has been collected on the possibility of an association between cannibal practices and the spread of kuru. As these practices vary considerably in the kuru region and in adjacent areas an attempt will be made to relate these findings to variations in kuru prevalence. The data collected from the borders of the kuru region are of particular interest, and these will be discussed in relation to the spread of kuru.

1962 and 1963 were peak years for the incidence of kuru in the South Fore, to which the Fore responded by holding great public meetings calling on the sorcerers to outlaw kuru sorcery and to persuade them to start a new life by confessing their past aggression. Although we felt that our investigation of the relationship between kuru and cannibalism was moderately discrete, a letter written to the ADO by Mildred

Cervenka, a missionary associated with World Missions, and living in the South Fore parish of Oriesa, illustrates the way in which misplaced rumours could spread, particularly during this time of turmoil for the South Fore.

> . . .The Doktaboi from Paiti is passing thro now on his way to Okapa. He is abt the only native I have met lately with any sense. His object in going is to find the basis of the "rumour" that has made all the native [*sic*] from Purosa on to Paiti, and Mentilesa too, all upset, all holding all kinds of 'kuru' meetings, etc.
>
> The facts as I cd gather them, from the natives here, and partly from Purosa are: that the couple working on kuru at Wanitabi called out for the LL's and TT's [Luluais and Tultuls] of all the villages, and said that the kiap [ADO] said they were to kill and eat all the women with kuru; then men and boys only wd be taken in an airplane by the kiap to a new locality where they cd have a fresh start without kuru, only that they wd have to swear that they wd not start working kuru nor poison in this new locality. The women and female children wd be left behind here . . .   (Orie, Saturday, December 22, 1962)

Throughout 1962 and 1963 we continued to gather data, and often spoke about the relationship of kuru and cannibalism with visitors to the Fore region. Richard Hornabrook, who we met during his preliminary visits to the area on May 4, 1963, recalls us discussing the topic with him at that time. MacFarlane Burnet also remembers the talk given by R.G. on the relationship of cannibalism and kuru to a group of medical investigators, of which he was part, during their visit to our South Fore fieldhouse (Burnet and White, 1972, p. 259). Our diary entries indicate that the occasion would have been on May 21, 1963, when we arranged a *mumu* (the earth oven cooking of vegetables in the Fore style) for the visitors, who included Drs Walsh, Dougherty, Jon Hancock, Norma McArthur, MacFarlane Burnet and Michael Alpers. (Michael Alpers, now the Director of the Papua New Guinea Institute for Medical Research, first arrived in Okapa at the end of 1961, and would soon make his own significant clinical and epidemiological contributions to kuru research.)

In the same year (October 8, 1963), shortly after we had left the field,Richard Sorenson, a historian, and at that time a photographer working with Carleton Gajdusek, made the following entry in his own field diary:

> I ran into another of Carleton's old acquaintances, Blu Russell, at the hotel. He tells me that kuru is caused by eating spiders and frogs and the reason men don't get it is that they leave these delicacies for the women and children. Everyone seems to have his own theory. Shirley Glass has just written a paper implying strongly that it is transmitted by cannibalism—particularly the eating of the brain of the deceased relative.

The paper Sorenson refers to is perhaps the manuscript drafted by R.G. in 1962, revised and reprinted by the NIH in 1963. R.G. subsequently presented a fuller account at the New York Academy of Science in 1967. The question of why we waited until 1967 before publishing our data on kuru and cannibalism was posed to S.L. during a taped interview by Hank Nelson, an Australian historian pursuing the kuru story in the 1970s. At that time, April 1978, Nelson asked:

H.N.:   Now, did you want to put it in print in stronger form before? Why didn't it
        get into print?

S.L: We were talking about it to everybody. Hornabrook . . .visited us and he
        says he can remember us talking to him about it . . . anthropologists didn't
        rush into print in those days. You came back, thought about the stuff you
        published and you published it later . . .

The analysis and reporting of research in anthropology differs from the procedures
that are typical in medical inquiry. Ethnographic data consist of both quantitative
and qualitative information. Neither kind of data is theory free (they are collected
with theoretical assumptions in mind, and they are analysed in light of theoretical
changes in the discipline). On return from the field, most anthropologists can provide
a superficial description of people's beliefs and ways of life, but serious consideration
of the ecology, economy, social organization, politics, ideology, ritual, symbolism and
so on, requires deeper theoretical analysis and interpretation. It is thus not unusual
to delay in publishing initial reports, and to continue to analyse field data for some
decades after its collection. (The luxury of such an approach may result also from
the fact that there are usually no other investigators carrying out similar research in
the same place).

As it happened, three months after returning from Papua New Guinea, we were
offered another opportunity to undertake fieldwork, and we departed for East Pakistan
(now Bangladesh), where for the next two years we carried out research on cholera.
It was not until our return to New York in 1966 that we began to further analyse
and write about the Fore material.

A little further in the interview in Canberra, S.L. mentions sharing some anthropological data with John Mathews, a physician who carried out his own kuru research,
and with whom we published an article on kuru and cannibalism in 1968 (Mathews
et al., 1968). S.L. adds that an epidemiologist she met later in Bangladesh expressed
surprise that we would share our data in this manner, since it was not something
one did in the medical world. This led to Nelson to reflect further on the different
approaches to research and publication in the natural and human sciences:

H.N.:   I think for them [the natural sciences] there can only be one first. For us,
        there are infinite varieties of truth. For them there is only that one reality.

Earlier in the interview, Nelson asked 'What was the reception to the cannibalism
idea?'

S.L.:   . . . I don't remember anybody actually saying it but I had the sense that
        they thought it was a piece of exotic information you might well expect from
        anthropologists. . . Nobody seemed to think it was all that likely.

This interpretation of events appears to be confirmed in Macfarlane Burnet's article
on kuru in the Encyclopaedia of Papua New Guinea:

The next significant investigation was a report by the Glasses that the existence
of kuru as a disease dates only from about 1920. This immediately implied that
an environmental as well as a genetic factor must be involved and Glasse's

suggestion that this may have been the development of cannibalism in the area. This suggestion has been much discussed. Some, like the present writer, initially found the suggestion incredible but must confess now to at least an open mind on the matter.

(Burnet, 1972, p. 587)[†]

ON CANNIBALISM AND ANTHROPOLOGY

Skepticism is, of course, a healthy state of mind for medical or anthropological research. We should perhaps take this opportunity to comment, however, on recent anthropological literature suggesting that, since no first-hand account of cannibalism as a socially approved custom exists, cannibalism is an invention of the anthropological (or missionary, or adventurer) imagination (Arens, 1979). The Fore case is unfortunately sometimes cited as an example of anthropological myth-making: unfortunate, since this pretended defence of 'primitive peoples' shows a lack of respect for the Fore's own documentation of the case. It is unfortunate in another respect also, because it is sometimes true that in speaking of cannibalism we are dealing, in some cases at least, with an instance of collective prejudice. The assumption of the cannibalistic nature of others is part of an ideology which attempts to discredit rivals. It is an aspect of the process by which societies attempt to create a conceptual order based on difference, and thus belongs to a category of disparaging allegations about the malevolence of 'others', who from time to time in human history have been identified as witches, satanists, heretics, or ethnic groups to which the accuser does not belong. We are dealing here with the seeds of racism. However, serious scholarship is hampered by the non sequitur that, because some ethnographic (or missionary, or adventurer) accounts of cannibalism are certainly fictitious, all reports of cannibalism belong to the field of mythology. Moreover, in this case, the Fore are not suggesting that 'others' who live in the next valley are eaters of human flesh and thus less than human. At the time of our fieldwork in the early 1960s, the Fore were speaking without self-criticism of their own behaviours, suppressed by government edict less than a decade earlier. Because the Fore could name for us (and for later investigators, such as John Mathews) those who had died of kuru and those who had participated in the consumption of the deceased person, a coherent account could be developed for the appearance of clinical disease, in certain instances four to twenty years after the ingestion of poorly cooked tissues containing the transmissible agent (Mathews *et al.*, 1968). That anthropologists are sometimes gullible is no doubt true, and is a

---

†In 1963, Gajdusek presented an outline of the status of kuru research after six years of medical surveillance. Stressing the genetic hypothesis of kuru pathogenesis, he also listed five other aetiological possibilities that had been considered. Among these, cannibalism was discussed in the context of a possible auto-immune hypersensitivity reaction. The hypothesis considered was that, after ingestion of human brain material, antigens absorbed from the gut might initiate an auto-immune reaction directed at the nervous system. This line of research was said to have led to negative results, although Sir Macfarlane Burnet was still examining thymus glands from autopsies performed on kuru victims. Later in his essay, Gajdusek observed that the remarkable age and sex distribution of the disease was an important clue, but interpreted this as likely to indicate a neuro-endocrine influence on pathogenesis.

notion that appears to find a receptive place in the popular media. However, all who have lived with the Fore for a significant amount of time during the past 30 years (anthropologists, medical investigators, linguists, missionaries, coffee planters, government administrators) have confidence in the detailed accounts of cannibalism that the Fore provide concerning their own past behaviour. (On a return visit to the Fore in 1991, S.L. tape recorded conversations with now elderly Fore men and women who still vividly recall seeing others consume certain deceased kin). The desire to believe that cannibal stories belong to the realm of fantasy would seem to be just as interesting as the notion that those who accept cannibal stories lack judgment.

The Fore rules of cannibalism seemed to fit the epidemiological evidence available to us at that time. Although cannibalism was no longer present in the 1960s, having been suppressed by the government and the missions, the first government patrols in the late 1940s reported cannibalism throughout the entire kuru region. By the 1950s (according to the anthropologists R. and C. Berndt) the government had put a stop to cannibalism in the North Fore, but it was still practiced surreptitiously in the south. The South Fore later told us that they indeed continued to hide and eat people until the mid-1950s, when a jeep road was constructed through the area. Thus, in the South Fore, the area with the highest incidence of kuru, cannibalism continued longer than it did in the north.

Little of a body was discarded. Maternal kin of the deceased dismembered a corpse, removing hands and feet, and cutting open the arms and legs to strip out the muscles. Opening the chest and belly, they avoided rupturing the bladder, whose bitter contents would ruin the meat. After severing the head, they fractured the skull to remove the brain. Meat, Viscera, and brain were all eaten. Marrow was sucked from cracked bones, and the pulverized bones were cooked and eaten with green vegetables. In the North Fore (but not in the South), the corpse was sometimes buried for several days, then exhumed and eaten when the flesh had ripened and the maggots could be cooked as a separate delicacy.

Little was wasted, but not all bodies were eaten. Fore did not eat people who died of dysentery, leprosy and possibly yaws, but kuru victims were viewed favourably, the layer of fat on those who died rapidly heightening the affinity of human flesh with pork. Most significantly, not all Fore were cannibals. Although cannibalism by adult males occurred more frequently among the North Fore, South Fore men rarely ate human flesh, and those who did said they avoided eating the bodies of women. (Contact with women was thought to deplete men's strength.) Small children residing in houses with their mothers ate what their mothers gave them. Initiated youths moved at about the age of 10 to the communal house with adult men, abandoning the world of immaturity, femininity and cannibalism. Cannibalism was thus largely limited to adult women, to children of both sexes, and to a few older men, matching again the epidemiology of kuru in the early 1960s.

Body parts from a corpse were due to those who received pigs and valuables by rights of kinship and friendship with the deceased (primarily maternal kin), and the gift had to be reciprocated. Pigs and humans were considered equivalent. Moreover, the Fore had not long been cannibals. Within the broad zone of cannibal peoples to the east and south of the government post at Goroka, they may have been among the last groups in the Eastern Highlands to include human flesh in their diet. Cannibalism was thus adopted by the Kamano and Keiagana before it became customary among the Fore. The North Fore say they were imitating these northern neighbours when they became cannibals at about the turn of the century, while in the South Fore people in the 1960s said that cannibalism had begun there as recently as 50 or 60 years before, that is, about a decade before the appearance there of kuru.

ANTHROPOLOGY AND MEDICINE

The ethnography of cannibalism was in some sense irrelevant until medical investigators provided the next piece of evidence. The anthropological and medical stories came together in 1966 when the chimpanzees, injected with brain material from Fore victims of the disease, exhibited a clinical syndrome astonishingly akin to human kuru (Gajdusek et al., 1966). This finding gave credence to the cannibalism hypothesis, which implied that kuru would not affect those born after the abandonment of the behaviour. This appears to have been substantiated by the disappearance of kuru among children (which was earlier in the North than in the South) and by the age of current victims (now approximately 35 years and older).

Although the medical puzzle seemed to have come to some kind of conclusion in the 1960s, we continued to fill out the ethnographic record based on the material we had gathered together during the 1960s, and on data from a return trip by R.G. in 1969 and S.L. in 1970. We turned our attention to the way in which the Fore responded to an eclipse of the sun (Glasse and Lindenbaum, 1967), to a further account of cannibalism (R.M. Glasse, 1968), to marriage in the South Fore (R.M. Glasse, 1969), Fore age mates (Lindenbaum and Glasse, 1969), South Fore politics (Glasse and Lindenbaum, 1969, 1971), sorcery and the structure of Fore society (Lindenbaum, 1971), sorcery in its social and symbolic aspects (Lindenbaum, 1975, 1976), the impact of kuru at Wanitabe (Glasse and Lindenbaum, 1976), the social and symbolic status of women (Lindenbaum, 1976), South Fore kinship (Glasse and Lindenbaum, 1979), and to changing images of the sorcerer in Papua New Guinea, which could be viewed as indigenous commentaries on the changing political, economic and social relations in different regions during the many decades of colonial experience (Lindenbaum, 1981, 1990).

The account of Fore social life, however, is not concluded. Just as the medical story of kuru and other degenerative diseases of humans and animals began to move from 'slow' viruses to the involvement of a prion protein as the possible infectious agent, an adequate account of contemporary life in the Eastern Highlands must now take account of the widespread involvement of the Fore in wage labour, the significant impact of mission activities on health and on ideas about the nature of the world, the experience of primary and (to a lesser degree) secondary education, and the radical effect of increased mobility as entrepreneurs provide daily transport for the shipment of people, ideas and pathogens back and forth across the highlands and from the highlands to the coast. The current generation of young Fore should join in telling the story during the next decade.

REFERENCES

Arens, W. (1979) *The Man-Eating Myth: Anthropology and Anthropophagy.* Oxford
    University Press, Oxford.

Bennett, J.H., Rhodes, F.A. and Robson, H.N. (1958) Observations on kuru: a possible
    genetic basis. *Aust. Ann. Med.* **7** 269.

Bennett, J.H., Rhodes, F.A. and Robson, H.N. (1959) A possible genetic basis for
    kuru. *Am.J. Hum. Genet.* **11** 169.

Berndt, R.M. (1962) *Excess and Restraint.* University of Chicago Press, Chicago, IL.

Blong, R.J. (1982) *The Time of Darkness.* Australian National University Press,
    Canberra.

Burnet, M. and White, D.O. (1972) *Natural History of Infectious Disease,* 4th ed.
    Cambridge University Press, Cambridge.

Burnet, M.F. (1972) Kuru. In: *Encyclopaedia of Papua New Guinea.* p.587.

Cervinka, M. (1962) Letter to ADO (Mert Brightwell), December 22, 1962.

Gajdusek, D.C. (1963) Kuru. *Trans. R. Soc. Trop. Med. Hyg.* **57** (3) 151–169.

Gajdusek, D.C. and Zigas, V. (1957) Degenerative disease of the central nervous
    system in New Guinea: the endemic occurrence of 'kuru' in the native population.
    N. Engl. J. Med. **257** 974.

Gajdusek, D.C. Zigas, V. and Baker, J. (1961) Studies on kuru 3. Patterns of kuru
    incidence: demographic and geographic epidemiological analysis. *Am. J. Trop.
    Med. Hyg.* **10** 599.

Gajdusek, D.C., Gibbs, C.J. and Alpers, M. (1966) Experimental transmission of a
    kuru-like syndrome to chimpanzees. *Nature (London)* **209** 794.

Glasse, R.M. (1962a) South Fore society: a preliminary report. Unpublished.

Glasse, R.M. (1962b) The spread of kuru among the Fore. Mimeograph, Department
    of Public Health, Territory of Papua New Guinea. (Reissued by National Institutes
    of Health, Bethesda, MD.)

Glasse, R.M. (1962c) South Fore cannibalism and kuru. Unpublished.

Glasse, R.M. (1963) Cannibalism in the Kuru Region. Mimeograph, Department of
    Public Health, Territory of Papua New Guinea. (Reissued by National Institutes
    of Health, Bethesda, MD.)

Glasse, R.M. (1967) Cannibalism in the Kuru region of New Guinea. *Trans. N.Y.
    Acad. Sci. Ser. 11,* **29** (6) 748–754.

Glasse, R.M. (1968) Cannibalism et Kuru chez les Fore de Nouvelle Guinee. *L'Homme*
    8 22–36.

Glasse, R.M. (1969) Marriage in the South Fore. In: Glasse, R.M. and Meggitt, M.J.
    (eds) *Pigs, Pearlshells, and Women* Prentice-Hall, Englewood Cliffs, NJ.

Glasse R.M. and Glasse, S. (1963) Report of fieldwork by R.M. and S. Glasse,
    November 1962–April 1963, Okapa, Eastern Highlands.

Glasse, R.M. and Lindenbaum, S. (1967) How New Guinea natives reacted to a total
    eclipse. *Transaction* **4** 46–52.

Glasse, R.M. and Lindenbaum, S. (1969) South Fore politics. *Anthropol. Forum* **2**
    308. (Reprinted in Berndt, R.M. and Lawrence, P. (eds), *Politics in New Guinea.*
    University of Western Australia Press.)

Glasse, R.M. and Lindenbaum, S. (1976) Kuru at Wanitabe. In: Hornabrook, R.W. (ed), *Essays on Kuru.* E.W.Classey Ltd., Farringdon, Berks.

Glasse, R.M. and Lindenbaum, S. (1979) South Fore kinship. In: Carroll, V. (ed.) *Blood and Semen.* University of Michigan Press.

Glasse, S. (1962a) Note on women in the South Fore. Unpublished.

Glasse, S. (1962b) Note on Fore medicine. Unpublished.

Glasse, S. (1962c) The social effects of Kuru. (Later published as Glasse, S. (1964) *Papua New Guinea Med. J.* **7** 36–47.)

Glasse, S. (1963a) The social life of women in the South Fore. Mimeograph, Department of Public Health, Territory of Papua New Guinea.

Glasse, S. (1963b) A note on Fore medicine and sorcery, with an ethno-botanical list. Mimeograph, Department of Public Health, Territory of Papua New Guinea (Reissued by National Institutes of Health, Bethesda, MD.)

Hartry, A.L., Keith-Lee, P. and Morton, W.D. (1964) Planaria: memory transfer through cannibalism reexamined. *Science* **146** 274–275.

Lindenbaum, S. (1971) Sorcery and structure in Fore society. *Oceania* **41** 277–288.

Lindenbaum, S. (1975) Sorcery and danger. *Oceania* **46** 68–75.

Lindenbaum, S. (1976) A wife is the hand of man. In: Brown, P. and Buchbinder, G. (eds) *Man and Woman in the New Guinea Highlands.* American Anthropological Association, Special Publication No.8.

Lindenbaum, S. (1979) *Kuru Sorcery. Disease and Danger in the New Guinea Highlands.* Mayfield Publications, Palo Alto, CA.

Lindenbaum, S. (1981) The image of the sorcerer in Papua New Guinea. In: Zelenietz, M. and Lindenbaum, S. (eds) *Social Analysis,* No.8. pp. 119–129.

Lindenbaum, S. and Glasse, R.M. (1969) Fore age mates. *Oceania* **39** 165–173.

Mathews, J.D., Glasse, R.M. and Lindenbaum, S. (1968) Kuru and cannibalism. *Lancet* **ii** 449–452.

McArthur, N. (1964) The age incidence of kuru. *Ann. Hum. Gen.* **27** 341–152.

McArthur, N. (1976) Cross-currents: a demographic study. In: Hornabrook, R.W. (ed.) *Essays on Kuru.* E.W. Classey Ltd., Farringdon, Berks.

Nelson, H. (1978) Transcript of Hank Nelson interview, Australian National University, April 5, 1978.

Sorenson, E.R. (n.d.) Expedition to the Kuru region. (Reprinted by National Institutes of Neurological Diseases and Blindness, Bethesda, MD, restricted circulation.)

Time (1962) *Time* May 18, 1962, p.48.

Part III   Human prion diseases

12

Molecular genetics of inherited, sporadic and iatrogenic prion disease

John Collinge and Mark S. Palmer

ABSTRACT

In addition to the transmissibility of Creutzfeldt–Jakob disease (CJD) and Gerstmann–Sträussler syndrome (GSS) to experimental animals by inoculation, missense and insertional mutations in the prion protein gene are associated with both familial CJD and GSS, genetic linkage studies demonstrating that they are also autosomal dominant diseases. The identification of pathogenic mutations has led to a clearer understanding of the phenotypic range of the human diseases, now more logically called prion diseases. However, 85% of CJD cases occur sporadically and are not known to be associated with prion protein (PrP) gene mutations. Their aetiology is unknown. In addition, rare CJD cases have arisen as a result of iatrogenic contamination. Human prion disease therefore has sporadic, iatrogenic and inherited forms. A number of lines of evidence strongly support the suggestion that an abnormal, partially protease-resistant isoform of PrP is an essential and possibly the sole constituent of the infective agent. Purified PrP fractions retain infectivity which is resistant to treatment with agents known to inactivate nucleic acids (Prusiner, 1989, for review). Furthermore, transgenic mice expressing PrP encoding a mutation analogous to the codon 102 leucine substitution seen in one form of inherited prion disease spontaneously develop spongiform neurodegeneration (Hsiao et al., 1990). The molecular difference between the disease-associated isoform of PrP ('PrP^{Sc}') and the normal cellular isoform ('PrP^{C}') is unknown, although the change is assumed to be post-translational as no differences in primary structure have been identified. Recent work in transgenic mice has suggested that this conversion of PrP^{C} to PrP^{Sc} involves interaction between homologous PrP molecules (Prusiner *et al.*, 1990) and that a conformational change may underlie this conversion. We have demonstrated that homozygosity with respect to a common protein polymorphism (either methionine or valine encoded at position 129) is involved in genetic susceptibility to both iatrogenic and sporadic CJD. Our work suggests that PrP valine 129 may be more susceptible to undergo transformation to the disease-related isoform (Collinge *et al.*, 1991) and also supports the idea that PrP homology is important in prion replication and implies that amino acid 129 is at an important position in PrP for interaction between PrP molecules (Palmer *et al.*, 1991). We discuss a model of prion propagation and argue that prion diseases of humans and animals may have been, prior to human intervention, essentially rare genetic diseases arising from germline or somatic mutation.

INTRODUCTION

The prion diseases are a group of neurodegenerative conditions of humans and animals previously described as the spongiform encephalopathies, slow virus diseases or transmissible dementias. The first disease of this group to be described was scrapie, a naturally occurring disease of sheep and goats recognized in the UK for over 200 years (McGowan, 1922). More recently recognized animal diseases include transmissible mink encephalopathy (Hartsough and Burger, 1965), chronic wasting disease of mule deer and elk (Williams and Young, 1980) and bovine spongiform encephalopathy (BSE) (Wells *et al.*, 1987). The recently described feline spongiform encephalopathy (Wyatt *et al.*, 1991) and spongiform encephalopathies of various zoo animals (Jeffrey and Wells, 1988; Kirkwood *et al.*, 1990) are very likely to be members of this group. It is generally accepted that the epidemic of BSE amongst British cattle resulted from the inclusion of dietary supplements prepared from the rendered carcases of scrapie-infected sheep and later BSE-infected cows (Wilesmith *et al.*, 1988). Spongiform encephalopathies of other captive animals have probably arisen also from such dietary exposure to transmissible agent. However, the mechanism of transmission of 'natural' sheep scrapie is still under debate (see Chapter 3).

The human diseases include kuru, Creutzfeldt–Jakob disease and Gerstmann–Sträussler syndrome and are, like the animal diseases, experimentally transmissible to various mammalian species by intracerebral or other routes of inoculation (Gibbs and Gajdusek, 1973). Horizontal transmission in man has resulted from ritualistic endocaniballism in the case of kuru which reached epidemic proportions amongst the Fore peoples of Papua New Guinea (Alpers, 1987), and rarely via iatrogenic innoculation. Such iatrogenic routes of transmission have included the use of inadequately sterilized intracerebral electrodes, dura mater and corneal grafts and treatment with human cadaveric pituitary derived growth hormone and gonadotrophin (Weller, 1989).

The spongiform encephalopathies are recognized neuropathologically by the classic triad of spongy degeneration (affecting any part of the cerebral grey matter), neuronal loss and the proliferation and hypertrophy of astrocytes (Beck and Daniel, 1987). These changes are accompanied by accumulation in the brain of an abnormal, partially protease-resistant, isoform of a host-encoded protein, prion protein, as amyloid plaques (Prusiner, 1987). A number of lines of evidence suggest that this abnormal, partly protease-resistant isoform of PrP is an essential and possibly the sole constituent of the transmissible agent. Purified PrP fractions retain infectivity which is resistant to treatment with agents known to inactivate nucleic acids (Prusiner, 1989, for review).

The term Creutzfeldt–Jakob disease was first used by Spielmeyer (1922) bringing together the original case report by Creutzfeldt (1920) with subsequent cases described by Jakob (1921a, b, c). The term was used in subsequent years to cover a range of neurodegenerative conditions. The successful transmission of CJD in 1968 to the chimpanzee by Gibbs and Gajdusek (Gibbs et al., 1968) and subsequent experience with transmission work of possible CJD cases that followed led to a reassessment of diagnostic criteria now validated by the criterion of transmissibility. The term transmissible spongiform encephalopathy was introduced by Gajdusek and Gibbs in 1969 to encompass scrapie, transmissible mink encephalopathy, kuru and CJD (Gibbs and Gajdusek, 1969).

CJD is currently recognized clinically by the occurrence of a rapidly progressive dementia with myoclonus often accompanied by pyramidal signs, cerebellar ataxia or extrapyramidal features. The clinical course is generally rapid and in most cases the duration is less than 12 months. Neuropathological confirmation of diagnosis relies on the demonstration of the classical features discussed above. However, atypical cases of CJD (transmissible to experimental animals) are well recognized. In 10% of cases clinical presentation is with ataxia (Gomori et al., 1973). Of particular interest is that 5–10% of CJD cases have a clinical course of 2 years or more (Brown et al., 1984). Routine laboratory investigations are unhelpful diagnostically. Neuroimaging reveals only varying degrees of cerebral atrophy. The electroencephalogram may show, however, characteristic pseudoperiodic sharp wave activity (Cathala and Baron, 1987).

Gerstmann–Sträussler syndrome was first described in 1928 by Gerstmann in a single patient (Gerstmann, 1928). A more detailed report followed in 1936 by Gerstmann, Sträussler and Scheinker of this individual and seven other affected cases from a single family (Gerstmann et al., 1936). The condition is also called Gerstmann–Sträussler–Scheinker disease. The classic description of GSS is of a chronic cerebellar ataxia; dementia occurs later in the clinical course, which is much more protracted than CJD. Clinical onset is usually in the 3rd to 4th decade and the mean duration of illness is 5 years. However, cases with a duration of up to 11 years are recorded in the literature. The pathological hallmark of GSS is the presence of multicentric amyloid plaques distributed throughout the brain. Neuronal loss, spongiform change and white matter loss are seen in most cases (Masters et al., 1981). GSS was transmitted to experimental animals in 1981 (Masters et al., 1981). Although horizontal transmission by various routes is well documented in man, such cases are extremely rare. About 85% of human cases occur as a sporadic illness without a history of exposure to infective agent. Of particular importance is that the incidence of sporadic CJD is remarkably uniform throughout the world, and (with two notable exceptions discussed later) does not show the clustering expected for an infectious disease (Brown et al., 1987). There is no evidence epidemiologically for a relationship between CJD and scrapie. However, around 15% of CJD cases occur in a familial context and GSS is nearly always described in a familial context, in both cases showing an autosomal dominant pattern of disease segregation (Masters *et al.*, 1979).

Partial amino acid sequencing of PrP isolated from affected brains permitted DNA probes to be generated in order to determine the source of PrP coding sequence. It was demonstrated that PrP was not a virally encoded protein but the product of a normal host cellular gene (Chesebro *et al.*, 1985; Oesch *et al.*, 1985; Basler *et al.*, 1986) which in man maps to the short arm of chromosome 20 (Sparkes *et al.*, 1986; Robakis *et al.*, 1986). All the spongiform encephalopathies are associated with accumulation in the brain of an abnormal isoform of PrP. Therefore the PrP gene was a strong candidate gene for genetic linkage studies in GSS and familial forms of CJD. The report by Hsiao *et al.* (1989) of genetic linkage between the PrP gene and GSS in two families was a milestone in spongiform encephalopathy research. This linkage study established GSS to be an autosomal dominant inherited condition in addition to being transmissible to experimental animals. The genetic marker used in this study was not, however, an anonymous DNA marker but a missense mutation at codon 102 of the PrP gene open reading frame resulting in a proline to leucine substitution in the PrP. The proline at codon 102 is highly conserved; all mammalian (including rodent) PrP genes which have been sequenced encode proline at the corresponding codon (Westaway and Prusiner, 1986). This mutation has now been confirmed in other GSS kindreds in Germany, US and Japan (Doh-ura *et al.*, 1989; Goldgaber *et al.*, 1989; Speer *et al.*, 1991) and in the original kindred reported by Gerstmann (Kretzschmar *et al.*, 1991). However, this mutation has not been seen in normal controls. It therefore seemed very likely on statistical grounds that this leucine substitution at codon 102 was a pathogenic mutation.

Another mutation in the PrP gene, consisting of a 144 base pair insertion, had previously been reported in a UK family with CJD (Owen *et al.*, 1989) and a number of other insertional and point mutations have since been described by a number of workers (see Fig. 1). Again these mutations have been seen only in affected or at-risk individuals in these families and not in normal controls and are almost certainly also pathogenic mutations.

Demonstration that the codon 102 mutation was indeed pathogenic was provided by Hsiao *et al.* (1990). Transgenic mice with multiple copies of a prion protein transgene encoding leucine at the corresponding murine codon spontaneously develop spongiform neurodegeneration. Furthermore, it has now been found that brain homogenates from these mice injected intracerebrally into Syrian hamsters transmit the disease (see Chapter 13). Hence a transmissible disease appears to have been induced by mutation of a protein. These experiments strongly suggest that PrP, or rather an abnormal isoform of PrP, is the transmissible agent in these diseases, as well as being the target for mutations in inherited disease.

INHERITED PRION DISEASE

Molecular genetics and phenotypic spectrum
The availability of gene markers for the inherited diseases has enabled direct examination of the phenotypic range of these disorders. Of particular interest in this

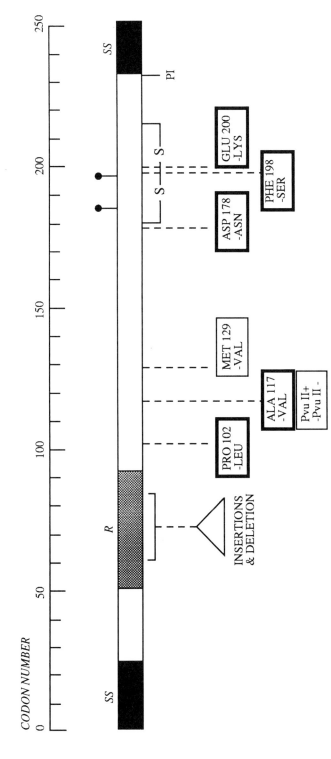

Fig. 1. Representation of PrP open reading frame showing the position of pathogenic mutations and polymorphisms relative to the codon number. Codon 1 is the translational start ATG. Long subdivided box is the amino acid sequence, shaded boxes indicating regions of interest within the protein. SS, amino- or carboxy-terminal signal sequences which are cleaved during maturation of the sequence; R, region of octapeptide repeats. Disulphide bond indicated by S—S. Glycosylation sites are represented by stick and ball attachments. PI, phosphatidyl inositol anchor. Pathogenic mutations are indicated in boxes with heavy outlines, polymorphisms are in boxes with light outlines. Insertions and deletions are located within the region of octapeptide repeats.

regard is the original kindred in which the 144 base pair insertion in the PrP gene was described (Owen *et al.*, 1989). The diagnosis of familial CJD in this kindred was based on an individual who died in the 1940s with both a clinical and neuropathological diagnosis of CJD. This case is described in the literature as a case of Heidenhain's variant of CJD (Meyer *et al.*, 1954). The clinical course was rapid with a duration of illness assessed as 6 months. At necropsy gross status spongiosus and astrocytosis affecting the entire cerebral cortex was described. Histology from this case is used to illustrate classic CJD pathology in Greenfield's *Neuropathology* textbook. However, another affected individual from this family had a duration of illness of over 4 years and only mild and subtle spongiform changes insufficient for a morphological diagnosis of CJD (Janota, personal communication). An additional family member has been affected for over 8 years. Hence a range of illness from that of classical subacute CJD to a GSS-like picture can occur in the same family and the classical neuropathology vary from gross to minimal. For this reason it was felt that this or other genetic variants of GSS–CJD may present atypically and an initial blind screen of 12 unrelated individuals with various familial dementias and ataxias (kindly provided by Professor A. E. Harding) was performed using the polymerase chain reaction (PCR). In this way a family unsuspected of having GSS–CJD was detected (Collinge *et al.*, 1989). The clinical diagnosis in this family was familial Alzheimer's disease; a diagnosis in a neurodegenerative disease was thus made by a direct gene test. A further screen of over 100 cases revealed 4 further families with an identical PrP gene insertion (Owen *et al.*, 1991) and subsequent genealogical investigation demonstrated that all of these cases form part of a single large kindred (Poulter *et al.*, 1992). Previous clinical diagnoses in these families included familial Alzheimer's disease, Huntington's disease, Pick's disease and familial presenile dementia (Collinge *et al.*, 1992). Of particular interest was a family which had carried previous clinical diagnoses of Huntington's disease and familial Alzheimer's disease in which one case was recently examined neuropathologically (Collinge *et al.*, 1990). The clinical features of this case were most unusual. The patient had longstanding antisocial personality traits and became increasingly aggressive in his early 20s and developed loss of balance and dysarthria. By age 27 he had obvious intellectual decline, memory loss and profound dyspraxia. At age 30 he had negligible abilities on verbal tasks, severe memory impairment, gross ataxia and episodes of clonic activity later developing generalized seizures. By age 36 he was permanently hypertonic, unable to talk, and died of bronchopneumonia. At necropsy there was mild cerebral atrophy but a remarkable absence of histological features of CJD, GSS or Alzheimer's disease. Despite the absence of classical histological features this case did, however, reveal evidence of abnormal immunostaining using anti-PrP antisera (Clinton *et al.*, 1990). This extreme variation in histological appearance does not seem to be restricted to genetic cases. A case of clinically diagnosed, apparently sporadic, CJD which lacked characteristic histological features has been transmitted to laboratory animals (Prusiner, personal communication).

It is therefore clear that individuals dying of a dementing or ataxic illness may have an inherited prion disease, as defined at the DNA level, despite the absence of the usually accepted diagnostic features of these conditions at necropsy. In the absence of sufficient neuropathological evidence to substantiate an alternative diagnosis in an individual dying of a dementing or ataxic illness, CJD–GSS cannot be excluded either on clinical or on conventional neuropathological grounds. Immunohistochemical staining with PrP antisera has aided diagnosis in cases where amyloid plaques are present (Nochlin et al., 1989), but plaques are absent in many cases. Diffuse PrP staining, reported in mouse scrapie (Bruce et al., 1989) may prove to be helpful in some cases but this requires formal evaluation in human case material. The rapid immunoblotting technique of Serban et al. (1990) utilizing the relative resistance of the abnormal isoform of PrP seen in CJD to protease digestion looks promising as a diagnostic test where fresh CNS tissue is available for study.

Diagnostic terms such as CJD, GSS, spongiform encephalopathy and possibly even transmissible dementia are therefore becoming less useful. Since an aberrant form of the prion protein and its gene play a central role in the aetiology of these conditions 'prion disease' with infectious, sporadic and inherited forms seems a more suitable term.

It is possible that non-transmissible forms of inherited prion disease occur. Transmission to experimental animals has been reported with all the point mutation types of inherited prion disease (except the codon 198 variety) and also with the 168 bp insertion type (Prusiner, 1991; Brown et al., 1991a). Transmission studies have not been attempted with the 216 bp type. Further transmission studies are required with the 144 bp insertion disease; limited transmission experiments in the marmoset have so far been negative (Poulter et al., 1992).

Classification of inherited prion diseases

The inherited prion diseases can now be subclassified according to mutation (Table 1), a nosology based on aetiological markers rather than descriptive terms. Clinical experience is limited with most of these subtypes so far. Many families have been ascertained as a result of their clinical features fulfilling established criteria for familial CJD or GSS; it is possible that screening of atypical cases for these mutations may reveal broader clinicopathological phenotypes for these mutations than hitherto realized.

Table 1   Human inherited prion diseases

Inherited prion disease (PrP leucine 102)
Inherited prion disease (PrP valine 117)
Inherited prion disease (PrP asparagine 178)
Inherited prion disease (PrP serine 198)
Inherited prion disease (PrP lysine 200)
Inherited prion disease (PrP 144 bp insertion)
Inherited prion disease (PrP 168 bp insertion)
Inherited prion disease (PrP 216 bp insertion)

Inherited prion disease (PrP 144 bp insertion)

This mutation has been demonstrated in a large family originating in south-east England (Poulter et al., 1992). As has been discussed above both the clinical and the pathological features of affected individuals in this family show marked variability. The clinical phenotype in this family varies from that of an individual with a rapidly progressive dementia similar in course to classical CJD to that of individuals in which the clinical course mimicked Alzheimer's disease and in. which a longstanding antisocial personality disorder preceded dementia and other features (Collinge et al., 1992). Information on the most recent two generations of this kindred, on which most detailed information is available, shows a more consistent picture of a personality disorder, followed in early or mid-adult life by a slowly progressive multifocal dementia preceded by behavioural changes involving aggression and depression. Additional features include a varying combination of cerebellar ataxia and dysarthria, pyramidal signs, myoclonus and occasionally extrapyramidal signs, chorea or seizures.

The personality disorder, which precedes onset of neurodegenerative features by many years in some individuals, has not been described in the other types of inherited prion diseases, but it is possible that this aspect of the history may not have been specifically sought in all these families. Such an observation raises the possibility that the insertion affects neurodevelopment in addition to leading to the development of the progressive neurodegenerative disease in early or mid-adult life.

Apart from DNA analysis, investigations were unhelpful diagnostically in this disorder. No case in which the EEG was performed had the characteristic pseudoperiodic activity seen in CJD although all recordings showed non-specific abnormalities.

Detailed histology is currently available on four individuals from this kindred and shows remarkable variability from classic spongiform encephalopathy to absence of histological features of specific neurodegenerative disease, as discussed above. Prion protein immunoreactive plaques were seen in one patient. Interestingly, beta A4 immunoreactive plaques have also been demonstrated in this patient (who died at age 43). However, neurofibrillary tangles, reported in inherited prion disease (PrP serine 198) were not seen (Collinge et al., 1992).

Interestingly, age at death of affected individuals in this family has been correlated with genotype at codon 129 of the prion protein gene (Baker et al., 1991). The significance of this finding is discussed later.

Inherited prion disease (PrP leucine 102)

This mutation was first described in two unrelated families from the US and the UK with neuropathologically confirmed GSS (Hsiao et al., 1989). It has subsequently been confirmed in several other unrelated families (from the US, Germany and Japan) with GSS (Doh-ura et al., 1989; Goldgaber et al., 1989; Speer et al., 1991) including the original Austrian family described by Gerstmann, Sträussler and Scheinker in 1936 (Kretzschmar et al., 1991). It has been suggested that this genetic subtype correlates with the ataxic type of GSS (Hsiao et al., 1989). However, the larger of the two original codon 102 mutation kindreds shows considerable phenotypic variability at the clinical and histological level (Adam *et al.*, 1983) and such variability has recently been confirmed in a German kindred (Brown *et al.*, 1991b). Doh-ura *et al.* (1990) report codon 102 mutations in individuals with what are described as 'CJD with kuru plaques', although these patients presented with spinocerebellar signs in addition to amnesia and disorientation and have a prolonged course.

Inherited prion disease (PrP valine 117)
This mutation was originally described in a French GSS patient (Doh-ura *et al.*, 1989) and subsequently in a US family of German descent (Hsiao *et al.*, 1991a) with presenile dementia associated with pyramidal signs and parkinsonism. This kindred had originally been described as familial Alzheimer's disease (Heston *et al.*, 1966) but was reclassified as GSS following the demonstration of PrP immunoreactive amyloid plaques (Nochlin *et al.*, 1989). Clinical features in the French (Alsatian) family were presenile dementia with pyramidal and pseudobulbar features, with cerebellar signs in some family members (Warter *et al.*, 1982).

Inherited prion disease (PrP lysine 200)
This subtype was originally described in a sibling pair with CJD (Goldgaber *et al.*, 1989) and interestingly accounts for the two main ethnogeographic clusters of CJD, in Slovakia (Goldfarb *et al.*, 1990a) and amongst Libyan Jews (Goldfarb *et al.*, 1990b). Patients with this subtype of prion disease present with a rapidly progressive dementia and myoclonus and pyramidal, cerebellar or extrapyramidal signs. The duration of illness is generally less than 12 months. The electroencephalogram shows the characteristic pseudoperiodic picture seen in sporadic CJD. Histological appearances are usually typical of CJD; plaques are absent but protease-resistant PrP has been demonstrated by immunoblotting techniques (Hsiao *et al.*, 1991b). Two UK families with inherited prion disease (PrP lysine 200) have recently been identified. One has Libyan Jewish ancestry but the second does not have either Libyan Jewish or Slovakian ancestry, suggesting that a separate UK focus on this type of inherited prion disease may exist (Collinge *et al.*, submitted).

Of particular interest with this variant of inherited prion disease is that incomplete penetrance has been reported. A codon 200 mutation has been reported in an apparently healthy 75 year old mother of an affected case (Goldfarb *et al.*, 1990). A possible contributing factor to this incomplete penetrance is discussed later.

Inherited prion disease (PrP asparagine 178)
This was first described in a Finnish kindred with CJD (Goldfarb *et al.*, 1991) and presented as a progressive presenile dementia with myoclonus, cerebellar and pyramidal features. The mean duration of illness was around 2 years and the characteristic EEG of CJD was not seen. Histological examination revealed spongiform change, neuronal loss and astrocytosis; plaques have not been reported. This mutation has subsequently been described in US kindreds of Dutch and Hungarian descent and in a French kindred (Nieto *et al.*, 1991).

Inherited prion disease (PrP serine 198)

This mutation has been recently described in a US kindred with presenile dementia, ataxia and parkinsonism (Prusiner, 1991). The duration of illness is around 6 years. Histological examination in this family demonstrated numerous neurofibrillary tangles, as seen in Alzheimer's disease, in addition to PrP immunoreactive plaques; mild spongiform change was also present (Ghetti *et al.*, 1989).

Other mutations

In addition to the 144 bp insertion, two different-sized insertions in the PrP gene have now been reported. As with the 144 bp insertion, these consist of different numbers of added repeats of an octapeptide. Goldfarb *et al.* (1990) have reported an insertion mutation consisting of seven extra octarepeats in a family described as having a bizarre family history of ill-defined neurological illness (Brown *et al.*, 1991a). Another insertion consisting of nine extra octarepeats has been described in an individual with a presenile dementia with myoclonus and extrapyramidal signs (Owen *et al.*, 1991).

Genetic counselling and presymptomatic testing

The availability of direct gene tests for subtypes of inherited prion disease allows presymptomatic testing in subjects at risk from families with known mutations as well as providing accurate diagnosis in affected families. Such tests have already been performed following careful genetic counselling (Collinge *et al.*, 1991). In many respects the counselling situation resembles that in Huntington's disease (HD) and patients were counselled according to established protocols developed for HD (Craufurd *et al.*, 1989; World Federation of Neurology Research Group on Huntingdon's Disease, 1989). In addition, the possibility of incomplete penetrance needs to be considered. This has already been reported in inherited prion disease (PrP lysine 200); reduced penetrance with the other variants cannot be excluded as experience of these conditions is still limited.

MOLECULAR GENETICS OF IATROGENIC AND SPORADIC PRION DISEASE

The demonstration of pathogenic mutations in the PrP gene in familial cases of CJD has been exploited as a powerful means of establishing whether individuals with familial neurodegenerative disorders can be diagnosed as having a prion disease. However, only 15% of cases of CJD are familial, the remainder being nearly all sporadic with a very small number of cases being iatrogenic in origin. Is there anything to be learned from the PrP gene that helps us to understand the mechanism or cause of these forms of CJD, or is the cellular gene only of importance with respect to the familial diseases? An interesting opportunity to pursue this question arose with the recognition that a group of individuals had developed CJD following earlier treatment with human cadaveric pituitary derived growth hormone. In the UK a total of 1908 individuals had received growth hormone from such a source and 6 have so far gone on to develop CJD (Buchanan *et al.*, 1991). Each batch of growth hormone had been pooled from as many as 3000 pituitary glands removed at routine autopsy, and each individual that subsequently developed CJD had received hormone from more than one batch. No single batch spanned all cases. Since it has not been possible to trace contamination to individual batches it is possible that all hormone recipients were exposed to CJD infectivity on at least one occasion. Although the six who developed CJD may have been exposed to a greater dose they may also have been more genetically susceptible to infection. Since it had been shown that genetic susceptibility or incubation time alleles existed in mice for experimental transmission of scrapie (Westaway et al., 1987), it was not unreasonable to look at the PrP gene for evidence of such susceptibility.

Seven samples were screened for PrP gene variations; six growth hormone cases and a single gonadotrophin-related case. These represented all cases described to date in the UK arising from treatment with human pituitary derived hormones. DNA was extracted from available tissues and the open reading frame of the PrP gene amplified by the polymerase chain reaction for analysis. None of the known pathogenic mutations was detected in any of the samples, and all were homozygous for a non-coding polymorphism at codon 117 identified by a Pvu II restriction site which is known to be present in 90% of alleles in the general population. However, an unexpected allele frequency was found for a coding polymorphism at codon 129 (Collinge et al., 1991). The amino acid at residue 129 is a methionine in 63% of alleles and valine in 37% of alleles in Caucasian populations. This gives a genotype frequency of 37% Met/Met, 51% Met/Val and 12% Val/Val (as estimated from a sample population of 106 normal controls). However, the seven CJD samples analysed had a significantly higher frequency of valine alleles. Four cases were Val/Val, two Met/Val and only one Met/Met. This gave a frequency of 71% valine to 29% methionine alleles, or 14% Met/Met, 29% Met/Val and 57% Val/Val. Although the number of cases is necessarily small the data suggested that there is genetic susceptibility to exogenous prion infection. Those individuals exposed to contaminated pituitary hormones who are valine homozygotes may be at an increased risk (albeit still a small risk) of developing CJD. The implications that this result has for the biology of prion diseases will be discussed below in the context of a general model for the role of PrP in these diseases.

The aetiology of sporadic cases of CJD has been less easy to address. It remains a possibility, despite the lack of epidemiological evidence to support such a route, that sporadic cases have arisen because of exposure to environmental prions. It follows from our observation on iatrogenic cases of CJD that if this were the case there might also be an excess of valine 129 homozygotes amongst sporadic cases of CJD.

Initially 22 cases of sporadic CJD were analysed for their genotypes at codon 129. Surprisingly only one case was heterozygous. Of the rest there were 16 methionine homozygotes and 5 valine homozygotes (Palmer et al., 1991). The excess of valine 129 homozygotes was not statistically significant, perhaps suggesting that most cases of sporadic CJD were not acquired from exposure to environmental prions. However, this was nevertheless a striking result; 95.5% were homozygotes (to either valine or methionine) giving a p value of 0.00017 against it being a chance occurrence. When the records of the single heterozygous case were investigated further it was found

MOUSE PRIONS HAMSTER PRIONS MOUSE PRIONS HAMSTER PRIONS

ILLNESS 140 days (>370 days) (>360 days) 90 days

DEATH 150 days – DEATH – 100 days

Fig. 2. Representation of the species barrier between mice and hamsters. Mice are susceptible to prions which have been propagated in other mice. Illness in this example is seen within about 140 days and death follows about 10 days later. When challenged with prions propagated in hamsters most mice do not succumb to illness. The few who develop disease do so after prolonged incubation times. Conversely hamsters are susceptible to prions propagated in hamsters but not to those from mice. Arrows above the animal in the figure represent inoculations with prions from the source indicated while arrows below the animal indicate the progression of the disease with times for incubation and death indicated in days at the side.

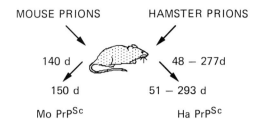

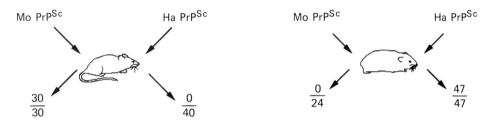

Fig. 3.   Abrogation of the species barrier in transgenic mice. The stippled mouse is transgenic for the normal hamster prion gene and is susceptible to challenge with prions propagated in both hamsters and mice. The incubation period following challenge with hamster prions varies depending on the copy number of the transgene; the higher the level of expression the shorter the incubation time. In the lower part of the figure normal non-transgenic mice or hamsters are inoculated with prions derived from the infected transgenic mouse. Mo PrP^{Sc} and Ha PrP^{Sc} represent infectious prion as isolated from the transgenic mouse after inoculation with mouse (Mo) or hamster (Ha) prions respectively. Numbers in the lower part of the figure represent the number of animals affected within experimental groups as reported by Prusiner *et al.* (1990). 30/30 indicates that all mice in a group of 30 developed the disease.

that the father had died with dementia (unfortunately further details were not available) so that the only heterozygote may in fact have been a familial rather than a sporadic CJD case.

A number of cases of suspected CJD, which had been notified to the United Kingdom CJD surveillance study, were also available (kindly provided by Dr R. G. Will). At the time of study it was likely that several of these cases would turn out, on subsequent investigation, not to be CJD. Of 23 such cases just 4 were heterozygous, with 13 methionine and 6 valine homozygotes. Further suspected CJD cases have now been studied, and an additional four heterozygotes identified. These cases were typed without knowing clinical details. Of these eight heterozygotes only one has proved to be a confirmed case of typical CJD with short duration (albeit with an atypical EEG). The other cases turned out to be either not CJD or were clinically atypical cases with a prolonged duration. (Collinge and Palmer, 1991; Collinge *et al.*, submitted). This supports the view that homozygotes are genetically more susceptible to sporadic CJD and that heterozygotes are to a certain extent protected; the heterozygotes that do develop sporadic CJD may have a longer duration 'atypical' disease.

Further evidence for the importance of the genotype relating to codon 129 comes from the analysis of the extended pedigree with a 144 bp insertion in the repeat region of the prion protein (Poulter *et al.*, 1992). This 144 bp insertion is carried on a methionine 129 allele so that affected individuals are either methionine 129 homozygotes or are heterozygotes with respect to codon 129. Affected individuals homozygous for codon 129 (although heterozygous for the insertion mutation) died significantly earlier than codon 129 heterozygotes (Baker *et al.*, 1991). It has also now been possible to correlate the age of onset with codon 129 genotype from a number of affected individuals. Strikingly all those who have an early age of onset, under the age of 40, were homozygous for methionine (the allele which bears the insert mutation), two individuals had an onset at age 40, one was a methionine homozygote, the other a heterozygote, and all those with a later age of onset were heterozygotes (Poulter *et al.*, 1992; Collinge *et al.*, 1992).

The observations on homozygosity at codon 129 lend support to the idea emerging from work on transgenic animals which implicates a role for protein–protein interaction in the establishment and progression of the disease (Prusiner *et al.*, 1990; Weissmann, 1991). Ordinarily there is a species barrier to transmission of disease across species as illustrated in Fig. 2. Mice are susceptible to prions originating from other mice but do not usually succumb to disease following inoculation with prions originating in hamsters. The occasional mouse that does succumb to infection following inoculation with hamster-derived prions does so after a prolonged incuba- tion period. Conversely hamsters are highly susceptible to hamster prions but not usually to mouse prions. In 1990 Prusiner *et al.* reported that they had constructed mice transgenic for the hamster prion gene. These mice expressed normal hamster PrP in the brain in quantities dependent upon the copy number of the transgene. Upon challenge with mouse-derived prions these transgenic animals were still susceptible to disease in the same way as wild-type mice. However, they were now also susceptible on challenge with hamster prions, the incubation period being inversely proportional to the level of expression of hamster PrP (Fig. 3). In addition, mice inoculated with hamster prions could be shown to generate prions infectious to wild-type hamsters (and not mice), whereas those inoculated with mouse-derived prions generated prions highly infectious for wild-type mice (but not hamsters). Since the only difference between the transgenic mice and normal mice is the expression of the hamster PrP, the abrogation of the species barrier in the transgenic animals can most easily be explained by a direct interaction between the cellular form of PrP (PrP^C), presumably on the cell surface, and the exogenous infectious form of the prion protein (PrP^{Sc}) with which the mice were inoculated. Since it is known from studies on the turnover of PrP in infected and non-infected murine neuroblastoma cells that PrP^{Sc} derives from PrP^C (Borchelt *et al.*, 1990) it is possible that the interaction between PrP^{Sc} and PrP^C drives the conversion of PrP^C into PrP^{Sc} (Fig. 4); whether this represents a conformational change in PrP as illustrated, a change in aggregation state or something else is not clear. The implication from the codon 129 genotype studies in the human diseases is that for optimum establishment and progression of the disease the interacting proteins need to be homologous (Fig. 5).

Westaway *et al.* (1987) demonstrated that murine strains with a long incubation time following experimental inoculation with prions encode a PrP (designated PrP^b) which differs at two residues from that seen in short incubation time strains (PrP^a)

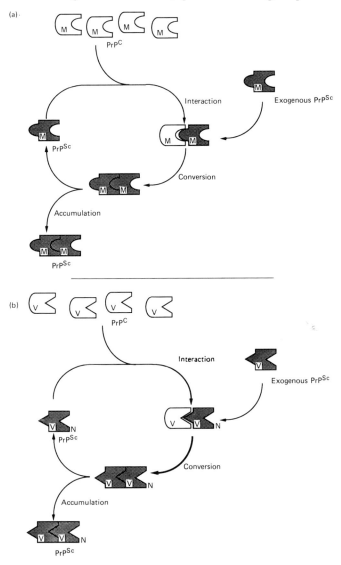

Fig. 4.   Representation of a possible mechanism by which the scrapie form of the prion protein PrP^{Sc} may lead to the progression of the disease by driving the conversion of the cellular form PrP^C into more PrP^{Sc}. Although the figure shows the conversion as a conformational change and the interaction as dimer formation this is for ease of representation; the nature of conversion of PrP^C into PrP^{Sc} is not known and could be a change in conformation, a change in aggregation state or some other state that renders it resistant to degradation. PrP^C molecules are represented as free-floating shaped boxes though it is known to be a membrane associated protein and these events are probably taking place in the presence of a lipid bilayer. Panels A and B represent molecules with a methionine or a valine at amino acid residue 129 respectively. Exogenous PrP^{Sc} introduced iatrogenically, by experimental routes or by oral routes interacts with cellular PrP. This drives the conversion of PrP^C into PrP^{Sc}. There is thus an increase of PrP^{Sc} some of which becomes associated with more PrP^C to fuel the cycle further, leading to a progressive accumulation of PrP^{Sc}. The darker arrow showing conversion of valine PrP^C to PrP^{Sc} follows from the data on genetic susceptibility to exogenous prions in iatrogenic cases and indicates that this conversion is more favourable than the conversion of the methionine variant.

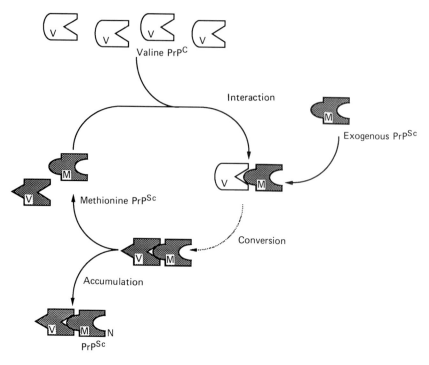

Fig. 5.   Representation of the reduced susceptibility of heterozygotes at residue 129 to develop disease. Differences in tertiary structure between methionine and valine variants reduce the efficiency of interaction and thus show a reduced level of conversion of heterologous types. For further details see legend to Fig. 4.

although it has not yet been unequivocally demonstrated that this variation, rather than variation at a separate but closely linked locus, is responsible for the incubation time differences. F₁ crosses between long and short incubation mice demonstrate that long incubation alleles are dominant. In an attempt to investigate this phenomenon further Westaway *et al.* (1991) constructed transgenic mice with multiple copies of a murine PrP transgenic construct derived from long incubation time mice. Paradoxically, those transgenic lines with marked overexpression of PrP^b had even shorter incubation times than non-transgenic short incubation time controls. A possible interpretation of this result is that it again indicates the importance of PrP homology with incubation times being inversely proportional to the level of PrP^b expression in an analogous fashion to the reduction in incubation time in the mice transgenic for hamster PrP discussed above.

    As different proteins, presumably valine 129 PrP and methionine 129 PrP will differ slightly in their propensity to undergo the transformation from PrP^C to PrP^{Sc}. It is therefore possible that valine 129 PrP molecules are more susceptible to undergoing the conformational (or other post-translational) change in response to exogenous prions (Fig. 6), accounting for the excess of valine 129 PrP alleles amongst

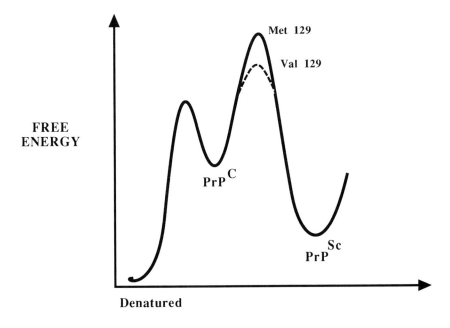

Fig. 6.   Theoretical energy level diagram for the different states of PrP adapted from Weissmann (1991). An energy barrier exists between the conformations of PrP^C and PrP^{Sc} such that the conversion of PrP^C to PrP^{Sc} is energetically unfavourable and uncommon, though is overcome by the interaction between PrP^{Sc} and PrP^C. The barrier to conversion is shown to be lower for the valine 129 polymorphic variant.

the iatrogenic cases (Collinge et al., 1991). However, homozygous cases are more likely to progress to the disease state since heterozygotes will be partially protected as a result of the unfavourable interaction between heterologous PrP molecules. Iatrogenic CJD in the individuals exposed to contaminated pituitary hormones then occurs most easily in valine 129 homozygotes. However, since dose of inoculation with prions is also a factor in transmission of the disease this would only be a relative protection for individuals with other genotypes. Indeed, it may be that iatrogenic CJD cases arising from intracerebral inoculation with prions rather than peripheral inoculation (e.g. dura mater grafting or intracerebral electrodes) may not show this effect. It will be interesting to see whether BSE occurs predominantly in cows homozygous with respect to bovine PrP polymorphisms, although selective breeding in cattle may make such analysis difficult to interpret.

According to this model, iatrogenic prion disease (or disease in animals following experimental inoculation) results from the introduction of exogenous PrP^{Sc} which interacts with host PrP^C and catalyses a chain reaction of transformation (whether conformational or other modification), resulting in accumulation of PrP^{Sc} and leading progressively towards the disease state. Both the initiation and the rate of progression

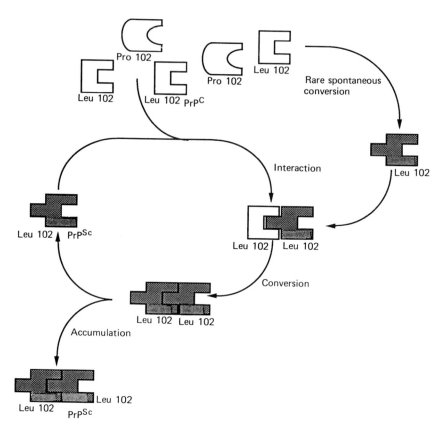

Fig. 7.   Representation of the spontaneous conversion of mutant variants of PrP into the scrapie form of PrP and the ensuing progression of the disease that leads to the accumulation of PrP^{Sc}. For fuller explanation of symbols see legend to Fig. 4. The conformation of the leucine 102 variant is indicated as being different to that of the wild-type proline form to represent the different property of spontaneous conversion and interactions as discussed in the text.

of this process are dependent on structural homology between inoculum PrP^{Sc} and host PrP^{C}. This could explain both the partial barrier to interspecies transmission seen experimentally (Pattison, 1965) (mammalian PrP molecules differing at a number of residues), and the protective effect of codon 129 heterozygosity in human disease which could be envisaged as an internal 'species barrier'.

In the inherited prion diseases, PrP^{C} containing one of the known pathogenic mutations presumably can spontaneously undergo the transformation to PrP^{Sc}, although such a change may still be an extremely rare event (Fig. 7).

Since the genotype associated with genetic susceptibility to infection with exogenous prions was not present to the same degree of excess in our sporadic CJD population as in the iatrogenic cases this might be interpreted as suggesting that most sporadic CJD cases are not environmentally acquired. Alternative explanations for the aetiology of sporadic CJD include new germ line mutations in a minority, somatic mutations resulting in the formation of PrP^{Sc} in the mutant cell clone, or perhaps the spontaneous conversion of normal sequence PrP^C to PrP^{Sc} as a rare stochastic event then triggering a disease causing chain reaction as before.

If sporadic disease arises in response to somatic mutation of the PrP gene all genotypes should be equally susceptible to mutation. The finding of a correlation between disease and homozygosity strongly suggests that it is the homology of the proteins that is permitting the disease to develop even though individuals who are heterozygous may also have developed somatic mutation. Those individuals who are heterozygous and nevertheless develop sporadic prion disease may show an atypical presentation of disease.

Such a model could account for the different clinical presentations of CJD and GSS, classical short duration CJD occurring in homozygotes, while GSS, which occurs as an autosomal dominant disease, always occurs in heterozygotes (heterozygous for the pathogenic mutation) and is therefore a longer duration illness. The interacting proteins can be even more heterologous if there is heterozygosity at codon 129 also. Sporadic disease in heterozygotes may result in a GSS-like illness or 'atypical CJD'. Clearly inherited prion disease (PrP lysine 200) does not fit into this neat picture. Individuals described so far with this condition present as a short duration CJD-like disease and homozygotes for the codon 200 mutation do not seem to differ clinically from heterozygotes (although it remains possible that more phenotypic variability, as is seen with other inherited prion diseases, may be identified when further cases are ascertained by genetic screening). How could this be explained? It may be that a mutation in PrP can have two different effects. Firstly it could act to destabilize the molecule leading to spontaneous production of PrP^{Sc} (i.e. a pathogenic mutation such as leucine 102); this could be called a 'type 1' effect.

Alternatively, a mutation could alter the structure of PrP such that it affects protein–protein interaction necessary for prion replication but does not alter the protein in a way that renders it more likely to undergo the transformation to PrP^{Sc}; this could be called a 'type 2' effect. According to such a model most of the known pathogenic mutations have both type 1 and type 2 effects. The codon 129 polymorphism has a type 2 effect (although presumably valine 129 has a slight effect in rendering the protein more liable to adopt the disease-related isoform but not sufficient to result in spontaneous disease). Lysine 200 would then have a type 1 effect but not a type 2 effect. This would in turn imply that residue 200 lies outside the region of PrP involved in dimerization (or other interaction). Of additional interest in this regard is the observation of incomplete penetrance of the codon 200 mutation. Lysine 200 may be somewhat less pathogenic than the other pathogenic mutations. Heterozygosity at codon 129 in lysine 200 individuals may be sufficient to delay onset of disease sufficiently to account for such non-penetrance. The two cases of inherited prion disease (PrP lysine 200) we have identified to date in the UK have both been homozygous at codon 129 (Collinge et al., submitted).

CONCLUSIONS

The transmissible spongiform encephalopathies are now more logically called prion diseases. Human prion diseases have inherited, sporadic and rare iatrogenic forms.

Inherited prion diseases are autosomal dominant genetic diseases caused by mutations in the prion protein gene. The full range of their phenotypes can now be directly examined and have still to be described. A powerful body of evidence, most notably from the production of transmissible spongiform neurodegeneration in transgenic mice encoding one of the human pathogenic mutations, indicates that an abnormal isoform of PrP is the essential component of the transmissible agent in these diseases. The disease process involves the normal cellular isoform of PrP being converted to the abnormal isoform. Work in mice transgenic for hamster PrP along with our work on the importance of PrP sequence homology in human prion disease strongly suggests that such a transformation may follow from direct interaction between PrP^{Sc} and PrP^{C} with PrP^{Sc} catalysing the conversion of PrP^{C} to PrP^{Sc} thereby resulting in a chain reaction leading progressively to the disease state. Presumably the pathogenic mutations seen in the human inherited diseases lead to the spontaneous production of PrP^{Sc} which then initiates a disease process transmissible to other individuals by inoculation. Sporadic prion disease most likely arises as a result of somatic mutation leading to the same process, although in this case it will not be vertically transmissible.

According to this emerging model, human prion diseases are (with the rare exception of iatrogenic disease and the previous transmission by cannibalism) essentially rare genetic diseases (Table 2). Because of the similarity of these diseases in different species and their ability to cross species it would not be unreasonable if the animal prion diseases were also originally genetic diseases. Transmission horizontally could then have resulted from human intervention, for instance abnormal feeding practices in which scrapie-infected sheep products are fed to captive or farmed animals (ranched mink, deer, cows and zoo animals). The use of BSE-infected cows in protein supplements fed to cows was of course a cannibalistic process (without species barrier) and resulted, like kuru, in epidemic disease. The remarkable resistance of the transmissible agent to inactivation is a vital aspect to this process. The mode of transmission of scrapie is still undecided but the practice of bringing ewes together for lambing and resulting ground contamination with placentae may well be relevant. Of course selective breeding may well have resulted in unusually genetically susceptible populations. Acquired prion diseases in animals, as well as in humans, may be man made.

Table 2.   Human prion diseases

Type	Syndrome	Aetiology
Acquired	Kuru Iatrogenic CJD ?Other	Cannibalism Accidental inoculation
Sporadic	Classical CJD Atypical CJD ?Other	?Somatic mutation ?Somatic mutation
Inherited	Familial CJD GSS Other*	Inherited germline mutation Inherited germline mutation

*Includes atypical dementias and fatal familial insomnia.

According to the model of prion replication discussed here, drugs designed with selective binding properties for the disease-related isoform of PrP may be expected to interfere with the disease process and may offer rational therapeutic approaches to human prion diseases.

REFERENCES

Adam, J., Crow, T.J., Duchen, L.W., Scaravilli, F. and Spokes, E. (1982) Familial cerebral amyloidosis and spongiform encephalopathy. *J. Neurol. Neurosurg. Psychiatry* **45** 37–45.

Alpers, M. (1987) Epidemiology and clinical aspects of kuru. In: Prusiner, S. B. and McKinley, M.P. (eds) *Prions: Novel Infectious Pathogens Causing Scrapie and Creutzfeldt–Jakob Disease.* Academic Press, San Diego, CA, pp. 451–465.

Baker, H.F., Poulter, M., Crow, T.J., Frith, C.D., Lofthouse, R., Ridley, R.M. and Collinge, J. (1991) Amino acid polymorphism in human prion protein and age at death in inherited prion disease. *Lancet* **337** 1286.

Basler, K., Oesch, B., Scott, M., Westaway, D., Walchli, M., Groth, D.F., McKinley, M.P., Prusiner, S.B. and Weissmann, C. (1986) Scrapie and cellular PrP isoforms are encoded by the same chromosomal gene. *Cell* **46** 417–428.

Beck, E. and Daniel, P.M. (1987) Neuropathology of transmissible spongiform encephalopathies. In: Prusiner, S.B. and McKinley, M.P. (eds) *Prions: Novel Infectious Pathogens Causing Scrapie and Creutzfeldt–Jakob Disease.* Academic Press, San Diego, CA, pp. 331–385.

Borchelt, D.R., Scott, M., Taraboulos, A., Strahl, N. and Prusiner, S.B. (1990) Scrapie and cellular prion proteins differ in their kinetics of synthesis and topology in cultured cells. *J. Cell Biol.* **110** 743–752.

Brown, P., Rodgers Johnson, P., Cathala, F., Gibbs, C.J., Jr, and Gajdusek, D.C. (1984) Creutzfeldt–Jakob disease of long duration: clinicopathological characteristics, transmissibility, and differential diagnosis. *Ann. Neurol.* **16** 295–304.

Brown, P., Cathala, F., Raubertas, R.F., Gajdusek, D.C. and Castaigne, P. (1987) The epidemiology of Creutzfeldt–Jakob disease: conclusion of a 15-year investigation in France and review of the world literature. *Neurology* **37** 895–904.

Brown, P., Goldfarb, L.G. and Gajdusek, D.C. (1991a) The new biology of spongiform encephalopathy: infectious amyloidoses with a genetic twist. *Lancet* **337** 1019–1022.

Brown, P., Goldfarb, L.G., Brown, W.T., Goldgaber, D., Rubenstein, R., Kascsak, R.J., Guiroy, D.C., Piccardo, P., Boellaard, J.W. and Gajdusek, D.C. (1991b) Clinical and molecular genetic study of a large German kindred with Gerstmann-Straussler–Scheinker syndrome. *Neurology* **41** 375–379.

Bruce, M.E., McBride, P.A. and Farquhar, C.F. (1989) Precise targeting of the pathology of the sialoglycoprotein, PrP, and vacuolar degeneration in mouse scrapie. *Neurosci. Lett.* **102** 1–6.

Buchanan, C.R., Preece, M.A. and Milner, R.D. (1991) Mortality, neoplasia, and Creutzfeldt–Jakob disease in patients treated with human pituitary growth hormone in the United Kingdom. *Br. Med. J.* **302** 824–828.

Cathala, F. and Baron, H. (1987) Clinical aspects of Creutzfeldt-Jakob disease. In: Prusiner, S. B. and McKinley, M. P. (eds) *Prions: Novel Infectious Pathogens Causing Scrapie and Creutzfeldt–Jakob Disease.* Academic Press, San Diego, CA, pp. 467–509.

Chesebro, B., Race, R., Wehrly, K., Nishio, J., Bloom, M., Lechner, D., Bergstrom, S., Robbins, K., Mayer, L., Keith, J.M., *et al.* (1985) Identification of scrapie prion protein-specific mRNA in scrapie-infected and uninfected brain. *Nature (London)* **315** 331–333.

Clinton, J., Lantos, P.L., Rossor, M., Mullan, M. and Roberts, G.W. (1990) Immunocytochemical confirmation of prion protein. *Lancet* **336** 515.

Collinge, J. and Palmer, M.S. (1991) CJD discrepancy. *Nature (London)* **353** 802.

Collinge, J., Harding, A.E., Owen, F., Poulter, M., Lofthouse, R., Boughey, A.M., Shah, T. and Crow, T.J. (1989) Diagnosis of Gerstmann–Straussler syndrome in familial dementia with prion protein gene analysis. *Lancet* **2** 15–17.

Collinge, J., Owen, F., Poulter, M., Leach, M., Crow, T.J., Rossor, M.N., Hardy, J., Mullan, M.J., Janota, I. and Lantos, P.L. (1990) Prion dementia without characteristic pathology. *Lancet* **336** 7–9.

Collinge, J., Palmer, M. S. and Dryden, A. J. (1991) Genetic predisposition to iatrogenic Creutzfeldt–Jakob disease. *Lancet* **337** 1441–1442.

Collinge, J., Brown, J., Hardy, J., Mullan, M., Rossor, M.N., Baker, H., Crow, T.J., Lofthouse, R., Poulter, M., Ridley, R., Owen, F., Bennett, C., Dunn, G., Harding, A.E., Quinn, N., Doshi, B., Roberts, G.W., Honavar, M., Janota, I., Lantos, P.L. (1992) Inherited prion disease with 144 base pair gene insertion. 2. Clinical and pathological features. *Brain* **115**, in press.

Craufurd, D., Dodge, A., Kerzin-Storrar, L. and Harris, R. (1989) Uptake of presymptomatic predictive testing for Huntington's disease. *Lancet* **334** 603–605.

Creutzfeldt, H.G. (1920) *Z. Gesamte Neurol. Psychiatr.* **57** 1–18.

Doh-ura, K., Tateishi, J., Sasaki, H., Kitamoto, T. and Sakaki, Y. (1989) Pro–leu change at position 102 of prion protein is the most common but not the sole mutation related to Gerstmann–Straussler syndrome. *Biochem. Biophys. Res. Commun.* **163** 974–979.

Doh-ura, K., Tateishi, J., Kitamoto, T., Sasaki, H. and Sakaki, Y. (1990) Creutzfeldt–Jakob disease patients with congophilic kuru plaques have the misssense variant prion protein common to Gerstmann–Straussler syndrome. *Ann. Neurol.* **27** 121–126.

Gerstmann, J. (1928) Über ein noch nicht beschriebenes Reflexphänomen bei einer Erkrankung des zerebellaren Systems. *Wien. Med. Wochenschr.* **78** 906–908.

Gerstmann, J., Sträussler, E. and Scheinker, I. (1936) Über eine eigenartige hereditär-familiäre Erkrankung des Zentralnervensystems. Zugleich ein Beitrag zur Frage des vorzeitigen lakalen Alterns. *Z. Neurol.* **154** 736–762.

Ghetti, B., Tagliavini, F., Masters, C.L. *et al.* (1989) Gerstmann–Straussler–Scheinker disease. II. Neurofibrillary tangles and plaques with PrP-amyloid coexist in an affected family. *Neurology* **39** 1453–1461.

Gibbs, C.J. and Gajdusek, D.C. (1969) In: Norris, F.H., Jr., and Kurland, L.T. (eds) *Motor Neuron Diseases: Research on Amyotrophic Lateral Sclerosis and Related*

Disorders, Neurology Symposia. Grune and Stratton, New York, pp. 269–279.

Gibbs, C.J. and Gajdusek, D.C. (1973) Experimental subacute spongiform encephalopathies in primates and other animals. *Science* **182** 67–68.

Gibbs, C.J., Gajdusek, D.C., Asher, D.M., Alpers, M.P., Beck, E., Daniel, P.M. and Matthews, W. B. (1968). Creutzfeldt–Jakob disease (spongiform encephalopathy): transmission to the chimpanzee. *Science* **161** 388–389.

Goldfarb, L.G., Brown, P., Goldgaber, D., Garruto, R.M., Yanagihara, R., Asher, D.M. and Gajdusek, D.C. (1990a) Identical mutation in unrelated patients with Creutzfeldt–Jakob disease. *Lancet* **336** 174–175.

Goldfarb, L.G., Korczyn, A.D., Brown, P., Chapman, J. and Gajdusek, D.C. (1990b) Mutation in codon 200 of scrapie amyloid precursor gene linked to Creutzfeldt–Jakob disease in Sephardic Jews of Libyan and non-Libyan origin. *Lancet* **336** 637–638.

Goldfarb, L.G., Haltia, M., Brown, P., Nieto, A., Kovanen, J., McCombie, W.R., Trapp, S. and Gajdusek, D.C. (1991) New mutation in scrapie amyloid precursor gene (at codon 178) in Finnish Creutzfeldt–Jakob kindred. *Lancet* **337** 425.

Goldgaber, D., Goldfarb, L.G., Brown, P. *et al.* (1989) Mutations in familial Creutzfeldt–Jakob disease and Gerstmann–Straussler–Scheinker's syndrome. *Exp. Neurol.* **106** 204–206.

Gomori, A.J., Partnow, M.J., Horoupian, D.S. and Hirano, A. (1973) The ataxic form of Creutzfeldt–Jakob disease. *Arch. Neurol.* **29** 318–323.

Hartsough, G.R. and Burger, D. (1965) Encephalopathy of mink. I. Epizootiologic and clinical observations. *J. Infect. Dis.* **115** 387–392.

Heston, L.L., Lowther, D.L.W. and Leventhal, C.M. (1966) Alzheimer's disease. A family study. *Arch. Neurol.* **15** 225–233.

Hsiao, K., Baker, H.F., Crow, T.J., Poulter, M., Owen, F., Terwilliger, J.D., Westaway, D., Ott, J. and Prusiner, S.B. (1989) Linkage of a prion protein missense variant to Gerstmann–Straussler syndrome. *Nature (London)* **338** 342–345.

Hsiao, K.K., Scott, M., Foster, D., Groth, D.F., DeArmond, S.J. and Prusiner, S.B. (1990) Spontaneous neurodegeneration in transgenic mice with mutant prion proteins. *Science* **250** 1587–1590.

Hsiao, K.K., Cass, C., Schellenberg, G.D., Bird, T., Devine Gage, E., Wisniewski, H. and Prusiner, S.B. (1991a) A prion protein variant in a family with the telencephalic form of Gerstmann–Straussler–Scheinker syndrome. *Neurology* **41** 681–684.

Hsiao, K., Meiner, Z., Kahana, E., Cass, C., Kahana, I., Avrahami, D., Scarlato, G., Abramsky, O., Prusiner, S.B. and Gabizon, R. (1991b) Mutation of the prion protein in Libyan Jews with Creutzfeldt–Jakob disease. *N. Engl. J. Med.* **324** 1091–1097.

Jakob, A. (1921a) *Z. Gesamte Neurol. Psychiatr.* **64** 147–228.

Jakob, A. (1921b) *Dtsch. Z. Nervenheilk.* **70** 132–146.

Jakob, A. (1921c) *Med. Klin.* **13** 372–376.

Jeffrey, M. and Wells, G.A. (1988) Spongiform encephalopathy in a nyala (*Tragelaphus angasi*). *Vet. Pathol.* **25** 398–399.

Kirkwood, J.K., Wells, G.A., Wilesmith, J.W., Cunningham, A.A. and Jackson, S.I. (1990) Spongiform encephalopathy in an arabian oryx (*Oryx leucoryx*) and a greater kudu (*Tragelaphus strepsiceros*). *Vet. Rec.* **127** 418–420.

Kretzschmar, H.A., Honold, G., Seitelberger, F., Feucht, M., Wessely, P., Mehraein, P. and Budka, H. (1991) Prion protein mutation in family first reported by Gerstmann, Straussler, and Scheinker. *Lancet* **337** 1160.

Masters, C.L., Harris, J.O., Gajdusek, D.C., Gibbs, C.J., Bernoulli, C. and Asher, D.M. (1979) Creutzfeldt–Jakob disease: patterns of worldwide occurrence and the significance of familial and sporadic clustering. *Ann. Neurol.* **5** 177–188.

Masters, C.L., Gajdusek, D.C. and Gibbs, C.J., Jr (1981) Creutzfeldt–Jakob disease virus isolations from the Gerstmann–Straussler syndrome with an analysis of the various forms of amyloid plaque deposition in the virus-induced spongiform encephalopathies. *Brain* **104** 559–588.

McGowan, J.P. (1922) Scrapie in sheep. *Scott. J. Agric.* **5** 365–375.

Meyer, A., Leigh, D. and Bagg, C.E. (1954) A rare presenile dementia associated with cortical blindness (Heidenhain's syndrome). *J. Neurol. Neurosurg. Psychiatr.* **17** 129–133.

Nieto, A., Goldfarb, L.G., Brown, P., McCombie, W.R., Trapp, S., Asher, D.M. and Gajdusek, D.C. (1991) Codon 178 mutation in ethnically diverse Creutzfeldt–Jakob disease families. *Lancet* **337** 622–623.

Nochlin, D., Sumi, S.M., Bird, T.D., Snow, A.D., Leventhal, C.M., Beyreuther, K. and Masters, C.L. (1989) Familial dementia with PrP-positive amyloid plaques: a variant of Gerstmann–Straussler syndrome. *Neurology* **39** 910–918.

Oesch, B., Westaway, D., Walchli, M. *et al.* (1985) A cellular gene encodes scrapie PrP 27-30 protein. *Cell* **40** 735–746.

Owen, F., Poulter, M., Lofthouse, R., Collinge, J., Crow, T.J., Risby, D., Baker, H.F., Ridley, R.M., Hsiao, K. and Prusiner, S.B. (1989) Insertion in prion protein gene in familial Creutzfeldt–Jakob disease. *Lancet* **1** 51–52.

Owen, F., Poulter, M., Collinge, J., Leach, M., Shah, T., Lofthouse, R., Chen, Y., Crow, T.J., Harding, A.E., Hardy, J. and Rossor, M.N. (1991) Insertions in the prion protein gene in atypical dementias. *Exp. Neurol.* **112** 240–242.

Palmer, M.S., Dryden, A.J., Hughes, J.T. and Collinge, J. (1991) Homozygous prion protein genotype predisposes to sporadic Creutzfeldt–Jakob protein. *Nature (London)* **352** 340–342.

Pattison, I.H. (1965) *Symposium on Slow, Latent and Temperate Virus Infections.* US National Institute of Neurological Diseases and Blindness, Monograph No. 2, p. 249.

Poulter, M., Baker, H.F., Frith, C.D., Leach, M., Lofthouse, R., Ridley, R.M., Shah, T., Owen, F., Collinge, J., Brown, G., Hardy, J., Mullan, M.J., Harding, A.E., Bennett, C., Doshi, R., Crow, T.J. (1992) Inherited prion disease with 144 base pair gene insertion. 1. Genealogical and molecular studies. *Brain* **115**, in press.

Prusiner, S.B. (1987) Prions and neurodegenerative diseases. *N. Engl. J. Med.* **137** 1571–1581.

Prusiner, S.B. (1989) Scrapie prions. *Annu. Rev Microbiol.* **43** 345–374.

Prusiner, S.B. (1991) Molecular biology of prion diseases. *Science* **252** 1515–1522.

Prusiner, S.B., Scott, M., Foster, D. *et al.* (1990) Transgenetic studies implicate interactions between homologous PrP isoforms in scrapie prion replication. *Cell* **63** 673–686.

Robakis, N.K., Devine Gage, E.A., Jenkins, E.C., Kascsak, R.J., Brown, W.T., Krawczun, M.S. and Silverman, W.P. (1986) Localization of a human gene homologous to the PrP gene on the p arm of chromosome 20 and detection of PrP-related antigens in normal human brain. *Biochem. Biophys. Res. Commun.* **140** 758–765.

Serban, D., Taraboulos, A., DeArmond, S.J. and Prusiner, S.B. (1990) Rapid detection of Creutzfeldt–Jakob disease and scrapie prion proteins. *Neurology* **40** 110–117.

Sparkes, R.S., Simon, M., Cohn, V.H. *et al.* (1986) Assignment of the human and mouse prion protein genes to homologous chromosomes. *Proc. Natl. Acad. Sci. USA* **83** 7358–7362.

Speer, M.C., Goldgaber, D., Goldfarb, L.G., Roses, A.D. and Pericak Vance, M.A. (1991) Support of linkage of Gerstmann–Straussler–Scheinker syndrome to the prion protein gene on chromosome 20p12-pter. *Genomics* **9** 366–368.

Spielmeyer, W. (1922) *Klin. Wochenschr.* **2** 1817–1819.

Warter, J.M., Steinmetz, G., Heldt, N., Rumbach, L., Marescaux, C., Eber, A.M., Collard, M., Rohmer, F., Floquet, J., Guedenet, J.C., Gehin, P. and Weber, M. (1982) Familial presenile dementia: Gerstmann–Straussler–Scheinker's syndrome. *Rev. Neurol. (Paris)* **138** 107–121 (in French).

Weissmann, C. (1991) The prion's progress. *Nature (London)* **349** 569–571.

Weller, R.O. (1989) Iatrogenic transmission of Creutzfeldt–Jakob disease. *Psychol. Med.* **19** 1–4.

Wells, G.A., Scott, A.C., Johnson, C.T., Gunning, R.F., Hancock, R.D., Jeffrey, M., Dawson, M. and Bradley, R. (1987) A novel progressive spongiform encephalopathy in cattle. *Vet. Rec.* **121** 419–420.

Westaway, D. and Prusiner, S.B. (1986) Conservation of the cellular gene encoding the scrapie prion protein. *Nucleic Acids Res.* **14** 2035–2044.

Westaway, D., Goodman, P.A., Mirenda, C.A., McKinley, M.P., Carlson, G.A. and Prusiner, S.B. (1987) Distinct prion proteins in short and long scrapie incubation period mice. *Cell* **51** 651–662.

Westaway, D., Mirenda, C.A., Foster, D., Zebarjadian, Y., Scott, M., Torchia, M., Yang, S.L., Serban, H., DeArmond, S.J., Ebeling, C., Prusiner, S.B. and Carlson, G.A. (1991) Paradoxical shortening of scrapie incubation times by expression of prion protein transgenes derived from long incubation period mice. *Neuron* **7** 59–68.

Wilesmith, J.W., Wells, G.A., Cranwell, M.P. and Ryan, J.B. (1988) Bovine spongiform encephalopathy: epidemiological studies. *Vet. Rec.* **123** 638–644.

Williams, E.S. and Young, S. (1980) Chronic wasting disease of captive mule deer: a spongiform encephalopathy. *J. Wildl. Dis.* **16** 89–98.

World Federation of Neurology Research Group on Huntington's Disease (1989) Ethical issues policy statement on Huntington's disease molecular genetics predictive tests. *J. Neurol. Sci.* **94** 327–332.

Wyatt, J.M., Pearson, G.R., Smerdon, T.N., Gruffydd-Jones, T.J., Wells, G.A.H. and Wilesmith, J.W. (1991) Naturally occurring scrapie-like spongiform encephalopathy in five domestic cats. *Vet. Rec.* **129** 233–236.

13

Genetic and transgenic studies of prion proteins in Gerstmann–Sträussler–Scheinker disease

K. K. Hsiao, D. Groth, M. Scott, S.-L. Yang, A. Serban, D. Rapp, D. Foster, M. Torchia, S. J. DeArmond and S. B. Prusiner

ABSTRACT

Gerstmann–Sträussler–Scheinker disease (GSS) is a rare, dominantly inherited neurodegenerative disease which can sometimes be transmitted to experimental animals through intracerebral inoculation of brain homogenates from patients. Substitution of leucine for proline at codon 102 of the prion protein gene has been found in several families with the disease; this mutation is genetically linked to GSS. Mice containing murine prion protein transgenes with this mutation spontaneously develop neurologic symptoms of ataxia, lethargy, and rigidity accompanied by spongiform degeneration throughout the brain. Thus many of the clinical and pathological features of GSS have been reproduced in this transgenic mouse paradigm; this study illustrates that a neurodegenerative illness similar to a human disease can be genetically modelled in animals.

INTRODUCTION

Gerstmann–Sträussler–Scheinker disease (GSS), also known as Sträussler's disease and Gerstmann–Sträussler syndrome, was first described by the Austrian neurologists J. Gerstmann and E. Sträussler and the neuropathologist I. Scheinker in 1936 (Gerstmann *et al.*, 1936). It has been defined as a 'spinocerebellar ataxia with dementia and plaque-like deposits' (Seitelberger, 1981). The original case of GSS and the majority of subsequently diagnosed cases were familial; about half the members of each generation were affected. A few sporadic cases resembling GSS clinically and pathologically have been reported (Masters *et al.*, 1981). Until recently there has been discussion over the exact classification of GSS and diseases resembling GSS (Kuzuhara *et al.*, 1983); however, the development of new immunohistologic and molecular genetic markers has greatly improved our ability to diagnose GSS.

The clinical manifestations of GSS may vary, even within the same pedigree (Masters *et al.*, 1981). Nevertheless, it is useful to examine a weighted composite of clinical features in several members of a given family; sometimes this approach enables investigators to decipher different clinical forms of GSS. In 'ataxic GSS', patients typically complain of difficulty walking and unsteadiness, sometimes accompanied by leg pains or paraesthesias in the early stage of disease. Later, mental and behavioural deterioration may occur. Examination usually reveals prominent cerebellar ataxia, dysarthria, ocular dysmetria, and hypo- or areflexia in the lower extremities with extensor plantar responses. Mild wasting or weakness and impairment of vibratory and proprioceptive sensations may be detected in the lower extremities. Dysphagia frequently develops and contributes to inanition in the final stages of disease. By contrast, in 'dementing GSS', patients typically complain of cognitive and behavioural changes in the early stages of disease (Warter *et al.*, 1982; Nochlin *et al.*, 1989). Later, Parkinsonism, pseudobulbar signs, and spasticity may develop. Ataxia is present in some cases.

The pathologic hallmark of GSS is the multicentric amyloid plaque which contains prion protein (PrP) (Kitamoto *et al.*, 1986; Roberts *et al.*, 1986; DeArmond *et al.*, 1987). The amyloid plaques are deposited mainly in the cerebral and cerebellar hemispheres. The amyloid plaques of GSS specifically stain with polyclonal antisera raised against PrP 27-30 isolated from scrapie-infected hamster brains. The diagnosis of GSS is most effectively established when immunohistologic examination of formalin-fixed brain embedded in paraffin blocks reveals multicentric PrP plaques. White matter degeneration resembling that of other system degenerations, such as Friedreich's ataxia, is a prominent feature in most cases of GSS (Seitelberger, 1981). Neuronal loss occurs in scattered areas throughout the brain and spinal cord. Spongiform change is variable in degree and extent. Neurofibrillary tangles, if present, are usually observed in quantity and locations consistent with the age of the patient; however, in one family with GSS they are visible in pathologic numbers (Azzarelli *et al.*, 1985; Farlow *et al.*, 1989; Ghetti *et al.*, 1989).

GSS belongs to a group of related conditions, including Creutzfeldt–Jakob disease (CJD) and kuru, which are clinically, pathologically, and biochemically similar to scrapie, a neurodegenerative disease naturally found in sheep and goats. These diseases can often be transmitted to experimental animals through intracerebral inoculation of brain homogenates from patients and animals with the disease. The transmissible agent, termed 'prion' to distinguish it from viruses and viroids, has been found to be largely composed of an aberrant form of a 33–35 kd host glycoprotein, called prion protein (PrP), and contains little or no nucleic acid (Prusiner, 1982). PrP is encoded in humans by a single-copy gene on chromosome 20 (Sparkes *et al.*, 1986). The entire 253-codon open reading frame is located in one exon (Basler *et al.*, 1986; Kretzschmar *et al.*, 1986; Puckett *et al.*, 1991). A more extensive discussion of the biochemistry

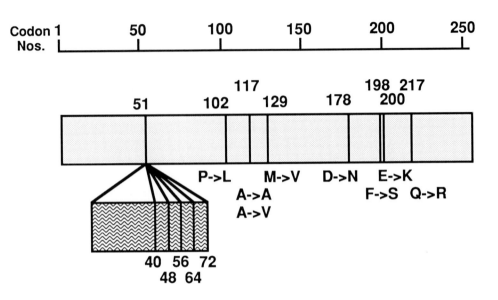

Fig. 1.   Human PrP open reading frame. The positions of known amino acid substitutions and insertions associated with inherited human prion diseases are shown. P, proline; L, leucine; A, alanine; V, valine; methionine; D, aspartic acid; N, asparagine; F, phenylalanine; S, serine; E, glutamic acid; K, lysine; Q, glutamine; R, arginine. Box with wavy pattern denotes inserts of 40 to 72 mino acids corresponding to 5 to 9 glycine-rich octarepeats.

and genetics of prions can be found elsewhere (Prusiner, 1987, 1991).

Assuming an autosomal dominant mode of inheritance, genetic linkage analysis of a leucine for proline substitution at PrP codon 102 yielded a lod score of > 3.2 in two unrelated Caucasian families (Hsiao *et al.*, 1989a). This evidence is consistent with GSS being inherited despite also being infectious. An affected descendent of the first patient with GSS described by Gerstmann, Sträussler, and Scheinker was shown to have this Leu_{102} substitution (Kretzschmar *et al.*, 1991). PrP Leu_{102} was also found in two Asian patients with GSS, thus supporting the hypothesis that the mutation is more than a linked genetic marker, possibly causing the disease (Doh-ura *et al.*, 1989; Hsiao *et al.*, 1989b). This and three other point mutations in the PrP open reading frame, Val_{117}, Ser_{198}, and Arg_{217}, have been found to segregate with disease in families with GSS (Doh-ura *et al.*, 1989; Hsiao *et al.*, 1992). Two point mutations, Asn_{178} and Lys_{200}, and several insertional mutations have been found to segregate with disease in families with CJD (Goldgaber *et al.*, 1989; Owen *et al.*, 1989; Goldfarb *et al.*, 1990, 1991a, b; Hsiao *et al.*, 1991; Brown *et al.*, 1991) (Fig. 1). There is a strong but not absolute correspondence of the various PrP mutations to different clinical and pathological phenotypes. The molecular basis for the pathogenic effects of these mutations is not understood. A more complete review of the molecular genetics of GSS and CJD can be found elsewhere (Hsiao and Prusiner, 1990; Brown *et al.*, 1991).

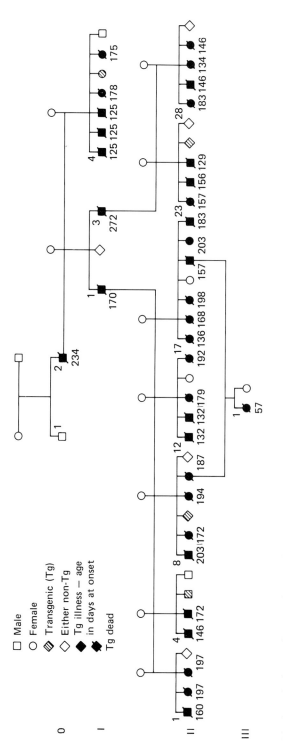

Fig. 2. Inheritance of a neurodegenerative disorder in Tg(GSSMoPrP)174 mice. The founding parents (FP) were both (C57BL/6 × SJL) F1 mice. Superscripts denote symbol numbers for each generation. Each of the filled symbols represents a transgenic (Tg) mouse, and the numbers below these symbols denote the age in days at which neurologic dysfunction was first observed. Stippled symbols represent transgenic mice who have not yet developed disease. Open symbols represent non-transgenic littermates. To compress the data, many stippled and open symbols represent multiple mice. Those denoting more than one mouse are as follows: I.2, $n = 8$; I.7, $n = 3$; II.3, $n = 7$; II.6, $n = 2$; II.7, $n = 3$; II.9, $n = 3$; II, $n = 2$; II.16, $n = 2$; II.22, $n = 5$; II.26, $n = 3$; II.27, $n = 2$; and II.31, $n = 3$. (Reprinted, with permission, from *Science* **250** 1587–1590 (1990).)

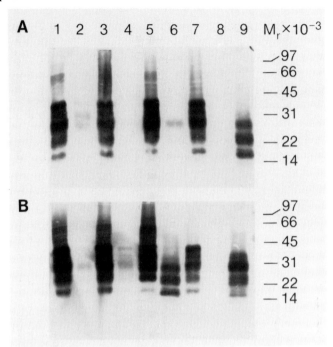

Fig. 3.   PrP immunoblots of brain extracts from Tg(GSSMoPrP)174 mice detected with a rabbit antiserum (R073). (A) Protease-resistant PrP in brains of Tg(GSSMoPrP)174 mice which exhibited neurologic dysfunction. Three transgenic mice are shown in the pedigree (Fig. 2): mouse 1.1, lanes 1 and 2; mouse 11.12, lanes 3 and 4; mouse 11.17, lanes 5 and 6. The variable amounts of protease-resistant PrP in Tg(GSSMoPrP)174 mice are shown in lanes 2, 4, and 6; lane 4 contained the lowest levels of protease-resistant PrP. Non-Tg(GSSMoPrP)174 littermate control, 128 days old, lanes 7 and 8. A Swiss CD-1 mouse with clinical signs of scrapie inoculated with the Chandler scrapie isolate, lane 9. (B) Protease-resistant PrP in brains of asymptomatic Tg(GSSMoPrP)174 mice. Two transgenic mice, 49 days old, lanes 1 to 4; Swiss CD-1 mouse with scrapie, lanes 5 and 6; non-Tg(GSSMoPrP)174 littermate control, lanes 7 and 8; Syrian golden hamster with scrapie, lane 9. (Reprinted, with permission, from *Science* **250** 1587–1590 (1990).)

Transgenic (Tg) mice were created to test the biologic activity of the PrP leucine substitution linked to GSS. Spontaneous ataxia, rigidity and lethargy with spongiform degeneration and gliosis developed in two of four lines of Tg(GSSMoPrP) mice containing murine PrP genes with a leucine substitution at codon 101 (homologous to codon 102 in humans) (Fig. 2) (Hsiao *et al.*, 1990; Hsiao, unpublished data). Little or no detectable protease-resistant PrP was seen on Western blots of brain homogenates from these Tg(GSSMoPrP) mice (Fig. 3). The residual PrP present after limited proteolysis with proteinase K is probably the result of increased substrate PrP in the Tg(GSSMoPrP) mice, which express four times more PrP than non-Tg littermates (Fig. 4).

Although protease-resistant PrP is associated with disease in inoculated animals with experimental scrapie (Jendroska *et al.*, 1991), there is thus far no evidence that this is so in Tg(GSSMoPrP) mice. In contrast to inoculated mice that develop scrapie, Tg(GSSMoPrP) mice develop clinical signs and spongiform degeneration in the

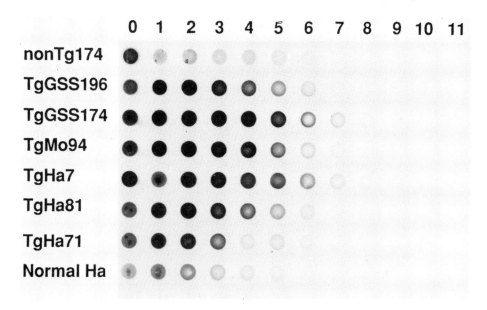

Fig. 4.   Brain PrP immunodot blots. Brain tissue from transgenic mice, non-transgenic control mice, and hamsters was disrupted in 0.32 M sucrose by passing it through a 20-gauge needle five times and through a 22-gauge needle ten times. The 10% (w/v) homogenate was centrifuged at 1600g for 5 min at 4°C. Supernatant protein measured by BCA dye binding was adjusted to 3 mg/ml in 10 mM Tris–acetate, pH 7.4, buffer containing 2% N-lauroyl-N- methylglycine (Sarkosyl). In column 0, 30 μg protein was applied to nitrocellulose membranes using a manifold filtration unit. Protein samples were serially diluted by a factor of 2 in columns 2–11. PrP was detected using a rabbit antiserum (R073) (Serban *et al.* 1990). Tg(GSSMoPrP)196 and Tg(GSSMoPrP)174 are two lines of transgenic mice which express murine PrP with a proline substitution at residue 101. Tg($Prn-p^h$)94 is a line of transgenic mice which expresses the $Prn-p^b$ variant of murine PrP. Tg(SHaPrP)7, Tg(SHaPrP)81, and Tg(SHaPrP)71 are three lines of transgenic mice which express hamster PrP.

absence of detectable levels of protease-resistant PrP. This suggests that an as-yet undefined abnormality in the metabolism of mutant PrP molecules contributes to or causes the disease in Tg(GSSMoPrP) mice. Histoblot analyses (Taraboulos *et al.*, 1992) currently being done with Tg(GSSMoPrP) brains could potentially reveal focal accumulations of protease-resistant PrP. However, if absent, then the likelihood that protease-resistant PrP plays a role in disease pathogenesis in Tg(GSSMoPrP) mice becomes more remote.

We now have preliminary studies suggesting that the disease produced in Tg(GSSMoPrP) mice can be sometimes transmitted through inoculation of brain extracts derived from affected Tg(GSSMoPrP) mice. Inoculation of a 10% brain homogenate derived from an offspring of the founder Tg(GSSMoPrP) mouse, Tg(GSSMoPrP)$_{170}$, ill at 170 days of age, produced neurologic symptoms of experimental scrapie in 7/10 Syrian golden hamsters at 221 $\pm$ 26 days, hence referred to as Tg(GSSMoPrP)$_{170}$ → Syr.

Neuropathologic studies of Tg(GSSMoPrP)$_{170}$ → Syr revealed prominent spongi-form degeneration in a novel topographic distribution, with striking white matter spongiosis and gliosis which is rarely if ever observed in Syrian hamsters inoculated with 139H or Sc237 hamster prion strains.

Serial passage and end-point titration of brain homogenate from Tg(GSSMoPrP)$_{170}$ → Syr into Syrian hamsters produced neurologic symptoms at 75 days and a high titre. Brain homogenates of Syrian hamsters contain protease-resistant PrP.

While prolonged incubation times can be indicative of contamination with low levels of scrapie prions, they are also consistent with the low levels of protease-resistant PrP in the brains of affected Tg(GSSMoPrP) mice. The apparent production of prions *de novo* in Tg(GSSMoPrP) mice argues that prions lack foreign nucleic acid, a conclusion which is consistent with many other lines of experimental evidence (Prusiner, 1991). The low or absent levels of detectable protease-resistant PrP and the low apparent titre of prions suggest that neuronal death in Tg(GSSMoPrP) mice may occur, at least in part, through abnormal metabolism of mutant PrP.

REFERENCES

Azzarelli, B., Muller, J., Ghetti, B., Dyken, M. and Conneally, P.M. (1985) Cerebellar plaques in familial Alzheimer's disease (Gerstmann–Sträussler–Scheinker variant?). *Acta Neuropathol. (Berlin)* **65** 235–246.

Basler, K., Oesch, B., Scott, M., Westaway, D., Wälchi, M., Groth, D.F., McKinley, M.P., Prusiner, S.B. and Weissmann, C. (1986) Scrapie and cellular PrP isoforms are encoded by the same chromosomal gene. *Cell* **46** 417–428.

Brown, P., Goldfarb, L. and Gajdusek, C. (1991) The new biology of spongiform encephalopathy: infectious amyloidoses with a genetic twist. *Lancet* **337** 1019–1022.

DeArmond, S.J., Kretzschmar, H.A., McKinley, M.P. and Prusiner, S.B. (1987) Molecular pathology of prion diseases. In: Prusiner, S.B. and McKinley, M.P. (eds) *Prions—Novel Infectious Pathogens Causing Scrapie and Creutzfeldt–Jakob Disease.* Academic Press, Orlando, FL, pp. 387–414.

Doh-ura, K., Tateishi, J., Sasaki, H., Kitamoto, T. and Sakaki, Y. (1989) Pro–Leu change at position 102 of prion protein is the most common but not the sole mutation related to Gerstmann–Sträussler syndrome. *Biochem. Biophys. Res. Commun.* **163** 974–979.

Farlow, M.R., Yee, R.D., Dlouhy, S.R., Conneally, P.M., Azzarelli, B. and Ghetti, B. (1989) Gerstmann–Sträussler–Scheinker disease. I. Extending the clinical spectrum. *Neurology* **39** 1446–1452.

Gerstmann, J., Sträussler, E. and Scheinker, I. (1936) Über eine eigenartige hereditär-familiäre Erkrankung des Zentralnervensystems zugleich ein beitrag zur frage des vorzeitigen lokalen alterns. *Z. Neurol.* **154** 736–762.

Ghetti, B., Tagliavini, F., Masters, C.L. *et al.* (1989) Gerstmann–Sträussler–Scheinker disease. II. Neurofibrillary tangles and plaques with PrP-amyloid coexist in an affected family. *Neurology* **39** 1453–1461.

Goldfarb, L., Korczyn, A., Brown, P., Chapman, J. and Gajdusek, D.C. (1990) Mutation in codon 200 of scrapie amyloid precursor gene linked to Creutzfeldt–Jakob disease in Sephardic Jews of Libyan and non-Libyan origin. *Lancet* **336** 637–638.

Goldfarb, L., Haltia, M., Brown, P., Nieto, A., Kovanen, J., McCombie, W.R., Trapp, S. and Gajdusek, D.C. (1991a) New mutation in scrapie amyloid precursor gene (at codon 178) in Finnish Creutzfeldt–Jakob kindred. *Lancet* **337** 425–426.

Goldfarb, L.G., Brown, P., McCombie, W.R., Goldgaber, D., Swergold, G.D., Wils, P.R., Cervenakova, L., Baron, H., Gibbs, C.J., Jr, and Gajdusek, D.C. (1991b) Transmissible familial Creutzfeldt–Jakob disease associated with five, seven, and eight extra octapaptide coding repeats in the *PRNP* gene. *Proc. Natl. Acad. Sci. USA* **88** 10926–10930.

Goldgaber, D., Goldfarb, L.G., Brown, P., *et al.* (1989) Mutations in familial Creutzfeldt–Jakob disease and Gerstmann–Sträussler–Scheinker's syndrome. *Exp. Neurol.* **106** 204–206.

Hsiao, K. and Prusiner, S.B. (1990) Inherited human prion diseases. *Neurology* **40** 1820–1827.

Hsiao, K., Baker, H.F., Crow, T. J., *et al.* (1989a) Linkage of a prion protein missense variant to Gerstmann–Sträussler syndrome. *Nature (London)* **338** 342–345.

Hsiao, K.K., Doh-ura, K., Kitamoto, T., Tateishi, J. and Prusiner, S.B. (1989b) A prion protein amino acid substitution in ataxic Gerstmann–Sträussler syndrome. *Ann. Neurol.* **26** 137 (Abstract).

Hsiao, K., Scott, M., Foster, D., Groth, D.F., DeArmond, S.J. and Prusiner, S.B. (1990) Spontaneous neurologic disease in transgenic mice expressing leucine mutant prion protein of Gerstmann–Sträussler syndrome. *Science* **250** 1587–1590.

Hsiao, K.K., Meiner, Z., Kahana, E., Cass, C., Kahana, I., Abraham, D., Scarlatto, G., Abramsky, O., Prusiner, A. and Gabizon, R. (1991a) Prion protein mutation in Libyan Jews with Creutzfeldt–Jakob disease. *N. Engl. J. Med.* **324** 1091–1097.

Hsiao, K.K., Cass, C., Schellenberg, G., Bird, T., Devine-Gage, E., Wisniewski, H. and Prusiner, S.B. (1991b) A prion protein variant in a family with the telencephalic form of Gerstmann–Sträussler–Scheinker syndrome. *Neurology* **41** 681–684.

Hsiao, K., Dlouhy, S.R., Farlow, M.R., Cass, C., Da Costa, M., Conneally, M.P., Hodes, M.E., Ghetti, B. and Prusiner, S.B. (1992) Mutant prion proteins in Gerstmann–Sträussler–Scheinker disease with neurofibrillary tangles. *Nature Genetics* **1** 68–71.

Jendroska, K., Heinzel, F.P., Torchia, M., Stowring, L., Kretzschmar, H.A., Kon, A., Prusiner, S.B. and DeArmond, S.J. (1991) Proteinase-resistant prion protein accumulation in Syrian hamster brain correlates with regional pathology and scrapie infectivity. *Neurology* **41** 1482–1490.

Kitamoto, T., Tateishi, J., Tashima, I. *et al.* (1986) Amyloid plaques in Creutzfeldt–Jakob disease stain with prion protein antibodies. *Ann. Neurol.* **20** 204–208.

Kretzschmar, H.A., Stowring, L.E., Westaway, D., Stubblebine, W. H., Prusiner, S. B. and DeArmond, S. J. (1986) Molecular cloning of a human prion protein cDNA. *DNA* **5** 315–324.

Kretzschmar, H., Honold, G., Seitelberger, F., Feucht, M., Wessely, P., Mehraein, P. and Budka. (1991) Prion protein mutation in family first reported by Gerstmann, Sträussler, and Scheinker. *Lancet* **337** 1160 (Letter).

Kuzuhara, S., Kanazawa, I., Sasaki, H., Nakanishi, T. and Shimamura, K. (1983) Gerstmann–Sträussler–Scheinker's disease. *Ann. Neurol.* **14** 216–225.

Masters, C.L., Gajdusek, D.C. and Gibbs, C.J. (1981) Creutzfeldt–Jakob disease virus isolation from the Gerstmann–Sträussler syndrome. *Brain* **104** 559–588.

Nochlin, D., Sumi, S.M., Bird, T.D., *et al.* (1989) Familial dementia with PrP-positive amyloid plaques: a variant of Gerstmann–Sträussler syndrome. *Neurology* **39** 910–918.

Owen, F., Poulter, M., Lofthouse, R., *et al.* (1989) Insertion in prion protein gene in familial Creutzfeldt–Jakob disease. *Lancet* **1** 51–52.

Prusiner, S.B. (1982) Novel proteinaceous infectious particles cause scrapie. *Science* **216** 136–144.

Prusiner, S.B. (1987) Prions and neurodegenerative diseases. *N. Engl. J. Med.* **317** 1571–1581.

Prusiner, S.B. (1991) Molecular biology of prion diseases. *Science* **252** 1515–1522.

Puckett, C., Concannon, P., Casey, C. and Hood, L. (1991) Genomic structure of human prion protein gene. *Am. J. Hum. Genet.* **49** 320–329.

Roberts, G.W., Lofthouse, R., Brown, R., Crow, T.J., Barry, P.A. and Prusiner, S.B. (1986) Prion protein immunoreactivity in human transmissible dementias. *N. Engl. J. Med.* **315** 1231–1233.

Seitelberger, F. (1981) Spinocerebelar ataxia with dementia and plaque-like deposits (Sträussler's disease). In: Vinken, P.J. and Bruyn, G.W. (eds) *Handbook of Clinical Neurology.* North-Holland, Amsterdam, pp. 182–183.

Serban, D., Taraboulos, A., DeArmond, S.J. and Prusiner, S.B. (1990) Rapid detection of Creutzfeldt–Jakob disease and scrapie prion proteins. *Neurology* **40** 110–117.

Sparkes, R.S., Simon, M., Cohn, V.H., *et al.* (1986) Assignment of the human and mouse prion protein genes to homologous chromosomes. *Proc. Natl. Acad. Sci. USA* **83** 7358–7362.

Taraboulos, A., Jendroska, K., Serban, D., Yang, S.-L., DeArmond, S.J. and Prusiner, S.B. (1992) Regional mapping of prion proteins in Brains. *Proc. Natl. Acad. Sci. USA*, in press.

Warter, J.M., Steinmetz, G., Heldt, N., *et al.* (1982) Familial presenile dementia: Gerstmann–Sträussler–Scheinker's syndrome. *Rev. Neurol. (Paris)* **138** 107–121.

14

Prion protein gene analysis and transmission studies of Creutzfeldt–Jakob disease

J. Tateishi, K. Doh-ura, T. Kitamoto, C. Tranchant, G. Steinmetz, J.M. Warter and J.W. Boellaard

ABSTRACT

DNA samples from French, German and Japanese patients with Creutzfeldt–Jakob disease (CJD), Gerstmann–Sträussler syndrome (GSS) and their family members were examined together with samples from control subjects of various ethnic backgrounds. A proline-to-leucine change at codon 102 of the prion protein (PrP) gene, an alanine-to-valine change at codon 117, a methionine-to-valine change at codon 129, a glutamate-to-lysine change at codon 200 and an insertion of 168 bp in the N-terminal region were confirmed among the patients and their family members. Val-129 was often carried by control Caucasians but seldom by Asiatics, except for a high incidence in Japanese GSS. French GSS patients carried both Val-117 and Val-129.

The inoculation of brain homogenates of the patients to mice resulted in a highly positive transmission from sporadic CJD patients without PrP gene variants, while it was negative from the few patients with the variants.

INTRODUCTION

Creutzfeldt–Jakob disease (CJD), Gerstmann–Sträusser syndrome (GSS) and kuru are characterized by the accumulation in the brain of an abnormal protease-resistant isoform of a host-encoded prion protein (PrP) (Prusiner, 1982). Human normal PrP is encoded by a single copy gene on chromosome 20 with entire 253-codon open reading frame within a single exon (Liao et al., 1986; Kretzschmar et al., 1986). Gene analysis has revealed that several PrP gene variants occur only in affected or at-risk individuals in families with GSS or CJD (Hsiao *et al.*, 1989; Doh-ura *et al.*, 1989; Goldgaber *et al.*, 1989; Owen *et al.*, 1989, 1990; Collinge *et al.*, 1990; Goldfarb *et al.*, 1990; Tateishi *et al.*, 1990). We screened DNA samples from a large number of patients with GSS, CJD and other dementing illnesses along with their family members, and controls from several ethnic backgrounds. The PrP variants were then compared with the disease types and with the results of transmission experiments from patients to mice. The methods of gene analysis and transmission experiments were described previously (Doh-ura *et al.*, 1989; Tateishi *et al.*, 1979).

RESULTS

As shown in Table 1, several PrP gene variants were carried by the individuals with GSS or CJD. Many of these genotypes correlated with particular clinical (Table 2) or pathological symptoms (Table 3), and some patients showed different results in the experimental transmission of their disease to mice (Table 4). Details are shown according to each genotype.

Table 1.   Genotypes for PrP polymorphism

Subjects EB (number)	Codon 102 P/P	Codon 102 P/L	Codon 117 A/A	Codon 117 A/V	Codon 200 E/E	Codon 200 E/K	Insertion −	Insertion +	Codon 129 M/M	Codon 129 M/V	Codon 129 V/V
GSS											
Leu-102 J (13)		13	13		13		13		12	1	
G (2)		2	2		2		2		2		
Val-117 F (15)	15			15	15		15			13	2
Val-129 J (9)	9		9		9		9			7	2
CJD											
Familial J (1)	1		1			1	1		1		
Sporadic J (17)	17		17		16	1	16	1	17		
Sporadic G (1)	1		1		1		1		1		
Controls											
Asiatic (198)	198		198		198		198		181	17	
Caucasian (157)	157		157		157		157		75	63	19

Eb, ethnic background (J, Japanese; G, German; F, French).

A proline-to-leucine change at codon 102 (Leu-102)
Twelve out of thirteen Japanese GSS patients exclusively showed this heterozygous mutation but only one combined with heterozygous Val-129. Two German GSS patients, belonging to the Tübingen family Sch (Boellaard and Schlote, 1980) carried only this mutation. Except for two patients belonging to the family Fuj (Tateishi *et al.*, 1979, 1988), all other Japanese patients belonged to unrelated families, in which other members died or were suffering from the same disease or had the same mutation.

Table 2.   Clinical data of each genotype

Genotype (number of patients)	Familial onset	Sex M/F	Age (years) at onset, M ± SD	Initial ataxia	Myoclonus or PSD	Duration (months) of illness, M ± SD
Leu-102 (15)	+ (11)	9/6	52 ± 11	15 (100)[a]	9 (60)[a]	66 ± 25
Val-117 (12)	+	7/5	40 ± 10	5 (42)	3 (25)	36 ± 13[b]
Val-129 (7)	+ (2)	2/5	53 ± 13	6 (86)	4 (57)	57 ± 32
Lys-200 (2)	+ (1)	1/1	47,54	0	2 (100)	12, < 3
Insertion						
168 bp (1)	NE	F	29	0	0	84
CJD without mutation (8)	–	3/5	66 ± 6	0	8 (100)	17 ± 8

[a] Percentage of positive patients.
[b] Eight patients have died, four are still living.
NE, not examined.

The mean age of the disease onset was younger than that of sporadic CJD patients without any known mutations (Table 2). The duration of the disease was far longer than that of sporadic CJD. An ataxic gait or uncoordinated movement were the initial symptoms of almost all patients. The symptoms progressed slowly, simulating cerebellar degenerative diseases. Brain stem symptoms, such as bulbar palsy, and cerebral involvement, such as dementia and pseudobulbar symptoms, appeared at different stages. About half the patients lacked either myoclonus or periodic synchronous discharge (PSD) in EEG, both of which were common to CJD patients without mutations. These clinical symptoms corresponded to those of the original patients reported by Gerstmann et al. (1936) and their family members were recently proven to have Leu-102 (Kretzschmar et al., 1991.)

The patients with Leu-102 had a common pathology; that was the presence of Congophilic amyloid masses called kuru plaques, which were stained with anti-PrP. Congophilic amyloid plaques had been found in patients with Leu-102, Val-117 and Val-129, but not in sporadic CJD patients without these mutations (Table 3). Kitamoto et al. (1991) were able to prove a peptide containing Leu-102 in the brain homogenate of a GSS patient by reverse phase HPLC. Therefore, Leu-102 is believed to correlate closely with the formation of kuru plaques in the brain. Other pathological changes, such as a spongiform change, loss of nerve cells and gliosis, were different from case to case, and a few patients showed minimal changes, except for kuru plaques (Tateishi et al., 1988).

Transmission experiments from these patients to mice are shown in Table 4. Five out of nine experiments were successful. Four of the successful experiments were done with a high incidence and within relatively short time periods. Transmission from one patient was difficult to mice, which were not confirmed by their mild sponginess but by PrP depositions after immunostaining. Two patients were negative and the

Table 3. Pathology of each genotype

Genotype	Brain weight (g) M ± SD (number of patients)	Spongiform change	Kuru plaque	Severe loss of nerve cells
Leu-102	1021 ± 158 (12)	9/15	15/15	11/15
Val-117	*ca* 950	0/1	1/1	0/1
Val-129	983 ± 173 (6)	5/7	7/7	4/7
Lys-200	880	1/1	0/1	1/1
Insert 168 bp	900	0/1	Unusual plaque, 1/1	0/1
CJD without mutation	914 ± 195 (6)	6/7	0/7	6/7

*Positive/total examined patients.

other two probably negative; however, with experiments still being continued 2 years later by repeated inoculations to several mouse strains.

An alanine-to-valine change at codon 117 (Val-117)
An A-to-G transition of the third letter of codon 117 abolishes polymorphic Pvu II site (Wu *et al.*, 1987; Hsiao *et al.*, 1989) but still codes for alanine. This silent mutation was found in one out of the 157 Caucasian controls but in none of the 198 Asiatic controls. An additional transition of C to T at the second letter of the codon 117 leads to an alanine-to-valine change. This two-point mutation was exclusively found in an Alsatian GSS family (Doh-ura *et al.*, 1989) that had at least 12 affected individuals spread over 4 generations (Tranchant *et al.*, 1991), and recently was confirmed in an American (Hsiao and Prusiner, 1990). From 47 blood samples and one frozen tissue sample, taken from patients and healthy members of the Alsatian family, 15 showed Val-117 and also Val-129.

Among these 15 members, one was autopsied, 4 were still living with neurologic symptoms and 10 were healthy without symptoms. Clinical and pathological findings of patients in this family were reported by Warter and his colleagues (Warter *et al.*, 1982; Tranchant *et al.*, 1991, 1992). Briefly, three patients in the first and second generations exclusively manifested dementia. The later generations showed a triad of pyramidal, pseudobulbar syndrome and dementia associated with symptoms indicating the spread of lesions to the spinal cord and cerebellum. These features were different from those of ordinary GSS and have been labelled by some authors as telencephalic GSS (Hsiao and Prusiner, 1990). The pathology was not uniform but commonly showed Congophilic amyloid plaques.

Western blot analysis of the protease-resistant PrP fraction from an autopsied brain showed weakly positive bands between 16 and 31 kd. Immunostaining of the brain tissue revealed a moderate number of kuru plaques densely decorated with anti-PrP (Tateishi *et al.*, 1990). Transmission experiments were repeated several times, inoculating homogenates taken from many areas of the autopsied brain to different mouse strains, but have been negative until now (Tateishi *et al.*, 1990).

Table 4. Transmission experiment

Genotype		Transmission		
		Positive	Negative	Negative?[a]
Leu-102	(9)[b]	5	2	2
Val-129	(4)	2		2
Val-117	(1)		1	
Insert 168 bp	(1)		1	
CJD without mutation	(8)	8		

[a]Negative but still continuing experiments.
[b]Number of patients examined for more than 2 years.

A methionine-to-valine change at codon 129 (Val-129)

This change was carried out 82 of 157 control Caucasians and 17 of 198 control Asiatics. The frequency of the PrP allele with this change showed a big difference between 0.04 in Asiatics and 0.32 in Caucasians, being eight times more frequent in the latter. Control studies on general populations with various ethnic backgrounds are thus very important concerning Val-129. Nine Japanese GSS patients carried only this mutation; seven heterozygously and two homozygously. Another Japanese patient had both Leu-102 and Val-129. Two Japanese patients showed familial onset; a brother of one patient and two sisters of another patient had similar clinical and pathological features and all were heterozygous for Val-129. All of the 15 members of the Alsatian family carrying Val-117 were also positive for Val-129, in which 13 were heterozygous and 2 homozygous. Of other 33 members of this family, 14 carried Met/Met, 14 carried Met/Val and 5 Val/Val at codon 129.

It is interesting that two out of three kuru patients had homozygous Val-129 (Goldfarb *et al.*, 1990) and out of nine GSS patients from the Indiana kindred with phenylalanine-to-serine mutation at codon 198, three were codon 129 Val/Val homozygotes and six were Met/Val heterozygotes (Dlouhy *et al.*, 1992). Homozygous Val-129 was reported in two US patients (Goldfarb *et al.*, 1990) and four UK patients who developed CJD after growth hormone treatment (Collinge *et al.*, 1991). The combined appearance of Leu-102 and Val-129 was reported in one UK family CBWU with GSS (Rosenthal *et al.*, 1976). The combination of Val-117 and Val-129 in the Alsatian family was also reported in the US family GCSA (Hsiao and Prusiner, 1990). Recently, Palmer *et al.* (1991) reported that homozygosity, Met/Met or Val/Val, at codon 129 confined a genetic predisposition to sporadic CJD in the United Kingdom. In Japanese patients, however, either homozygous or heterozygous Val-129 related to the phenotype of GSS, and 15 sporadic CJD patients with an acute clinical course and the triad symptoms did not carry this mutation.

The age at onset and duration of the illness in Japanese patients with Val-129 resembled those of GSS patients with Leu-102, but were different from those of sporadic CJD patients. Clinical symptoms were similar to those of patients with Leu-102, beginning with ataxia, slowly progressive dementia, and often lacking myoclonus and PSD. The pathology of this group also resembled that of a group with Leu-102, namely the Congophilic kuru plaques in all seven autopsied patients. Therefore, patients with Val-129 should be categorized as GSS. Transmission experiments from many patients to mice are ongoing but were easily done in two patients, while the transmissions from a Japanese GSS patient with both Leu-102 and Val-129 and from an Alsatian patient with both Val-117 and Val-129 have been negative, thus far.

Glutamate-to-lysine change at codon 200 (Lys-200)

This mutation was found in two Japanese CJD patients, one of which had an uncle and an aunt who died from similar symptoms as well as four other healthy members carrying the same mutation. The clinical and pathological feature of this patient is as follows. A 47-year-old housewife complained of decreasing visual acuity in March 1990, followed by occipitalgia and forgetfulness in June, and difficulties in articulation and walking appeared in early July. She showed somnolence and myoclonus in late July, and akinetic mutism since August. An EEG study revealed PSD and CT showed generalized atrophy of the brain. She died at the end of February 1991, one year after the onset of the illness. At autopsy, the brain weighed 880 g. Severe loss of nerve cells, spongiform change and glial proliferation were seen in the cerebral cortex, basal ganglia, thalamus and cerebellar cortex. Kuru plaques were not found and immunostaining showed fine deposits of PrP diffusely in the grey matter. These clinicopathological findings corresponded well to those seen in sporadic CJD patients. Transmission experiments from this patient to mice are now ongoing.

Another patient with Lys-200 is a 54-year-old male without any known family history. He complained of forgetfulness in the beginning of May 1991 and dementia progressed rapidly, accompanied by PSD. At present, 3 months after onset, he is in a vegetative state. Lys-200 has been reported in CJD patients in Czechoslovakia (Goldfarb *et al.*, 1990), and in Libyan Jews (Hsiao *et al.*, 1991). This mutation may associate with typical CJD symptoms with familial onset, regardless of regional or racial differences. The distribution pattern of PrP in the brain of our patient was identical to those of sporadic CJD and different from kuru plaques in GSS patients with Leu-102, Val-117 and Val-129.

An insert of 168 bp at codon 53

In our efforts to analyse the PrP gene from autopsy cases with progressive dementia, one patient showed a new insertion of 168 bp. A 29-year-old female retired from her office job, because of a decreasing capacity of memory and calculation. As a result of an increasing state of dementia and unsteadiness of gait, she was admitted to a chronic hospital, where she often fell down and eventually died from a sudden subdural haematoma at 36 years old. The laboratory examination did not reveal any particular abnormalities. At autopsy, the brain weighed 900 g, showing diffuse atrophy. The microscopic examination disclosed minimal change, such as moderate gliosis in the cerebral white matter and slight loss of Purkinje cells. A definite diagnosis was not established, but it was included in a category of progressive subcortical gliosis (Neumann). Recently improved methods of immunostaining disclosed a few plaque-type deposits of PrP in the cerebellar cortex, which were not stained with Congo red. Experimental transmission has also been negative after repeated inoculations of the brain homogenate into numerous mouse strains and SD rats.

A PrP gene study of her family members has not been done. The insert codes for 56 amino acids corresponding to seven extra uninterrupted repeats of proline–glycine-rich octapeptide (PHGGGWGQ); this was inserted into the domain of imperfect five proline–glycine-rich repeats in the N-terminal region, beginning at codon 53. Insertion of 96, 144 and 216 bp, or deletion of 24 bp has been reported at codon 53. A 144 bp insert was found in a British family with CJD (Owen et al., 1989), a familial neurodegenerative illness in another British patient (Collinge et al., 1989), a British family with presenile dementia (Collinge et al., 1990), American families with CJD and Slovakian Czecks with CJD (Goldgaber et al., 1989). Inserts of 96 bp and 216 bp were found in patients with CJD (Goldfarb et al. (1989), cited in Prusiner (1991), whereas a deletion of one octapeptide or four additional octarepeats has been identified in individuals without the neurologic disease cited in Prusiner (1991). As Collinge et al. (1990) stated, patients with insertion at codon 53 have difficulties in differential diagnosis as did our patient who had clinicopathological features quite different from those of prion diseases. Western blot analysis of our patient brain showed several larger molecular weight bands ranging up to about 41 kD, in addition to the ordinary PrP bands (unpublished data). This may be due to an insert at codon 53, which is located outside the protease-resistant portion of PrP. The disease of this patient could not be transmitted to mice, but further experiments from other patients to other animal species may be important in determining the pathophysiological meaning of the insertion or deletion at codon 53.

Sporadic CJD patients without known mutations

As a comparison, we analysed the PrP gene of sporadic CJD patients and confirmed an absence of any known mutations in one German and eight Japanese patients. The Japanese patients had rather uniform clinicopathological features, including a later onset of the disease than mutation groups; clinical symptoms such as dementia, myoclonus and PSD in EEG progressed rapidly to a vegetative stage within a shorter period than did mutation groups; spongiform change, severe loss of nerve cells, gliosis and a lack of Congophilic kuru plaques were common pathological traits. These features were also common to many sporadic CJD patients whose PrP gene was not analysed. One German patient with sporadic CJD was a 73-year-old man who died 2 months after the onset of dementia. His brain showed extensive amyloid angiopathy and minimal sponginess, probably because of the short duration of the illness. Western blotting and immunohistochemistry revealed an abnormal isoform of PrP in the brain.

Experimental transmission from these sporadic CJD patients to mice were highly positive. From four Japanese patients, transmission was done between 200 and 300 days, while from the other four including the German patient, longer periods of over

400 days were required. Albeit there were some differences, transmission from other sporadic CJD patients whose PrP gene was not studied was almost 100% positive (Kitamoto *et al.*, 1989). Though many experiments are still ongoing, different results have been obtained in the Leu-102 and Val-129 groups. Transmission from a French GSS patient with both Val-117 and Val-129, and a Japanese GSS patient with both Leu-102 and Val-129, was not successful. A negative transmission from a patient with 168 bp insertion may indicate the non-infectious nature of this patient.

Our results are summarized in Table 5, and indicate that PrP genotypes correlate closely with phenotypes of prion diseases, especially with clinical symptoms, the rapidity of worsening and the distribution pattern of PrP in the brain, as well as with the transmissibility to laboratory animals. Transmission studies suggest that more infectious factors cause the disease in sporadic CJD patients without PrP variations, while less infectious and more genetic factors tend to cause the disease in subjects with variant PrP genes.

Table 5.   Genotypes, phenotypes and transmission

Genotype	Phenotype	PrP deposits	Transmission Positive/total
Leu-102	Ataxic GSS	Plaque type	5/9
Val-117	Telencephalic GSS	Plaque type	0/1
Val-129	Ataxic GSS	Plaque type	2/4
Lys-200	Familial CJD	Synapse type	Ongoing
Insertion 168 bp	Undefined dementia	Unusual plaque type	0/1
Without mutation	Sporadic CJD	Synapse type	8/8

ACKNOWLEDGEMENTS

We thank the members of other affiliated laboratories for their materials and advice; K. Beppu for preparing the manuscript; M. Yoneda and K. Hatanaka for photographic and laboratory assistance. This work was supported in part by a Grant-in-Aid from the Ministry of Education, Science and Culture, from the Ministry of Health and Welfare, and from the Science and Technology Agency, Japan.

REFERENCES

Boellaard, J.W. and Schlote, W. (1980) Subakute spongiforme Encephalopathie mit multiformer Plaquebildung. Eigenartige familiär-hereditäre Krankheit des Zentralnervensystems [Spinocerebellare Atrophie mit Demenz, Plaques und plaqueähnlichen Ablagerungen im Klein- und Großhirn (Gerstmann, Sträussler, Scheinker)]. *Acta Neuropathol. (Berlin)* **49** 205–212.

Collinge, J., Harding, A.E., Owen, F., Poulter, M., Lofthouse, R., Boughey, A.H., Shah, T. and Crow, T.J. (1989) Diagnosis of Gerstmann–Sträussler syndrome in familial dementia with prion protein gene analysis. *Lancet* **2** 15–17.

Collinge, J., Owen, F., Poulter, M., Leach, M., Crow, T.J., Rossor, M.N., Hardy, J., Mullan, M.J., Janota, I. and Lantos, P.L. (1990) Prion dementia without characteristic pathology. *Lancet* **336** 7–9.

Collinge, J., Palmer, M.S. and Dryden, A.J. (1991) Genetic predisposition to iatrogenic Creutzfeldt–Jakob disease. *Lancet* **337** 1441–1442.

Dlouhy, S.R., Hsiao, K., Farlow, M.R., Foroud, T., Conneally, P.M., Johnson, P., Prusiner, S.B., Hodes, M.E. and Ghetti, B. (1992) Linkage of the Indiana kindred of Gerstmann–Sträussler–Scheinker disease to the prion protein gene. *Nature Genetics* **1** 64–67.

Doh-ura, K., Tateishi, J., Sasaki, H., Kitamoto, T. and Sakaki, Y., (1989) Pro→Leu change at position 102 of prion protein is the most common but not the sole mutation related to Gertsmann–Sträussler syndrome. *Biochem. Biophys. Res. Commun.* **163** 974–979.

Gerstmann, J., Sträussler, E. and Scheinker, I. (1936) Über eigenartige hereditär-familiäre Erkrankung des Zentralnervensystems. *Z. Neurol.* **154** 736–762.

Goldfarb, L., Korczyn, A., Brown, P., Chapman, J. and Gajdusek, D.C. (1990) Mutation in codon 200 of scrapie amyloid precursor gene linked to Creutzfeldt–Jakob disease in Sephardic Jews of Libyan and non-libyan origin. *Lancet* **336** 637–638.

Goldgaber, D., Goldfarb, L.G., Brown, P., Asher, D.M., Brown, W.T., Lin, S., Teener, J.W., Feinstone, S.M., Rubenstein, R., Kascsak, R.J., Boellaard, J.W. and Gajdusek, D.C. (1989) Mutations in familial Creutzfeldt–Jakob disease and Gerstmann–Sträussler–Scheinker's syndrome. *Exp. Neurol.* **106** 204–206.

Hsiao, K. and Prusiner, S.B. (1990) Inherited human prion diseases. *Neurology* **40** 1820–1827.

Hsiao, K., Baker, H.F., Crow, T.J., Poulter, M., Owen, F., Terwilliger, J.D., Westaway, D., Ott, J. and Prusiner, S.B. (1989) Linkage of a prion protein missense variant to Gerstmann–Sträussler syndrome. *Nature (London)* **338** 342–345.

Hsiao, K.K., Cass, C., Schellenberg, G.D., Bird, T., Devine-Gage, E., Wisniewski, H. and Prusiner, S.B. (1991) A prion protein variant in a family with the telencephalic form of Gerstmann–Sträussler–Scheinker syndrome. *Neurology* **41** 681–684.

Hsiao, K., Meiner, Z., Kahana, E., Cass, C., Kahana, I., Avrahami, D., Scarlato, G., Abramsky, O., Prusiner, S.B. and Ruth, G. (1991) Mutation of the Prion protein in Libyan Jews with Creutzfeldt–Jakob disease. *New Engl. J. Med.* **324** 1091–1097.

Kitamoto, T., Tateishi, J., Sawa, H. and Doh-ura, K. (1989) Positive transmission of Creutzfeldt–Jakob disease verified by murine kuru plaques. *Lab. Invest.* **60** 507–512.

Kitamoto, T., Yamaguchi, K., Doh-ura, K. and Tateishi, J. (1991) A missense variant of prion protein is the major component of kuru plaque in patients with Gerstman–Sträussler syndrome. *Neurology* **41** 306–310.

Kretzschmar, H.E., Stowring, L.E., Westaway, D., Stubblebine, W.H., Prusiner, S.B. and DeArmond, S.J. (1986) Molecular cloning of a human prion protein cDNA. *DNA* **5** 315–324.

Kretzschmar, H.A., Honold, G., Seitelberger, F., Feucht, M., Wessely, P., Mehraein, P. and Budka, H. (1991) Prion protein mutation in family first reported by Gerstmann, Sträussler, and Scheinker. *Lancet* **337** 1160.

Liao, J.Y., Lebo, R.V., Clawson, G.A. and Smuckler, E.A. (1986) Human prion protein cDNA: molecular cloning, chromosomal mapping, and biological implications. *Science* **233** 364–367.

Owen, F., Poulter, M., Lofthouse, R., Collinge, J., Crow, T.J., Risby, D., Baker, H.F. and Ridley, R.M. (1989) Insertion in prion protein gene in familial Creutzfeldt–Jakob disease. *Lancet* **1** 51–52.

Owen, F., Poulter, M., Shah, T., Collinge, J., Lofthouse, R., Baker, H., Ridley, R., McVey, J. and Crow, T.J. (1990) An in-frame insertion in the prion protein gene in familial Creutzfeldt–Jakob disease. *Mol. Brain Res.* **7** 273–276.

Palmer, M.S., Dryden, A.J., Hughes, J.T. and Collinge, J. (1991) Homozygous prion protein genotype predisposes to sporadic Creutzfeldt–Jakob disease. *Nature (London)* **352** 340–342.

Prusiner, S.B. (1982) Novel proteinaceous infectious particles cause scrapie. *Science* **216** 136–144.

Prusiner, S.B. (1991) Molecular biology of prion diseases. *Science* **252** 1515–1522.

Rosenthal, N.P., Keesey, J., Crandall, B. and Brown, W.J. (1976) Familial neurological diseases associated with spongiform encephalopathy. *Arch. Neurol.* **33** 252–259.

Tateishi, J., Ohta, M., Koga, M. and Sato, Y. (1979) Transmission of chronic spongiform encephalopathy with kuru plaques from humans to small rodents. *Ann. Neurol.* **5** 581–584.

Tateishi, J., Kitamoto, T., Hashiguchi, H. and Shii, H. (1988) Gerstman–Sträussler–Scheinker disease: immunohistological and experimental studies. *Ann. Neurol.* **24** 35–40.

Tateishi, J., Kitamoto, T., Doh-ura, K., Sakaki, Y., Steinmetz, G., Tranchant, C., Warter, J.M. and Heldt, N. (1990) Immunochemical, molecular genetic, and transmission studies on a case of Gerstmann–Sträussler–Scheinker syndrome. *Neurology* **40** 1578–1581.

Tranchant, C., Doh-ura, K., Steinmetz, G., Chevalier, Y., Kitamoto, T., Tateishi, J. and Warter, J.M. (1991) Mutation du codon 117 du gène du prion dans une maladie de Gerstmann–Sträussler–Scheinker. *Rev. Neurol. (Paris)* **147** 274–278.

Tranchant, C., Doh-ura, K., Warter, J.M., Steinmetz, G., Chevalier, Y., Hanauer, A., Kitamoto, T. and Tateishi, J. (1992) Gerstman–Sträussler–Scheinker disease in an Alsatian family: clinical and genetic studies. *J. Neurol., Neurosurg. Psychiatry* **55** 185–187.

Warter, J.M., Steinmetz, G., Heldt, N., Rumbach, L., Marescaux, C., Eber, A.M., Collard, M., Rohmer, F., Floquet, J., Guedenet, J. C., Gehin, P. and Weber, M. (1982) Familial presenile dementia: Gerstman–Sträussler–Scheinker's syndrome. *Rev. Neurol. (Paris)* **138** 107–121.

Wu, Y., Brown, W.T., Robakis, N.K., Dobkin, C., Devine-Gage, E., Merz, P. and Wisniewski, H.M. (1987) A PvuII RFLP detected in the human prion protein (PrP) gene. *Nucleic Acids Res.* **15** 3191.

15

The molecular genetics of human transmissible spongiform encephalopathy

Lev G. Goldfarb, Paul Brown and D. Carleton Gajdusek

ABSTRACT

We have studied 120 patients with human transmissible spongiform encephalopathies to determine structural changes in the coding region of the PRNP gene. Each of the 63 tested familial CJD patients from 56 affected families had one of four mutations: 178^{Asn} mutation was identified in 7 families, 200^{Lys} mutation in 45 families, extra-repeat insertions in 3 families, and a single-repeat deletion in one family. Genetic linkage to CJD was shown for the 178^{Asn} mutation, and a strong association demonstrated for the 200^{Lys} and insert mutations. We have not so far found mutations in patients with sporadic CJD, of whom three were fully sequenced. Of five patients with iatrogenic CJD following native human growth hormone therapy (all five fully sequenced) two were homozygous for 129^{Val} and one had a single-repeat deletion. Twelve patients from six families with GSS all had the 102^{Leu} mutation. Our data suggest that familial forms of spongiform encephalopathy are primarily genetic in nature, in which the point mutations, extra-repeat insertions, and the deletion are causative or predisposing factors.

INTRODUCTION

Soon after it was shown that mutations in the PrP gene control scrapie incubation time in mice (Westaway et al., 1987), mutations were also found in the analogous human PRNP gene in patients with familial transmissible spongiform encephalopathies (Owen et al., 1989; Hsiao et al., 1989; Doh-ura et al., 1989; Goldgaber et

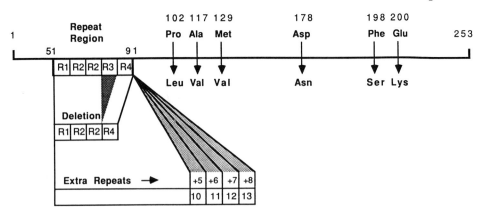

Fig. 1.   Schematic diagram of the PRNP coding region. Codon numbers are shown above line and octapeptide coding repeats below line. Point mutations at codons 178 and 200, extra-repeat insertions, and the deletion are associated with familial CJD. Point mutations at codons 102, 117, and 198 are associated with GSS.

al., 1989; Goldfarb *et al.*, 1991). Fig. 1 summarizes updated information on types of mutations that are associated with familial CJD (point mutations at codons 178 and 200, and extra-repeat insertions or a single-repeat deletion between codons 51 and 91) and GSS (point mutations at codons 102, 117, and 198). The fact that different sets of mutations are associated with these syndromes suggests that each mutation has its own biological significance.

Our Laboratory of Central Nervous System Studies at the NIH has been a referral centre for spongiform encephalopathies for the last 30 years. Specimens for more than 500 patients with confirmed, probable, or possible diagnoses of CJD or GSS have been forwarded to this laboratory for experimental transmission or molecular genetic studies, and our systematic storage of frozen tissues and clinical files provide a unique opportunity to determine the prevalence of mutations in different forms of human transmissible spongiform encephalopathies.

RESULTS AND DISCUSSION

Familial CJD

178^{Asn} mutation
This mutation was first identified by direct sequencing of the PRNP coding region in two members of a Finnish CJD kindred (Goldfarb *et al.*, 1991) and a member of the US–Dutch family (Nieto *et al.*, 1991). Fig. 2 shows the GAC-to-AAC substitution on one allele resulting in an amino acid change of aspartic acid to asparagine. Using the *Tth III* 1 restriction analysis, we have now identified a total 7 CJD families with 64 patients from Finland, Hungary, Holland, Canada, England, and France as carrying

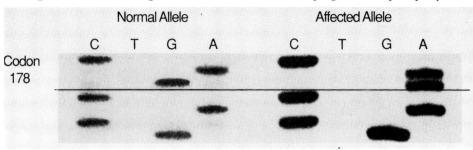

Fig. 2.   A nucleotide sequencing autoradiogram from the area of codon 178 in the PRNP coding region of a CJD patient from the Finnish family. The normal sequence GAC is changed to AAC in the affected allele, which results in a substitution of asparagine for aspartic acid.

the same 178^{Asn} mutation (Table 1). Twenty-eight patients were neuropathologically verified and six experimentally transmitted. All 17 patients available for genetic screening showed the mutation, as well as 10 of 26 first degree relatives, but none of 83 unrelated controls. Statistical analysis confirmed a strong linkage (lod score of 5.3) between the mutation and CJD in these families.

Table 1.   Genetic screening of families with the 178^{Asn} mutation

				Number positive/number tested			
Country of residence	Country of origin	Number of affected members	Confirmed/ transmitted cases[a]	CJD patients	First-degree relatives	Unrelated controls	P[b]
Finland	Finland	14	4/1	8/8	3/12	0/12	<0.001
US	Hungary	9	5/2	3/3	2/4		
US	Holland	10	4/2	1/1	3/4		
US	Canada	6	4/0	1/1	nt[c]	0.69	<0.001
US	England	8	4/0	2/2	nt		
France	France	14	6/1	1/1	2/4	0/2	
France	France	3	1/0	1/1	0/2		
Total		64	28/6	17/17	10/26	0.83	

[a] Neuropathologically confirmed/experimentally transmitted.
[b] Comparing the frequency of the 178^{Asn} mutation in CJD patients and unrelated controls: $\chi^2 = 20$ for the Finnish family, and $\chi^2 = 74$ for the US families.
[c] nt, none tested.

200^{Lys} *mutation*

This mutation was first found in a family from Poland (Goldgaber *et al.*, 1989), seemingly sporadic patients from Slovakia and Chile, and in a Sephardic Jewish family from Greece (Goldfarb *et al.*, 1990a). It was later established that all patients from a CJD cluster in Slovakia (Goldfarb *et al.*, 1990c), all tested patients from a similar cluster in Libyan Jews living in Israel (Goldfarb *et al.*, 1990b; Hsiao *et al.*, 1991), patients in all other Sephardic families originating in Greece and Tunisia

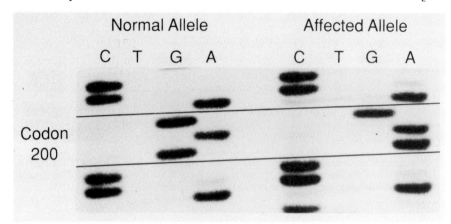

Fig. 3. Sequencing autoradiogram from the area of codon 200 in the PRNP coding region of a CJD patient from the Slovakian cluster. The normal codon sequence GAG is changed to AAG in the affected allele that results in amino acid substitution of lysine for glutamic acid.

(Brown et al., 1991b), and all tested familial CJD patients in Chile (Brown et al., in preparation) carried the same 200^{Lys} mutation.

An autoradiogram in Fig. 3 illustrates the GAG-to-AAG change in the affected allele that was found by sequencing in both Polish siblings and two Slovakian CJD patients. This change resulted in a substitution of glutamic acid to lysine in the encoded protein. Restriction endonuclease analysis with Bsm A1 and single nucleotide extension reaction (Kuppuswamy et al., 1991) were used for screening for this mutation.

The current annual mortality rate of CJD in the Slovakian clusters is approximately 200 per million population (in some Northern villages it approaches 2000 per million); the Libyan Jewish population in Israel is characterized by rates close to 100 per million; and the frequency of CJD in some populations in Chile is 18 per million. The codon 200^{Lys} mutation was detected in 34 families with 54 known CJD cases from the cluster areas (Table 2) by testing CJD patients and/or their first-degree

Table 2. Genetic screening of families with the 200^{Lys} mutation in geographic CJD clusters

Country	Number of families	Number of affected members	Confirmed/ transmitted cases[a]	Number positive/number tested			P[b]
				CJD patients	First-degree relatives	Unrelated controls	
Slovakia	20	32	26/1	17/17	23/68	1/80	<0.001
Libya	8	11	6/1	7/7	1/3	0/23	<0.001
Chile	6	11	6/5	6/6	nt[c]	nt	
Total	34	54	38/7	30/30	24/71	1/103	

[a] Neuropathologically confirmed/experimentally transmitted.
[b] Comparing the frequency of the 200^{Lys} mutation in CJD patients and unrelated controls: $\chi^2 = 90$ for Slovakian families, and $\chi^2 = 34$ for Libyan families.
[c] nt, none tested.

relatives. Thirty-eight patients were neuropathologically verified and seven experimentally transmitted. All 30 tested CJD patients and 24 of 71 first-degree relatives, but only 1 of 103 unrelated healthy control individuals from the same populations, had the mutation. A very strong association was demonstrated between the mutation and disease by comparing frequencies of the mutation in patients and control individuals ($\chi^2 = 90$ for Slovakia and $\chi^2 = 34$ for Libya). The positive control individual comes from a severely affected North Slovakian village, and may eventually be found to be related to a local CJD family.

Eleven families emigrated at different times from the cluster areas to Western countries (Table 3). Of 33 patients in these families, 3 were themselves emigrants. Thirty patients belong to the second, third, or the fourth generations of these families

Table 3.   Genetic screening of families with 200^{Lys} mutation emigrating from cluster areas to other countries

| Country of residence (origin) | Number of families | Affected members born in the country of | | C/T^a | Number positive/number tested | | |
		origin	residence		CJD patients	First-degree relatives	Unrelated controls
US, Canada (Slovakia, Poland Germany, Greece)	8	3	16	8/5	7/7	8/19	0/102
France (Tunisia, Greece)	3	0	14	5/4	3/3	13/19	0/2
Total	11	3	30	13/9	10/10	21/38	0/104

^{a} Neuropathologically confirmed/experimentally transmitted.

after emigration, and they have never revisited the country of origin. Thirteen patients, at least one per family, were neuropathologically verified and nine experimentally transmitted. Ten CJD patients and 21 of 38 first degree relatives, but none of 104 control individuals in the country of residence, had the 200^{Lys} mutation. The fact that branches of some families migrating from cluster areas to other countries continue to have CJD over several generations argues against a role of local environmental factors and supports the view that familial CJD is a primarily genetic disorder, in which the 200^{Lys} mutation is responsible for disease.

The codon 200^{Lys} mutation probably originated in Spain and was dispersed in the middle ages by mass migration of expelled Sephardic Jews to North African and European countries, and migration of Spanish people to Latin American countries, including Chile. The map in Fig. 4 shows routes of migration and places of settlement of the Sephardim in the 16th and 17th centuries. Historical data suggest that Sephardic Jewish communities were established in the areas of Cracow, Vienna, and Prague,

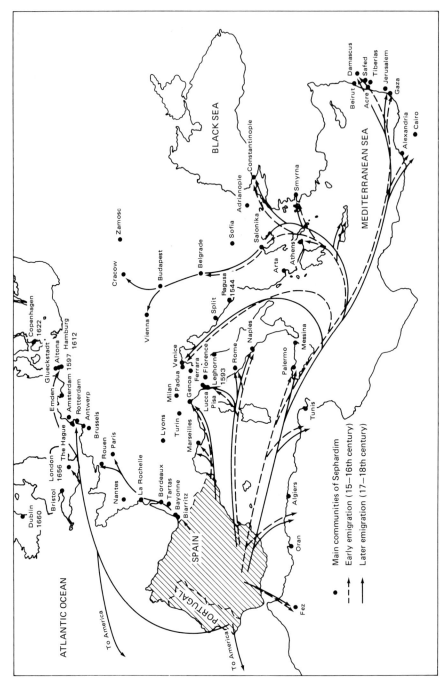

Fig. 4. Dispersion of Jews originating from Libya in the 20th century. The Tripoli Jewish community emigrated about 1950. Most of its members emigrated to Israel while others left Libya for Greece or Italy. The Djerban community (large bold arrow) emigrated mostly to the city of Tunis and from there to Israel or France.

Map labels:

ATLANTIC OCEAN

BLACK SEA

MEDITERRANEAN SEA

SPAIN

PORTUGAL

To America

Dublin 1660
Bristol
London 1656
The Hague
Emden
Glueckstadt 1597
Copenhagen 1622
Altona
Amsterdam
Hamburg 1612
Rotterdam
Antwerp
Brussels
Rouen
Paris
Nantes
La Rochelle
Bordeaux
Tartas
Bayonne
Biarritz

Fez
Oran
Algiers
Tunis

Lyons
Turin
Milan
Padua
Venice
Genoa
Ferrara
Marseilles
Lucca
Florence
Pisa
Leghorn 1593
Rome
Naples
Palermo
Messina

Cracow
Zamosc
Vienna
Budapest
Belgrade
Split
Ragusa 1544
Arta
Athens
Sofia
Adrianople
Constantinople
Salonika
Smyrna

Beirut
Damascus
Acre
Safed
Tiberias
Jerusalem
Gaza
Alexandria
Cairo

Legend:
● Main communities of Sephardim
- - - → Early emigration (15—16th century)
——→ Later emigration (17—18th century)

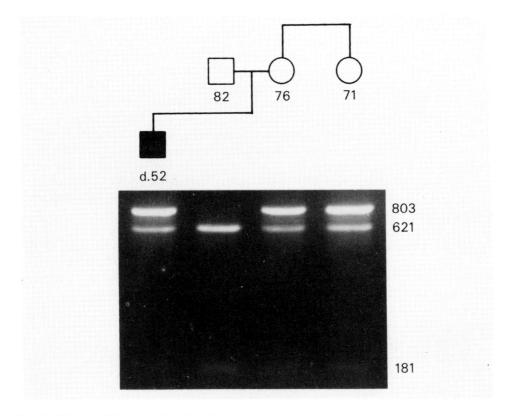

Fig. 5.   Pedigree of Slovak family 'Bal' and electrophoresis of PCR-amplified and endonuclease-digested DNA fragments containing the PRNP coding region. Lane 1, CJD proband; lane 2, his unaffected father; lane 3, his unaffected mother; lane 4, his mother's unaffected sister. *Bsm* A1 endonuclease normally cleaves the 803 bp fragment into two smaller fragments of 621 and 182 bp. The G-to-A substitution in codon 200 abolishes this cleavage site. Thus the DNA fragment from the normal allele breaks into 621 and 182 bp fragments, whereas the mutated allele is unchanged. As a result, heterozygous individuals carrying the 200^{Lys} mutation show an additional band (lanes 1, 3 and 4).

surrounding Slovakia, and it is also known that Sephardim settled in Tunisia, Libya, and Greece.

Unlike 178^{Asn}, the codon 200^{Lys} mutation often occurs in families with 'skipped' generations, with cases limited to a single generation, or even in apparently sporadic cases. Fig. 5 presents the pedigree of a seemingly sporadic case, in which the proband, a 200^{Lys} mutation carrier, died of neuropathologically confirmed CJD at the age of 52. Both his parents are still alive. The patient's 76-year old mother and her 71-year old sister also show the 200^{Lys} mutation, but no signs of CJD. We found two more mutation-positive individuals over the age 70, and five in their 60s (the average age of CJD onset in the 200^{Lys} families is 55). From calculations based on analysis of 65 mutation carriers the disease penetrance was found to be 0.56, i.e. the presence of the 200^{Lys} mutation results in disease in approximately half of the carriers.

Extra-repeat insertions

Owen *et al.* (1989, 1990) first reported and described an insert in the area of octapeptide coding repeats in the PRNP gene associated with CJD and atypical dementias. We confirmed in our work that the area of tandem 24 bp sequences between codons 51 and 91 is unstable. Normally, it consists of five variant repeats. Of a total of 532 individuals screened for the number of repeats, six members of three CJD families and one non-neurological control patient were identified as having extra-repeats in this region (Table 4). The control patient, who died at age 63 of advanced micronodular cirrhosis, had no family history of neurological disease, no clinical or pathological signs of spongiform encephalopathy, and brain tissue did not transmit disease to experimental primates. The PRNP coding region was completely sequenced, and nine repeats were found in this individual instead of the normal five. No irregular nucleotide substitutions were seen.

Table 4.   Octapeptide coding repeat insertions in the PRNP gene found in CJD and non-neurological control patients

Number of repeats	Medical condition	Order of repeats
5	Normal	R1,R2,R2,R3,R4
9	Cirrhosis	R1,R2,R2,R3,R2,R3,R2,R3,R4
10	CJD	R1,R2,R2,R3,R2,R3g,R2,R2,R3,R4
11	CJD	R1,R2,R2,R2,R3,R2,R3g,R2,R2,R3,R4
12	CJD	R1,R2,R2c,R3,R2,R3,R2,R3,R2,R3g,R3,R4
13	GSS–GSS	R1,R2,R2,R3,R2,R2,R2,R2,R2,R2,R2,R2a,R4

Irregular repeats

CCC	CAT		GGT	GGT	GGC	TGG	GGG	CAG		R3g
pro	his		gly	gly	gly	trp	gly	gln		
CCT	CAT		GGC	GGT	GGC	TGG	GGG	CAG		R2c
pro	his		gly	gly	gly	trp	gly	gln		
CGT	CAT		GGT	GGT	GGC	TGG	GGA	CAG		R2a
pro	his		gly	gly	gly	trp	gly	gln		

The patient from American family 'Kel' with 10 repeats and an irregular nucleotide substitution at the 6th repeat (Tables 4 and 5) had a 15-year illness with progressive dementia, abnormal behaviour, cerebral signs, tremor, rigidity, and myoclonus. Neuropathological examination showed spongiform change, and inoculated brain suspension transmitted spongiform encephalopathy to a squirrel monkey. The patient's father had a similar disease that was neuropathologically verified as spongiform encephalopathy.

The mother and two daughters in family 'Ald' showing 12 repeats with irregular nucleotide substitutions in the 3rd and 10th repeats (Tables 4 and 5) had a prolonged disease with abnormal behaviour, clumsiness, cerebellar signs, overt dementia, and myoclonus. Severe spongiform change and gliosis were found in the mother, but no spongiosis was seen in the brain of the deceased daughter. Inoculated brain suspension from the mother transmitted disease to a chimpanzee, squirrel and capuchin monkeys.

The patient in family 'Che', with 13 repeats and an irregular substitution in the 12th repeat, had a 3-month illness with abnormal behaviour, mutism, cerebellar signs,

Table 5. Characteristics of familial CJD and GSS patients with extra-repeat insertions

Family	Sex, age at onset	Duration (years)	Number of repeats	Confirmed/ transmitted[a]	Atypical neuro- pathology
'Kel'	M31	15	10	+/+	
	M45	5		+/na[b]	
'Ald'	F31	11	12	+/+	
	F23	10	12	+/na	No spongiosis
	F35	>13 (alive)	12		
'Che'	M35	1/4	13	+/+	

[a] Neuropathologically confirmed/experimentally transmitted.
[b] na, not attempted.

and myoclonus (Tables 4 and 5). His neuropathology was typical of GSS, and inoculated brain suspension transmitted disease to a chimpanzee. The patient's mother and seven other members of this family had similar symptoms but over a much more protracted course (2–5 years). Six unaffected individuals, all children of affected members, had an identical insertion.

The occurrence of 10 or more octapeptide coding repeats (5 or more inserted repeats) in the PRNP gene is thus associated with neuropathologically verified and experimentally transmitted familial CJD and GSS with an unusually early age of onset and (in most of the cases) prolonged disease.

One-repeat deletions

Deletion of a 24-nucleotide sequence was first incidentally found in a human brain cDNA library and HeLa cell culture (Puckett *et al.*, 1991). We independently detected this same deletion in a 63-year old patient having a 4-year illness with dementia, pyramidal signs, rigidity, and myoclonus. Her mother died of a similar illness. The deletion occurs in R3 and R4 repeats resulting in an irregular R4c repeat (Table 6). An identical change was found in a 26-year old patient with no known family history of neurological disease who died of a neuropathologically confirmed iatrogenic CJD following a course of pituitary derived growth hormone inoculations. A similar (but not identical) one-repeat deletion was detected in patients from two unrelated families having prolonged illnesses with cerebellar signs and no dementia. One patient's unaffected brother and two of four children also show deletions.

Phenotypic expression of different mutations in familial CJD

The age of onset in familial CJD varied depending on the type of mutation (Table 7). The patients were younger in extra-repeat and 178^{Asn} families than in 200^{Lys} and sporadic cases. The disease duration has nearly the opposite distribution: it was much longer in repeat insertion and 178^{Asn} patients than in 200^{Lys} or sporadic cases.

Table 6. Octapeptide coding repeat deletion in the PRNP gene identified in familial and iatrogenic CJD patients

Number of repeats	Medical condition	Order of repeats
4	Familial CJD	R1, R2, R2, R3, R4c
4	Iatrogenic CJD	R1, R2, R2, R3, R4c

Irregular repeat

CCC	CAT	GGT	GGT	GGC	TGG	GGT	CAA	R4c
pro	his	gly	gly	gly	trp	gly	gln	

The 178^{Asn} mutation patients were also distinctive in that the initial sign of disease was invariably an insidious memory loss, and during the course of illness the triphasic periodic slow waves characteristic for all other familial and sporadic CJD patients were never seen.

Data on experimental transmission of the disease to squirrel monkeys (Table 8) show that the rate of transmission of familial CJD with extra-repeat insertions, 178^{Asn} and 200^{Lys} mutations (24 of 28 attempted) is comparable with that of sporadic CJD cases. Incubation time of experimental disease was significantly shorter in transmissions from familial CJD patients carrying the 178^{Asn} mutation than in sporadic cases. This finding suggests that inoculated PrP molecules with the 178^{Asn} change are most efficient in transforming the host PrP precursor molecules.

Sporadic CJD

We have been so far unable to find any mutations in sporadic CJD patients, of whom 35 were screened for known mutations, and 3 were fully sequenced. In view of the recent suggestion that codon 129 homozygosity may be disproportionately common in sporadic CJD (Palmer et al., 1991), we are currently re-examining the frequencies of alleles in these patients.

Iatrogenic CJD

There are 35 reported cases of iatrogenic CJD with a known source of infection: inoculation of contaminated pituitary hormone in 22 patients, dura mater implants in 6 patients, other types of neurosurgery in 4 patients, corneal transplant in 1 patient, and stereotactic EEG in 2 patients (P. Brown, unpublished data). We recently reported that some patients with iatrogenic CJD were homozygous for the 129^{Val} mutation (Goldfarb et al., 1989). This result was later confirmed and extended by the study of seven more iatrogenic CJD patients, of whom four had the same change (Collinge et al., 1991). We have now completely sequenced the PRNP coding region in five patients with iatrogenic CJD following inoculation of growth hormone. Two patients were

Table 7.   Phenotypic expression of mutations associated with familial CJD

	Familial CJD			
	Extra-repeats	178^{Asn}	200^{Lys}	Sporadic CJD
Number of cases studied	9	33	42	211
Age at onset (years; mean ± SD)	38 ± 10	46 ± 7	55 ± 8	62 ± 9
Duration of illness (months; mean ± SD)	86 ± 5	22 ± 13	8 ± 18	6 ± 8
Periodic EEG activity (%)	14	0	74	56

homozygous for 129^{Val}, one was heterozygous, and two patients were homozygous for 129^{Met}. One of the last two patients had a 24-nucleotide deletion identical to that in a familial CJD patient (Table 6). No other mutations were found in any of these patients.

Considering that only 10% of the US–Canadian population is normally homozygous for the 129^{Val} allele, homozygosity for 129^{Val} in two patients and the presence of a comparatively rare deletion in a third patient with iatrogenic CJD (of five tested) suggests a predisposing role of mutations or polymorphisms in this infectious form of spongiform encephalopathy.

Table 8.   Incubation time of experimental encephalopathy in squirrel monkeys following intracerebral inoculation of 10% homogenates of frozen human brain tissue

	Familial CJD			
	Extra-repeats	178^{Asn}	200^{Lys}	Sporadic CJD
Number of cases inoculated	3	10	15	211
Number of cases transmitted	3	6	15	188
Incubation time (months; mean $\pm$ SD)	33	18.6 ± 2.9	21.3 ± 3.5	25.4 ± 4.4

GSS

Linkage between codon 102^{Leu} mutation and GSS was first established in US and English families (Hsiao *et al.*, 1989), and confirmed in Japanese and German families (Doh-ura *et al.*, 1989; Goldgaber *et al.*, 1989; Brown *et al.*, 1991a). Subsequent studies

Table 9.   Genetic screening of GSS families with 102^{Leu} mutation

Family	Number of affected members	Confirmed/ transmitted cases[a]	Number positive/number tested		
			GSS patients	First-degree relatives	Unrelated controls
German	13	4/1	3/3	3/12	
Hungarian	3	3/0	3/3	nt[b]	
Italian	1	(alive)	1/1	nt	
French	3	1/0	1/1	nt	
American	2	(alive)	1/1	nt	
US–Dutch	3	1/0	3/3	5/11	
Total	25	9/1	12/12	8/23	0/23

[a] Neuropathologically confirmed/experimentally transmitted.
[b] nt, none tested.

resulted in discoveries of two new mutations associated with GSS: 117^{Val} (Doh-ura *et al.* 1989; Tranchant *et al.*, 1991) and 198^{Ser} (Prusiner, 1991). The human 102^{Leu} mutation was shown to produce encephalopathy in transgenic mice (Hsiao *et al.*, 1990).

  We studied 6 families with 25 members affected with GSS (Table 9). Nine patients were neuropathologically verified and one experimentally transmitted. In all 12 patients from these families, including the pathologically confirmed and transmitted case, the 102^{Leu} mutation was identified. Eight of 23 tested first-degree relatives had the same substitution. Linkage of this mutation to GSS (lod score 4.52) was confirmed by using pooled data from the German, American, and British families (Speer *et al.*, 1991).

CONCLUSION

Mutations in the PRNP gene—point mutations in codons 178 and 200, extra-repeat insertions and, perhaps, deletions—are all associated with familial CJD. The presence of a single heterozygous mutation in each studied case of familial CJD and in approximately half of the first-degree relatives, the absence of these mutations in unrelated control individuals, and the evidence for linkage between the individual mutations and the disease, indicate that familial CJD is a genetic disorder with an autosomal-dominant pattern of inheritance. Occurrence of the disease in several generations in branches of families emigrating from cluster areas to different countries and continents argues against a role for local environmental factors and strongly favours the contention that mutations are the cause of familial CJD. There is an emerging overall correlation between individual mutations and clinical patterns in each of the genetically different subsets of spongiform encephalopathy. A predisposing role of mutations or polymorphisms may also be important in iatrogenic forms of CJD.

ADDENDUM

After this paper was sent for publication it became known that the 200^{Lys} mutation-positive individual who was considered control from Slovakia is in fact a first-degree relative of a newly discovered CJD patient.

REFERENCES

Brown, P., Goldfarb, L.G., Brown, W.T., Goldgaber, D., Rubenstein, R., Kascsak, R.J., Guiroy, D.C., Piccardo, P., Boellaard, J.W. and Gajdusek, D.C. (1991a) Clinical and molecular genetic study of a large German kindred with Gerstmann–Sträussler–Scheinker syndrome. *Neurology* **41** 375–379.

Brown, P., Goldfarb, L.G., Cathala, F., Vrbovská, A., Sulima, M., Nieto, A., Gibbs, C.J., Jr, and Gajdusek, D.C. (1991b) The molecular genetics of familial Creutzfeldt–Jakob disease in France. *J. Neurol. Sci.* **105** 240–246.

Collinge, J., Palmer, M.S. and Dryden, A.J. (1991) Genetic predisposition to iatrogenic Creutzfeldt–Jakob disease. *Lancet* **337** 1441–1442.

Doh-ura, K., Tateishi, J., Sasaki, H., Kitamoto, T. and Sakaki, Y. (1989) *Pro–leu* change at position 102 of prion protein is the most common but not the sole mutation related to Gerstmann–Sträussler syndrome. *Biochem. Biophys. Res. Commun.* **163** 974–979.

Goldfarb, L.G., Brown, P., Goldgaber, D., Asher, D.M., Strass, N., Graupera, G., Piccardo, P., Brown, W.T., Rubenstein, R., Boellaard, J.W. and Gajdusek, D.C. (1989) Patients with Creutzfeldt–Jakob disease and kuru lack the mutation in the

PRIP gene found in Gerstmann–Sträussler–Scheinker syndrome, but they show a different double-allele mutation in the same gene. *Am. J. Hum. Genet.* **45** (Supplement) A189.

Goldfarb, L.G., Brown, P., Goldgaber, D., Garruto, R.M., Yanagihara, R., Asher, D.M. and Gajdusek, D.C. (1990a) Identical mutations in unrelated patients with Creutzfeldt–Jakob disease. *Lancet* **336** 174–175.

Goldfarb, L.G., Korczyn, A.D., Brown, P., Chapman, J. and Gajdusek, D.C. (1990b) Mutation in codon 200 of scrapie amyloid precursor gene linked to Creutzfeldt–Jakob disease in Sephardic Jews of Libyan and non-Libyan origin. *Lancet* **336** 637–638.

Goldfarb, L.G., Mitrová, E., Brown, P., Toh, B.H. and Gajdusek, D.C. (1990c) Mutation in codon 200 of scrapie amyloid protein gene in two clusters of Creutzfeldt–Jakob disease in Slovakia. *Lancet* **336** 514–515.

Goldfarb, L.G., Haltia, M., Brown, P., Nieto, A., Kovanen, J., McCombie, W.R., Trapp, S. and Gajdusek, D.C. (1991) New mutation in scrapie amyloid precursor gene (at codon 178) in Finnish Creutzfeldt–Jakob kindred. *Lancet* **337** 425.

Goldgaber, D., Goldfarb, L.G., Brown, P., Asher, D.M., Brown, W.T., Linn, W.S., Teener, J.W., Feinstone, S.M., Rubenstein, R., Kascsak, R.J., Boellaard, J.W. and Gajdusek, D.C. (1989) Mutations in familial Creutzfeldt–Jakob disease and Gerstmann–Sträussler–Scheinker's syndrome. *Exp. Neurol.* **106** 204–206.

Hsiao, K., Baker, H.F., Crow, T. ., Poulter, M., Owen, F., Terwilliger, J.D., Westaway, D., Ott, J. and Prusiner, S.B. (1989) Linkage of a prion protein missense variant to Gerstmann–Sträussler syndrome. *Nature (London)* **338** 342–345.

Hsaio, K.H., Meiner, Z., Kahana, E., Cass, C., Kahana, I., Avrahami, D., Scarlato, G., Abramsky, O., Prusiner, S.B. and Gabizon, R. (1991) Mutation of the prion protein in Libyan Jews with Creutzfeldt–Jakob disease. *New Engl. J. Med.* **324** 1091–1097.

Hsiao, K.K., Scott, M., Foster, D., Groth, D.F., DeArmond, J., Prusiner, S.B. (1990) Spontaneous neurodegeneration in transgenic mice with mutant prion protein. *Science* **250** 1587–1590.

Kuppuswamy, M.N., Hoffmann, J.W., Kasper, C.K., Apitzer, S.G., Groce, S.L., and Bajaj, S.P. (1991). Single nucleotide primer extension to detect genetic diseases: experimental application to hemophilia B (factor IX) and cystic fibrosis genes. *Proc. Natl. Acad. Sci USA* **88** 1143–1147.

Nieto, A., Goldfarb, L.G., Brown, P., Chodosh, H.L., McCombie, W.R., Trapp, S., Asher, D.M. and Gajdusek, D.C. (1991) Codon 178 mutation in ethnically diverse Creutzfeldt–Jakob disease families. *Lancet* **337** 622–623.

Owen, F., Poulter, M., Lofthouse, R., Collinge, J., Crow, T.J., Risby, D., Baker, H.F., Ridley, R.M., Hsiao, K. and Prusiner, S.B. (1989) Insertion in prion protein gene in familial Creutzfeldt–Jakob disease. *Lancet* **1** 51–52.

Owen, F., Poulter, M., Shah, T., Collinge, J., Lofthouse, R., Baker, H., Ridley, R., McVey, J. and Crow, T.J. (1990) An in-frame insertion in the prion protein gene in familial Creutzfeldt–Jakob disease. *Mol. Brain Res.* **7** 273–276.

Palmer, M.S., Dryden, A.J., Hughes, J.T. and Collinge, J. (1991) Homozygous prion protein genotype predisposes to sporadic Creutzfeldt–Jakob disease. *Nature*

(*London*) **352** 340–342.

Prusiner, S.B. (1991) Molecular biology of prion diseases. *Science* **252** 1515–1522.

Puckett, C., Concannon, P., Casey, C. and Hood, L. (1991) Genomic structure of the human prion protein gene. *Am. J. Hum. Genet.* **49** 320–329.

Speer, M.C., Goldgaber, D., Goldfarb, L. G., Roses, A. D. and Pericak-Vance, M. A. (1991) Support of linkage of Gerstmann–Sträussler–Schneiker syndrome to the prion protein gene on chromosome 20p12-pter. *Genomics* **9** 366–368.

Tranchant, C., Doh-ura, K., Steinmetz, G., Chevalier, Y., Kitamoto, T., Tateishi, J. and Warter, J.M. (1991) Mutation du codon 117 du gène du prion dans une maladie de Gerstmann–Sträussler–Scheinker. *Rev. Neurol.* (*Paris*) **147** 274–278.

Westaway, D., Goodman, P.A., Mirenda, C.A., McKinley, M.P., Carlson, G.A. and Prusiner, S. B. (1987) Distinct prion proteins in short and long scrapie incubation period mice. *Cell* **51** 651–662.

16

Indiana variant of Gerstmann–Sträussler–Scheinker disease

B. Ghetti, F. Tagliavini, K. Hsiao, S. R. Dlouhy, R. D. Yee, G. Giaccone,
P. M. Conneally, M. E. Hodes, O. Bugiani, S. B. Prusiner, B. Frangione and M. R.
Farlow

ABSTRACT

The Indiana variant of Gerstmann–Sträussler–Scheinker disease is an autosomal
dominant neurological disorder clinically characterized by onset with gradual loss of
short-term memory and ataxia, followed by rigidity, bradykinesia and dementia.
Death occurs within 5–7 years of onset of ataxia. The established kindred is large;
at the time of this writing there are 10 living affected, 164 fully at risk, and 240 at
half risk family members.

Neuropathologic features include neurofibrillary tangles, uni- and multicentric
amyloid deposits in cerebral cortex and hippocampus as well as multicentric amyloid
deposits in cerebellar cortex and basal ganglia. Most amyloid deposits in cerebral
cortex and hippocampus are surrounded by a neuritic component. Amyloid cores
immunoreact with antisera to the hamster prion protein (PrP) 27-30. Neurofibrillary
tangles, composed of paired helical filaments, are morphologically indistinguishable
from those seen in Alzheimer disease.

The major component of the amyloid is an 11 kDa PrP fragment, the N-terminus
of which corresponds to residue 58 of the amino acid sequence deduced from the
human PrP cDNA.

In affected patients, a thymine-to-cytosine transition has been found in the second
position of codon 198 of the human prion protein gene (PRNP), resulting in
substitution of serine for phenylalanine.

INTRODUCTION

An adult-onset, autosomal dominant neurological disorder characterized clinically by cerebellar ataxia, Parkinsonism and dementia, has been studied in a large Indiana kindred (IK) (Azzarelli et al., 1985; Farlow et al., 1989). The clinical signs of affected individuals are similar to those observed in patients from an Austrian family ('H' family) originally described by Gerstmann et al. (1936) and further characterized by von Braunmühl (1954), Seitelberger (1962, 1981), Budka et al. (1991) and Kretzschmar et al. (1991). In the 'H' family, the salient clinical features are adult-onset ataxia, pyramidal signs and dementia. The disease in the 'H' family is inherited as an autosomal dominant trait. Neuropathologically, amyloid deposits in cerebellum and cerebrum as well as atrophy in spinal cord, brain stem and cerebellum, have been consistently observed.

Other familial and sporadic cases, that have many of the clinical and pathologic features of the patients from the 'H' family, have been given the eponym Gerstmann–Sträussler–Scheinker (GSS) disease or syndrome; for review see Farlow et al. (1992). However, the nosography of this group of disorders is complicated by the fact that, in some cases, spongiform changes similar to those seen in Creutzfeldt–Jakob disease (CJD) are present and that laboratory animals inoculated with brain tissue obtained from such patients develop a spongiform encephalopathy (Masters et al., 1981). Further, other transmissible spongiform encephalopathies, such as Creutzfeldt–Jakob disease and kuru in humans and scrapie in animals, have in common the presence of proteinase-resistant prion protein (PrP) and, in some cases, amyloid deposits immunoreactive with antisera to PrP; for review see Hsiao and Prusiner (1990) and Prusiner (1991).

The pathologic phenotype of the IK is also characterized by amyloid deposits immunoreactive with anti-PrP antisera (Ghetti et al., 1989); however, one of the most striking features is the consistent presence of neurofibrillary tangles in many areas of the central nervous system (Azzarelli et al., 1985; Ghetti et al., 1989; Giaccone et al., 1990, 1991a). These lesions are comparable with those seen in Alzheimer disease and make the disease of the IK unique in the spectrum of GSS disease and among those belonging to the group of prion diseases.

Recent progress in understanding clinical, pathological and molecular genetic characteristics of the IK will be reviewed.

RESULTS AND DISCUSSION

The pedigree of the IK spans eight generations and includes 2226 members, with 73 known to have been affected (Fig. 1). There is no sexual predominance and inheritance is autosomal dominant, with an apparently high degree of penetrance. There are currently ten known affected living individuals, 164 fully at-risk (children of affected individuals) and 240 one-half at-risk (children of at-risk subjects) living family members. Fifty-two at-risk and one-half at-risk family members have been examined over the last five years, with ten of these determined to be affected.

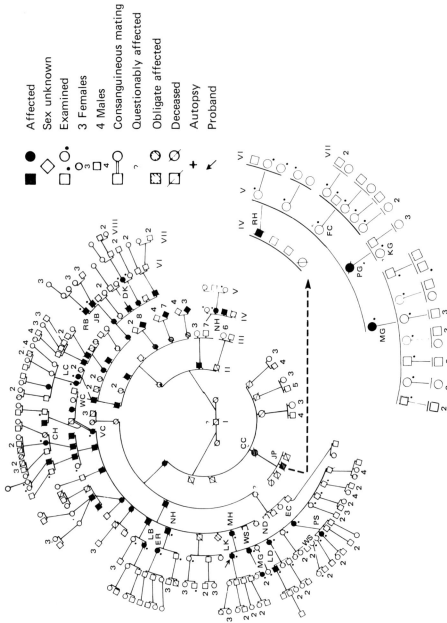

Fig. 1. Pedigree of the Indiana kindred.

Clinical studies

The age of onset of symptoms varies from the mid-30s to the early 60s. From neurological examinations in affected patients, a composite clinical picture emerges: initial symptoms are a gradual loss of short-term memory and progressive clumsiness in walking, exaggerated or first noticed when the individual is under stress. These symptoms may progress slowly over five years or rapidly over as little as one year. Rigidity and bradykinesia generally occur late in the disease; dementia worsens with their onset. Psychotic depression has been seen in several patients. Tremor is mild or absent. Without carbidopa–levodopa treatment, rapid weight loss and death usually occur within one year of onset of Parkinsonism. Bradykinesia and rigidity improve with treatment; however, death still occurs from pneumonia or other illness within two years after the onset of Parkinsonian signs. Early in the course, testing reveals abnormalities of short-term memory, eye movement, cognition, as well as mild cerebellar incoordination. Later, there is increasingly severe bradykinesia, rigidity, difficulty with gait and dysarthria. In the last stage, patients have severe extrapyramidal abnormalities, dysphagia and global dementia (Farlow et al., 1989).

Magnetic resonance imaging (MRI) studies

We have studied six patients by MRI of the brain and preliminary correlations between MRI abnormalities and clinical signs and symptoms have been made (Farlow et al., 1990). These patients show mild to moderate cerebellar atrophy and ataxia. Marked decrease in T2 signal intensity is seen in the basal ganglia, particularly the putamen and caudate nucleus, in three of the six imaged individuals. These findings are associated with moderate to severe Parkinsonian signs. Based on MRI and neuropathologic findings (see below), it is suggested that iron accumulation takes place in the caudate nucleus and substantia nigra.

Eye movements

Five affected and 11 unaffected at-risk family members have been studied (Yee et al., 1991). The affected individuals show abnormal eye movements that are characteristic of cerebellar damage. They include (1) gaze-evoked and rebound nystagmus, (2) upbeat nystagmus, (3) saccade hypometria (undershooting), (4) impaired smooth pursuit and optokinetic nystagmus, and (5) abnormal visual–vestibular interactions, including inability to suppress vestibulo-ocular responses by fixation.

Three of the five affected persons could not produce voluntary upward saccades of more than 30° in amplitude. However, oculocephalic manoeuvres increase upward eye movements. Supranuclear palsy of upward gaze is similar to that seen in Parkinsonian patients.

Most of the at-risk subjects have normal eye movements. However, two have pathologic nystagmus, although they did not have definite abnormalities in the remainder of their neurologic examination. One of them showed progression of eye movement abnormalities over a six-month period. Three other at-risk members have slight abnormalities of pursuit, optokinetic nystagmus, or visual–vestibulo-ocular responses. Progression of abnormalities has been found in one out of three at-risk individuals, that were examined at a 6 to 10 months interval.

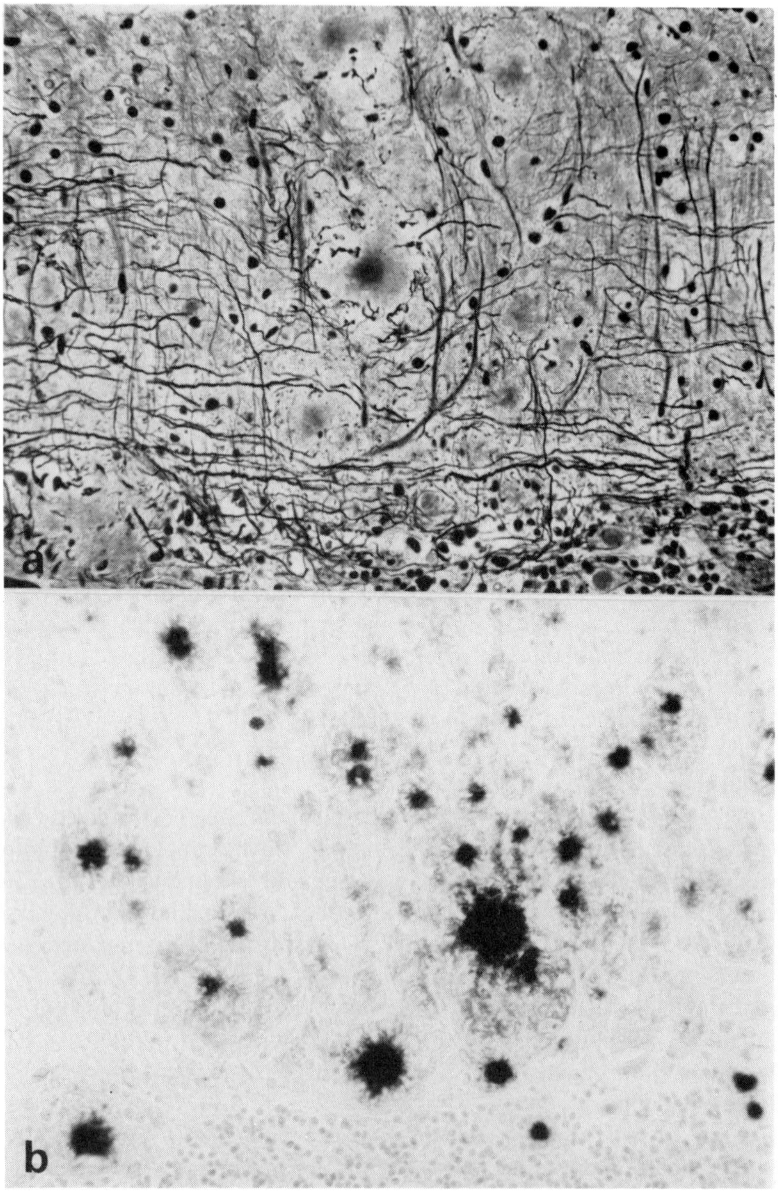

Fig. 2. Cerebellar cortex. (a) Amyloid deposits in the molecular layer (Bodian stain). (b) Amyloid deposits of the molecular and the Purkinje cell layers are immunolabelled with anti-PrP antibodies.

Neuropathology in affected individuals

The central nervous system of the six patients studied so far has consistently shown moderate cerebral atrophy and moderate to severe cerebellar atrophy. Other main macroscopic findings are marked loss of pigment of both the substantia nigra and the locus coeruleus. Microscopically, the following lesions are observed: amyloid deposits with or without a neuritic component, neurofibrillary tangles, nerve cell loss, gliosis, mild spongiform changes, and iron deposition (Azzarelli *et al.*, 1985; Ghetti *et al.*, 1989).

Amyloid deposits (Figs 2, 3(a)) are distributed throughout the gray structures of the cerebrum, cerebellum, and midbrain. They are intertwined at the periphery with astrocytic processes. Microglial elements and macrophages are present. The size of the amyloid deposits ranges between 10 and 100 μm in the cerebellum, and between 15 and 160 μm in the cerebral cortex. One or more amyloid cores are observed in each plaque. In thioflavine-S treated sections, amyloid deposits are strongly fluorescent when observed under ultraviolet light. Fluorescent deposits are not seen in or around the walls of blood vessels; however, they may be present in the parenchyma adjacent to vessels. In the cerebral cortex, the highest concentration of amyloid deposits of any type is found in the deeper layers; however, deposits in clusters are often found in subpial regions of the cerebral and cerebellar cortex (Ghetti *et al.*, 1989). In the cerebellar and cerebral cortex, poorly circumscribed fluorescent deposits are also seen. Their fluorescence is less intense than that of amyloid cores. These deposits may correspond to an early stage of amyloid formation.

In the cerebral cortex, amyloid deposits are in most instances associated with a crown of degenerating neurites, so that when the lesions are studied with classical stains they are morphologically indistinguishable from the neuritic plaques of Alzheimer disease (Fig. 3(a)). They are invariably found in most areas of the cerebral cortex (Fig. 3(a)) and hippocampus of symptomatic individuals.

The region showing most amyloid deposits is the molecular layer of the cerebellar cortex (116 amyloid deposits/mm^2) (Fig. 2). The mean number of amyloid deposits in vertical strips of cerebral cortex comprising the entire cortical thickness is 16/mm^2 in the parahippocampal gyrus, 12/mm^2 in the insular gyrus, and 11/mm^2 in the superior frontal gyrus. Fewer amyloid deposits are found in the hippocampus and they are rare in the calcarine cortex. The subcortical nuclei and the brainstem (mesencephalic tegmentum and periaqueductal gray matter) are severely involved.

By electron microscopy, the amyloid deposits appear to be composed of bundles of fibrillar structures, measuring 9–10 nm, and radiating out from a central zone. Surrounding the mass of fibrils, neuritic processes are found in the cerebral cortex, but are generally absent in the cerebellum.

The cerebral and cerebellar amyloid deposits, with and without a neuritic component, are immunoreactive with antisera raised against the protease-resistant core of the prion protein (PrP) designated PrP 27–30, a glycosylated peptide of 27–30 kDa that is derived by limited proteolysis from a neuronal sialoglycoprotein of 33–35 kDa (Fig. 2(b)). Immunolabelling with anti-PrP antisera is not detected in the vessel walls; however immunolabelled deposits are occasionally found adjacent to the Virchow–Robin spaces (Ghetti *et al.*, 1989). Using antisera to synthetic peptides

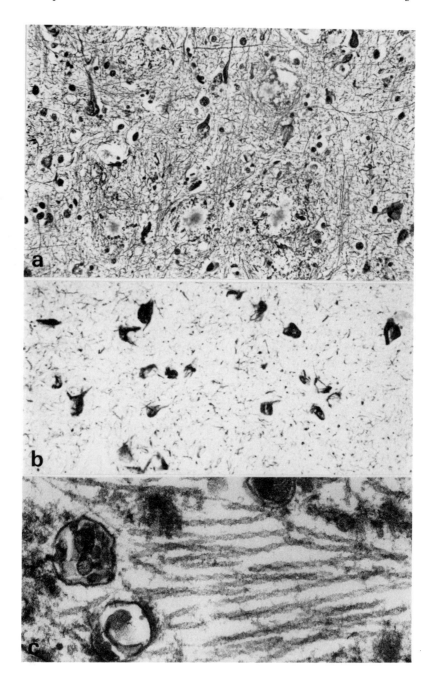

Fig. 3.   Cerebral cortex. (a) Amyloid deposits and neurofibrillary tangles in the deep cortical layers (Bodian stain). (b) Neurofibrillary tangles and neuropil threads in the parahippocampal gyrus are immunolabelled with Alz50. (c) Paired helical filaments from a neurofibrillary tangle (electron micrograph).

corresponding to residues 15–40 (P_2), 90–102 (P_1), and 220–232 (P_3) of the cDNA-deduced amino acid sequence of hamster PrP respectively, the amyloid deposits are strongly labelled by anti-P_1, and not by anti-P_2 and anti-P_3. With anti-P_2 and anti-P_3 antisera, a ring of immunopositivity is seen around the centre of the amyloid deposits (Giaccone et al., 1992).

Immunolabelling is not detected when sections from temporal cortex and cerebellum of patients who died at 49, 51, 55, and 58 years of age are exposed to anti-SP28, an antiserum to a 28-residue synthetic peptide homologous to the NH_2-terminal region of the β-protein. However, in one patient with onset of GSS signs at age 66 and death at 73, rings of anti-β-protein immunoreactivity around cores of anti-PrP immunoreactivity were frequently found (Giaccone et al., 1991b).

Neurofibrillary tangles (Fig. 3) are numerous in the second (Fig. 3(b)) and fifth layers of the parahippocampal cortex. Other cortical regions particularly rich in neurofibrillary tangles are the cingulate gyrus and the insular cortex, where neurons of layer V are most involved. In the remaining cortical regions, neurofibrillary tangles are present but less numerous. In the hippocampus, neurofibrillary tangles are not found in the CA1 through CA3 areas, but are numerous in CA4. The subcortical nuclei show a variable degree of involvement with tangles, the nucleus basalis being severely involved. In midbrain and pons, the substantia nigra, the griseum centrale, and the locus coeruleus have high numbers of neurofibrillary tangles (Ghetti et al., 1989).

Electron microscopic studies show that paired helical filaments (PHFs) are the main constituents of the neurofibrillary tangles found in cell bodies and in neurites (Fig. 3(c)). Each member of the pair is a filament about 10 nm in diameter. The pair of filaments measures about 22–24 nm at its maximum width and the helical twist has a period of about 70 nm (Azzarelli et al., 1985; Ghetti et al., 1989).

Many nerve cell bodies are strongly immunolabelled by Alz50, a monoclonal antibody which recognizes a 68 kDa protein present in large amounts in AD brains, and two patterns of perikaryal staining are recognized (Giaccone et al., 1990). In most labelled neurons the reaction product is circumscribed in a portion of the cytoplasm, which shows a globular or flame-shaped fibrillary appearance and contains neurofibrillary tangles. On the other hand, in a few neurons, the cytoplasm is diffusely immunolabelled by Alz50, but no neurofibrillary tangles can be detected. Anti-PHF, anti-PHF absorbed with tau, and anti-ubiquitin antibodies also immunostain the neurofibrillary tangles.

All these antibodies reveal the neuritic component, which surrounds the amyloid deposits in the cerebral cortex. Moreover, Alz50 labels neuropil threads that are not spatially associated with amyloid deposits or neurofibrillary tangles and are particularly abundant in the parahippocampal gyrus. Double anti-PrP/Alz50 and anti-SP28/Alz50 immunohistochemical staining clearly demonstrates that anti-PrP-positive and anti-SP28-negative amyloid deposits in the cerebral cortex are surrounded by Alz50-positive dystrophic neurites. With regard to the cerebellum, no immunolabelling with Alz50 and anti-PHF antibodies is detected around the amyloid deposits.

Spongiform changes: rarely, areas of the hippocampal pyramidal layer show vacuolation of the neuropil.

Nerve cell loss is observed in almost all cortical regions. In the cerebellar cortex, rarefaction of Purkinje cells is severe; swollen axons of Purkinje cells are found in the granule cell layer. Neuronal loss is evident in the red nucleus, substantia nigra, dentate nucleus, inferior olivary nucleus, and various other nuclear groups of the medulla. The oculomotor nuclei and the nucleus hypoglossus are apparently normal. In the substantia nigra, abundant pigment is found in the neuropil.

Perls stain shows that deposition of iron is prominent in globus pallidus, caudate nucleus, putamen, red nucleus and substantia nigra.

Neuropathology in at-risk individuals

Recently, we found a previously unexplored branch of the IK pedigree. A 52-year-old patient from this branch has typical clinical signs. A neurologically asymptomatic sister of this patient died suddenly at age 42. The neuropathologic examination shows that amyloid deposits in the cerebellar molecular and granule cell layers are already prominent. The amyloid deposits are immunolabelled by antibody to PrP. Neurofibrillary tangles are not found (Farlow *et al.*, 1991).

Transmission

Inoculation of brain tissue from one affected individual into hamsters was carried out in the laboratory of Dr. E. Manuelidis in 1987 and no transmission has occurred. Tissue from another affected member of the IK has been subsequently inoculated in hamsters and mice in Dr. S. Prusiner's laboratory and no transmission has occurred 450 days after inoculation.

Amyloid protein isolation and characterization

Amyloid cores were isolated from cerebral cortex and basal ganglia of two patients from the IK. Proteins were extracted with formic acid and fractionated by gel filtration. Two major peaks were obtained: the void volume (fraction 1) and a peak centred at 11 kDa on SDS–PAGE (fraction 6). Moreover, four minor intermediate peaks (fractions 2 to 5) and a lower molecular weight peak were present in the chromatograms. These fractions were subjected to immunoblot analysis using an antiserum to PrP 27-30, purified from scrapie-infected hamster brain, and four antisera to synthetic peptides corresponding to residues 15–40 (P_2), 90–102 (P_1), 140–172 (P_5) and 220–232 (P_3) of cDNA-deduced amino acid sequence of hamster PrP respectively. The immunochemical analysis showed that the 11 kDa peptide, the major constituent of the amyloid extract, was strongly labelled by the antiserum to PrP 27-30 and by anti-P_1; conversely it was weakly detected by anti-P_5 and was non-reactive with anti-P_2 and anti-P_3, suggesting that the amyloid protein was an internal fragment of PrP. In addition, higher molecular weight species of PrP-related peptides with apparently intact *N*-termini were present in fractions 2–5. The 11 kDa amyloid protein was further purified by HPLC and aliquots were digested with endoproteinase Lys-C. Automated sequence analysis of the intact 11 kDa peptide and of peptides generated by enzymatic digestion showed that the amyloid protein has a *N*-terminal lysine corresponding to codon 58 of the amino acid sequence deduced from human PrP cDNA and appears to end at codon 150 (Fig. 4) (Tagliavini *et al.*, 1991).

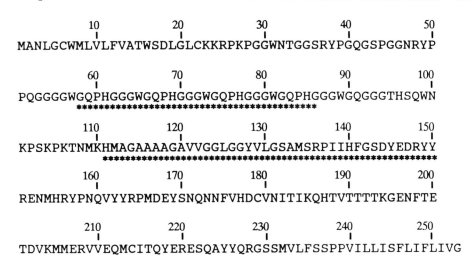

Fig. 4.   Amino acid sequence of 11 kDa amyloid protein isolated from plaques of the IK. Letters indicate the 'full-length' human PrP sequence as deduced from the cDNA. Asterisks (***) indicate the 11 kDa fragment. Note that the 11 kDa contains a V (valine) at position 129.

PRNP gene in the IK

DNA from affected patients was subjected to polymerase chain amplification of the PrP open reading frame (ORF). Both total amplified product and individually subcloned amplified *PrP* ORFs have been sequenced. Some patients were heterozygous (met/val) for a normal intragenic variant at codon 129, thus enabling identification of both PrP alleles in those individuals. Other patients were homozygous (val/val) at codon 129. In all affected individuals sequenced, a thymine (T) to cytosine (C) transition was found at codon 198 of one allele of the PrP ORF (Figs 5 and 6).

These studies demonstrate that Gerstmann–Sträussler–Scheinker disease in the Indiana kindred represents a unique phenotypic expression in the spectrum of GSS disease and of prion diseases. The missense mutation found at codon 198 of PRNP, resulting in a substitution of a serine for a phenylalanine, is likely to be the disease-causing mutation in this family (Hsiao *et al.*, 1992), a conclusion also supported by a lod score of greater than 6 ($\theta=0$) obtained in genetic linkage analysis (Dlouhy *et al.*, 1992). How the mutation triggers amyloid deposition is unknown. Since the amyloid represents only an 11 kDa fragment of the full-length PrP, our findings suggest that the disease process leads to proteolytic cleavage of PrP 33-35, generating an amyloidogenic peptide that polymerizes into insoluble fibrils. However, it is interesting that the mutation lies outside of the sequence that codes for the portion of PrP that was isolated as a major component of the amyloid. Thus, the exact relationship, if any, between the codon 198 mutation, PrP cleavage and amyloid deposition must be elucidated. Also, it is tempting to speculate that the 198 mutation, alone or in combination with other factors, may have a significant role in the pathogenesis of neurofibrillary tangle formation.

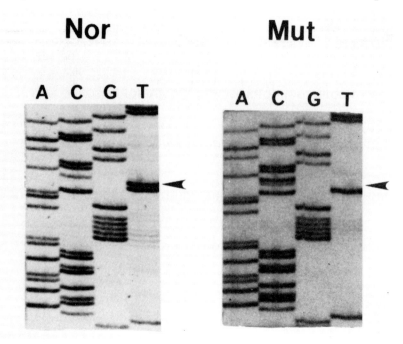

Fig. 5.   DNA sequencing gels demonstrating the codon 198 mutation in the Indiana kindred. The arrow indicates the position of a T that is present in the normal (Nor) PRNP gene but which has been replaced by a C in the mutant (Mut). Each panel represents a single PRNP allele that has been sequenced by cloning into M13.

The pathologic phenotype in the Indiana variant is unique in view of the coexistence of amyloid deposits, neuritic plaques in the cerebral cortex and neurofibrillary tangles. The topography of amyloid deposition in the cerebellum, basal ganglia and cerebral cortex and that of neurofibrillary tangles in cortex, nucleus basalis, substantia nigra and pontine tegmentum correlates well with ataxia, Parkinsonism and dementia. Specific clinicopathological correlations are needed to interpret best the pathogenetic mechanisms of ocular movement abnormalities.

At this time, there are three additional mutations in the PRNP gene that are associated with GSS disease. The mutations are at codons 102, 117, and 217. The mutation at codon 102 has been found in several families, including the 'H' family (Hsiao et al., 1989; Kretzschmar et al., 1991). Neuropathologically, patients with mutation at codon 102 show the presence of PrP-amyloid deposits in cerebrum and cerebellum, and absence of neurofibrillary tangles. Patients with mutation at codon 117 have mild or no cerebellar involvement; neuropathologically, PrP plaques in cerebrum and rare neurofibrillary tangles are seen (Nochlin et al., 1991; Hsiao et al., 1991). The mutation at codon 217 appears to be characterized by late onset GSS disease with cerebral and cerebellar involvement (Hsiao et al. 1992). Neuropathologically some neurofibrillary tangles and association of PrP amyloid with β-amyloid have been found in the only two cases of one family (Ikeda et al., 1991).

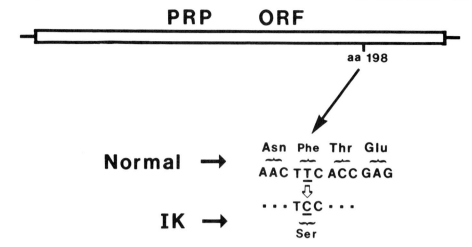

Fig. 6.   Schematic representation of PRNP gene with the IK mutation.

CONCLUSIONS

The following are the characteristics of the Indiana variant of Gerstmann–Sträussler–Scheinker disease:

(1) autosomal dominant inheritance;
(2) ataxia, Parkinsonism, and presenile dementia;
(3) 2–7 years duration from the onset of ataxia;
(4) PrP-immunopositive amyloid deposits in cerebrum and cerebellum; deposition of PrP-immunopositive amyloid occurs in the cerebellum before the onset of clinical signs;
(5) the major component of the amyloid is an 11 kDa fragment of prion protein with an N-terminal glycine at codon 58;
(6) PrP immunopositive amyloid deposits of the cerebral cortex are surrounded by abnormal neurites;
(7) Alzheimer's neurofibrillary tangles in cerebral cortex, hippocampus, nucleus basalis, and other subcortical nuclei;
(8) PrP amyloid deposition may be associated with β-protein in aged patients;
(9) mutation in the PRNP gene, codon 198.

ACKNOWLEDGMENTS

We thank Ms Patricia Johnson for DNA sequencing and Mr Joseph Demma for his photographic expertise.

REFERENCES

Azzarelli, B., Muller, J., Ghetti, B., Dyken, M. and Conneally, P.M. (1985) Cerebellar plaques in familial Alzheimer's disease (Gerstmann–Sträussler–Scheinker variant?). *Acta Neuropathol.* **65** 235–246.

Budka, H., Seitelberger, F., Feucht, M., Wessely, P. and Kretzschmar, H.A. (1991) Gerstmann–Sträussler–Scheinker syndrome (GSS): rediscovery of the original Austrian family. *Clin. Neuropathol.* **10** 99.

Dlouhy, S.R., Hsiao, K., Farlow, M.R., Foroud, T., Conneally, P.M., Johnson, P., Prusiner, S.B., Hodes, M.E. and Ghetti, B. (1992) Linkage of the Indiana kindred of Gerstmann–Sträussler–Scheinker disease to the prion protein gene. *Nature Genetics* **1** 64–67.

Farlow, M.R., Yee, R.D., Dlouhy, S.R., Conneally, P.M., Azzarellli, B. and Ghetti, B. (1989) Gerstmann–Sträussler–Scheinker disease. I. Extending the clinical spectrum. *Neurology* **39** 1446–1452.

Farlow, M.R., Edwards, M.K., Kuharik, M., Giaccone, G., Tagliavini, F., Bugiani, O. and Ghetti, B. (1990) Magnetic resonance imaging in the Indiana kindred of Gerstmann–Sträussler–Scheinker disease. *Neurobiol. Aging* **11** 265.

Farlow, M.R., Bugiani, O., Giaccone, G., Tagliavini, F. and Ghetti, B. (1991) Neuropathology of presymptomatic Gerstmann–Sträussler–Scheinker disease of the Indiana kindred. *Neurology* **41** (Suppl. 1) 119.

Farlow, M.R., Tagliavini, F., Bugiani, O. and Ghetti, B. (1992) Gerstmann–Sträussler–Scheinker disease. In: de Jong, J.M.B.V. (ed.) *Handbook of Clinical Neurology*, Vol. 16 (60). Elsevier, Amsterdam.

Gerstmann, J., Sträussler, E. and Scheinker, I. (1936) Über eine eigenartige hereditär-fimiliäre Erkrankung des Zentralnervensystems. Zugleigh ein Beitrag zur Frage des vorzeitigen lokalen Alterns. *Z. Neurol. Psych.* **154** 736–762.

Ghetti, B., Tagliavini, F., Masters, C.L., Beyreuther, K., Giaccone, G., Verga, L., Farlow, M.R., Conneally, P.M., Dlouhy, S.R., Azzarelli, B. and Bugiani, O. (1989) Gerstmann–Sträussler–Scheinker disease. II. Neurofibrillary tangles and plaques with PrP-amyloid coexist in an affected family. *Neurology* **39** 1453–1461.

Giaccone, G., Tagliavini, F., Verga, L., Frangione, B., Farlow, M.R., Bugiani, O. and Ghetti, B. (1990) Neurofibrillary tangles of the Indiana kindred of Gerstmann–Sträussler–Scheinker disease share antigenic determinants with those of Alzheimer disease. *Brain Res.* **530** 325–329.

Giaccone, G., Tagliavini, F., Verga, L., Fragione, B., Farlow, M.R., Bugiani, O. and Ghetti, B. (1991a) Indiana kindred of Gerstmann–Sträussler–Scheinker syndrome: neurofibrillary tangles and neurites of plaques with PrP amyloid share antigenic determinants with those of Alzheimer's disease. In: Iqbal, K., McLachlan, D.R.C., Winblad, B. and Wisniewski, H.M. (eds) *Alzheimer's Disease: Basic Mechanisms, Diagnosis and Therapeutic Strategies*. Wiley, New York.

Giaccone, G., Tagliavini, F., Bugiani, O., Frangione, B., Farlow, M.R. and Ghetti, B. (1991b) PrP and β-amyloid coexist in plaques of an aged patient of the Indiana kindred of Gerstmann–Sträussler–Scheinker disease. *Neurology* **41** (Suppl. 1) 155.

Giaccone, G., Verga, L., Bugiani, O., Frangione, B., Serban, D., Prusiner, S.B., Farlow,

M.R., Ghetti, B. and Tagliavini, F. (1992) Prion protein preamyloid and amyloid deposits in Gerstmann–Sträussler–Scheinker disease, Indiana kindred. *Proc. Natl. Acad. Sci.. USA*, in press.

Hsiao, K. and Prusiner, S. B. (1990) Inherited human prion diseases. *Neurology* **40** 1820–1827.

Hsiao, K., Baker, H.F., Crow, T.J., Poulter, M., Owen, F., Terwillinger, J.D., Westaway, D., Ott, J. and Prusiner, S.B. (1989) Linkage of a prion protein missense variant to Gerstmann–Sträussler syndrome. *Nature* (*London*). **338** 342–345.

Hsiao, K.K., Cass, C., Schellenberg, G.D., Bird, T., Devine-Gage, E., Wisniewski, H. and Prusiner, S.B. (1991) A prion protein variant in a family with the telencephalic form of Gerstmann–Sträussler–Scheinker syndrome. *Neurology* **41** 681–184.

Hsiao, K., Dlouhy, S., Ghetti, B., Farlow, M., Cass, C., DaCosta, M., Conneally, P. M., Hodes, M.E. and Prusiner, S.B. (1992) Mutant prion proteins in Gerstmann–Sträussler–Scheinker disease with neurofibrillary tangles. *Nature Genetics* **1** 68–71.

Ikeda, S., Yanagisawa, N., Allsop, D. and Glenner, G.G. (1991) A variant of Gerstmann–Sträussler–Scheinker disease with βB-protein epitopes and dystrophic neurites in the peripheral regions of PrP-immunoreactive amyloid plaques. In: Natvig, J. B., Førre, Ø., Husby, G., *et al.* (eds) *Amyloid and Amyloidosis 1990*. Kluwer, Dordrecht, pp. 737–740.

Kretzschmar, H.A., Honold, G., Seitelberger, F., Feucht, M., Wessely, P., Mehraein, P. and Budka, H. (1991) Prion protein mutation in family first reported by Gerstmann, Sträussler, and Scheinker. *Lancet* **337** 1161.

Masters, C., Gajdusek, D. and Gibbs, C. (1981) Creutzfeldt–Jakob disease virus isolations from the Gerstmann–Sträussler syndrome with an analysis of the various forms of amyloid plaque deposition in the virus induced spongioform encephalopathies. *Brain* **104** 559–588.

Nochlin, D., Sumi, S.M., Bird, T.D., Snow, A.D., Leventhal, C.M., Beyreuther, K. and Masters, C.L. (1991) Familial dementia with PrP positive amyloid plaques: a variant of Gerstmann–Sträussler syndrome. *Neurology* **39** 910–918.

Prusiner, S.B. (1991) Molecular biology of prion diseases. *Science* **252** 1515–1522.

Seitelberger, F. (1962) Eigenartige familiär-hereditäre Krankheit des Zentralnervensystems in einer neiderösterreichischen Sippe. *Wien. Klin. Wochenschr.* **74** 687–691.

Seitelberger, F. (1981) Straubler's disease. *Acta Neuropathol.* (*Berlin*) **7** 341–343.

Tagliavini, F., Prelli, F., Ghiso, J., Bugiani, O., Serban, D., Prusiner, S.B., Farlow, M.R., Ghetti, B. and Frangione, B. (1991) Amyloid protein of Gerstmann–Sträussler–Scheinker disease (Indiana kindred) is an 11 kd fragment of prion protein with an *N*-terminal glycine at codon 58. *EMBO J.* **10** 513–519.

von Braunmühl, A. (1954) Über eine eigenartige hereditär-familiäre erkrankungh des Zentralnervensystems. *Arch. Psychiatr. Z. Neurol.* **191** 419–449.

Yee, R.D., Farlow, M.R., Suzuki, D.A., Betelak, K.F. and Ghetti, B. (1992) Abnormal eye movements in Gerstmann–Sträussler–Scheinker disease. *Arch. Opthalmol.* **110** 68–74.

17

Inherited prion disease in Libyan Jews

Ruth Gabizon, Esther Kahana, Karen Hsiao, Stanley B. Prusiner and Zeev Meiner

ABSTRACT

A focus of Creutzfeldt–Jakob disease (CJD) among Jews from Libyan origin was identified in Israel 20 years ago. The incidence of the disease in this ethnic group is about 100 times more than in the worldwide population. The consumption of lightly cooked sheep brain has been invoked to explain the high incidence of CJD in this community. The discovery of mutations in the PrP gene that segregate with other familial prion diseases such as Gerstmann–Sträussler–Scheinker syndrome (GSS) led us to perform a molecular genetic study and to compare it with an epidemiological survey of the Libyan Jewish community. The epidemiological data suggest a very high familial incidence of CJD in this population and a molecular genetic research elucidated that CJD segregates with a point mutation at codon 200 of the PrP gene resulting in the substitution of lysine for glutamate. This mutation was found in ~40 CJD patients of Libyan origin and not found in one Moroccan Jew suffering from CJD. It was also absent in almost 100 healthy Libyan controls above age 60. This result strongly supports a genetic aetiology for CJD pathogenesis in the Libyan Jewish community and overturns the culinary hypothesis. The disease is vertically transmitted with an autosomal dominant inheritance of unknown penetrance. All of our patients were heterozygous for the mutation except for one homozygous patient. The course of the disease in this patient was identical to that in the heterozygous patients, strongly arguing that inherited CJD displays complete phenotypic dominance. The Libyan Jewish cluster does not appear to be connected with the Czechoslovakian cluster of CJD patients associated with the same codon 200 mutation, suggesting this mutation developed independently in two geographically separated populations.

INTRODUCTION

Creutzfeldt–Jakob disease (CJD) has a worldwide annual incidence of 0.5 to 1 case per million population; 5–10% of the cases are familial (Brown *et al.*, 1987). Several clusters of the disease have been identified in different countries such as Czechoslovakia and Chile (Galvez *et al.*, 1983; Mitrova, 1987). The highest incidence of the disease, 31 per million, had been reported in 1974 in the Libyan-born Jews living in Israel (Kahana *et al.*, 1974), and at that time was attributed to their culinary or cultural habits (Herzberg *et al.*, 1974; Alter, 1974; Goldberg *et al.*, 1979). The clinical presentation and features of CJD in the Libyan born patients are similar to Israeli patients of other ethnic origins. Some higher frequency of myoclonic jerks, more pronounced periodic EEG changes and a rapidly progressive course of shorter duration in the Libyan patients have been observed (Kahana *et al.*, 1991).

EPIDEMIOLOGICAL DATA ON CJD IN LIBYAN JEWS

The first epidemiological survey of CJD in Israel disclosed 29 cases with onset between 1963 and 1972, among which 13 were Jewish immigrants from Libya (Kahana *et al.*, 1974). This study confirmed previous observations of the unusual high incidence of CJD in this community (Goldhammer *et al.*, 1972; Behar *et al.*, 1969). The latest calculated incidence of CJD in Libyan Jews as compared with other ethnic groups in Israel, 52.6 per million, is the highest ever reported for the disease (Table 1) (Zilber *et al.*,

Table 1. Incidence of CJD in Israel (1962–1987) by place of birth

Place of birth	Number of cases		
	Observed (*O*)	Expected (*E*)	Incidence rate (*O/E*)
Libya	40	0.76	52.6
Other North African countries	13 (2)[a]	5.95	2.2
Israel	7 (1)[a]	5	1.4
Asia[b]	9	8.3	1.1
Europe or America	19	29	0.65

E is the expected number of cases in each ethnic population as calculated by the general incidence rate and the age adjusted number of individuals. *O* is the observed number of cases.
[a] Individuals with relatives of Libyan origin.
[b] All countries except Israel.

1991). Whereas no familial cases were identified in the first epidemiological studies, explanations were sought among the cultural and culinary habits of this community (Alter, 1974; Alter and Kahana, 1976). The similarity between scrapie, a transmissible neurodegenerative disease of sheep, and CJD (Parry, 1983a), as well as the evidence that ritualistic cannibalism is responsible for the spread of kuru among the Fore people of New Guinea (Gajdusek, 1977), led to the assumption that transmission of CJD occurred through consumption of scrapie-infected sheep brains. This hypothesis was reinforced by the fact that Libyan Jews consumed a considerable amount of lightly cooked sheep brain in their diet (Goldberg et al., 1979). The consumption of sheep eyeballs was also proposed as a mode CJD transmission (Herzberg et al., 1974).

Scrapie appears largely confined to flocks in Europe and North America and there is no evidence for its existence among the flocks of sheep in Libya or the Middle East (Parry, 1983b). Although non-Libyan Mediterranean Jews also consume a large amount of sheep brain, the incidence of CJD found in these communities was not elevated. Indeed, some of the CJD patients born in other North African countries were descendants of the Libyan Jewish community (Table 1). Within the Libya Jewish community, there was no correlation between consuming sheep brain and the appearance of CJD (Goldberg et al., 1979).

The high incidence of CJD does not seem to be related to the country of origin since the incidence of the disease among non-Jewish inhabitants of Libya appears to be similar to the incidence worldwide (Radhakrishnan and Mousa, 1988). The latest epidemiological study which includes all cases diagnosed prior to 1987 does not report any correlation between the age of emigrés from Libya and the incidence of the disease, suggesting the time of exposure to a hypothetical infectious is irrelevant (Zilber et al., 1991). In addition, two Israeli-born CJD patients from Libyan families have been identified recently.

The genetic aetiology of CJD was not seriously contemplated at first by any of the investigators in the field since, as mentioned above, in the first epidemiological publication all the patients were labelled as sporadic (Kahana et al., 1974). Lack of reliable medical information on previous generations, the fear from stigmata of dementia in the family and early deaths due to infectious diseases prevented accurate determination of the percentage of familial cases.

Improper diagnosis was another factor that interfered with attainment of accurate epidemiological data concerning CJD. For instance, lack of imaging equipment until recent years led to the erroneous diagnosis of brain tumours in cases of CJD. A special problem was the limited number of autopsies performed in Israel for religious reasons, In recent years, as a result of better medical organization and equipment, increased longevity and improved epidemiological surveys, more patients have been diagnosed accurately. The percentage of familial cases in Israel is now about 64% (Table 2). Most of the affected individuals appear to be members of a limited number of large pedigrees with many consanguinities, strongly consistent with genetic aetiology for CJD in this community (Fig. 1) (Hsiao et al., 1991).

MOLECULAR GENETICS OF CJD IN LIBYAN JEWS

The discovery of the prion protein and its role in CJD pathogenesis led to the search for mutations in the PrP gene as a possible cause for inherited prion diseases (Hsiao and Prusiner, 1990). The first point mutation in the PrP gene to be identified in humans was a codon 102 leucine substitution; it was shown to be linked to ataxic

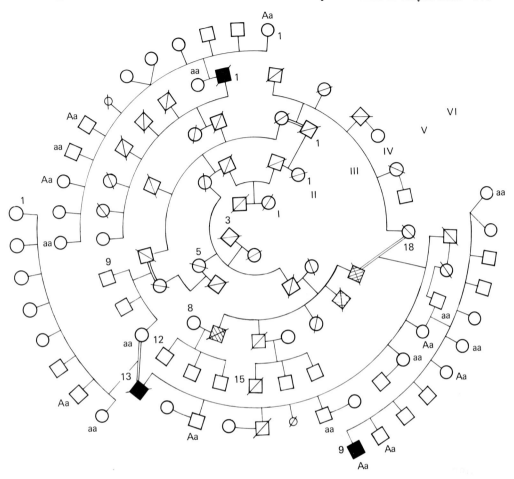

Fig. 1.   Pedigree of Libyan Jewish family M with Creutzfeldt–Jakob disease. Circles denote female subjects, square male subjects, and symbols with diagonals dead subjects. Solid symbols denote subjects in whom Creutzfeldt–Jakob disease was confirmed, and hatched symbols subjects in whom it was suspected. On the basis of the predicted amino acid sequence, *a* indicates a glutamate at codon 200 of the prion protein and *A* indicates a lysine at codon 200.

Gerstmann–Sträussler–Scheinker disease with a lod score of approximately 3.2 (Hsiao *et al.*, 1989). Other mutations have also been found to be associated with familial CJD (Goldgader *et al.*, 1989; Owen *et al.*, 1989; Doh-ura *et al.*, 1989). Earlier, polymorphisms in the PrP gene were found to be linked to a scrapie incubation time gene in mice (Carlson *et al.*, 1986).

With this background, we searched for a mutation in the PrP gene of Libyan Jews suffering from CJD. The method used was amplification of the PrP gene open reading frame (ORF) by PCR using primers previously described (Hsiao *et al.*, 1989). Sequencing the ORF from one patient revealed a G-to-A change in the first position

Table 2. Libyan familial cases in different epidemiological studies

	Cases	Familial cases	Familial percentage
Kahana *et al.* (1974)	13	0	
Alter and Kahana (1976) and			
Goldberg *et al.* (1979)	21	2	9.5
Neugut *et al.* (1979)	23	8	34.8
Zilber *et al.* (1991)	49	20	40.8
To date	83	54	64

of codon 200 in one allele resulting in the substitution of lysine for glutamate (Fig. 2) (Hsiao *et al.*, 1991). No other mutations were found. This mutation eliminates a unique BsmA1 restriction site with the PrP gene ORF but digestion with this enzyme is frequently incomplete. Therefore, other CJD patients as well as family members and unrelated controls were probed for the mutation by allele-specific oligonucleotide hybridization with two oligonucleotides, one for the mutant allele and one for the wild type (Fig. 3). The same mutation was discovered concomitantly in brains of Mediterranean Sephardic Jews with CJD, including four of Libyan origin, using the BsmA1 restriction enzyme method (Goldfarb *et al.*, 1990a).

Our original survey included 11 Libyan Jewish CJD patients; one from Italy and the others referred to us by different Israeli physicians. All of the patients had the codon 200 lysine substitution while 37 healthy Libyan Jews of age 60 or greater who served as controls and one Moroccan Jew with CJD tested negative for the mutation. A χ^2 analysis of the codon 200 lysine haplotype revealed a significant association with CJD in the Libyan Jews ($\chi^2 = 42$, $p < 0.001$). All the patients except one were heterozygous for this substitution. Subsequently, we have enlarged our database by testing for the mutation in family members of patients with CJD as well as older healthy individuals from Libyan community without a family history of CJD (Table 3). Twenty-eight CJD patients have the codon 200 mutation, either by direct testing of DNA or by inference from positive offspring and a negative spouse. In 17 other patients, positive offspring were identified but the spouse was unavailable for testing, suggesting that these patients also carried the mutation. Ninety healthy Libyan controls were found to be negative for the mutation. Forty-six healthy Libyan Jews were identified as mutation carriers, three of them above age 65.

Four individuals originally tested as controls proved to be positive for the mutation. In each of these cases, a close relative who died in the past from a CJD-like disease was identified and affiliation to a known CJD pedigree was documented. This experience suggests that there are a significant number of misdiagnosed CJD cases.

Our results argue that CJD among the Libyan community is vertically transmitted in an apparent autosomal dominant inheritance with unknown penetrance. Pedigree

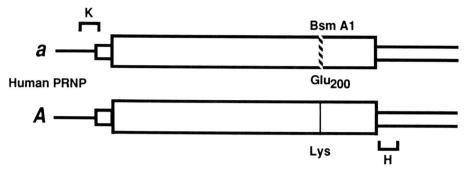

Fig. 2.   Prion protein gene mutation at codon 200 in Libyan Jews with Creutzfeldt–Jakob disease. In the mutant allele, designated *A*, lysine (Lys) is substituted for glutamate (Glu) at codon 200. A BsmA1 restriction site is abolished by the mutation. K and H show the positions of the primers used to amplify DNA in the polymerase chain reaction. The open reading frame is indicated in the large box, the mRNA untranslated region by the narrower box, and the intron by a solid line.

analyses currently suggest incomplete penetrance. However, only long-term follow-up establish the probability of a specific mutation carrier acquiring the disease. Penetrance determination is hindered by the wide range in age for disease onset (35–73 years old); therefore, determining whether a given individual is above an upper age limit for the appearance of CJD is problematic. Only a minority of mutation carriers are elderly, yet in one family two healthy individuals were identified as carriers for the mutation at the ages of 68 and 66 years old, respectively (subjects V-15 and V-23 in Fig. 1). In a few other cases, individuals aged 65–70 that were not available for testing were designated obligate carriers for the mutation. Whether genetic factors other than codon 200 mutation in the PrP gene may play a role in determining the expression of disease remains to be established.

An interesting phenomenon in some CJD families is the occurrence of initial symptoms of the disease at younger ages in successive generations. This phenomenon, denominated anticipation, has been observed in other autosomal dominant diseases such as Huntington's disease and myotonic dystrophy but its molecular mechanism is unknown (Ridley *et al.*, 1988; Howeler *et al.*, 1989).

HOMOZYGOUS PATIENTS

One of our patients, a woman 42 years old, was found to be homozygous for the PrP gene codon 200 lysine mutation. No hybridization with the wild-type allele-specific oligonucleotide probe was found in her case. Her four children tested positive (heterozygous) for the codon 200 lysine mutation while her husband was negative. We have identified two other possible homozygous patients among the historical cases, since all of their offspring were positive for the mutation but the spouses negative. Unfortunately no tissue was available for direct examination of these cases. No differences in the clinical course, age of appearance and laboratory findings were observed in these cases as compared with heterozygous CJD patients. These observations argue that CJD associated with the codon 200 lysine mutation is a true

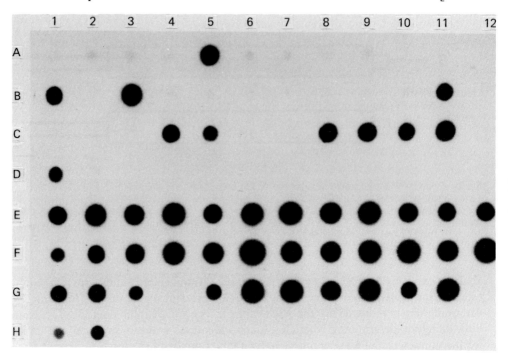

Fig. 3.　Allele-specific oligonucleotide hybridization. The prion protein open reading frame in DNA from peripheral leukocytes was amplified with the polymerase chain reaction. Denatured, amplified DNA was immobilized on nitrocellulose and hybridized to radiolabeled oligonucleotide probes encoding either lysine (mutant probe; rows A to D) or glutamate (wild-type probe; rows E to H) at codon 200 of the prion protein. Samples from normal subjects reacted only with the wild-type probe, whereas those from patients reacted with both the wild-type and mutant probe except one sample (dots C4 and G4), which reacted only with the mutant probe, suggesting the patient was homozygous for the mutation.

dominant disorder, and displays complete phenotypic dominance like Huntington's disease (Wexler *et al.*, 1987). It also demonstrates that a wild-type copy of the PrP gene is not essential for foetal viability or for normal growth and development. It argues as well against the possibility that the disease is triggered in heterozygous patients when a recessive, somatic mutation disables the wild-type prion protein allele.

ORIGIN OF THE CODON 200 MUTATION IN LIBYAN JEWS

The Libyan Jewish community in Israel has ∼50 000 members mostly living in the cities of Ashkelon and Natania.

The first Jews are thought to have arrived in Djerba about 600 BC. Later waves of Jewish immigration to North Africa were created by their expulsion during the Spanish inquisition. For many centuries, the Djerban Jewish community remained isolated from other North African communities.

Mostly of the Israeli CJD patients with the codon 200 mutation originated from either Tripoli or the island of Djerba, a Tunisian island geographically close to Tripoli.

Table 3.   Current data on codon 200 mutation in Libyan Jews

	Verified	Suspected
CJD patients	28	17[a]
Healthy mutation carriers	46	—
Mutation carriers age 65–70	3	5[b]
Unrelated Libyan controls	90	—

[a] Positive offspring but untested spouse.
[b] Obligated carriers.

To avoid assimilation with the Arab inhabitants, there were many consanguineous marriages within this small and isolated community, and therefore the fact that homozygous individuals for the codon 200 mutation were identified in this population has no bearing on the prevalence of the mutation. About 1950, most of the Jews in Tripoli emigrated to Israel, but some went to Italy and Greece. Djerban Jews immigrated to Tunis and from there to Israel and France (Fig. 4). Consequently CJD patients with the codon 200 mutation were also found in the Libyan Jewish communities of these European countries (Goldfarb et al., 1990a) (Fig. 4).

Two other CJD foci are associated with the codon 200 mutation; a large cluster of CJD patients in Czechoslovakia (Goldfarb et al., 1990b), as well as the CJD cluster in Chile (Paul Brown, personal communication). The fact that geographically and culturally distant communities manifest the same mutation suggests that this mutation must have occurred independently since there appears to be no common ancestry between these three communities.

The codon 200 mutation may have resulted from deamination of deoxycytidine-phosphate-deoxyguanosine and the subsequent formation of deoxythimidine-phosphate-deoxyguanosine on the negative strand. This mutation, which occurs frequently in mammalian DNA (Barker et al., 1984), could explain how the mutation at codon 200 of the PrP gene developed independently in three geographically separate populations.

CONCLUSIONS

The detection of a specific mutation that segregates with CJD in Libyan Jews argues persuasively for a genetic aetiology and suggests that the previous culinary hypothesis should be discarded (Hsiao et al., 1991). The majority of CJD cases in Israel are familial. CJD is vertically transmitted through autosomal dominant pattern of inheritance but the penetrance is still unknown. The apparent incidence of the disease has increased in the last decade, partially as a result of increased longevity which enables late onset cases to be diagnosed. The young generation of Libyan Jews are now marrying outside their ethnic group; this social practice will result in the spread of the codon 200 mutation to the general Israeli population. How these changes in genetic background will influence penetrance as well as age of onset of familial CJD is unknown.

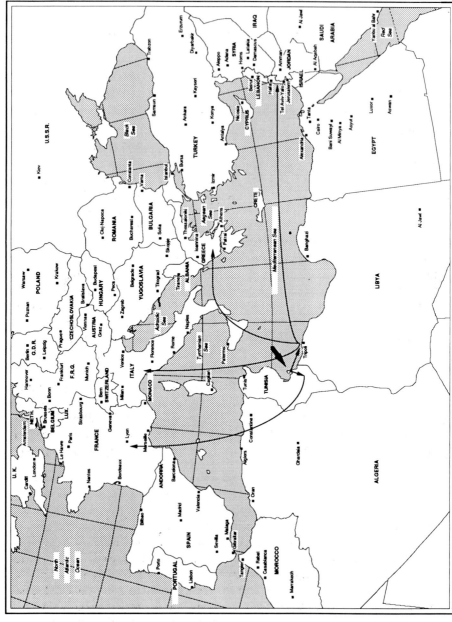

Fig. 4. Dispersion of Jews originating from Libya in the 20th century. The Tripoli Jewish community emigrated about 1950. Most of its members emigrated to Israel while others left Libya for Greece or Italy. The Djerban community (large bold arrow) emigrated mostly to the city of Tunis and from there to Israel or France.

The adult onset of an autosomal dominant inherited disease creates special ethical problems (Huggins *et al.*, 1990). In another autosomal disorder, Huntington's disease, it is now possible to predict the likelihood of acquiring the disease for a given individual, although the gene responsible for the disease is unknown (Gusella *et al.*, 1983). In CJD caused by the codon 200 mutation, the penetrance is still unknown and thus cannot predict which of the mutation carriers will manifest the disease. This dilemma impacts on decisions regarding prenatal diagnosis of CJD, an examination which should be considered as a possible measure to control the spread of the codon 200 mutation within and outside the Libyan Jewish community. More information will have to be collected before any of these ethical issues can be resolved.

REFERENCES

Alter, M., (1974) Creutzfeldt–Jakob disease: hypothesis for high incidence in Libyan Jews in Israel. *Science* **186** 848.

Alter, M. and Kahana, E. (1976) Creutzfeldt–Jakob disease among Libyan Jews in Israel. *Science* **192** 428.

Barker, D., Schafer, M. and White, R. (1984) Restriction sites containing CpG show a higher frequency of polymorphism in human DNA. *Cell* **36** 131–138.

Behar, M., Sroka, C., Elian, E., Kott, E., Korczyn, A.D., Bornstein, B. and Sandbank, U. (1969) Creutzfeldt–Jakob disease and its relation to pre-senile dementia. *Harefua* **77** 275–279.

Brown, P., Cathala, F., Raubertas, R.F., Gajdusek, D. C. and Castaigne, P. (1987) The epidemiology of Creutzfeldt–Jakob disease: conclusion of a 15-years investigation in France and review of the world literature. *Neurology* **37** 895–904.

Carlson, G.A., Kingsbury, D.T., Goodman, P.A., Coleman, S., Marshal, S.T., DeArmond, S.J., Westaway, D. and Prusiner, S.B. (1986) Linkage of prion protein and scrapie incubation times genes. *Cell* **46** 503–511.

Doh-rua, K., Tateishi, J., Sasaki, H., Kitamoto, T. and Sakaki, Y. (1989) Pro–Leu change at position 102 of prion protein is the most common but not the sole mutation related to Gerstmann–Straussler syndrome. *Biochem. Biophys. Res. Commun.* **163** 974–979.

Gajdusek, D.C. (1977) Unconventional viruses and the origin and disappearance of kuru. *Sicnece* **197** 943–957.

Galvez, S., Masters, C. and Gajdusek, D.C. (1980) Descriptive epidemiology of Creutzfeldt–Jakob disease in Chile. *Arch. Neurol.* **37** 11–14.

Goldberg, H., Alter, M. and Kahana, E. (1979) The Libyan Jews focus of Creutzfeldt–Jakob disease: a search for the mode of natural transmission. In: Prusiner, S.B. and Hadlow, W.J. (eds) *Slow Transmissible Disease of the Nervous System*, Vol 1. Academic Press New York, pp. 195–211.

Goldfarb, L.G., Korczyn, A.D., Brown, P., Chapman, J. and Gajdusek, D.C. (1990a) Mutation in codon 200 of scrapie amyloid precursor gene linked to Creutzfeldt–Jakob disease in Sepharadic Jews of Libyan and non-Libyan origin. *Lancet* **336** 637–638.

Goldfarb, L.G., Mitrova, E., Brown, P., Toh, B.H. and Gajdusek, D.C. (1990b) Mutation in codon 200 of scrapie amyloid protein gene in two clusters of Creutzfeldt–Jakob disease in Slovakia. *Lancet* **336**, 514–515.

Goldgaber, D., Goldfarb, L.G., Brown, P., Asher, D.M., Brown, W.T., Lin, S., Teener, J.W., Feinstone, S.M., Rubenstein, R., Kascsak, R.J., Boellaard, J.W. and Gajdusek, D.C. (1989) Mutation in familial Creutzfeldt–Jakob disease and Gerstmann–Straussler–Scheinker's syndrome. *Exp. Neurol* **106** 204–206.

Goldhammer, Y., Bubis, J.J., Sarova-pinhas, I. and Braham, J. (1972) Subacute spongiform encephalopathy and its relation to Jakob–Creutzfeldt disease: report on six cases. *J. Neurol. Neurosurg. Psychiatry* **35** 1–10.

Gusella, J.F., Wexler, N.S., Conneally, M.P., Naylor, S.L., Anderson, M.A., Tanzi, R.E., Watkins, P.C., Ottina, K., Wallace, M.R., Sakaguchi, A.Y., Young, A.B., Shoulson, I., Bonilla E. and Martin, J.B. (1983) A polymorphic DNA marker genetically linked to Huntington's disease. *Nature (London)* **306** 234–239.

Herzberg, L., Herzberg, B.N., Gibbs, C.J., Jr, Sullivan, W., Amyx, H. and Gajdusek, D.C. (1974) Creutzfeldt–Jakob disease: hypothesis for high incidence in Libyan Jews in Israel. *Science* **186** 848.

Howeler, C.J., Busch, H.F., Geraedts, J.P., Niermeijer, M.F. and Staal, A. (1989) Anitcipation in myotonic dystrophy: fact or fiction? *Brain* **122** 779–797.

Hsiao, K. and Prusiner, S.B. (1990) Inherited human prion dieases. *Neurology* **40** 1820–1827.

Hsiao, K., Baker, H.F., Crow, T.J., Poulter, M., Owen, F., Terwilliger, J.D., Westaway, D., Ott, J. and Prusiner, S.B. (1989) Linkage of a prion protein missense variant to Gerstmann–Straussler syndrome. *Nature (London)* **338** 342–345.

Hsiao, K., Meiner, Z., Kahana, E., Cass, C., Kahana, I., Avrahami, D., Scarlato, G., Abramsky, O., Prusiner, S.B. and Gabizon, R. (1991) Mutation of the prion protein in Libyan Jews with Creutzfeldt–Jakob disease. *N. Engl. J. Med.* **324** 1091–1097.

Huggins, M., Bloch, M., Kanani, S., Quarrell, O.W.J., Theilman, J., Hedrick, A., Dickens, B., Lynch, A. and Hayden, M. (1990) Ethical and legal dilemmas arising during predictive testing for adult-onset disease: the experience of Huntington's diease. *Am. J. Hum. Genet.* **47** 4–12.

Kahana, E., Alter, M., Braham, J. and Sofer, D. (1974) Creuzfeldt–Jakob disease: focus among Libyan Jews in Israel. *Science* **183** 90–91.

Kahana. E., Zilber, N. and Abraham, M. (1991) Do Creutzfeldt–Jakob disease patients of Jewish Libyan origin have unique clinical features? *Neurology*, **41** 1390–1392.

Lowenstein, D.H., Butler, D.A., Westaway, D., McKinley, M.P., DeArmond, S.J. and Prusiner, S.B. (1990) Three hamster speacies with different scrapie incubation times and neuropathological features encode distinct prion proteins. *Mol. Cell Biol.* **10** 1153–1163.

Mitrova, E. (1987) Epidemiological analysis of Creutzfeldt–Jakob disease in Slovakia (1972–1985). In: Court, L. (ed.) *Virus Non Conventional du Systeme Nerveux Central.* Masson, Paris.

Neugut, R.H., Neugut, A.I., Kahana, E., Stein, Z. and Alter, M. (1979) Creutzfeldt–Jakob disease clustering among Libyan born Israelis. *Neurology* **29**, 225–231.

Owen, F., Poulter, M., Lofthouse, R., Collinge, J., Crow, T.J., Risby, D., Baker, H.F.,

Ridley, R.M., Hsiao, K. and Prusiner, S.B. (1989) Insertion in prion protein gene in familial Creutzfeldt–Jakob disease. *Lancet* **1** 51–52.

Parry, H.B. (1983) *Scrapie Disease in Sheep.* Academic Press, New York, 192 pp.

Parry, H.B. (1988) Records of occurrence of scrapie from 1750. In: Oppenheimer, D.R. (ed.) *Scrapie Disease in Sheep* Academic Press, New York, pp.31–59.

Radhakishnan, K. and Mousa, E. M. (1988) Creutzfeldt-Jakob disease in Benghazi, Libya. *Neuroepidemiology* **7** 42–43.

Ridley, R.M., Firth, C.D., Crow, T.J. and Conneally, P.M. (1988) Anticipation in Huntington's disease is inherited through the male line but may originate in the female. *J. Med. Genet.* **25** 589–595.

Wexler, N.S., Young, A.B., Tanzi, R.E., Travers, H., Starosta-Rubinstein, S., Penney, J.B., Snodgrass, S.R., Shoulaon, I., Gomez, F., Ramos Arroyo, M.A., Penchaszadh, G.K., Moreno, H., Gibbons, K., Faryniarz, A., Hobbs, W., Anderson, M.A., Bonilla, E., Conneally, P.A. and Gusella, J.F. (1987) Homozygotes for Huntington's disease. *Nature (London)* **326** 194–197.

Zilber, N., Kahana, E. and Abraham, M. (1991) The Libyan Creutzfeldt–Jakob disease (CJD) focus in Israel— an epidemiological evaluation. Neurology, **41** 1385–1389.

18

Fatal familial insomnia: a prion disease with a mutation in codon 178 of the prion protein gene—study of two kindreds

Rossella Medori, Hans-Juergen Tritschler, Andréa C. LeBlanc, Federico Villare, Valeria Manetto, Pasquale Montagna, Pietro Cortelli, Patrizia Avoni, Mirella Mochi, Lucila Autilio-Gambetti, Elio Lugaresi, Pierluigi Gambetti

ABSTRACT

We present clinical, histopathological, immunochemical and genetic studies of members of two large Italian kindreds affected by a disease that we originally named fatal familial insomnia (FFI).

FFI, which is apparently inherited as an autosomal dominant trait, begins between the ages of 35 and 61 years and has a course of 7 to 36 months. It is characterized by progressive loss of the ability to sleep and to generate electroencephalographic (EEG) sleep patterns, as well as by dysautonomia, endocrine dysfunction and motor signs including ataxia, myoclonus and dysarthria. The major histological findings are severe and selective atrophy of thalamic nuclei and variable atrophy of inferior olives as well as gliosis of cerebral and cerebellar cortices. Spongiform degeneration and EEG periodic spike activity has been found only in two of the nine cases examined.

Immunochemical analysis has revealed the presence of proteinase-K-resistant prion protein (PrP^{Sc}) fragments which differed in size from those of the sporadic form of Creutzfeldt–Jakob disease (CJD).

Following amplification and sequencing, a mutation in PrP codon 178 resulting in the substitution of asparagine for aspartic acid has been found in all five affected members tested and in 12 non-affected members, three of which were 61 to 63 years old. Twenty non-affected members, eight of which were 60 to 78 years old, lacked the mutation.

Three additional kindreds likely to be affected by FFI are reviewed. The FFI disease phenotype is compared with that of three kindreds carrying the same mutation in PrP codon 178. The clinical and histopathological differences between the FFI and the other three kindreds underline the need to clarify the relationships between disease phenotype and genotype.

INTRODUCTION

> ... he gathered together the heads of families to explain to them what he knew about the sickness of insomnia ...
>
> Gabriel Garcia Marquez. *One Hundred Years of Solitude.*

Since our original report on a disease that we named fatal familial insomnia (FFI) (Lugaresi *et al.*, 1986), we have examined clinical and histopathological features and carried out immunochemical and molecular genetic analyses on affected and non-affected members of two unrelated Italian kindreds (Manetto *et al.*, 1991; Medori, *et al.*, 1991, unpublished data). The findings indicate that FFI is a distinctive disease entity which belongs to the group of the prion diseases.

CLINICAL FEATURES

The pedigree of the first FFI kindred (FFI-1) includes 288 members spanning six generations; 29 members died of a disease consistent with FFI; in 8 of these individuals, FFI was proved with histopathological examination. The second FFI kindred (FFI-2) has a total of 162 members from seven generations; 13 members were likely affected by FFI; in 1 of these, the disease was proved with histological examination. In both kindreds, the disease affected both men and women in consecutive generations, as expected of autosomal dominant traits. The mean age of onset for the nine histopathologically proved cases from both kindreds was 50 ± 8 years (range 35–61 years) and the mean duration 15 ± 10 months (range 7–36 months) (Table 1).

The major clinical features involved sleep and vigilance as well as autonomic, endocrine and motor systems. Disturbances of sleep and vigilance were characterized by progressive loss of the ability to sleep demonstrated in four patients by polysomnographic tests. Twenty-four hour recording revealed a decrease and eventually loss of electrical activities associated with sleep such as slow wave and rapid eye movement. These sleep disturbances were associated at later stages with hallucinations and dream-like behaviours. Testing revealed mild and selective impairment of attention, vigilance, visuomotor performances and memory. Stupor and coma followed at terminal stages. Major dysautonomic signs were hyperhydrosis, pyrexia, tachycardia, hypertension and irregular respiration with sympathetic over-activity revealed by autonomic tests. The abnormalities of the endocrine system included decreased ACTH and increased cortisol levels as well as abnormal circadian rhythms of growth hormone, prolactin and melatonin. Ataxia, tremor, spontaneous

Table 1.   FFI and other kindreds with PrP codon 178 mutation:
similarities and differences

	FFI[a]	Other[a]
Similarities		
Age of onset	50[a] ± 8 (35–61)	46[c] (38–53)
Duration	15[b] ± 10 (7–36)	16[c] (10–36)
Ataxia	+ +	+ +
Myoclonus	+ +	+ +
Dysarthria	+ +	+ +
EEG spike activity	±	−
Differences		
Memory deficit	+	+ + +
Insomnia	+ + +	−
Dysautonomia	+ + +	±
Endocrine disorders	+ +	?
Thalamic atrophy	+ + +	±
Spongiosis	±	+ + +
Cerebellar and olivary atrophy	+ +	−

[a]Data for FFI are based on the two Italian kindreds, FFI-1 and FFI-2; for 'other kindreds' data are based on the Hungarian–Romanian, Finnish and French kindreds reviewed by Medori *et al.* (1991).
[b]Mean ± SD (range).
[c]Approximate values.

and evoked myoclonus and dysarthria were the major motor signs. The mode of presentation of the disease varied. Sleep impairment and dysautonomia were the presenting signs in three of the five patients which received multiple clinical examinations from early stages of the disease. The other two patients presented with ataxia and dysarthria. Two of the nine patients had periodic or pseudoperiodic EEG activities while all the others had only slowing of the background activity.

HISTOPATHOLOGICAL FEATURES

Thalamus, cerebral cortex, cerebellum and inferior olives were the four anatomical regions of the brain consistently affected in the nine cases examined.

Thalamus.   The anterior ventral and mediodorsal nuclei showed severe neuronal loss with reactive astrogliosis in all the cases, whereas atrophy of other thalamic nuclei was variable. Morphometric analyses of thalamic nuclei, carried out in two cases, showed 80% to 95% neuronal loss in the anterior ventral and mediodorsal nuclei with no significant involvement of ventral lateral, pulvinar and centre median nuclei (Lugaresi *et al.*, 1986).

Cerebral cortex.   All cases had variable degrees of astrogliosis in the deep cortical layers and/or in the subcortical white matter. Spongiform degeneration was present in two cases.

Cerebellum showed various degrees of torpedo formation in the Purkinje cell axons, and mild to minimal loss of Purkinje and granular cells with reactive astrogliosis.

Inferior Olives were variably atrophic.

ANATOMOCLINICAL CORRELATIONS

No clear correlation emerged between histopathological changes and clinical parameters such as age of onset and duration, except for the spongiform degeneration of the cerebral cortex. This change occurred only in the two cases with a course of 25 and 36 months, which is much longer than the 12 month mean course of the seven cases with no spongiosis. The severity and consistency of the involvement of the anterior ventral and mediodorsal thalamic nuclei suggests that these nuclei are the main, and possibly initial, site of the disease process. This conclusion is supported by data obtained with position emission tomography (PET) scanning (Lenzi *et al.*, unpublished data).

OTHER CASES OF FFI

Following our first description, Julien *et al.* (1990) reported a French kindred with six members affected by a condition identified as FFI. Detailed clinical evaluation carried out in three patients, two of which were also examined histologically, revealed a disease phenotype very similar to the one we described. The age of onset was between 40 and 52 years with a duration of 6 to 11 months. A sleep disorder was noted in two cases; in one, lack of EEG sleep patterns and presence of endocrine dysfunctions were searched for and demonstrated. Myoclonus, ataxia, dysautonomia and selective intellectual impairment were also present. At histopathological examination, the anterior ventral and mediodorsal thalamic nuclei in both cases, and the pulvinar in one, were markedly atrophic. Segmental atrophy of the inferior olives was also seen while other thalamic nuclei, cerebral cortex and cerebellum were reported as normal. There was no periodic or pseudoperiodic EEG activities, or spongiform degeneration.

Other familial disease phenotypes similar to FFI have been reported by Little *et al.* (1986). Sleep disorders, although not polygraphically proved, ataxia, myoclonus and dysautonomia were observed in one or more of these cases. Atrophy of the anterior ventral and mediodorsal and, less consistently, of other thalamic nuclei, as well as atrophy of the inferior olives and gliosis of the cerebral cortex without spongiosis were the most common histopathological changes.

IMMUNOHISTOCHEMISTRY

The finding of spongiform degeneration in the cerebral cortex of two cases raised the suspicion that FFI might be a familial prion disease. This possibility was investigated

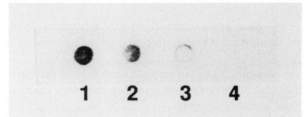

Fig. 1.   Dot blots of proteinase-K-treated brain homogenates: blot 1, Creutzfeld–Jakob disease (CJD); blot 2, fatal familial insomnia kindred 1 (FFI-1) case V-58 (Manetto *et al.*, 1991); 3, FFI-1 case IV-37 (Manetto *et al.*, 1991); blot 4, normal control. Immunoreaction was consistently stronger in case V-58 than in case IV-37. The former had a longer course (25 months) than the latter (18 months). A case from FFI-2 with an 8 month course showed no consistent immunoreaction (data not shown). The dots were immunoreacted with antiserum R073 provided by S. Prusiner

by searching for the proteinase-K-resistant isoform of the prion protein (PrP^{Sc}) in frozen brains from three cases. PrP^{Sc} was definitely detected in two of the three cases examined by dot blot analyses using antibodies to PrP generously provided by Dr S. Prusiner (Fig. 1). The presence of PrP^{Sc} provided strong evidence that FFI is a prion disease. Western blotting showed that PrP^{Sc} fragments of FFI differed in size from those of sporadic Creutzfeldt–Jakob disease (CJD). While three PrP^{Sc} fragments of approximately 29, 25 and 21 kDa were obtained from CJD tissue, those from FFI were of 29 and 27 kDa. This finding suggests that the PrP isoform of FFI is different from that of sporadic CJD (Medori *et al.*, 1991). Moreover, the case with apparently higher amount of PrP^{Sc} was the one with the longer duration suggesting that, as is the case for spongiform degeneration, deposition of PrP^{Sc} correlates with the duration of the disease.

GENE ANALYSIS

Direct sequencing, as well as sequencing following cloning, of the PCR-amplified coding region of the PrP gene showed a GAC → AAC mutation resulting in the substitution of asparagine for aspartic acid at codon 178 and in the elimination of the Tth111 I restriction site in both kindreds (Fig. 2) (Medori *et al.*, 1991; unpublished data). This mutation was found in all five affected members tested and in twelve non-affected members, three of which were between 61 and 63 years of age (Fig. 3). Twenty non-affected members, nine of which were between 60 and 78 years of age, as well as twenty unrelated Italian individuals, did not carry the mutation. The presence of non-affected members which carried the mutation and were at the age of disease onset, indicates that the penetrance of FFI is likely to be lower than 100%. Fig. 2 and 3

DISCUSSION

The finding that a disease such as FFI is a prion disease with a mutation at PrP codon 178 raises several issues.

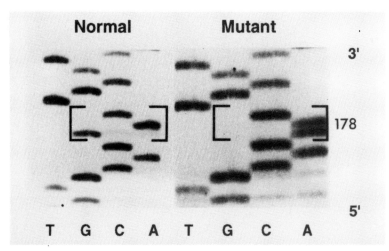

Fig. 2.   Sequencing gel of portion of the PrP open reading frame (ORF) from one of the confirmed cases of FFI. One of the alleles shows the G-to-A mutation at codon 178 resultingin the substitution of aspartic acid with asparagine. PrP ORF sequencing in two additional FFI cases, showed the same mutation.

The first is whether FFI is a distinct disease entity, different from the known prion diseases. The combination of distinctive clinical and pathological features of FFI, namely loss of ability to sleep, dysautonomia, endocrine disorders and selective atrophy of thalamic nuclei, observed in two Italian kindreds (Lugaresi et al., 1986; Manetto et al., 1991; Medori et al., 1991, unpublished data) and in a French kindred (Julien et al., 1990) have never been described before. Since loss of ability to sleep, dysautonomia and endocrine disorder are difficult to demonstrate unless they are searched for, it is likely that the other familial cases of thalamic atrophy reported by Little et al (1986) are indeed cases of FFI. Moreover, since at least five sporadic cases with a disease phenotype similar to FFI are on record, it is likely that FFI may occur also in sporadic form (see Manetto et al, 1991, for review).

The identification of four kindreds with proved or probable FFI (Lugaresi et al., unpublished data; Julien et al., 1990; Little et al., 1986) since the original report of 1986 (Lugaresi et al., 1986), along with the difficulty of identifying FFI even after histopathological examination unless special tests are carried out, suggests that this condition is not exceedingly rare. Its prevalence might actually exceed that of Gertsmann–Sträussler–Scheinker syndrome (GSS) of which 13 kindreds have been reported since the first complete description in 1936 (Farlow et al., 1991; Schiffer 1990).

The finding that FFI is a not exceedingly rare prion disease, as well as the recent discovery of several disorders with heterogeneous clinicopathological phenotypes associated with mutations in the PrP gene (Baker et al., 1991; Brown et al., 1991; Owen et al., 1990), widens the spectrum of the prion diseases. Moreover, FFI and the other new conditions challenge the identity of well-established prion diseases such as CJD and GSS, and underlines the need for a reclassification of the prion diseases.

The GAC → AAC mutation at codon 178 of the PrP gene present in two FFI kindreds has been previously reported in six kindreds of various ethnic origins (see Chapter 15). Available clinical and histopathological information on a total of 12

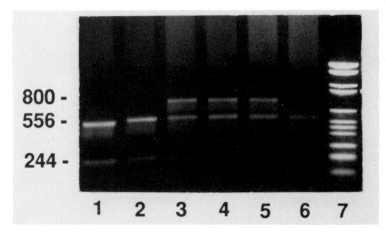

Fig. 3.   Fragments of PrP coding regions generated by Tthlll I digestion: lanes 1–4, members of FFI-2; lane 5, members of FFI-l; lane 6, unrelated normal control; lane 7, DNA size markers, Bgl I and Hinf I digested pBR322. Members carrying the mutation (lane 3–5) which abolishes the Tthlll I restriction site in one allele show three fragments of 800, 556 and 244 bp. Individuals lacking the mutation (lanes 1, 2, 6) display only two fragments since both alleles are digested.

members of these three kindreds shows that they share a similar disease phenotype (Medori *et al.*, 1991). This phenotype, however, has remarkable clinical and his-topathological differences from that of FFI (Table 1). The differences in disease phenotype, despite the common mutation in PrP codon 178, raise the important question of the relationship between disease genotype and phenotype. Studies directed at clarifying this issue are in progress in several laboratories.   In human prion diseases, it has been shown that homozygosity at codon 129 increases the risk of acquiring the sporadic and iatrogenic forms of CJD and that it also influences the age of onset and/or duration of the genetic forms of prion diseases with an 144 bp insertion in the PrP gene (Baker *et al.*, 1991; Collinge *et al.*, 1991; Palmer *et al.*, 1991). Experimental evidence indicates that *Sinc* and *Sip*, the scrapie incubation time genes in mouse and sheep, may also modulate the pathology of the disease (see Chapters 24 and 41). Therefore, genetic polymorphism of PrP and related genes, as well as other genetic factors (Wexler *et al.*, 1991), might explain the phenotypic differences between FFI and the other conditions that share the mutation at PrP codon 178.

ACKNOWLEDGEMENT

The authors are very grateful to Stanley Prusiner and Dan Serban for the generous supply of antibodies and invaluable advice. We thank Hsiao Ying Chen, Run Xue, Barbara Schaezle and Jennifer Secki for technical help and Sandra Bowen for secretarial expertise. We are deeply indebted to the members of the kindreds studied and to Ignazio Roiter for their thoughtful and generous cooperation.

    Division of Neuropathology, Institute of Pathology, Case Western Reserve University, Cleveland, Ohio, USA
Neurological Institute, University of Bologna Medical School, Bologna Italy
Address correspondence: Pierluigi Gambetti, M.D., Division of Neuropathology,

Institute of Pathology, Case Western Reserve University, 2085 Adelbert Road, Cleveland, Ohio 44106.
Telephone: 216-844-1808; Fax: 216-844-1810
NINCDS NS 14509-13, NIA ADRC AG-08012-03, NIH NIA 1 R01 AGNS08155-02, AG08992, and the Britton Fund.

REFERENCES

Baker, H.F., Poulter, M., Crow, T.J., *et al.* (1991) Amino acid polymorphism in human prion protein and age at death in inherited prion disease. *Lancet* **337** 1286.

Brown, P., Goldfarb, L.G., Gajdusek, D.C. (1991) The new biology of spongiform encephalopathy: infectious amyloidoses with a genetic twist. *Lancet* **337** 1019–1022.

Collinge, J., Palmer, M.S., Dryden, A.J. (1991) Genetic predisposition to iatrogenic Creutzfeldt–Jakob disease. *Lancet* **337** 1441.

Farlow, M.R., Tagliavini, F., Bugiani, O., Ghetti, B. (1991) Gerstmann–Sträussler–Scheinker disease. In *Handbook of Cinical Neurology*, Vol 16(60) *Hereditary Neuropathies and Spino-Cerebellar Atrophies*. J.M.B.V. De Jong (ed.), Elsevier Science Publisher.

Garcin, R., Brion, S., Khochneviss, A.A. (1963) Le syndrome de Creuzfeldt–Jakob et les syndromes cortico-striés du presenium (a l'occasion de 5 observations anatomo-cliniques). *Rev. Neurol. (Paris)* **109** 419–441.

Julien, J., Vital, C., Delepanque, B. *et al.* (1990) Atrophie thalamique subaigue familiale. Troubles mnésques et insomnie totale. *Rev. Neurol. (Paris)* *146*(3) 173–178.

Little, B.W., Brown, P.W., Rodgers-Johnson, P., *et al.* (1986) Familial myoclonic dementia masquerading as Creutzfeldt–Jakob disease. *Ann. Neurol.* **20** 231–239.

Lugaresi, E., Medori, R., Montagna, P., *et al.* (1986) Fatal familial insomnia and dysautonomia with selective degeneration of thalamic nuclei. *New Engl. J. Med.* **315** 997–1003.

Manetto, V., Medori, R., Cortelli, P., *et al.* (1992) Fatal familial insomnia clinical and pathological study of five cases. *Neurology* **42** 312–319.

Martin, J.J. (1975) Thalamic degeneration. In: Vinken, P.J, Bruyn, G.W. (eds). *Handbook of Clinical Neurology*, Vol. 21. North-Holland, Amsterdam, pp. 587–604.

Medori, R., Tritschler, H.J., LeBlanc, A.C., *et al.* (1992) Fatal familial insomnia is a prion disease with a mutation at codon 178 of the prion gene. *New Engl. J. Med.* **326** 444–449.

Owen, F., Poulter, M., Shah, T., *et al.* (1990) An in-frame insertion in the prion protein gene in familial Creutzfeldt–Jakob disease. *Mol. Brain Res.* **7** 273–276.

Palmer, M.S., Dryden, A.J., Hughes, T., Collinge, J. (1991) Homozygous prion protein genotype predisposes to sporadic Creutzfeldt–Jakob disease. *Nature (London)* **352** 340–342.

Prusiner, S.B. (1991) Molecular biology of prion diseases. *Science* **252** 1515–1522.

Schiffer, D. (1990) Gertsmann–Sträussler syndrome in a familial case with amyotrophy. *Proc. 3rd European Meeting of Neuropathology*, 1988, p. 207.

Wexler, N.S., Rose, E.A., and Housman, D.E. (1991) Molecular approaches to hereditary diseases of the nervous system: Huntington's disease as a paradigm. *Annu. Rev. Neurosci.* **14** 503–529.

19

Creutzfeldt–Jakob disease epidemiology

R.G. Will, T.F.G. Esmonde and W.B. Matthews

ABSTRACT

The body of epidemiological evidence precludes case-to-case transmission as a major pathogenic mechanism in Creutzfeldt–Jakob disease (CJD) and no alternative mechanism of natural transmission of the unconventional causative agent has been identified.

A close correlation between disease and mutations of the prion protein gene has been established in some pedigrees of CJD and Gerstmann–Sträussler syndrome (GSS) but a genetic influence on susceptibility remains to be established in sporadic cases of CJD. Public concern regarding the possibility of transmission of bovine spongiform encephalopathy (BSE) to the human population has prompted the reinstitution of epidemiological surveillance of CJD in the United Kingdom. Preliminary results, including an analysis of the incidence and occupational distribution of CJD, indicate no change in the pattern of CJD up to the present.

INTRODUCTION

The epidemiology of Creutzfeldt–Jakob disease (CJD) has been under investigation in England and Wales since the 1970s. The incidence and geographical distribution of cases have been established for the years 1970 to 1984 (Cousens *et al.*, 1990) and a case-control study between 1980 and 1984 (Harries-Jones *et al.*, 1988) provided no significant evidence of an environmental influence on causation. The occurrence of bovine spongiform encephalopathy (BSE) has led to concern about the possibility of the spread of this novel condition to man and has prompted the reinstitution of epidemiological surveillance of CJD in the United Kingdom. The primary aim of this study is to identify any change in the epidemiological, clinical or pathological characteristics of CJD that might be attributable to BSE. The systematic accumulation of information on all identified cases on a national basis may also provide evidence for the relative importance of genetic and environmental influences on the aetiology of CJD.

CLINICAL FEATURES
The clinical presentation of CJD is relatively stereotyped in the great majority of cases, with subacute cognitive decline evolving rapidly to a state of akinetic mutism over a period of weeks or months. There is myoclonus in over 80% of cases and a characteristic electroencephalogram (EEG) recording in the majority (Will and Matthews, 1984). This combination of clinical and investigative features is readily identifiable, and previous investigation suggests that a high percentage of cases are accurately diagnosed, including those individuals who are initially referred to psychiatric institutions (Will and Matthews, 1984). The presumption that there is a high level of case ascertainment is crucial to the epidemiological surveillance of CJD. The close correlation of findings in systematic surveys of CJD, despite differing methodologies, suggests that typical cases are consistently identified. However, atypical presentations of CJD are well documented (Brown et al., 1984) and it has been proposed that many such cases are missed in epidemiological surveys. (Editorial, 1990).

In the UK studies, cases of CJD are identified from death certificates or by direct notification by clinicians. As a safety net, all neuropathologists are asked to notify cases confirmed at post-mortem, whether or not the diagnosis was clinically suspect in life. By examining the clinical features of all pathologically confirmed cases of CJD, cases with an atypical clinical presentation can be identified. In the 1970s in England and Wales less than 10% of pathologically certified cases of CJD had an atypical clinical presentation (Will and Matthews, 1984), with duration of illness greater than one year (Table 1).

Table 1. Creutzfeldt–Jakob disease: England and Wales 1970–1979

121 pathologically confirmed cases	
12 clinically atypical	
Dementia and myoclonus terminally	8 cases
Characteristic EEG	2 cases
No suggestive clinical features	4 cases

In these 12 cases the clinical diagnosis of CJD was made in one patient and is likely to have been suspected in other cases, including the two patients with a characteristic EEG. In four cases there were no suggestive clinical features and it is unlikely that the diagnosis of CJD would have been considered.

A retrospective survey of CJD in the UK between 1985 and April 1990 has now been completed and in this study it was possible to determine from case notes whether or not the diagnosis of CJD was suspected in life. Fifty-seven pathologically confirmed cases of CJD were identified and of these adequate clinical information has been obtained on 53 cases (Table 2).

Table 2. Creutzfeldt–Jakob disease in the UK: 1985–April 1990

	Diagnosis suspected	Diagnosis unsuspected
Clinically typical	47	1
Clinically atypical	2	3
Total	49	4

The diagnosis of CJD was unsuspected in one clinically typical case, admitted under the care of a general physician and not seen by a neurologist. This reinforces the validity of the assumption that only rarely are typical cases not identified by current surveillance procedures. The diagnosis of CJD was not suspected in three clinically atypical cases, defined as cases with a duration of illness of greater than 1 year, or cases in which there were no suggestive features such as myoclonus, akinetic mutism or the characteristic EEG. One of these patients was the first human growth hormone recipient in the UK who died in 1985 prior to the recognition of an association between human pituitary growth hormone therapy and CJD. One patient died suddenly early in the course of a subacute dementing illness and the third patient exhibited clinical features indistinguishable from Alzheimer's disease. It is of note that in two cases of relatively long duration the diagnosis of CJD was suspected pre-mortem.

This evidence suggests that clinically unrecognized CJD is rare and that a clinical presentation mimicking Alzheimer's disease is exceptional. Cases of dementia which come to post-mortem are, of course, selected with a bias towards younger age groups, and the possibility remains that CJD is underdiagnosed in the elderly, although the available evidence does not suggest that this occurs frequently. (Brown and Cathala, 1979). The clinical features of CJD in some pedigrees may be very variable, with some individuals exhibiting the expected clinical course while in others the clinical course may be so atypical as to make even the suspicion of CJD unlikely. (Collinge *et al.*,1990). The analysis of clinical features in the consecutive series of confirmed cases described above indicates that the evidence from highly selected and unusual familial aggregates of CJD may not be applicable to sporadic cases.

Inevitably the systematic identification of all suspect cases of CJD requires that some cases of CJD are classified by clinical criteria alone, with no subsequent pathological confirmation. The possibility of misdiagnosis is an important consideration in this group. In the study of CJD in England and Wales 1980–1984 there was only one case fulfilling all the criteria for 'probable' CJD which came to post-mortem and in which the diagnosis was not confirmed pathologically. In the study in France there were four such cases over a 15 year period (Cathala and Baron,1987), underlining the importance of the strict application of clinical criteria to the classification of such cases. The EEG is an important criterion in the categorization of clinical cases and, although generalized periodic complexes mimicking CJD may occur in other conditions, the clinical context usually allows the correct diagnosis. However, the EEG may evolve in parallel to the clinical course in CJD, with suggestive tracings becoming characteristic on subsequent recordings. Such suggestive, but not characteristic, tracings may occur in conditions other than CJD, for example in a recent confirmed case of Lewy body dementia. Formal criteria for the classification of the EEG in CJD would be a useful adjunct to epidemiological surveillance and would allow more accurate comparison between studies.

EPIDEMIOLOGY

The incidence of CJD in systematic surveys is 0.5–1 case per million per year (Brown *et al.*,1987). Contact between sporadic cases is exceptional and there is no convincing evidence of significant geographical aggregation of cases. This largely excludes the possibility of case-to-case transmission of CJD, although some form of cross-contamination, perhaps by minor medical or dental procedure early in life, is difficult to exclude by current epidemiological methods. Although iatrogenic transmission of CJD is well recognized (Buchanan *et al.*, 1991), this cannot be the explanation for CJD in the great majority of cases. The occurrence of familial cases of CJD and the absence of evidence for any mechanism of natural transmission led to the proposition that CJD might be a genetic disorder. (Baker *et al.*,1985). The identification of six separate mutations of the prion protein (PrP) gene in pedigrees of familial CJD or Gerstmann–Sträussler syndrome (Brown *et al.*, 1991) has provided convincing evidence of a correlation between genetic anomalies and disease. The paradoxically high incidence of CJD in certain areas of Slovakia and in Libyan-born Israelis has now been linked to a high frequency of the codon 200 mutation of the PrP gene in these populations. (Goldfarb *et al.*, 1990; Hsiao *et al.*, 1991). Although genetic anomalies have been identified in only a small number of sporadic cases, it has been proposed, on the basis of this accumulating evidence, that the human spongiform encephalopathies should be reclassified according to molecular biological criteria. (Alzheimer's Disease Research Group,1991). An important issue in this context is the frequency of PrP mutations in CJD and how this relates to the proportion of cases that are sporadic or familial.

   In the study of CJD between 1980 and 1984 only 1 out of 122 definite or probable cases was familial and between 1985 and April 1990 there was 1 familial case out of 114 definite or probable cases. The implication of this evidence is that the mutations of the PrP gene so far identified are unlikely to be present in a high proportion of cases, because so few are familial. There are, however, difficulties in the classification of cases of CJD as familial or sporadic because of the problems in the clinical categorization of family members dying of dementia or senility. Only rarely is there sufficient information to be certain of the diagnosis of CJD in individuals who may have died many years previously, but inaccuracies will be inevitable if any history of dementia in a family member is regarded as being due to CJD. Other cases of dementia such as Alzheimer's disease or cerebrovascular disease are common. On the other hand, the application of rigid criteria for the diagnosis of CJD in the relatives of patients is likely to result in missed cases of familial CJD. This dilemma is difficult to resolve but the systematic examination of the PrP gene in a consecutive series of patients with CJD may provide some indication of the frequency of familial cases, as mutations of the PrP gene are known to occur in the majority of familial cases so far studied, but in only a small proportion of sporadic cases.

Blood has been taken for molecular analysis in all suspected cases of CJD in the UK from May 1990 onwards. Analysis by Dr J. Collinge has revealed three mutations in the PrP gene in the first 61 suspect cases, of which approximately half have so far been classified as definite or probable CJD. The relative infrequency of PrP mutations in this consecutive series of cases suggests that large numbers of familial cases are not being misclassified as sporadic cases, and either that other genetic anomalies are yet to be identified or that genetic factors are not crucial to the development of CJD in sporadic cases. Detailed investigation of kuru has provided little evidence for a genetic influence on susceptibility. Furthermore in the UK study we have identified one identical twin pair discordant for CJD and three dizygotic discordant twin pairs.

If sporadic cases of CJD are not due to germline mutations an alternative is that disease may develop following a somatic mutation in the PrP gene, leading to progressive accumulation of the modified form of PrP (Hsiao and Prusiner, 1990). Epidemiological evidence cannot verify or exclude this possibility. However, rarely CJD can present clinically as a unilateral disorder. In one case in England, the clinical presentation was truly unilateral with progressive hemiparesis and unilateral periodic complexes in the EEG up to the time of death. If CJD is caused by a somatic mutation in the PrP gene the tissue distribution of PrP may provide clues as to the kinetics of pathogenesis. In laboratory models of the spongiform encephalopathies replication of the agent occurs in the reticuloendothelial system before neuro-invasion occurs and this is usually symmetrical. (Kimberlin, 1979). In contrast, direct intraocular inoculation of the agent in mice results in initial unilateral pathological changes in the contralateral cerebral hemisphere. (Jeffrey, 1991).

An alternative explanation for the rarity and isolation of sporadic cases of CJD is that the infectious agent is widely distributed, resulting in disease only rarely, perhaps in relation to host genetic factors. Preliminary evidence of a genetic influence on susceptibility in sporadic cases of CJD has recently been described (Palmer, 1991), with complementary findings in UK growth hormone recipients. (Collinge, 1991). A genetic influence on susceptibility is well recognised in scrapie(Kimberlin,1990) and polymorphisms have been identified in the bovine PrP gene (Goldmann, 1991), but in both scrapie and BSE the epidemiological evidence indicates an environmental source of infection. The incidence of scrapie, with an estimated 30% of UK flocks affected (Morgan, 1990), and BSE, with 3.9/1000 cattle affected, contrasts with the rarity of CJD, with the implication that CJD is unlikely to be due to a rare environmental source of infection. Although there must be some doubt about the efficiency of case identification, feline spongiform encephalopathy has, however, proved to be a rare disorder with an annual incidence of approximately 1.7 cases per million per year. The likelihood is that FSE was caused by contamination of feed with the BSE agent, thus indicating that environmental contamination with an unconventional agent can nonetheless result in a rare disease.

The potential for the agent of BSE to cross species barriers after dietary exposure has been one of the major sources of concern in relation to human health. The possibility of a change in agent characteristics following species to species transmission (Kimberlin, 1990) indicates that the BSE agent may have different properties from the scrapie agent, and there is a remote possibility that the BSE agent might be pathogenic to man. Experimental evidence indicates the occurrence of differing strains of scrapie agent (Bruce *et al.*, 1991) which influence pathogenesis independently of the host genome. The implication is that, even should CJD be primarily a genetic disorder, the BSE agent, having crossed the species barrier, could cause disease in man. In the unlikely event of this occurring, occupational exposure to high risk tissue such as brain or spinal cord is potentially an area of risk, because of the possibility of inoculation of material through cuts or abrasions. Analysis of the occupational distribution of cases and of temporal changes in the incidence of CJD are important parameters in assessing any change potentially related to BSE.

OCCUPATION AND CJD

The occupation in patients dying of CJD has been examined in previous studies in the UK. In the retrospective survey 1970–1979 in England and Wales, occupation at death was ascertained largely from death certificates and showed no excess of cases in specific occupational groups (Table 3).

The potentially extended incubation period of CJD, exemplifed by iatrogenic cases, suggests that past occupation may be more relevant than occupation at death. A significant proportion of individuals with CJD are classified on death certificates as 'retired' and female patients are often classified according to the spouse's occupation.

Between 1980 and 1984 in England and Wales an occupational history was obtained in 122 patients. Occupations which might hypothetically be associated with an increased risk of developing CJD are listed in Table 4.

This classification includes individuals who were employed at any time in specific occupational groups, even if this employment was short lived or transient. One patient had been employed as a dentist and the close proximity of his surgery to other cases has previously been described. (Will and Matthews, 1982). The doctor worked as a general practitioner and had no known contact with any case of CJD. Five individuals had worked in the nursing profession often early in life and for short periods and nine patients had worked in the farming industry, many years before the advent of BSE. For comparison, in Table 5 are listed those individuals who worked in the mining industry or in munitions production; professions with no theoretical causative link to CJD. Occupations listed as 'Other' demonstrate the extraordinary variation in types of employment in patients with CJD.

Table 3.   Occupation of CJD patients by patient and spouse: 1970–1979

	Patient	Spouse		Patient	Spouse
Foodhandlers			Storeman	2	0
Grocer	1	1	Civil servant	2	0
Kitchen hand	3	0	Chartered accountant	2	0
Housekeeper	1	0	Building trade	1	3
Baker	0	1	Electrician	1	1
Butcher	0	1	Gardener	1	1
			Private investigator	1	0
Medical			Laboratory technician		
Nurse	1	0	(Ford Motors)	1	0
Hospital domestic	2	0	Shopkeeper	1	1
Optician	1	0	Mariner	1	0
Dental technician	1	0	Security administrator	1	0
Doctor	0	1	Woollen weaver	1	0
			Messenger	1	0
Other			Bricklayer	1	0
Manager–			Teacher	1	0
company director	9	4	Draper	1	0
Housewife	8	0	Police–fire officer	0	2
Engineer	8	6	Administrative officer	0	2
Factory worker	5	0	Decorator	0	2
Metal worker	4	0	Butler–chauffeur	0	2
Representative	4	3	Carpenter	0	1
Secretary	4	1	Hall porter	0	1
Miner	3	4	Refuse collector	0	1
Clerk	3	3	Clothing trade	0	1
Domestic	3	0	Armed forces	0	1
Driver	2	2	Forestry	0	1
Labourer	2	6	Window cleaner	0	1
Lecturer	2	2			
Total	86	56			

An apparent excess of individuals with CJD employed in medically related professions was described by Masters *et al.*, (1979), in a review of the worldwide incidence of CJD. This was not a systematic survey, and the possibility of bias in diagnosis and notification of cases was a possible confounding factor. In the study of CJD in France over a 15 year period, no significant association between occupation at diagnosis and CJD was discovered. (Cathala and Baron, 1987). In the UK, between 1980 and 1984, the initial impression is that there may be an excess of individuals employed in the medical or farming professions or as butchers. However, it is important to stress that the tabulation includes individuals who were employed at any time in these occupations. Similar numbers of individuals had been employed in the professions listed in Table 5, in which any link to CJD is difficult to envisage. This evidence does not suggest a correlation between CJD and occupation and initial results from the case-control study of CJD from May 1990 onwards show no difference in occupational history between cases and controls (Table 6). Detailed enquiry in individual cases failed to establish any contact with cases of CJD in those employed in medically related professions. The nursing auxillary who had worked in the operating theatre was employed in a district hospital in which neurosurgical

Table 4.　Creutzfeldt–Jakob disease 1980–1984: Occupational history

Medical	Years employed	Butcher	Years employed
Dentist	1925–1977	Butcher	1930–1933
Doctor	1957–1983	(also milkman)	
Nurse	1950–some years	Butcher	1936
Nurse	1946–1954	Butcher	1921–1979
Nurse	1942–a few years	Butcher's	
Nurse	1950–1951	apprentice	
Nurse	1952–a few years		
Farming	*Years employed*	*Foodhandlers*	*Years employed*
Farmworker	1958	Chef	1945–1978
Farm labourer	1944–1947	Cafe manageress	1959–1963
(also butcher's shop)			
Farmer's wife	1940–1948	Cook	1949–1981
Farm labourer	1937–1952	Cafe proprietress	
Farmer's wife		Cook	
Farmer	1923–1973	Kitchen assistant	
Farmer's wife			
Farmer		*Other*	*Years employed*
Farmer	1946–1960	Skin cleaner	1929–1936
		(leather)	

procedures were never carried out. The farmhand had no possible contact with animals affected by BSE.

The analysis of occupation at diagnosis or occupation throughout life in large numbers of individuals with CJD can exclude any major correlation between type of employment and disease. However, the possibility that CJD may rarely be caused by occupational contact with the agent, perhaps by accidental inoculation, cannot be excluded. One neurosurgeon has died of CJD (Schoene *et al.*, 1981) and two laboratory workers who may possibly have processed CJD tissue have been identified. (Miller, 1988). It is impossible to know whether these individuals developed CJD because of occupational contact with the infectious agent, but it is inevitable that extended surveillance will result in the identification of individuals with CJD from a wide range of occupational groups even if there is no causative link.

INCIDENCE

Data on the incidence of CJD in the UK is available since 1970 in England and Wales and from 1985 for the whole United Kingdom. Table 7 shows the incidence

Table 5.   Creutzfeldt–Jakob disease 1980–1984: occupational history

Miner–quarryman	Years employed	Ammunition worker	Years employed
Collier	1947–1977	Royal Ordnance	1939–1941
Miner	1935–1936	Munitions factory	1939–1942
Quarryman	1936–1941	Ammunition worker	1940–1941
Coal miner		Munitions	
Underground		Ammunition maker	
electrician			
Miner			
Colliery railway worker			
Other			
Tree felling	Vegetable stall	Flower arranger	
Lithograph engineer	Yacht club stewardess	Pylon maintenance	
Labelling tinned fruit	War Office Intelligence	Vagrant	

of CJD in various epochs and varies mainly in relation to the methodology of case ascertainment.

Although the incidence of CJD is higher in the prospective than in the retrospective surveys, overall there has been no significant change in the incidence with time.

The geographical distribution of cases in the UK from 1985 onwards parallels population density, with no obvious spatiotemporal aggregation of cases. In particular there is no evidence of an excess of cases in rural districts or a bias towards the South of England which might suggest a differential regional incidence in relation to BSE. Preliminary analysis of epidemiological data and patient characteristics does not therefore indicate any change in the pattern of CJD in comparison with previous studies. However, epidemiological surveillance will have to continue for many years before any change attributable to BSE can be excluded.

CONCLUSION

The remarkable developments in the study of the molecular biology of human spongiform encephalopathies have led to a resolution of some of the epidemiological paradoxes of CJD. However, the cause of CJD in the great majority of cases remains unknown and in the United Kingdom the possibility of transmission of BSE to the human population has led to the establishment of long-term epidemiological surveillance of CJD. It is to be hoped that developments in basic science and in particular molecular biology may in the future obviate the need for continued surveillance, perhaps by the discovery of an *in vivo* test for the presence of an unconventional agent.

Table 6. Creutzfeldt–Jakob disease May 1990–August 1991: occupational history

Cases	Controls
Nurse	Nurse
Auxillary nurse	Nursing auxillary
(also sausage factory worker)	
Nursing auxillary (theatres)	
Hospital domestic	
Doctor's receptionist	
Farmhand	Farm labourer
Pork butchers (1970–1975)	Butcher's assistant
Milk delivery	Farm worker
	Dairymaid
Wool warehouseman	Wool warehouseman
School dinner lady	School dinner lady
	Canteen worker

Table 7. Incidence of CJD

	Type of study	Incidence[a]
1970–1979 England and Wales	Retrospective	0.32
1980–1984 England and Wales	Prospective	0.49
1985–1990 UK	Retrospective	0.39
1990–1991 UK	Prospective	0.51

[a] Cases per million per year.

ACKNOWLEDGEMENTS

The national CJD Surveillance Programme is funded by the Department of Health. The authors wish to thank Miss J. Mackenzie for her invaluable help in preparing the manuscript and administering the study.

REFERENCES

Alzheimer's Disease Research Group (1991) Molecular classification of Alzheimer's disease. *Lancet* **337**, 1342–1434.

Baker, H.F., Ridley, R.M. and Crow, T.J. (1985) Experimental transmission of autosomal dominant spongiform encephalopathy: does infectious agent originate in human genome? *Br. Med. J.* **291**, 299–302.

Brown, P. and Cathala, F. (1979) Creutzfeldt–Jakob disease in France. In: *Slow Transmissible Diseases of the Nervous System* Vol. 1, Academic Press, New York, pp 213–227.

Brown, P., Rodgers–Johnson, P., Cathala, F., Gibbs, C.J., Jr., and Gajdusek, D.C. (1984) Creutzfeldt–Jakob disease of long duration: clinicopathological characteristics, transmissibility, and differential diagnosis. *Ann. Neurol.* **16**, 295–304.

Brown, P., Cathala, F., Raubertas, R.F., Gajdusek, D.C. and Cataigne, P. (1987) The epidemiology of Creutzfeldt–Jakob disease: conclusion of a 15-year investigation in France and review of the world literature. *Neurology* **37**, 895–904.

Brown, P., Goldfarb, L.G. and Gajdusek, D.C. (1991) The new biology of spongiform encephalopathy: infectious amyloidoses with a genetic twist. *Lancet* **337**, 1019–1022.

Bruce, M.E., McConnell, I., Fraser, H. and Dickinson, A.G. (1991) The disease characteristics of different strains of scrapie in sinc congenic mouse lines: implications for the nature of the agent and host control of pathogenesis. *J. Gen. Virol.* **72**, 595–604.

Buchanan, C.R., Preece, M.A. and Milner, R.D.G. (1991) Morality, neoplasia, and Creutzfeldt–Jakob disease in patients treated with human pituitary growth hormone in the United Kingdom. *B. Med. J.* **302**, 824–828.

Cathala, F. and Baron, H. (1987) Clinical aspects of Creutzfeldt–Jakob disease. In: Prions: *Novel Infectious Pathogens causing Scrapie and Creutzfeldt–Jakob Disease.* (eds. Prusiner, S.B. and McKinley, M.P.), Academic Press, London, pp. 467–509.

Collinge, J., Owen, F., Poulter, M., Leach, M., Crow, T.J., Rossor, M.N., Hardy, J., Mullan, M.J., Janota, I. and Lantos, P.L. (1990) Prion dementia without characteristic pathology. *Lancet* **336**, 7–9.

Collinge, J., Palmer, M.S. and Dryden, A.J. (1991) Genetic predisposition to iatrogenic Creutzfeldt–Jakob disease. *Lancet* **337**, 1441–1442.

Cousens, S.N., Harries-Jones, R., Knight, R., Will, R.G., Smith, P.G. and Matthews, W.B. (1990) Geographical distribution of cases of Creutzfeldt–Jakob disease in England and Wales 1970–1984. *J. Neurol. Neurosurg. Psych.* **53**, 459–465.

Editorial (1990) Prion disease: spongiform encephalopathies unveiled. *Lancet* **336**, 21–22.

Goldfarb, L.G., Mitrova, E., Brown, P., Toh, B.H. and Gajdusek, D.C. (1990) Mutation in codon 200 of scrapie amyloid protein gene in 2 clusters of Creutzfeldt–Jakob disease in Slovakia. *Lancet* **000** 514–515.

Goldmann, W., Hunter, N., Martin, T., Dawson, M. and Hope, J. (1991) Different forms of the bovine PrP gene have 5 or 6 copies of a short G–C-rich element within the protein coding exon. *J. Gen. Virol.* **72**, 201–204.

Harries-Jones, R., Knight, R., Will, R.G., Cousens, S., Smith, P.G. and Matthews.

W.B. (1988) Creutzfeldt–Jakob disease in England and Wales 1980–1984: a case-control study of potential risk factors. *J. Neurol. Neurosurg. Psych.* **51**, 1113–1119.

Hsiao, K. and Prusiner, S.B. (1990) Inherited human prion diseases. *Neurology* **40**, 1820–1827.

Hsiao, K., Meiner, Z., Kahana, E., Cass, C., Kahana, I., Avrahami, D., Scarlatto, G., Abramsky, O., Prusiner, S.B. and Gabizon, R. (1991) Mutation of the prion protein in Libyan Jews with Creutzfeldt–Jakob disease. *New Engl. J. Med.* **324**, 1091–1097.

Jeffrey, M., Scott, J.R. and Fraser, H. (1991) Scrapie inoculation of mice: light and electron microscopy of the superior colliculi. *Acta Neuropathol.* **81**, 562–571.

Kimberlin, R.H. (1979) Early events in the pathogenesis of scrapie in mice: biological and biochemical studies. In: Prusiner, S.B. and Hadlow, W.J. (eds.) *Slow Transmissible Diseases of the Nervous System.* Academic Press, New York, pp. 33–54.

Kimberlin, R.H. (1990) Transmissible encephalopathies in animals. *Can. J. Vet. Res.* **54**, 30–37.

Kimberlin, R.H., Cole, S. and Walker, C.A. (1990) Temporary and permanent modifications to a single strain of mouse scrapie on transmission to rats and hamsters. *J. Gen. Virol.* **68**(7), 1875–1881.

Masters, C.L., Harris, J.O., Gajdusek, D.C., Gibbs, C.J., Jr., Bernoulli, C. and Asher, D.M. (1979) Creutzfeldt–Jakob disease: patterns of worldwide occurrence and the significance of familial and sporadic clustering. *Ann Neurol* **5**, 177–188.

Miller, D.C. (1988) Creutzfeldt–Jakob disease in histopathology technicians. *New Engl. J. Med.* **318**, 853–854.

Morgan, K.L., Nicholas, K., Glover, M.J. and Hall, A.P. (1990) A questionnaire survey of the prevalence of scrapie in sheep in Britain. *Vet. Rec.* **127**, 373–376.

Palmer, M.S., Dryden, A.J., Hughes, J.T. and Collinge, J. (1991) Homozygous prion protein genotype predisposes to sporadic Creutzfeldt–Jakob disease. *Nature* **352**, 340–341.

Schoene, W.C., Masters, C.L., Gibbs, C.J., Jr, Gajdusek, D.C., Tyler, H.R., Moore, F.D. and Dammin, G.J. (1981) Transmissible spongiform encephalopath (Creutzfeldt–Jakob disease) Atypical clinical and pathological findings. *Arch. Neurol.* **38**(8), 473–477.

Will, R.G. and Matthews, W.B. (1982) Evidence for case-to-case transmission of Creutzfeldt–Jakob disease. *J. Neurol. Neurosurg. Psych.* **45**, 235–238.

Will, R.G. and Matthews, W.B. (1984) A retrospective study of Creutzfeldt–Jakob disease in England and Wales 1970–1979 I: clinical features. *J. Neurol. Neurosurg. Psych.* **47**, 134–140.

20

Familial dementia in relation to the 144 bp insert and its implications

T. J. Crow, M. Poulter, H. F. Baker, C. D. Frith, M. Leach, R. Lofthouse,
R. M. Ridley, T. Shah, F. Owen, J. Collinge, G. Brown, J. Hardy, M. J. Mullan,
A. E. Harding, C. Bennett and R. Doshi

ABSTRACT

A 144 bp insert was identified in the prion protein gene in the region encoding the octapeptide repeats in four families with early onset autosomal dominant dementia. These families derive from four children of a couple born in the late 18th century in South East England. Duration of illness (between 1 and 19 years) and the presence of characteristic brain changes are variable; progressive dementia is sometimes preceded by longstanding personality aberrations. Within the pedigree the presence of the insert is closely linked to disease (lod score 11.02 at zero recombination); age at death is significantly lower when the normal prion protein gene codes for a methionine (as does the insert-carrying allele) rather than a valine at the polymorphic codon 129. The interaction between the products of the normal and abnormal prion protein alleles appears to be critical for disease progression.

INTRODUCTION

Creutzfeldt–Jakob disease (CJD) and Gerstmann–Sträussler–Scheinker syndrome (GSS) are fatal neurodegenerative diseases that have in common spongiform encephalopathy, deposition of extracellular amyloid plaques containing prion protein (PrP) in the brain and experimental transmissibility (Masters *et al.*, 1981; Adam *et al.*, 1982).

CJD is defined as a rapidly progressive dementia (duration less than 2 years) with variable additional features such as myoclonus, pyramidal tract signs and cerebellar ataxia. GSS refers to a more slowly progressive cerebellar ataxia (with a duration of up to 10 years or more) with a varying degree of dementia (Masters *et al.*, 1981; Will and Matthews, 1984).

Approximately 15% of cases of CJD and almost all those of GSS occur in a familial form with autosomal dominant inheritance (Masters *et al.*, 1979, 1981) provoking the question of how a disease that has been shown to be transmissible to animals can also be inherited. Either what is inherited is a strong predisposition to an environmental agent, or the agent itself must be assumed to arise from the host genome (Baker *et al.*, 1985).

A major advance in understanding these diseases occurred with the identification of the prion protein gene (Cheseboro *et al.*, 1985; Oesch *et al.*, 1985). This highly conserved gene, present as a single copy on chromosome 20 in man (Sparkes *et al.*, 1986), codes for a ~ 253 amino acid glycosylated cell membrane protein with an as-yet unknown function. The normal cellular isoform is degraded by proteinase K treatment *in vitro*, whereas in prion diseases a proteinase K resistant form accumulates in affected tissue (for review see Prusiner, 1987). This pathological isoform, referred to as PrP^{CJD} or PrP^{GSS} in the case of the human prion diseases, copurifies with infectivity indicating that it is likely to be an integral part of the transmissible agent. Numerous experiments have failed to identify an infectivity-related nucleic acid suggesting that PrP^{Sc} (the scrapie-related protein) or PrP^{CJD} and PrP^{GSS} (in human disease) may be the only constituent of the infectious agent. One possibility is that the disease-related isoform (PrP^{Sc}, PrP^{CJD} or PrP^{GSS}) carries the ability to convert the normal protein PrP^{C} into the proteinase K resistant form.

Of particular interest is the recent discovery (Owen *et al.*, 1989; Hsiao *et al.*, 1989) that some families with CJD and GSS carry mutations within the open reading frame (ORF) of the PrP gene. A number of such gene mutations have now been described (Table 1).

Table 1.   Mutations within the PrP gene in cases of CJD and GSS

Codon	Changes	Reference
51–91	144 bp insertion	Owen *et al.* (1989, 1990)
51–91	168 bp insertion	Brown *et al.* (1991)
51–91	216 bp insertion	Owen *et al.* (1991b)
102	Pro → leu	Hsiao *et al.* (1989)
117	Ala → Val	Doh-ura *et al.* (1989)
178	Asp → Asn	Nieto *et al.* (1991)
198	Phe → Ser	Prusiner (1991)
200	Glu → Lys	Goldgaber *et al.* (1989), Goldfarb *et al.* (1990a,b)

Table 2. Normal and disease-related alleles

Wild-type allele

CCT	CAg	GGc	GGT	GGt	ggc	TGG	GGG	CAG	R$_1$
Pro	Gln	Gly	Gly	Gly	Gly	Trp	Gly	Gln	
CCT	CAT	GGT	GGT	GGC	—	TGG	GGG	CAG	R$_2$
Pro	His	Gly	Gly	Gly	—	Trp	Gly	Gln	
CCT	CAT	GGT	GGT	GGC	—	TGG	GGG	CAG	R$_2$
Pro	His	Gly	Gly	Gly	—	Trp	Gly	Gln	
CCc	CAT	GGT	GGT	GGC	—	TGG	GGa	CAG	R$_3$
Pro	His	Gly	Gly	Gly	—	Trp	Gly	Gln	
CCT	CAT	GGT	GGT	GGC	—	TGG	GGt	CAa	R$_4$
Pro	His	Gly	Gly	Gly	—	Trp	Gly	Gln	

Mutant allele

CCT	CAg	GGc	GGT	GGt	ggc	TGG	GGG	CAG	R$_1$
Pro	Gln	Gly	Gly	Gly	Gly	Trp	Gly	Gln	
CCT	CAT	GGT	GGT	GGC	—	TGG	GGG	CAG	R$_2$
Pro	His	Gly	Gly	Gly	—	Trp	Gly	Gln	
CCT	CAT	GGT	GGT	GGC	—	TGG	GGG	CAG	R$_2$
Pro	His	Gly	Gly	Gly	—	Trp	Gly	Gln	
CCT	CAT	GGT	GGT	GGC	—	TGG	GGG	CAG	R$_2$
Pro	His	Gly	Gly	Gly	—	Trp	Gly	Gln	
CCc	CAT	GGT	GGT	GGC	—	TGG	GGa	CAG	R$_3$
Pro	His	Gly	Gly	Gly	—	Trp	Gly	Gln	
CCT	CAT	GGT	GGT	GGC	—	TGG	GGG	CAG	R$_2$
Pro	His	Gly	Gly	Gly	—	Trp	Gly	Gln	
CCc	CAT	GGT	GGT	GGC	—	TGG	GGG	CAG	R$'_3$
Pro	His	Gly	Gly	Gly	—	Trp	Gly	Gln	
CCT	CAT	GGT	GGT	GGC	—	TGG	GGG	CAG	R$_2$
Pro	His	Gly	Gly	Gly	—	Trp	Gly	Gln	
CCT	CAT	GGT	GGT	GGC	—	TGG	GGG	CAG	R$_2$
Pro	His	Gly	Gly	Gly	—	Trp	Gly	Gln	
CCc	CAT	GGT	GGT	GGC	—	TGG	GGa	CAG	R$_3$
Pro	His	Gly	Gly	Gly	—	Trp	Gly	Gln	
CCT	CAT	GGT	GGT	GGC	—	TGG	GGt	CAa	R$_4$
Pro	His	Gly	Gly	Gly	—	Trp	Gly	Gln	

(for notation see text)

DEMENTIA CASES WITH 144 bp INSERT

A 144 bp insert was previously identified (Owen *et al.*, 1989) in the prion protein gene in the region encoding the octapeptide repeats in a family that included a member with classical CJD as well as a proband with a dementing illness of early onset and long duration.

DNA was extracted from whole blood from a control and affected pedigree member. The entire ORF of the PrP gene was amplified using the polymerase chain reaction (PCR), as described in Owen *et al.* (1990). Restriction mapping had previously localized the insertion to between the Dde1 site at codon 51 and the PvuII site at codon 118. Primers were synthesized to allow sequencing of the entire insert using the method of double strand sequencing of Chen and Seeburg (1985).

The difference between the normal and the disease-related allele is presented in Table 2. The normal allele has 5 uninterrupted repeat sequences between codons 51 and 91, whereas the mutant allele has 11 uninterrupted, in-frame, repeat sequences. The repeats are presented with invariant bases in upper and variant bases in lower case letters. Apart from the first repeat the amino acid sequence of both the normal and the mutant alleles is identical in each of the remaining repeats. The repeats can, however, be distinguished from each other on the basis of their nucleotide sequences. Thus the repeats of the normal allele can be abbreviated to R_1, R_2, R_3, R_4 and of the mutant allele to R_1, R_2, R_2, R_2, R_3, R_2, R_3', R_2, R_2, R_3, R_4. R_3' differs from R_2 and R_3 by one base in each case, and is not found in the normal allele. It could, however, be reconstructed from R_2 and R_3.

The insertion presumably arises from an unequal crossing over but the six extra repeats cannot have been generated by a single recombinational event. Complex duplication and perhaps deletion may have occurred to achieve the present arrangement. Repeat sequences in the N-terminal region of the prion protein comprising two short repeats of GG(N/S)RYP followed by five longer uninterrupted repeats of the form p(H/Q)GGG(–/G)WGQ are highly conserved across species which suggests that they are important in the normal function of the PrP.

ORIGIN OF THE 144 pb INSERT

As noted above the insert was first identified in a family that included a number of members with degenerative neurological disease with an onset in middle age, including one with a diagnosis of the Heidenhain variant (with particular involvement of the visual cortex) of Creutzfeldt–Jakob disease. The affected members in this family, with case note diagnoses and ages at death, are shown in Fig. 1. In three members (two of whom survive) the presence of the insert has been established. Dementia or cerebral atrophy was diagnosed in five others and may well also have been present in the individual who died before his fortieth year in an asylum.

A study of other cases of early onset dementia referred from other neurological and psychiatric centres in London revealed other families with the 144 bp insert. In one family (Fig. 2), referred to St Mary's Hospital, Paddington, two members (one still alive) have been shown to have the insert. Five members in preceding generations had a diagnosis of dementia, one of whom received a diagnosis of CJD.

Ho

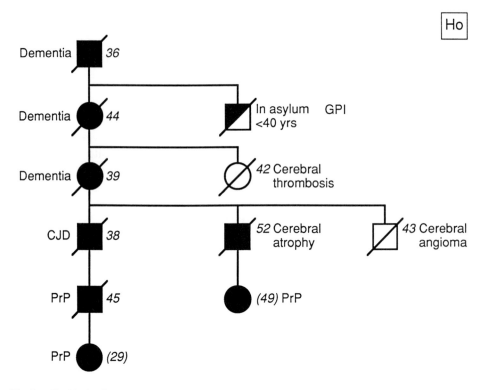

Fig. 1. The Ho family showing only affected males (■) and females (●) or possibly affected males (□) and females (○). /, dead. Numbers refer to age at death or, in parentheses, current age. PrP, 144 bp insert detected.

A further family referred to the National Hospital at Queen Square (Fig. 3) includes three members still alive with the prion protein gene insert, a further nine members with a diagnosis of dementia, cerebral atrophy, or brain disease or with a death at an early age in an asylum, and two members with relatively early deaths and diagnoses (nephritis and toxaemia due to pressure sores) that may have been accompaniments of neurodegenerative disease.

A fourth family was first seen at the Brook General Hospital in South East London. This family includes two sisters and a cousin now known to have the insert in the prion protein gene, two members who died with known mental impairment in their forties in asylums, two who died at a similar age with dementia, and another demented patient who died at the age of 53 of accidental burns. A sibling of one of the cases shown to have the insert had already died with a diagnosis of Huntington's chorea (Fig. 4).

Family lineages were traced back from affected probands. Birth certificates were obtained for each from the office of the Registrar General of Births, Deaths and Marriages (St Catherine's House, London), thus identifying the parents and enabling a search to be made of the marriage index. The marriage certificate gave information on ages, residence, condition, occupations, women's maiden name and also the

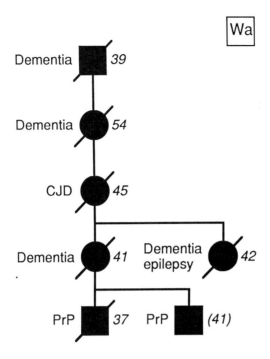

Fig. 2.   The Wa family; see Fig. 1 for key.

proband's grand-parental (male) names and their occupations. Searches could then be made for births of siblings of the original proband and deaths of the parents. Death certificates gave information on age, place of death, cause of death and informant's description and residence. The age at death enabled searches to recommence with the birth index. Working from this and subsequent information, lineages were traced back to 1837 when registration commenced. Local parish records were then consulted for information prior to this date. Use was also made of the census records (1841–1881 at ten-yearly intervals) currently available for inspection at the Public Records Office, Chancery Lane, London. These list the names, ages (only approximate in 1841), relationships, occupations and places of birth for all individuals staying at a particular address on census night. These were useful for identifying sibling's names and some family relationships not formally registered. Case notes were obtained wherever individuals died in hospital, and these provided some information on other family members. Other sources of information included county archivists, medical records offices and coroners' reports.

These genealogical investigations demonstrated that the four families, originally thought to be independent, did in fact derive from a sibship born to a couple living in South East England in the early nineteenth century. Dementing illness has also been tracked forward in this pedigree to reveal new cases (see Fig. 5). Overall the family includes at least 53 affected or probably affected members in 7 generations. It

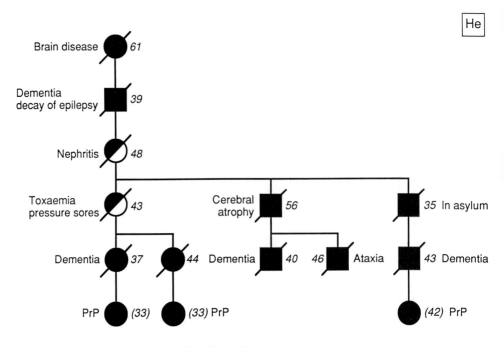

Fig. 3.   The He family; see Fig. 1 for key.

has not proved possible to deduce which of the potential founders I1 or I2 was affected. I1 died at age 50 from 'inflammation of the chest'. His wife, I2, did not appear on the 1841 census with him or on later returns suggesting that she may have died or been institutionalized between 1834 (the birth of her youngest child) and 1841 (aged approximately 40 years).

CLINICAL FEATURES AND PATHOLOGY

From the accumulated data within this family (Collinge *et al.*, 1992) mean age at onset is 35.2 (± 7.3 sd) years and mean age at death is 45.4 (± 7.3) years with a mean duration of illness (in those in whom both onset and age at death are recorded) being 7.9 (± 4.3) years. Dementia appears to be present in all cases at the time of death. Pyramidal tract signs were present or probably present in 16 of 21 cases in which neurological examination is recorded, cerebellar ataxia or dysarthria in 20 out of 22 cases, and extrapyramidal features in 6 of 12 cases. Chorea was noted in three cases, athetosis, homonymous hemianopia, puerperal psychosis, and cerebral thrombosis in one case each. An interesting feature is that personality disturbance of sufficient severity to have been recorded in the case notes is mentioned in nine cases as having been a lifelong feature. In a further eight cases personality change (including aggression and depression) heralded onset of illness.

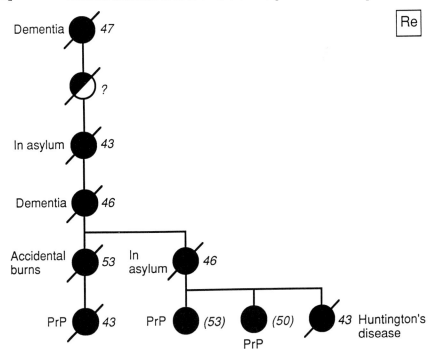

Fig. 4.   The Re family; see Fig. 1 for key.

Pathological examination of brain tissue obtained at post-mortem was available in four cases (Collinge *et al.*, 1992). Atrophy was marked in one case but only mild in degree in the other three. Status spongiosus was clearly present in two cases, equivocal in the third, and absent in the fourth (Collinge *et al.*, 1990). Neuronal loss and astrocytosis were clearcut in two cases but equivocal or absent in the other two. By immunohistochemical staining, prion protein plaques were detected in one out of three cases, this case being the one in which beta amyloid plaques of the type seen in Alzheimer's disease were also present.

LINKAGE ANALYSIS

Genetic linkage analysis was carried out using the MLINK program from the LINKAGE package (Lathrop *et al.*, 1984) assuming a dominant mode of transmission, age-dependent penetrance and a disease gene frequency of 1 in 10^6. A maximum lod score of 11.02 at a recombination fraction of zero was obtained that remained above 10 with the assumptions that the disease frequency is 100 times higher, that the frequency of the insert within the population was as high as 1 in 100, or that the age-independent phenocopy rate is 1 in 100 (Poulter *et al.*, 1992).

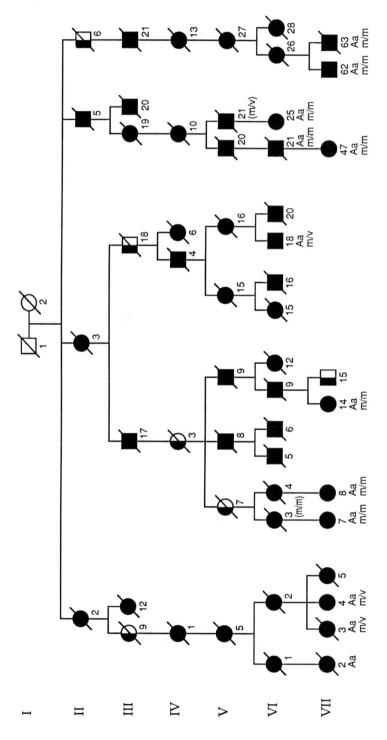

Fig. 5. The extended pedigree carrying the 144 bp insert; see Fig. 1 for key. Again unaffected individuals and spouses are not shown. The tree was constructed following detailed genealogical investigation and study of clinical notes and death certificates. Genotypes: A = normal allele; a = mutant allele; m = methionine–129; v = valine–129 Codon 129 genotypes in parenthesis are inferred from types offspring.

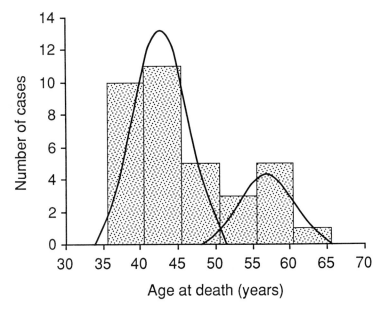

Fig. 6.   The age at death distribution is best described by two normal distributions with equal standard deviations: see text.

AGE AT DEATH AND CODON 129 POLYMORPHISM

A polymorphism at codon 129 in the prion protein gene has been described that is manifest in the amino acid sequence as either a methionine, with a population frequency of 0.68, or a valine with a frequency of 0.32 (Owen *et al.*, 1991a). Within this pedigree the insert is carried within an allele encoding methionine at codon 129. In 13 affected individuals tested, the other (normal) allele was found to code for methionine at codon 129 in 9 cases and a valine in 4 cases.

An analysis was made of age at death in the 35 individuals for whom death was thought to have occurred when the disease had run its full course, i.e. excluding suicide and accidental death. Using maximum likelihood estimate procedures the best fit for the age at death distribution was obtained from two normal distributions with equal standard deviations (Fig. 6), such a distribution giving a significantly better fit than a unimodal one ($\chi^2 = 13.1$, df 2, $p < 0.01$). The 'late' age at death distribution has a mean age at death of 56.9 ± 3.47 sd and accounted for 26% of cases, while the 'early' age at death distribution had a mean age at death of 42.7 ± 3.47 sd years and accounted for 74% of cases.

All four cases with a valine-129 encoding normal allele fell into the 'late' death group (two died over the age of 50 and two are alive but over the age of 50). Of the nine with a methionine-129 encoding normal allele, six were assigned to the 'early' death group, three having died under the age of 50 and three being currently below the age of 35 and severely ill. This association between methionine/valine genotype and age at death is significant (Fisher's exact $p = 0.0096$, two-tailed) (Baker *et al.*,

1991). The remaining three affected individuals with a methionine-encoding normal allele were aged between 35 and 50 and were not included in this analysis because they could possibly survive beyond 50 years. There was also an association between methionine/valine genotype at codon 129 and age at onset, three cases with a valine genotype having an onset after their fortieth year, and eight cases with a methionine genotype having an onset before this age. One case with each genotype had an onset at 40 years. The difference between the distributions was significant (Fisher's exact $p=0.014$, two-tailed). The data are insufficient to establish whether age at onset is the primary determinant of earlier age at death in the codon 129 methionine group or whether this group also has a more rapid course of illness.

IMPLICATIONS FOR THE NATURE OF THE PATHOGEN

The paradox of the familial prion-related dementias is that some at least of these conditions are both inherited and infectious. For example, the degenerative neurological disease transmitted within a family as an autosomal dominant as described by Adam *et al.* (1982) has been shown both to be transmissible by intracerebral inoculation to the marmoset (Baker *et al.*, 1985) and to be linked to a proline-to-leucine substitution at codon 102 in the prion gene (Hsiao *et al.*, 1989). Linkage of the 144 bp insert to neurological disease in the present family has been securely demonstrated (lod score in excess of 10). Although experimental transmission in this family has not yet been demonstrated (marmosets remain well 5 years after having been injected with cortical material taken at autopsy from case VI21), there are reports (Goldfarb *et al.*, 1991) that disease associated with other inserts in the repeat region of the prion protein gene is transmissible to primates.

These findings have implications for the general problem of the nature of the infectious agent in prion-related disease. It is increasingly recognized that the agent that transmits scrapie to experimental animals includes a protein component but does not require a nucleic acid (Alper *et al.*, 1967: Pattison and Jones, 1967; Prusiner *et al.*, 1980, 1982; Brown *et al.*, 1990). One possibility is that a particular form of the prion protein (that at present appears to be closely associated with protease resistance) is able to induce a conformational change in the normal form of the protein, and that it is this sequentially transmitted conformational change that causes disease. A possible precedent for such an interaction has been reported in the case of the p53 protein (Milner and Metcalf, 1991). According to this view the species barrier in transmission experiments relates to interspecies differences in structure of the prion protein. Consistent with this interpretation are the findings of transgenic experiments in the mouse with the hamster prion gene (Hsiao *et al.*, 1990).

The findings with respect to the methionine–valine polymorphism in this pedigree are consistent with this hypothesis—some aspect of pathology (although whether disease initiation or progression is unclear) is faciliated by homozygosity at codon 129. Earlier we suggested that inherited and sporadic forms of the transmissible dementias might be governed by similar age-related genetic mechanisms (Ridley *et*

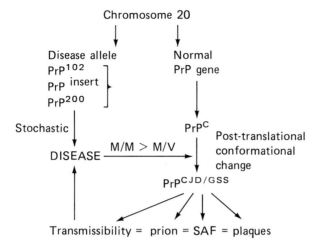

Fig. 7.   Scheme illustrating possible relationships underlying the pathogenesis of prion diseases: see text.

al., 1986). For example, the transition from the normal or cellular (PrP^C) form of the prion protein to the abnormal and disease-related forms (PrP^{CJD} or PrP^{GSS}) may be understandable as a stochastic process with a finite but very low probability of occurring in a normal individual in the course of a lifetime. This could account for the low, relatively uniform and age-dependent incidence of the sporadic forms of the prion-related dementias. In the familial forms, it is envisaged that the presence of a mutation in the prion protein gene increases the probability of the transition from normal to abnormal forms of the protein so that, in the presence of a mutation, the probability that disease will occur within the normal lifespan is much increased (Fig. 7).

According to this concept once the critical transition has occurred disease progression will be more rapid when the interaction between abnormal and normal forms of the protein is between protein structures that are similar in all relevant respects, e.g. at codon 129. A prediction of this hypothesis is that had the insert been on a valine-encoding allele age at death would have been earlier in codon 129 valine–valine homozygous individuals.

Consistent with this view is the report of Palmer *et al.* (1991) that 21 of 22 cases of sporadic CJD and 19 of 23 cases of suspected sporadic CJD were homozygous at codon 129. This is what might be expected if homozygosity at this site predisposes, after the occurrence of some as-yet unspecified random event, to rapid progression of disease. However, it should be noted that this sample of cases was defined by relatively restrictive criteria for typical CJD. Less typical cases, e.g. of longer duration, would have been excluded. It remains possible therefore that variants of prion-related dementia, of longer duration and perhaps later age at death, occurring in individuals heterozygous at codon 129, have yet to be described.

REFERENCES

Adam, J., Crow, T.J., Duchen, L.W., Scaravilli, F. and Spokes, E. (1982) Familial cerebral amyloidosis and spongiform encephalopathy. *J. Neurol. Neurosurg. Psychiatry* **45** 37–45.

Alper, T., Cramp, W.A., Haig, D.A. and Clarke, M.C. (1967) Does the agent of scrapie replicate without a nucleic acid? *Nature (London)* **214** 764–766.

Baker, H.F., Ridley, R.M. and Crow, T.J. (1985) Experimental transmission of an autosomal dominant spongiform encephalopathy: does the infectious agent originate in the human genome? *Bri. Med. J.* **291** 299–302.

Baker, H.F., Poulter, M., Crow, T.J., Frith, C.D., Lofthouse, R., Ridley, R.M. and Collinge, J. (1991) Aminoacid polymorphism in human prion protein and age at death in inherited prion disease. *Lancet* **337** 1286.

Brown, P., Liberski, P.P., Wolff, A. and Gajdusek, D.C. (1990) Conservation of infectivity in purified fibrillary extracts of scrapie-infected hamster brain after sequential enzymatic digestion or polyacrylamide gel electrophoresis. *Proc. Natl. Acad. Sci.* **87** 7240–7244.

Brown, P., Goldfarb, L.G. and Gajdusek, D.C. (1991) The new biology of spongiform encephalopathy: infectious amyloidosis with a genetic twist. *Lancet* **337** 1019–1022.

Chen, E.Y. and Seeburg, P.H. (1985) Supercoil sequencing: a fast and simple method for sequencing plasmid DNA. *DNA* **4** 165–170.

Cheseboro, B., Race, B., Wehrly, K., Nishio, J., Bloom, M., Lechner, D., Bergstrom, S., Robbins, K., Mayer, L., Keith, J.M., Garon, C. and Haase, A. (1985) Identification of scrapie prion-specific mRNA in scrapie infected and uninfected brain. *Nature (London)* **315** 331–333.

Collinge, J., Owen, F., Poulter, M., Leach, M., Crow, T.J., Rossor, M.N., Hardy, J., Mullan, M.J., Janota, I. and Lantos, P.L. (1990) Prion dementia without characteristic pathology. *Lancet* **336** 7–9.

Collinge, J., et al. (1992) Inherited prion disease with 144 base pair gene insertion II: clinical and pathological description. *Brain* (in press).

Doh-ura, K., Tateishi, J., Sasaki, H., Kitamoto, T. and Sakaki, Y. (1989) Pro → Leu change at position 102 of prion protein gene is the most common but not the sole mutation related to Gerstmann–Sträussler syndrome. *Biochem. Biophys. Res. Commun.* **163** 974–979.

Goldfarb, L.G., Korczyn, A.D., Brown, P., Chapman, J. and Gajdusek, D.C. (1990a) Mutation in codon 200 of scrapie amyloid precursor gene linked to Creutzfeldt–Jakob disease in Sephardic Jews of Libyan and non-Libyan origin. *Lancet* **336** 637–638.

Goldfarb, L.G., Mitrova, E., Brown, P., Tom, B.H. and Gajdusek, D.C. (1990b) Mutation in codon 200 of scrapie amyloid gene in two clusters of Creutzfeldt–Jakob disease in Slovakia. *Lancet* **336** 514–515.

Goldgaber, D., Goldfarb, L.G., Brown, P., Asher, D.M., Brown, W.T., Scott, L., Teener, J.W., Feinstone, S.M., Rubinstein, R., Kascsak, R.J., Boellaard, J.W. and Gajdusek, D.C. (1989) Mutations in familial Creutzfeldt–Jakob disease and Gerstmann–Sträussler–Scheinker syndrome. *Exp. Neurol.* **106** 204–206.

Hsiao, K., Baker, H.F., Crow, T.J., Poulter, M., Owen, F., Terwilliger, J.D., Westaway, D., Ott, J. and Prusiner, S.B. (1989) Linkage of a prion missense variant to Gerstmann–Sträussler syndrome. *Nature (London)* **338** 342–345.

Hsiao, K.K., Scott, M., Foster, D., Groth, D., DeArmond, S.J. and Prusiner, S.B. (1990) Spontaneous neurodegeneration in transgenic mice with mutant prion protein. *Science* **250** 1587–1589.

Lathrop, G.M., Lalouel, J.M., Julier, C. and Ott, J. (1984) Strategies for multilocus linkage analysis in humans. *Proc. Natl. Acad. Sci. USA* **81** 3443–3446.

Masters, C.L., Harris, J.O., Gajdusek, D.C., Gibbs, C.J., Bernoulli, C. and Asher, D. M. (1979) Creutzfeldt–Jakob disease: patterns of worldwide occurrence and the significance of familial and sporadic clustering. *Ann. Neurol.* **5** 177–188.

Masters, C.L., Gajdusek, D.C. and Gibbs, C.J. (1981) Creutzfeldt–Jakob disease virus isolations from the Gerstmann–Sträussler syndrome. *Brain* **104** 559–588.

Milner, J. and Metcalf, E.A. (1991) Co-translation of activated mutant p53 with wild type drives the wild-type p53 protein into mutant conformation. *Cell* **65** 765–774.

Nieto, A., Goldfarb, L.G., Brown, P., McCombie, W.R., Trapp, S., Asher, D.M. and Gajdusek, D.C. (1991) Codon 178 mutation in ethnically diverse Creutzfeldt–Jakob disease families. *Lancet* **337** 622–623.

Oesch, B., Westaway, D., Walchli, M., McKinley, M.P., Kent, S.B.H., Debersold, R., Barry, R.A., Tempst, P., Teplow, D.B., Hood, L. E., Prusiner, S.B. and Weissman, C. (1985) A cellular gene encodes scrapie PrP27-30 protein. *Cell* **40** 735–746.

Owen, F., Poulter, M., Lofthouse, R., Collinge, J., Crow, T.J., Risby, D., Baker, H.F., Ridley, R.M., Hsiao, K. and Prusiner, S.B. (1989) Insertion in the prion protein gene in familial Creutzfeldt–Jakob disease. *Lancet* **i** 51–52.

Owen, F., Poulter, M., Shah, T., Collinge, J., Lofthouse, R., Baker, H.F., Ridley, R.M., McVey, J. and Crow, T.J. (1990) An in-frame insertion in the prion protein gene in familial Creutzfeldt–Jakob disease. *Mol. Brain Res.* **7** 273–276.

Owen, F., Poulter, M., Collinge, J. and Crow, J.J. (1991a) Codon 129 changes in the prion protein gene in Caucasians. *Am. J. Hum. Genet.* **46** 1215–1216.

Owen, F., Poulter, M., Collinge, J., Leach, M., Shah, T., Lofthouse, R., Chen, Y., Crow, T.J., Harding, A., Hardy, J. and Rossor, M. (1991b) Insertions in the prion protein gene in atypical dementias. *Exp. Neurol.* **112** 240–242.

Palmer, M.S., Dryden, A.J., Hughes, T.J. and Collinge, J. (1991) Homozygous prion protein genotype predisposes to sporadic Creutzfeldt–Jakob disease. *Nature (London)* **352** 340–342.

Pattison, I.H. and Jones, K.M. (1967) The possible nature of the transmissible agent of scrapie. *Vet. Rec.* **80** 2–9.

Poulter, M., Baker, H.F., Frith, C.D., Leach, M. Lofthouse, R., Ridley, R.M., Shah, T., Owen, F., Collinge, J., Brown, G., Hardy, J.A., Mullan, M. J., Harding, A.E., Bennet, C., Doshi, R. and Crow, T.J. (1992) Inherited prion disease with 144 base pair insertion: I. Genealogical and molecular studies. *Brain* (in press).

Prusiner, S.B. (1987) Prions and neurodegenerative disease. *New Engl. J. Med.* **317** 1517–1581.

Prusiner, S.B. (1991) Molecular biology of prion diseases. *Science* **252** 1515–1522.

Prusiner, S.B., Groth, D.F., Cochran, S.P., Masiarz, F.R., McKinley, M.P. and

Martinez, H.M. (1980) Molecular properties, partial purification, and assay by incubation period measurements of the hamster scrapie agent. *Biochemistry* **19** 4883–4991.

Prusiner, S.B., Bolton, D.C., Groth, D.F., Bowman, K.A., Cochran, S.P. and McKinley, S.P. (1982) Further purification and characterization of scrapie prions. *Biochemistry* **21** 6942–6950.

Ridley, R.M., Baker, H.F. and Crow, T.J. (1986) Transmissible and non-transmissible neurodegenerative disease: similarities in age of onset and genetics in relation to aetiology. *Psychol. Med.* **16** 199–207.

Sparkes, R.S., Simon, M., Cohn, V.H., Fournier, R.E.K., Lem, J., Klisuk, I., Heinzmann, C., Blatt, C., Lucero, M., Mohandas, T., DeArmond, S.J., Westaway, D., Prusiner, S.B. and Weiner, L.P. (1986) Assignment of the mouse and human prion protein genes to homologous chromosomes. *Proc. Natl. Acad. Sci. USA* **83** 7358–7362.

Will, R.G. and Matthews, W.B. (1984) A retrospective study of Creutzfeldt–Jakob disease in England and Wales 1970–79 I: clinical features. *J. Neurol. Neurosurg. Psychiatry* **47** 134–140.

21

Prion disease: the spectrum of pathology and diagnostic considerations

Gareth W. Roberts and Joanne Clinton

ABSTRACT

Prion disease encompasses scrapie in sheep and bovine spongiform encephalopathy (BSE) in cattle as well as a wide range of other animals and Creutzfeldt–Jakob disease (CJD) and Gerstmann–Sträussler syndrome (GSS) in humans. The core pathological process involves an aberrant form of a normal cell protein—prion protein (PrP). PrP is a cell surface constituent of neurons and is encoded by a single gene. During the disease the normal PrP isoform is converted into an abnormal protease-resistant isoform (PrP^{Sc}). The accumulation of PrP^{Sc} is pathognomonic of prion disease. Altered PrP gene structure appears to make this conversion happen spontaneously in some families. In humans point mutations in the PrP gene segregate with inherited prion diseases. Thus, the structure of the PrP gene and prion protein is of considerable importance in relation to the onset, course and pathology of prion disease.

The neuropathology of prion disease is marked by the prominent spongiform appearance of the cerebral cortex, astrocytosis and possible prion protein-positive plaques in the cerebellum and cortex. Other neuronal abnormalities include swelling and loss of dendrites and axons, a marked loss of synaptic contacts and accumulation of abnormal cytoskeletal protein. As a result of these processes, a massive neuronal loss occurs, especially in the cerebral and cerebellar cortices and striatum.

The broadening spectrum of neuropathology and the equally varied clinical symptoms in prion disease make diagnosis an increasingly difficult task. Molecular biology offers the surest method of diagnosis.

INTRODUCTION

Prion disease is a new nosological entity derived from research which has uncovered a common molecular pathology underlying the disparate collection of syndromes previously known as the spongiform encephalopathies, unconventional viral infections or transmissible dementias (Table 1).

Table 1.   Prion diseases

Species	Common name
Humans	Creutzfeldt–Jakob disease (sporadic)
	Creutzfeldt–Jakob disease (familial)
	Gerstmann–Sträussler syndrome
	Kuru
	Atypical dementia
Sheep	Scrapie
Cow	Bovine spongiform encephalopathy
Cat	Feline spongiform encephalopathy
Mule deer	Chronic wasting disease
Mink	Transmissible mink encephalopathy
Kudu	Transmissible encephalopathy
Nyala	Transmissible encephalopathy
Mice	Transmissible encephalopathy
Monkey	Transmissible encephalopathy

Their collective name of spongiform encephalopathy derives from the characteristic vacuolar, spongy appearance of the brain seen at postmortem. In 1959, the pathological similarities between CJD, kuru, and the sheep disease scrapie were first noted (Hadlow, 1959; Klatzo et al., 1959). Following these observations, essentially identical diseases (going by different names) are now recognized to occur in many species (Table 1). For 30 years these disorders have been the subject of research interest out of proportion to their apparent rarity because of a number of unique features. Most notably, the human forms have been shown to be transmissible by inoculation, both to other species (Gadjusek et al., 1966; Gibbs et al., 1968; Baker et al., 1985; Manuelidis et al., 1985), and also iatrogenically from person to person after transplantation, neurosurgery and therapeutic procedures (e.g. treatment with human growth hormone (Brown et al., 1985; Weller, 1989; Preece et al., 1991)). In addition the human forms can also be inherited in an autosomal dominant fashion (Masters et al., 1981a,b; Will and Mathews, 1984; Brown et al., 1987; Owen et al., 1989; Hsiao et al., 1989).

Prion disease shows an array of clinical presentations and neuropathological findings, apparently overlapping not only with each other but also with other neurodegenerative disorders (Flament-Durand and Couck, 1979; Ball, 1980; Mancardi et al., 1982; Smith et al., 1987; Hirano et al., 1972; Brown et al., 1990; Arendt et al., 1984; Gray, 1986; Brown, 1989; Masters et al., 1981b; Brown et al., 1986; Irving et al., 1990). Until recently, there was no way unequivocally to separate one 'disease' from another, since neither the clinical nor neuropathological findings are pathognomonic. However, advances from molecular biology and immunocytochemistry have provided a means for a simpler and more rational grouping of these disorders based on their underlying features. These advances also suggest that the diseases may be commoner than hitherto suspected, increasing their potential clinical significance.

Nomenclature

In this chapter the normal cellular isoform of the prion protein shall be designated PrP^C and PrP^{Sc} will be used to designate the pathological abnormal isoform of PrP found in all the natural and experimental prion diseases.

NEUROPATHOLOGY

Pathology

No important pathological changes occur outside the nervous system. The neuropathological features vary considerably and may be influenced by the length of the illness.

When death is rapid the brain will show little or no macroscopic change and the weight will probably be normal. In patients surviving several years a moderate, or even considerable, degree of cerebral atrophy is more likely and the weight may fall to as low as 850 g. No gross evidence of inflammation is present and the vessels and meninges usually appear normal for the age. At times the atrophy may affect the corpus striatum and thalamus; the brain stem itself may appear considerably diminished in size; as a rule the white matter shows no abnormality (Tomlinson and Corsellis, 1984).

Histopathology

The histopathology of prion disease encompasses a triad of features, spongiform change, neuronal loss and gliosis (Fig. 1), which have been well described in standard works of reference (Tomlinson and Corsellis, 1984; Beck and Daniel, 1987).

Spongiform change

The sponginess, which cannot be seen with the naked eye, comes in two forms which can usually be distinguished (Masters and Richardson, 1978). The first is known as 'spongiform change' and is made up of many small, usually rounded or oval, vacuoles in the neuropil. Although they do not appear under the light microscope to be confined within a cell's limits, electron microscopy has shown that they lie in neuronal dendrites or axons and in glial processes. The smallest vacuoles measure as little as 1–2 μm across but they may coalesce into confluent areas up to 50 μm or more across. Electron microscopy of the vacuoles (Chou et al., 1980) has shown them to contain

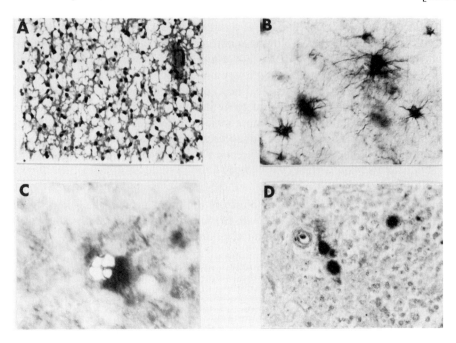

Fig. 1.   The standard histological changes observed in prion disease. (A) Spongiform change in the cortex, stained with haematoxylin and eosin. (B) Astrocytosis. Astrocytes labelled with glial fibrillary acidic protein. (C) Prion plaque, showing the typical maltese cross birefringence with polarized light when stained with Congo red. (D) Prion plaques in the granular layer of the cerebellum, stained with antibody to prion protein.

alterations in the membranes in which ulceration, focal thickening and splitting may occur. Small blisters containing particulate matter have also been described, but the possible relation of these structures to any transmitting agent was left, and remains, open. The vacuolation is seen in all layers of the cortex and may spread almost uniformly through the thalamus, the striatum and the grey matter of the brain stem, as well as the molecular layer of the cerebellum. Most often it is patchy with stretches of apparently normal cortex or grey matter which alternate and intermingle with discrete patches of typical vacuolation (Beck and Daniel, 1987).

Neuronal loss
Neuronal loss is also present in the great majority of cases. It can be difficult to detect, particularly in those in whom the illness had been rapidly fatal or when a cortical biopsy has been taken early in the course of the illness. In either instance, areas of neuronal loss may be visible, but slight. When spongiform change is not readily detectable diagnosis is difficult. Postmortem studies reveal a variable degree of neuronal loss in the cortex, thalamus and striatum, from severe and widespread (prominent in the third and deeper layer of the cortex) to irregular patchy loss of

cells in the different layers. Groups of abnormally shrunken, and often pale-staining, ghost-like remnants of cells are often seen amongst other apparently normal neurons (Tomlinson and Corsellis, 1984; Beck and Daniel, 1987).

Gliosis

This loss of nerve cells is accompanied by the third type of change: the proliferation of astrocytic glial cells (Fig. 1). This may well be the dominant feature, particularly in long-standing cases. The affected areas are largely occupied by hypertrophied and often gemistocytic astrocytes among which vacuolation is commonly seen, interspersed with small numbers of surviving neurons, which may be shrunken or swollen and show cytoplasmic vacuoles. When the astrocytic proliferation is severe, it may spread over into areas of nerve cell which do not show obvious neuronal destruction or vacuolation and may include adjacent parts of the white matter.

Plaques

About 15% of prion disease brains contain extracelluar amyloid plaques, usually restricted to the cerebellum (Masters et al., 1981b; Roberts et al.,1988). These plaques are Congophilic and show birefringence with polarized light (Fig. 1). These plaques are usually found to have a 'halo' of reactive glial processes (Masters et al., 1981a,b; Beck and Daniel, 1987; Roberts, 1988). Although these plaques can be morphologically indistinguishable from the senile plaques of Alzheimer's disease (AD), immunocytochemistry has shown that they are formed from prion protein, rather than β-amyloid protein of Alzheimer's disease (Roberts et al., 1986; Tateishi et al., 1986; Roberts et al., 1988; Nochlin et al., 1989). This observation forms the basis of methods for diagnosing cases of prion disease. The plaques exhibit a range of morphology from the discrete 'Kuru' plaques to the 'multicentric' plaques in GSS (Roberts et al., 1988) (see Fig. 2).

Other changes

In spite of the transmissible nature of the disorder evidence of an inflammatory reaction is strikingly absent. The leptomeninges are normal and the very occasional cluster of perivascular lymphocytes in the occasional case is no more common than similar minor changes which occur in ageing generally (Eikelenboom et al., 1991). Even microglial proliferation is not prominent at any stage of the disease although an occasional microglial star or cluster can be found, particularly in the deeper layer of the cortex (Eikelenboom et al., 1991).

Other neuronal abnormalities include swelling and loss of dendrites and axons and accumulation of abnormal cytoskeletal protein (Baker et al., 1990; DeArmond et al., 1987). The pathology present in axons and dendrites is associated with a marked loss of spines and synaptic contacts (Landis et al., 1981; Gray, 1986). Presumably as a result of these processes, a massive neuronal loss occurs, especially in the cerebral and cerebellar cortices and striatum, whereas pathology in the hippocampus can be relatively light. The importance of synaptic pathology has been highlighted recently by Clinton et al (1991). They immunocytochemically re-examined a case of inherited prion disease with no obvious pathology (Collinge et al., 1990; Clinton et al., 1990)

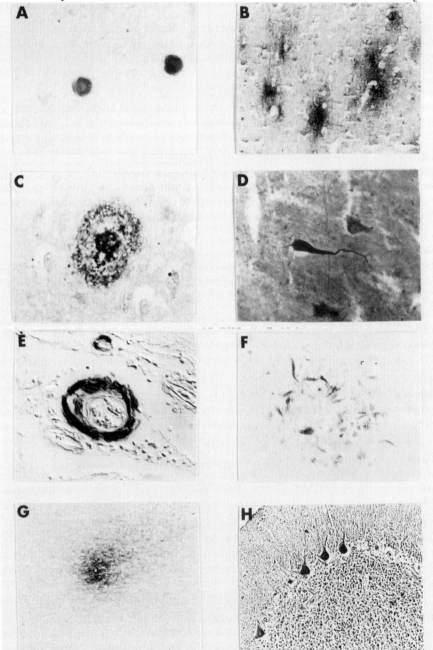

Fig. 2.   The range of pathological histological features now observed in prion disease: (A) small, compact, discrete prion protein positive kuru-type plaques in the cerebellum; (B) multicentric, more diffuse-type plaque observed in GSS; (C) β/A4 plaque occasionally observed in prion disease; (D) a silver-stained neurofibrillary tangle (NFT); (E) a Nomarski image demonstrating the presence of Congophilic angiopathy; (F) ubiquitin-positive plaque. (G) a plaque-like synaptic lesion in the hippocampus labelled with an antibody to synaptophysin; (H) prion protein-positive Purkinje cells in the cerebellum.

and discovered aberrant patterns of synaptic organization and the presence of synaptophysin-positive plaques (Fig. 2G). It is hypothesized that this disorganization may be an early manifestation of pathological change previously overlooked. The assessment and extent of synaptic disorganization might give a practical index of the degree of neuronal disorganization occurring during the disease.

Molecular pathology

It is now probable that the spongiform encephalopathies occurring in various species, including the three human variants, are all caused by a single common agent. What is surprising is that increasing evidence points to the transmissible factor being a protein, called a prion (proteinaceous infectious agent) (Prusiner, 1982). Conversely, although not completely excluded, evidence for the involvement of DNA or RNA in the infective process—as would be axiomatic for a conventional infection of any kind—is lacking (see Chapter 44).

Immunocytochemistry and immunoblotting have provided evidence of the central role of PrP in the transmissible dementias. Antibodies raised against the protease-resistant PrP^{Sc} found in scrapie-infected mouse or hamster brain cross-react with similar proteins seen in affected brains of other species, including CJD, GSS and kuru in humans (Roberts et al., 1986, 1988; Farquhar et al., 1989). This antigenic similarity is strong evidence for the unitary nature of the transmissible dementias across and within species, and emphasizes the central role of this protein. This evidence has been further highlighted by the recent advances of Doi-Yi et al., (1991), who demonstrated that protein denaturation treatments such as guanidine thiocyanate, trichloroacetate and phenol, enhance the immunoreactivity of prion plaques.

Making the connection between the tentative evidence that PrP might be involved in the disease and the presence of PrP^{Sc} immunoreactivity as an established pathological feature of the transmissible dementias awaited the application of molecular biological techniques.

The prion hypothesis came of age in the mid-1980s when it was established that PrP is encoded by a normal cellular gene (Oesch et al., 1985; Kretschmar et al., 1986a) located on human chromosome 20 (Sparkes et al., 1986). Other experiments showed that the PrP gene is expressed and its encoded protein synthesized by cells including neurons and glia throughout life (Kretschmar et al., 1986b; DeArmond et al., 1987; McKinley et al., 1987; Brown et al., 1990; Piccardo et al., 1990). These data provided unequivocal evidence that PrP production is a normal event in a healthy brain. Therefore, any pathological potential of PrP must result from an alteration in some aspect of its sequence, expression or properties.

Another important finding was that the prion gene in scrapie-affected animals had the same nucleotide sequence as in unaffected individuals, indicating that scrapie (at least in these cases) is not caused by a mutation (Basler et al., 1986). Moreover, prion gene expression is not increased during the disease (Cheseboro et al., 1985; Kretschmar et al., 1986b), suggesting that the deposition of PrP^{Sc} in amyloid plaques which occurs in disease is not simply the result of excessive production. Neither are variants of PrP likely to arise from differential processing of its messenger RNA, since the protein is encoded on a single exon (Basler et al., 1986). Further analysis of the gene sequence indicated PrP to be a 'housekeeping' gene, necessary for basic functioning of cells, in keeping with its marked conservation between organisms as diverse in evolution as fruitflies and humans (Basler *et al.*, 1986). These data were also suggestive that PrP is a membrane protein, agreeing with the biochemical localisation of PrP (and infectivity) to membranes (Hunter, 1972; Meyer *et al.*, 1986; Safar *et al.*, 1990a).

Research into kuru and scrapie showed that the infectivity of affected brain tissue is not removed by procedures which are routinely used to destroy nucleic acids. Even traditional disinfectants and preservatives are often not effective—exemplified by the iatrogenic spread of CJD by intracranial EEG electrodes despite sterilization (Weller, 1989). The agent can also pass through fine-pored filters, showing that it is much smaller than most conventional infective agents such as viruses of the kinds which cause other neurodegenerative disorders (Mathews, 1981; Prusiner, 1982). Infectivity could, however, be countered by proteolytic treatments, such as strong alkali or autoclaving, suggesting instead that proteins might be central to the process (Gajdusek, 1977; Prusiner, 1982).

Cellular fractionation experiments have supported this view, since infectivity of different fractions generally coincides closely with the presence of the protease-resistant PrP (Gabizon and Prusiner, 1990). Importantly, several recent, independent techniques have also now failed to separate PrP^{Sc} from infectivity or to produce any evidence for the association of nucleic acid (or other non-PrP component) with infectivity (Brown *et al.*, 1990b; Ceroni *et al.*, 1990; Safar *et al.*, 1990a,b).

These findings answered some perplexing questions, but raised several more. The normal presence of PrP provided an explanation for the lack of an immune response during disease, since the disease-producing form of the protein is not sufficiently different to be recognized as an antigen by the host. It also removed the requirement for PrP to be a self-replicating protein to account for its proposed infectivity; its infectious properties could now be explained (much more plausibly) in terms of a disruption of pre-existing cellular PrP metabolism (see Chapter 36).

Indeed, the main feature which distinguishes PrP^C from PrP^{Sc} is the resistance to proteolysis of the latter; this change in property is likely to be a correlate of such post-translational modifications. The presence of PrP^{Sc} in these cases could occur as a result of an uncharacterized gene defect or from aberrant processing of nascent PrP. Newly synthesized peptides often undergo post-translational processing and additions of various moieties before they can adopt their final conformation and position within the cell. Slight changes in any one of these steps, even when the primary amino acid sequence specified by the gene is normal, can produce a protein with modified properties leading to alterations in its functioning and antigenicity of the type seen in the prion diseases (Gabizon and Prusiner, 1990).

For example, a specific glycolipid must be added to the carboxy terminal of PrP to permit its attachment to the cell membrane (Stahl *et al.*, 1987), whilst other data show that the membrane attachment and transmembrane transfer of PrP is altered in prion diseases, although the mechanisms are unclear (Lopez *et al.*, 1990; Stahl *et al.*, 1990; Yost *et al.*, 1990).

Further support for the central role of PrP in the transmissible dementias is provided by studies investigating factors which determine the incubation period of scrapie. It has long been known that the incubation time is affected by the strain both of the host and of the infectious agent. Various theories for the basis of this phenomenon had been advanced (Hope and Kimberlin, 1987; Westaway *et al.*, 1989), but it became likely from linkage data that the scrapie incubation gene was either identical to the prion gene, or very tightly linked to it (Carlson *et al.*, 1986; Race *et al.*, 1990). Soon afterwards, it was demonstrated that the two genes are almost certainly the same, by showing that mice and sheep with long incubation periods differed from those with short incubation times by variation in the prion gene sequence (Westaway *et al.*, 1987; Goldmann *et al.*, 1991). Specifically, they varied in the amino acid coded for at either codon 108 or 189 in mice and codons 112, 130 and 147 in sheep. It is thought that these changes produce a significant effect on the extracelluar part of PrP and its potential for post-translational modification. These differences in the host-encoded PrP demonstrate another central role for the prion gene in the disease: not only does it encode the protein associated with infectivity, it also controls the incubation period for the disease. Confirmation of these findings was given when it was shown that forms of the PrP encoded by these alleles affected incubation time (Carlson *et al.*, 1989), and especially that introducing different prion genes into transgenic mice resulted in different incubation periods (Scott *et al.*, 1989), and altered susceptibility to the infection and the pattern of pathological features (Lowenstein *et al.*, 1990).

A similar effect on incubation time has now been shown to occur in humans. In a large family with CJD and a 144 bp insert in the prion gene, age of disease onset appears to be linked to valine (V) or methionine (M) homozygosity at codon 129 (Collinge *et al.*, 1991). Patients who are VV at codon 129 have an early age of onset and patients who are MM have a late age of onset.

Together, these findings provide explanations for many aspects of these disorders for which nucleic acid in the infectious agent had been assumed to be essential (Carlson *et al.*, 1989). It is now postulated that allelic heterogeneity within the prion gene, as this sequence variability is called, can explain differences in individual susceptibility to, and features of, prion-related diseases in humans (Scott *et al.*, 1989). What is not clear is whether the mutations now known to be associated with cases of GSS and CJD are simply other allelic variants equivalent to those described in mice, or whether they have different relationships to the occurrence of disease. For example, the codon 200 mutation has now been discovered in some unaffected members of the previously described pedigree (Goldfarb *et al.*, 1990); it remains to be seen whether these individuals subsequently develop CJD, or whether this particular mutation is not invariably associated with disease. Further study of prion gene sequences in both humans and animals is needed to clarify these points.

MECHANISMS OF PRION PATHOGENICITY

There is uncertainty as to the molecular events by which the inheritance of an abnormal prion gene or acquisition of abnormal PrP results in the clinical and pathological stigmata of the spongiform encephalopathies. This question is relevant whether or not prions are the sole infectious agent. A number of mechanisms have been proposed (Prusiner, 1982; Harrison and Roberts, 1991; Prusiner, 1991). If PrP[Sc] is indeed the cause, it could produce more of itself either by reverse translation (whereby a protein gives rise to its encoding messenger RNA) or by promoting synthesis of further aberrant PrP molecules through a variety of indirect pathways. Diedrich *et al.*, (1991) have implicated astrocytes in the formation of PrP^{Sc} scrapie and suggest that this cell type might also be involved in the replication of PrP^{Sc}. This is based on evidence that PrP^{Sc} accumluates in astrocytes prior to the standard pathological changes of astrocytosis, vacuolization, neuron loss and amyloid deposition. At present, there are no data to indicate which of these candidate cellular and molecular processes actually occur to explain how the inheritance of an abnormal prion gene, or contraction of infection, results in production of pathology. One possibility currently under investigation to account for the acquired form is that PrP^{Sc} serves as a template for the catalytic conversion of normally produced PrP into more PrPp^{Sc} (Harrison and Roberts, 1991; Prusiner, 1991).

In a series of experiments McKinley and colleagues (1990) have demonstrated that PrP^{Sc} is localized in secondary lysosomes, possibly primary lysosomes and within the cytoplasm of the cell. PrP^{Sc} was not seen in the nucleus, nuclear membrane, endoplasmic reticulum or Golgi stack. Treatment of cells with brefeldin A, which destroys Golgi apparatus, did not alter the pattern of distribution. These results indicate that the conversion of PrP^C into PrP^{Sc} is a post-translational event that occurs downstream of the endoplasmic reticulum and the mid Golgi apparatus. Previous studies have shown that the PrP^{Sc} isoform derives from a phospholipase C sensitive precursor (Stahl *et al.*, 1990). Together these studies suggest that the site of conversion is located either on the external surface of the membrane or in the early stages of endosome formation. The localization of PrP^{Sc} in putative primary lysosomes reveals a possible pathway for the entry of PrP^{Sc} into the cell by hitchhiking on the cells own transport system. This may give a vital clue to the mechanism of cellular pathology. It is possible that whilst it is on the membrane surface PrP^{Sc} does not affect cellular metabolism. However, once internalized (Taraboulos *et al.*, 1990) by the cells own transport mechanism the protease resistance of the abnormal isoform interferes with lysosomal function, causing a breakdown in lysosome structure and thus leading to a cellular catastrophe (Roberts and Collinge, 1991).

PRION GENE MUTATIONS

Genetic linkage analysis has confirmed that prion mutations occur in some familial cases and that they are tightly linked to disease. These data provide the strongest support yet for the prion hypothesis. The first mutation was described by Owen and colleagues (Owen *et al.*, 1989, 1990a), who found an insertion of an extra 144 base pairs at codon 52 in the prion gene in affected members of a family with CJD. Interestingly, the same insertion was then found in a case clinically diagnosed as familial AD (Collinge *et al.*, 1990), and has been identified in 5 (out of 101) cases of atypical dementias of various diagnoses but in which CJD and GSS had not been suspected (Owen *et al.*, 1990b).

Soon afterwards, Hsaio *et al.*, (1989) reported a pedigree of GSS in which the prion gene contained a single base mutation, leading to an amino acid substitution (at codon 102) of proline to leucine. This mutation was confirmed in other families with GSS, and an additional mutation at codon 189 (changing alanine to valine) was reported in another pedigree. The link between GSS and CJD was emphasized when the same group identified the codon 102 mutation in a familial CJD case (Doh-ura *et al.*, 1990); further mutations have been identified. It is assumed that these mutations are the direct cause of disease in these families. It is also assumed that such mutations are pathogenic by virtue of the encoded protein having altered properties which result in cellular dysfunction and ultimately neurological disorder.

PRION DISEASE CLINICAL AND PATHOLOGICAL OBSERVATIONS

Clinical Features

The current clinical diagnostic criteria for the best known incarnation of prion diseases, namely CJD, are shown in Table 2.

Table 2.   Current clinical diagnostic criteria

CJD	A history of rapidly advancing dementia, abnormal EEG, varying combinations of myoclonus, other movement disorders, cortical blindness, cerebellar ataxia, rigidity and akinetic mutism (Will *et al.* 1984).
GSS	Familial autosomal dominant cerebellar disorder, mild dementia, other diagnostic criteria as for CJD excluding abnormal EEG.
Kuru	The same criteria as the CJD, except that it only occurs in the Fore tribe of Papua New Guinea.

Since CJD was originally described in the early 1920s, it has become apparent that CJD has a distinctive typical picture (Irving *et al.*, 1990), but that it also has variants which merge into several other conditions and which are the source of much nosological and scientific discussion.

A typical case of CJD is that of a presenile dementia in the absence of a family history, occurring in the sixth decade and progressing rapidly to death within a year. A prodromal stage lasting weeks or months and characterized by neurasthenic symptoms can occur. The patient complains of fatigue, insomnia, anxiety and depression, and shows a gradual change towards mental slowness and unpredictability of behaviour. At this stage there may be evidence of impaired memory and concentration, and the limbs may appear to be weak and the gait unsteady. Frequently, neurological signs are lacking and a functional psychiatric disorder is suspected. This is especially likely in cases in which the early symptoms remit for several weeks at a time.

Intellectual deterioration or neurological defects soon become prominent. Although variable they are liable to involve motor functions, speech and/or vision. In addition, ataxia of cerebellar type, spasticity of limbs with progressive paralysis, extrapyramidal rigidity, tremor or choreoathetoid movements may also be found. Involvement of the anterior horn cells of the cord may lead to muscular fibrillation and atrophy, especially of the small hand muscles, resembling amyotrophic lateral sclerosis. Speech disturbances are common with dysphasia and dysarthria, likewise parietal lobe symptoms such as right–left disorientation, dyscalculia and finger agnosia. Vision may be severely affected with rapidly progressive cortical blindness. Apart from this, sensory changes are usually absent. Brain-stem involvement may lead to nystagmus, dysphagia, or bouts of uncontrollable laughing and crying. Myoclonic jerks are frequently seen and epileptic fits may occur. Intellectual deterioration follows or appears along with the neurological defects and evolves with great rapidity. An acute organic picture may be present initially with clouding of consciousness or frank delirium. Auditory hallucinations and delusions may be marked, and confabulation is often seen.

Ultimately a state of profound dementia is reached, accompanied by gross rigidity or spastic paralysis and often a decorticate or decerebrate posture. Repetitive myoclonic jerking of muscle groups is often still evident late in the disease. Emaciation is usually profound by the time death occurs (Tomlinson and Corsellis, 1984; Masters *et al.*, 1981a; Will and Mathews, 1984).

The course is a good deal more rapid than with most other primary dementing illnesses. Over half of the patients are dead within nine months and the great majority within two years (Will and Mathews, 1984). Death is usually preceded by a period of deepening coma which lasts for several weeks. There is no treatment able to prevent or retard development of prion disease.

Clinical and pathological variability

The emphasis on particular clinical features of marked severity in a given case has led to many subtypes being recognized. A familial form exists, accounting for 6–15% of cases (Masters *et al.*, 1981a; Will and Mathews, 1984; Brown *et al.*, 1987). These tend to be younger than sporadic cases but survive longer. Familial cases may also be less transmissible to primates (64% of cases compared with over 90% of sporadic CJD (Brown *et al.*, 1984, 1986). Prion disease can take different clinical forms; for example, there are some in which ataxia and cerebellar signs are prominent and a variety named after Heidenhain (1929) with severe occipital cortical involvement. However, a close examination of these and other cases demonstrates that these variants lie within the broad spectrum of the disorder. Considerable overlapping of pathological and clinical features occurs from case to case. Under these circumstances the prominence given to subtypes in the clinical or pathological descriptions of the disorder would appear to be unjustified.

The true extent of clinical and pathological variability is only now being appreciated since the advent of molecular biological techniques have allowed the accurate diagnosis of cases of familial prion disease. Gerstmann–Sträussler syndrome (GSS) is rarer than CJD and is an autosomal dominant cerebellar disorder, in which dementia tends to be mild. Mean age of death is earlier than CJD (around 50 years old) but duration of illness is longer (4–5 years) (Masters *et al.*, 1981b; Ridley *et al.*, 1986). Most pathological features are similar to those of CJD, but amyloid plaques in the cortex and cerebellum (see histopathology) are invariably present (Masters *et al.*,

1981b; Roberts *et al.*, 1988). Like CJD, the pathology of GSS may sometimes overlap with that of AD (De Courten-Myers and Mandybur, 1987; Pearlman *et al.*, 1988; Farlow *et al.*, 1989; Ghetti *et al.*, 1989; Nochlin *et al.*, 1989) and the clinical and pathological pictures of GSS and CJD merge into each other (Masters et al 1981b; Nochlin *et al.*, 1989). Moreover, there is considerable individual clinical and pathologic variation among affected members within GSS kinships (Table 3), as is also true in CJD families.

Table 3.   The variation of pathology in selected cases in one large family of inherited prion disease with the 144 base pair gene insertion

Case	Sex	Age	Brain weight (g)	Histology
V.20	M	38	1200	Severe cortical spongiform change most prominent in occipital lobe. Severe neuronal loss and gross astrocytic proliferation. Mild atrophy, no prion or β/A4 plaques and no NFTs.
VI.21	M	45	1220	Mild and patchy spongiform change. Mild neuronal loss and increase in number of astrocyte nuclei in the cerebral cortex. Mild atrophy, no prion or β/A4 plaques and no NFTs.
VII.2	f	43	804	Moderate general spongiform change, but severe in deep cortical laminae. Moderate neuronal loss and astrocytosis, except in the basal ganglia, peri-aqueductal grey matter and colliculi where astrocytosis was severe. Severe atrophy and prion plaque. No β/A4 plaques or NFTs.
VII.63	M	36	1315	No cortical spongiform change, neuronal loss or astrocytosis. Moderate astrocytosis in the globus pallidus. Mild atrophy and patchy loss of Purkinje cells in cerebellum. No prion or β/A4 plaques and no NFTs.

In a German kindred with the 102 codon mutation (Brown *et al.*, 1991) no single clinical sign was invariably present in all affected members; dementia, behavioural disturbances, cerebellar and corticospinal signs, and abnormal movements all varied from patient to patient, as did age at onset (third to seventh decade) and the duration of illness (3 to 10 years). Even the pathology was variable; although all patients appeared to have prominent multicentric plaques composed of prion amyloid protein, the spongiform changes varied from severe to absent, and the Indiana kindred had widespread neurofibrillary tangles with minimal spongiform change (Ghetti *et al.*, 1989). Collinge *et al.*, (1991) report similar widespread clinical and pathological variability in a large kindred of inherited prion disease (GSS) with the 144 base pair gene insert. None of these pathologic features appear to be correlated with the clinical course (or with experimental transmissibility) in these or other reported families, but the number of patients for which such comparisons can be made is still very limited (Brown *et al.*, 1991).

DIAGNOSING PRION DISEASE

When the triad of presenile dementia, myoclonus and abnormal EEG are present, clinical diagnosis of prion disease is usually reliable and accurate; conversely, the absence of both of the latter two features virtually excludes prion disease (Brown *et al.*, 1986).

Unfortunately, these conditions are not always satisfied, and the occurrence of clinical variants can make definite diagnosis impossible in the absence of biopsy confirmation (Jones *et al.*, 1985), neuropathology or demonstration of experimental transmissibility. To compound matters, the atypical clinical forms are less likely to show helpful EEGs (Will and Mathews, 1984). In some cases, therefore, the differential diagnosis includes all causes of dementia and ataxia.

On probability grounds, the commonest misdiagnosis is that of AD, especially for those later onset cases of CJD (about a third have an onset between 60 and 70 years (Tsuji and Kuroiwa, 1983)), or the 10–25% which have a prolonged course over two years (Tsuji and Kuroiwa, 1983; Brown *et al.*, 1984; Davanipour *et al.*, 1988). Helpful differentiating features, apart from speed of the progression of illness, include cerebellar, pyramidal or extrapyramidal signs and early visual disturbances, all of which are more common in prion disease (Will and Mathews, 1984; Brown *et al.*, 1986).

With the exception of the EEG, routine investigations are generally unhelpful in both conditions; CT scans tend to remain normal in prion disease, whereas in AD cortical atrophy is common (Galvez and Cartier, 184; Will and Mathews, 1984). The clinical overlap between prion disease and AD is mirrored pathologically; AD can show spongiform changes (Flament-Durand and Couck, 1979; Ball, 1980; Mancardi *et al.*, 1982; Smith *et al.*, 1987), whilst prion disease brains may contain Alzheimer-type senile plaques (Hirano *et al.*, 1972; Brown, 1989), neurofibrillary tangles (Ghetti *et al.*, 1989; Brown *et al.*, 1990a), or show cell loss in the cholinergic basal nucleus (Arendt *et al.*, 1984) and in both cases sulfated gylcosaminoglycans colocalize with the amyloid protein found in AD plaques and NFTs and the amyloid protein found in prion plaques (Guiroy *et al.*, 1991). However, an interesting difference is the absence of an inflammatory response in the vicinity of the plaques found in prion disease (Eikelenboom *et al.*, 1991).

Moreover, a third of familial AD cases are clinically indistinguishable from prion disease, yet show unequivocal AD pathology and are not transmissible to animals (Brown, 1989). Other diseases which have been confused clinically with prion disease include tumours, Huntington's chorea, supranuclear palsy, amyotrophic lateral sclerosis, Leigh's disease, corticostriate degeneration and progressive multifocal leucodystrophy (Masters *et al.*, 1981b; Brown *et al.*, 1986; Irving *et al.*, 1990). In addition, Byrne and Lowe (1991) recently pointed out some clinical and pathological overlap between diffuse Lewy body (DLB) disease and prion disease. They point out that this may give rise to a number of cases of DLB disease in which prion disease is suspected because of heightened clinical awareness and thus cause considerable confusion. Most of these diagnoses are more likely to be considered if motor or bizarre neurological symptoms are prominent relative to dementia, and such cases are often seen and managed by neurologists.

Molecular biology offers the surest method of diagnosis. Detection of the abnormal isoform of prion protein in CNS tissue by immunological techniques (eg Western blots, (Brown et al., 1986) or immunochemistry, (Roberts et al., 1986, 1988) or mutations and inserts in the prion gene are diagnostic (see Chapter 12).

When such tests are not available or cannot be completed diagnosis is based on the exclusion of similar conditions such as Huntington's chorea and Alzheimer's disease. In general the speed of disease progression is one of the best indicators. Confirmation of a clinical diagnosis by neuropathological examination is essential.

Within the rubric of the prion diseases, a subgrouping could be based on the origin of the PrP^{Sc} in terms of genetic, sporadic–idiopathic or iatrogenic factors. The disease could then be classified further in terms of the specific molecular causes (e.g. the exact prion mutation) as such details become apparent. This approach supersedes the inevitably inconclusive debates as to the correct nomenclature for each clinicopathological variant, or the relationship between one syndrome and another.

IMMUNO-DETECTION OF PrP

Reliable detection of PrP^{Sc} in plaques by immunostaining (Roberts et al., 1986, 1988), in brain tissue by Western blotting (Brown et al., 1986) or by localization of specific defects in the prion gene via polymerase chain reactions and Southern blotting (Hsiao et al., 1989; Owen et al., 1989a,b; Collinge et al., 1989) can detect the disease and be used as specific diagnostic tests.

The clinical overlap with other atypical dementias would remain a problem but could be unequivocally resolved by prion gene or protein analysis, irrespective of confounding clinical or pathological features. At present, the identification of PrP^{Sc} can only be carried out on brain tissue at either biopsy or autopsy. However, genetic analysis is rapidly becoming a diagnostic tool through regional centres, and it is hoped that improvements in immunological techniques should allow a test to be developed for demonstration of PrP^{Sc} in serum (Serban et al., 1990).

The value of PrP testing over conventional diagnostic approaches to possible transmissible dementias is illustrated by a recent extreme example. A familial CJD patient has been described who shared the prion gene mutation seen in other affected members of his family, yet at post-mortem the brain did not show significant pathology (Collinge et al., 1990). Moreover, the patient had satisfied clinical diagnostic criteria for AD. Subsequent PrP immunostaining demonstrated the presence of an abnormal pattern of PrP^{Sc} immunoreactivity despite the lack of any other neuropathological features of a spongiform encephalopathy (Clinton et al., 1990) and further investigations have indicated pathology of synaptic organization (Clinton et al., 1991).

In this instance, therefore, conventional clinical and pathological criteria would have been inadequate to make the correct diagnosis.

Analysis of PrP protein may also clarify the prion positivity (and thus potential transmissibility) of other disorders, such as amyotrophic lateral sclerosis with dementia (Salazar *et al.*, 1983) and other atypical dementias (e.g. Knopman *et al.*, 1990; Neary *et al.*, 1990), whose nosological and aetiological status is unclear.

The presence of PrP^{Sc} molecules can be used to define and classify prion disease. This molecular classification of prion diseases could take its place within a broader grouping of several disorders conceptualized on similar grounds according to the nature of their cerebral amyloidosis (Roberts *et al.*, 1988); it represents an important example of the shift towards classification of neuropsychiatric disorders according to their pathogenesis.

EPIDEMIOLOGY

The true extent of prion disease (as opposed CJD or GSS etc.) in the population is unknown at present. This is due to its recent emergence as a single nosological entity. A guide to the epidemiology of prion disease can be obtained by looking at the figures for CJD. CJD is found all over the world and has an incidence of 1 to 2 cases per million of the population (Will and Mathews, 1984). Approximately equal numbers of men and women are affected and around 10% of cases are familial. The age at onset is usually between 55 and 75 years of age with a peak frequency in the 60s (Fig. 3). Rare cases with onset as young as 16 or as old 85 have been reported. Higher rates than this have been recorded in some regions or populations such as amongst Libyan Jews (Hsiao *et al.*, 1991), and in areas of central Slovakia and Hungary and eastern England. It is now known that these high local rates are accounted for by large families with mutation in the prion protein gene.

The disease is generally diagnosed using the criteria for CJD. However, since the clinical phenotype is known to be very variable these criteria mean that cases might be misdiagnosed particularly in older patients (65 + years). This point is illustrated by the marked fall in mortality rates from CJD seen after the age of 65 in England and Wales in the periods 1970–1979 and 1980–1984 (Cousens *et al.*, 1990) (see Fig. 3). It should be noted, however, that the decline in mortality after 65 years is less pronounced in the period 1980–1984. It is possible that this is due to the increased awareness of the confusion between prion disease (CJD) and other neurodegenerative diseases, mainly AD.

It is unfortunate that, because of exaggerated fears of the infection risk, autopsies are rarely performed on atypical dementia cases. This is a major obstacle to pathological and epidemiological understanding of the spongiform encephalopathies.

Partly as a result of the low autopsy rate, and the tendency to certify 'Alzheimer's disease' as the cause of death for any dementia lacking tissue diagnosis, the true prevalence of the prion diseases (as opposed to classical CJD and GSS) is unknown. However, it is likely to be higher than the often-quoted figure of 1 per million (corresponding to 30 cases per year in the UK) (Will and Mathews, 1984). For example, some 10% of dementia cases have no definitive neuropathological diagnosis after

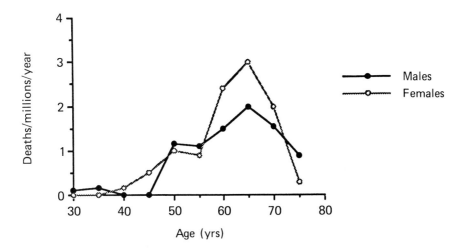

Mortality rates from Creutzfeldt-Jakob disease in England and Wales in the period 1980–1984.

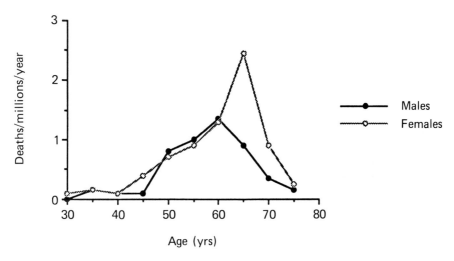

Mortality rates from Creutzfeldt-Jakob disease in England and Wales in the period 1970–1979

Fig. 3.   This figure demonstrates the marked decline in mortality rates from CJD after the age of 65 years. Note that it was more pronounced in the period from 1970 to 1979.

investigation. It is likely that some of these cases will prove to be prion disease. A consideration of this fact gives a possibility that prion disease might account for up to 9000 cases of dementia per year in the UK, i.e. 10% of undiagnosed dementia cases (Roberts, 1990a). Undoubtedly, this theoretical figure is an overestimate. However, even if the latter range overestimates the true prevalence by a factor of ten, the prion diseases would remain seriously underdiagnosed at present. Such uncertainties must be taken into account when interpreting epidemiological surveys or assessing apparent changes in disease incidence over the next few years.

ACCIDENTAL TRANSMISSION OF PRION DISEASES

Once the fundamental similarity between animal and human types of spongiform encephalopathy is appreciated, the experimental evidence for transmissibility of these disorders within and between species leads to concern. Several points can be made.

The recent spread of prion disease to cows through scrapie-infected bone meal and cats in the UK, and the experimental transmission to pigs, all species not previously known to be affected, underlines the potential of all animals who posses a prion gene to develop prion disease when challenged with the abnormal isoform of the prion protein, PrP^{Sc} (Roberts, 1990b).

This argument has been underlined by the recent reports documenting the occurrence of prion disease in cats. The striking similarity of the disease and pathology in cats to those previously described transmissible spongiform encephalopathies in other animals (Table 1) leaves little doubt that the affected cats had prion disease. The relatively widespread geographical distribution and pattern of emergence of these cases is similar to that witnessed at the start of the BSE epidemic (Wilesmith et al., 1988). This pattern is consistent with the disease in cats resulting from exposure to a common source of infection (possibly contaminated petfood), as has been shown to be the case for BSE (Wilesmith et al., 1988, 1991). In the BSE epidemic, studies have indicated that effective exposure began suddenly in 1981–1982 and was due to ingestion of contaminated cattlefeed (Wilesmith et al., 1991). The survival of scrapie infectivity after 3 years interment (Brown and Gajdusek, 1991) has implications for enviromental contamination of areas which are known to contain infective material, such as landfill sites used to dispose of BSE-infected carcasses. However, it is important to put this risk into perspective. For example, the extensive scrapie research is reassuring: there is no epidemiological association between the prevalence of classically diagnosed CJD and that of scrapie (Masters et al., 1979), nor a clear association with 'high risk' occupations such as pathology technicians and butchers (Brown et al., 1987) who may handle contaminated human or animal material, despite anecdotal reports (Mathews, 1990) and occasional findings to the contrary (Kondo and Kuroiwa, 1982; Davanipor et al., 1985). The occurrence of CJD in a lifelong vegetarian may also provide comfort to meat eaters (Mathews and Will, 1981). More importantly, experimental oral transmission is calculated to be 10^9 times less efficient than intracerebral inoculation (although 1 g of infected tissue may be sufficient for transmission to animals) and often fails to produce disease at all (Gibbs et al., 1980; Prusiner et al., 1985).

CONCLUSIONS

Improvements in our understanding of the molecular pathology of scrapie have led to the grouping together of the spongiform encephalopathies as a single nosological entity—prion disease.

The identification of families with autosomal dominant prion disease has dramatically widened our appreciation of the spectrum of pathology seen in cases of prion disease.

Although the potential for diagnostic confusion is substantial screening for gene defects and immunoreactivity to PrP^{Sc} after a way to diagnostic certainty.

The adoption of improved methods of diagnosis will improve our understanding of the epidemiology of prion disease and enable us to evaluate the risks of infection in humans.

ACKNOWLEDGEMENTS

This work was supported by the AFRC and the MHF.

REFERENCES

Arendt, T., Bigl, V. and Arendt, A. (1984) Neurone loss in the nucleus basalis of Meynert in Creutzfeldt–Jakob disease. *Acta Neuropathol. (Berlin)*, **65** 85–88.

Baker, F., Duchen, L.W., Jacobs, J.M. and Ridley, R.M. (1990) Spongiform encephalopathy transmitted experimentally from Creutzfeld Jakob and familial Gerstmann–Sträussler–Scheinker diseases. *Brain,* **113** 1891–1909.

Baker, H.F., Ridley, R.M. and Crow, T.J. (1985) Experimental transmission of an autosomal dominant spongiform encephalopathy: does the infectious agent originate in the human genome? *Br. Med. J.* **291** 299–302.

Ball, M.J. (1980) Features of Creutzfeldt–Jakob disease in brains of patients with familial dementia of the Alzheimer type. *Can. J. Neurol. Sci.* **7** 51–57.

Basler, K., Oesch, B., Scott, M., Westaway, D., Walchi, M., Groth, D.F., McKinley, M.P., Prusiner, S.B. and Weissman, C. (1986) Scrapie and cellular PrP isoforms are encoded by the same chromosomal gene. *Cell* **46** 417–428.

Beck, E. and Daniel, P.M. (1987) Neuropathology of transmissible spongiform encephalopathies. In: Prusiner, S.B. and McKinley, M.P. (eds) *Prions: Novel Infectious Pathogens Causing Scrapie and Creutzfeldt–Jakob Disease.* Academic Press, San Diego, CA, pp. 331–385.

Brown, P. (1989) Central nervous system amyloidoses: a comparison of Alzheimer's disease and Creutzfeldt–Jakob disease. *Neurology,* **39** 1103–1105.

Brown, P. and Gajdusek, C.D. (1991) Survival of scrapie virus after 3 years' interment. *Lancet,* **337** 269–270.

Brown, P., Rodgers-Johnson, P., Cathala, F., Gibbs, C.J.,Jr, and Gadjusek, D.C. (1984) Creutzfeldt–Jakob disease of long duration: clinicopathological characteristics, transmissibility, and differential diagnosis. *Ann. Neurol.* **16** 295–304.

Brown, P., Gajdusek, D.C., Gibbs, C.J. and Asher, D.M. (1985) Potential epidemic of Creutzfeldt–Jakob disease from human growth hormone therapy. *N. Engl. J. Med.* **313** 728–731.

Brown, P., Cathala, F., Castaigne, P. and Gajdusek, D.C. (1986) Creutzfeldt–Jakob disease: clinical analysis of a consecutive series of 230 neuropathologically verified cases. *Ann. Neurol.* **20** 597–602.

Brown, P., Cathala, F., Raubertas, R.F., Gajdusek, D.C. and Castiagne, P. (1987) The epidemiology of Creutzfeldt–Jakob disease: conclusions of a 15 year investigation in France and review of the world literature. *Neurology* **37** 895–904.

Brown, H.R., Goller, N.L., Rudelli, R.D., Merz, G.S., Wolfe, G.C., Wisniewski, H.M. and Robakis, N.K. (1990a) The mRNA encoding the scrapie agent is present in a variety of non-neuronal cells. *Acta Neuropath. (Berlin)*, **80** 1–6.

Brown, P., Jannotta, F., Gibbs, C.J., Baran, H., Guiroy, D.C. and Gadjusek, D.C. (1990b) Co-existence of Creutzfeldt–Jakob disease and Alzheimer's disease in the same patient. *Neurology* **40**, 226–228.

Brown, P., Liberski, P.P., Wolff, A. and Gdjusek, D.C. (1990c) Conservation of infectivity in purified fibrillary extracts of scrapie-infected hamster brain after sequential enzymatic digestion or polyacrilamide gel electrophoresis. *Proc. Natl. Acad. Sci. USA* **87** 7240–7244.

Brown, P., Goldfarb, L.G. and Gadjdusek D.C. (1991) The new biology of spongiform encephalopathy: infectious amyloidoses with a genetic twist. *Lancet* **337** 1019–1022.

Byrne, E.J., and Lowe, J. (1991) Transmissible Dementias. *Br. J. Psychiatry* **158** 291.

Carlson, G.A., Kingsbury, D.T., Goodman, P.A., Coleman, S. and Marshall, S.T. (1986) Linkage of prion protein and scrapie incubation time genes. *Cell* **46** 503–511.

Carlson, G.A., Westaway, D., DeArmond, S.J., Peterson-Torchia, M. and Prusiner, S.B. (1989) Primary structure of prion proteins may modify scrapie isolate properties. *Proc. Natl. Acad. Sci. USA* **86** 7475–7479.

Ceroni, M., Piccardo, P., Safar, J., Gajdusek, C.D. and Gibbs, C.J., Jr (1990) Scrapie infectivity and prion protein are distributed in the same pH range in agarose isoelectric focusing. *Neurology* **40**, 508–513.

Cheseboro, B., Race, R., Wehrly, K., *et al.*, (1985) Identification of scrapie prion protein-specific mRNA in scrapie-infected and uninfected brain. *Nature (London)* **315** 331–333.

Chou, S.M., Payne, W.N. and Gibbs, C.J. (1980) Transmission and scanning electron microscopy of spongiform change in Creutzfeldt–Jakob disease. *Brain* **103** 885–904.

Clinton, J., Lantos, P.L., Rossor, M., Mullan, M. and Roberts, G.W. (1990) Immunocytochemical confirmation of prion protein. *Lancet* **336** 515.

Clinton, J., Lantos, P.L., Doey, L. and Roberts, G.W. (1991) Synaptic degeneration in prion disease: a primary neuropathological feature. *Neuroreport* (submitted)

Collee, J.G. (1990) Foodborne illness. Bovine spongiform encephalopathy. *Lancet,* **336**, 1300–1303.

Collinge, J., Harding, A.E., Owen, F., Poulter, M., Lofthouse, R., Boughey, A.M., Shuh, T. and Crow, T.J. (1989) Diagnosis of Gerstmann–Sträussler syndrome in familial dementia with prion protein gene analysis. *Lancet* **336** 15–16.

Collinge, J., Owen, F., Poulter, M., Leach, M., Crow, T.J., Rossor, M.N., Hardy, J.,

Mullan, M., Janota, I. and Lantos, P.L. (1990) Prion dementia without characteristic pathology. *Lancet* **336** 7–9.

Collinge, J., Palmer, M. and Dryden, A.J. (1991) Genetic predisposition to iatrogenic Creutzfeld–Jakob disease. *Lancet* **337** 1441–1442.

Cousens, S.N., Harries-Jones, R., Knight, R., Will, R.G., Smith, P.G. and Mathews, W.B. (1990) Geographical distribution of cases of Creutzfeldt–Jakob disease in England and Wales 1970–84. *J. Neurol. Neurosurg. Psuychiatry*. **53** 459–465.

Davanipour, Z., Alter, M., Sobel, E., Asher, D. and Gadjdusek, D.C. (1985) Creutzfeldt–Jakob disease: possible medical risk factors. *Neurology* **35** 1483–1486.

Davanipour, Z., Alter, M., Coslett, H.B., Sobel, E., Kundu, S. and Hoeing, E.M. (1988) Prolonged progressive dementia with spongiform encephalopathy: a variant of Creutzfeldt–Jakob disease? *Neuroepidemiology* **7** 56–65.

DeArmond, S.J., Mobley, W.C., DeMott, D.L., Barry, R.A., Beckstead, J.H. and Prusiner, S.B. (1987) Changes in the localization of brain prion proteins during scrapie infection. *Neurology* **37** 1271–1280.

De Courten-Myers, G. and Mandybur, T. (1987) Atypical Gerstmann–Sträussler syndrome or familial spinocerebellar ataxia and Alzheimer disease? *Neurology* **37** 269–275.

Diedrich, J.F., Bendheim, P.E., Kim, Y.S.M., Carp, R.I. and Haase, A.T. (1991) Scrapie-associated prion protein accumulates in astrocytes during scrapie infection. *Proc. Natl. Acad. Sci. USA* **88** 375–379.

Doh-ura, K., Tateishi, J., Kitamoto, T. and Sakaki, Y. (1990) Creutzfeldt–Jakob disease patients with congophilic kuru plaques have the missense variant prion protein common to Gerstmann–Sträussler syndrome. *Ann. Neurol.* **27** 121–126.

Doi-Yi, R., Kitamoto, T. and Tateishi, J. (1991) Immunoreactivity of cerebral amyloidosis is enhanced by protein denaturation treatments. *Acta Neuropathol.* **82** 260–265.

Farlow, M.R., Yee, R.D., Dlouhy, S.R., Conneally, P.M., Azzarelli, B. and Ghetti, B. (1989) Gerstmann–Sträussler–Scheinker disease. I. Extending the clinical spectrum. *Neurology* **39** 1446–1452.

Farquhar, C.F., Somerville, R.A. and Ritchie, L.A. (1989) Post-mortem immunodiagnosis of scrapie and bovine spongiform encephalopathy. *J. Virol. Methods* **24** 215–222.

Flament-Durand, J. and Couck, A.M. (1979) Spongiform alterations in brains of presenile dementia. *Acta Neuropathol. (Berlin)* **46** 159–162.

Gabizon, R. and Prusiner, S.B. (1990) Prion liposomes. *Biochem. J.* **266** 1–14.

Gajdusek, D.C. (1977) Unconventional viruses and the origin and disappearance of kuru. *Science* **197**, 943–960.

Gajdusek, D.C., Gibbs, C.J. and Alpers, M. (1966) Experimental transmission of a kuru-like syndrome to chimpanzees. *Nature* **209** 794–796.

Gajdusek, D.C., Gibbs, C.J., Asher, D.M., *et al.* (1977) Precautions in medical care of and in handling material from patients with transmissible virus dementias (Creutzfeldt–Jakob disease). *N. Engl. J. Med.* **297** 1253–1258.

Galvez, S. and Cartier, L. (1984) Computerised tomography in 15 cases of Creutzfeldt–Jakob disease with histological verification. *J. Neurol. Neurosurg. Psychiatry* **47**

1244–1246.

Ghetti, B., Tagliavini, F., Masters, C.L., Beyreuther, K., Giaccone, G. and Verga, L. (1989) Gerstmann–Sträussler–Scheinker disease. II. Neurofibrillary tangles and plaques with PrP-amyloid coexist in an affected family. *Neurology* **39** 1453–1459.

Gibbs, C.J., Gajdusek, D.C., Asher, D.M. and Alters, M.P. (1968) Creutzfeldt–Jakob disease (spongiform encephalopathy): transmission to the chimpanzee. *Science* **161** 388–389.

Gibbs, C.J., Amyx, H.L., Bacote, A., Masters, C.L. and Gadjusek, D.C. (1980) Oral transmission of kuru, Creutzfeldt–Jakob disease and scrapie to non-human primates. *J. Infect. Dis.* **142** 205-208.

Goldfarb, L.G., Mitrova, E., Brown, P., Toh, B.K. and Gadjdusek, D.C. (1990) Mutation in codon 200 of scrapie amyloid protein gene in two clusters of Creutzfeldt–Jakob disease in Slovakia. *Lancet* **336** 514–515.

Goldmann, W., Hunter, N., Benson, G., Foster, J.D. and Hope, J. (1991) Different scrapie-associated fibril proteins (PrP) are encoded by lines of sheep for different alleles of the *Sip* gene. *J. Gen. Virol.* **72** 2411–2417.

Gray, E.G. (1986) Spongiform encephalopathy: a neurocytologist's viewpoint with a note on Alzheimer's disease. *Neuropathol. Appl. Neurobiol.* **12** 149–172.

Guiroy, D.C., Yanagihara, R., and Gajdusek, D.C. (1991) Localisation of amyloidogenic proteins and sulfated glycoaminoglycans in nontransmissible and transmissible cerebral amyloidosis. *Acta Neuropathol.* **82** 87–92. Hadlow, W.J. (1959) Scrapie and kuru. *Lancet* **ii** 289–290.

Harrison, P.J. and Roberts, G.W. (1991) Life Jim, but not as we know it. *Br. J. Psychiatry* **158** 457–470.

Hirano, A., Ghatak, N.R., Johnson, A.B., *et al.* (1972) Argentophilic plaques in Creutzfeldt–Jakob disease. *Arch. Neurol.* **26** 530–542.

Hope, J. and Kimberlin, R.H. (1987) The molecular biology of scrapie: the last two years. *Tins.* **10** 149–151.

Hsaio, K., Baker, H.F., Crow, T.J., Poulter, M., Owen, F., Terwilliger, J.D., Westaway, D., Oh, J. and Prusiner, S.B. (1989) Linkage of a prion protein missense variant to Gerstmann–Sträussler syndrome. *Nature (London)* **338** 342–345.

Hsiao, K., Meiner, Z., Kahana, E., Cass, C., Kahana, I., Arrahami, D., Scarlato, G., Abramsky, O., Prusiner, S.B and Gabizon, R. (1991) Mutation of the prion protein in Libyan jews with Creutzfeld–Jakob disease. *New Engl. J. Med.* **324** 1091–1097

Hunter, G.D. (1972) Scrapie: a prototype slow infection. *J. Infect. Dis.* **125** 427–440

Irving, W.L., Crimmins, D.S., Masters, C.L. and Cunningham, A.L. (1990) Creutzfeldt–Jakob disease and slow infections: a review. *Aust. NZ Med. J.* **20** 283–290.

Jones, H.R., Hedley-Whyte, E.T., Friedberg, S.R. and Baker, R.A. (1985) Ataxic Creutzfeldt–Jakob disease: diagnostic techniques and neuropathologic observations in early disease. *Neurology* **35** 254–257.

Klatzo, I., Gajdusek, D.C. and Zigas, V. (1959) Pathology of kuru. *Lab. Invest.* **8** 799–847.

Knopman, D.S., Mastri, A.R., Frey, W.H., Sung, J.H. and Rustan, T. (1990) Dementia lacking distinctive histological features. A common non-Alzheimer degenerative dementia. *Neurology* **40** 251–256.

Kondo, K. and Kuroiwa, Y. (1982) A case control study of Creutzfeldt–Jakob disease: association with physical injuries. *Ann. Neurol.* **11** 377–381.

Kretschmar, H.A., Stowring, L.E., Westaway, D., Stubblebine, W.H., Prusiner, S.B. and DeArmond, S.J. (1986a) Molecular cloning of a human prion protein cDNA. *DNA* **5** 315–324.

Kretschmar H.A., Prusiner, S.B., Stowring, L.E. and DeArmond, S.J. (1986b) Scrapie prion proteins are synthesized in neurons. *Am. J. Pathol.* **122** 1–5.

Landis, D.M.S., Williams, R.S. and Masters, C.L. (1981) Golgi and electromicroscopic studies of spongiform encephalopathy. *Neurology* **31** 538–549.

Lopez, C.D., Yost, C.S., Prusiner, S.B., Meyers, R.M. and Lingappa, V.R. (1990) Unusual topogenic sequence directs prion protein biosynthesis. *Science* **248** 226–229.

Lowenstein, D.H., Butler, D.A., Westaway, D., McKinley, M.P., DeArmond, S.J. and Prusiner, S.B. (1990) Three hamster species with different scrapie incubation times and neuropathological features encode distinct prion proteins. *Mol. Cell. Biol.* **10** 1153–1163.

Mancardi, G.L., Mandybur, T.I. and Liwnicz, B.H. (1982) Spongiform-like changes in Alzheimer's disease: an ultrastructural study. *Acta Neuropathol. (Berlin)* **56** 146–150.

Manuelidis, E.E., Kim, J.H., Mericangas, J.R. and Manuelidis, L. (1985) Transmission to animals of Creutzfeldt–Jakob disease from human blood. *Lancet* **ii** 896–897.

Masters, C.L. and Richards, E.P. Jr (1978) Sub-acute spongiform encephalopathy (Creutzfedt–Jakob disease). The nature and progression of spongiform change. *Brain* **101** 333–344.

Masters, C.L., Harris, J.O., Gadjusek, D.C., Gibbs, C.J.Jr, Bernoulli, C. and Asher, D.M. (1979) Creutzfeldt–Jakob disease: patterns of world-wide occurrence and the significance of familial and sporadic clustering. *Ann. Neurol.* **5** 177–188.

Masters, C.L., Gajdusek, D.C. and Gibbs, C.J. (1981a) The familial occurrence of Creutzfeldt–Jakob disease and Alzheimer's disease. *Brain* **104** 535–558.

Masters, C.L., Gajdusek, D.C. and Gibbs, C.J. (1981b) Creutzfeldt–Jakob disease virus isolations from the Gerstmann–Sträussler syndrome. With an analysis of the various forms of amyloid plaque deposition in the virus-induced spongiform encephalopathies. *Brain* **104** 559–588.

Mathews, W.B. (1981) Slow virus infections. *J. R. Coll. Physicians* **15** 109–112.

May, W.W., Itabishi, H.H. and Dejong, R.N. (1968) Creutzfeldt–Jakob disease. II. Clinical, pathological and genetic study of a family. *Arch. Neurol.* **19** 137–149.

McKinley, M.C., Hay, B., Lingappa, V.R., Lieberburg, I. and Prusiner, S.B. (1987) Developmental expression of prion protein gene in brain. *Dev. Biol.* **121** 105–110.

Meyer, R.K., McKinley, M.P., Bowman, K.A., Braunfield, M.B., Barry, R.A. and Prusiner, S.B. (1986) Separation and properties of cellular and scrapie prion proteins. *Proc. Natl. Acad. Sci. USA* **83** 2310–2314.

Neary, D., Snowden, J.S., Mann, D.M.A., Nothern, B., Goulding, P.J. and MacDermott, N. (1990) Frontal lobe dementia and motor neuron disease. *J. Neurol. Neurosurg. Psychiatry* **53** 23–32.

Nochlin, D., Sumi, S.M., Bird, T.D., Snow, A.D., Leventhal, C.M., Beyreuther, K. and

Masters, C.L. (1989) Familial dementia with PrP-positive plaques: a variant of Gerstmann–Sträussler syndrome. *Neurology* **39** 910–918.

Oesch, B., Westaway, D., Walchli, M., McKinley, M.P., Kent, S.B. and Aebersold, R. (1985) A cellular gene encodes scrapie PrP 27–30 protein. *Cell* **40** 735–745.

Owen, F., Poulter, M., Lofthouse, R., Collinge, J., Risby, D., Baker, H.F., Ridley, R.M., Hsiao, K. and Prusiner, S.B. (1989) Insertion in prion protein gene in familial Creutzfeldt–Jakob disease. *Lancet* **i** 51–52.

Owen, F., Poulter, M., Shah, T., Collinge, J., Lofthouse, R., Baker, H., Ridley, R., McVey, J. and Crow, T.J. (1990a) An in-frame insertion in the prion protein gene in familial Creutzfeldt–Jakob disease. *Mol. Brain Res.* **7** 273–276.

Owen F., Poulter, M., Collinge, J., *et al.* (1990b) Insertions in the prion protein gene in atypical dementias. *Exp. Neurol.* (submitted).

Pearlman, R.L., Towfighi, J., Pezeshkpour, G.H., Tenser, R.B. and Turel, A.P. (1988) Clinical significance of types of cerebellar amyloid plaques in human spongiform encephalopathies. *Neurology* **38** 1249–1254.

Piccardo, P., Safar, J., Ceroni, M., Gajdusek, D.C. and Gibbs, C.J., Jr (1990) Immunohistochemical localization of prion protein in spongiform encephalopathies and normal brain tissue. *Neurology* **40** 518–522.

Preece, M.A. (1991) Creutzfeldt–Jakob disease following treatment with human pituitary hormones. *Clin. Endocrinol.* **34** 527–529.

Prusiner, S.B. (1982) Novel proteinaceous infectious particles cause scrapie. *Science* **216** 136–144.

Prusiner, S.B., Cochran, S.P. and Alpers, M.P. (1985) Transmission of scrapie in hamsters. *J. Infect. Dis.* **152** 971–978.

Race, R.E., Graham, K., Ernst, D., Caughey, B. and Cheseboro, B. (1990) Analysis of linkage between scrapie incubation period and the prion protein gene in mice. *J. Gen. Virol.* **71** 493–497.

Ridley, R.M., Baker, H.F. and Crow, T.J. (1986) Transmissible and non-transmissible neurodegenerative disease: similarities in age of onset and genetics in relation to aetiology. *Psychiatry Med.* **16** 199–207.

Roberts, G.W. (1990a) Prion disease—spongiform encephalopathies unveiled. *Lancet* **336** 21–22.

Roberts, G.W. (1990b) Bovine spongiforn encephalopathy (BSE). In: *Agricultural Committee, Fifth Report.* HMSO, London, pp. 35–39.

Roberts, G.W. and Collinge, J. (1991) Playing clue with prion disease. *Lab. Invest.* (in press).

Roberts, G.W., Lofthouse, R., Brown, R., Crow, T.J., Barry, R.A., and Prusiner, S.B. (1986) Prion protein immunoreactivity in human transmissible dementias. *N. Engl. J. Med.* **315** 1231–1233.

Roberts, G.W., Lofthouse, R., Allsop, D., Landon, M., Kidd, M., Prusiner, S.B. and Crow, T.J. (1988) CNS amyloid proteins in neurodegenerative diseases. *Neurology* **38** 1534–1540.

Safar, J., Ceroni, M., Piccardo, P., Liberski, P.P., Miyazaki, M., Gadjusek, D.C. and Gibbs, C.J., Jr (1990a) Subcellular distribution and physicochemical properties of scrapie-associated precursor protein and relationship with scrapie agent. *Neurology* **40** 503–508.

Safar, J., Wang, W., Padgett, M.P., Ceroni, M., Piccardo, P., Zopf, D., Gadjdusek, D.C. and Gibbs, C.J., Jr (1990b) Molecular mass, biochemical composition, and physicochemical behaviour of the infectious form of the scrapie precursor protein monomer. *Proc. Natl. Acad. Sci. USA* **87** 6373–6377.

Salazar, C., Masters, C.L. Gajdusek, D.C. and Gibbs, C.J., Jr (1983) Syndrome of amyotrophic lateral sclerosis and dementia: relationship to transmissible Creutzfeldt–Jakob disease. *Ann. Neurol.* **14** 17–26.

Scott, M., Foster, D., Mirenda, C. *et al.* (1989) Transgenic mice expressing hamster prion protein produce species-specific scrapie infectivity and amyloid plaques. *Cell* **581** 847–857.

Serban, D., Taraboulos, A., DeArmond, S.J., and Prusiner, S.B. (1990) Rapid detection of Creutzfeldt–Jakob disease and scrapie prion proteins. *Neurology* **40** 110–117.

Smith, T.W., Anwer, U., DeGirolami, U. and Drachman, D.A. (1987) Vacuolar change in Alzheimer's disease. *Arch. Neurol.* **44** 1225–1228.

Sparkes, R.S., Simon, M., Cohn, V.H., Fournier, R.E., Lem, J., Klisak, I., Heinzmann, C., Blatt, C., Lucero, M., Mohandus, T., *et al.* (1986) Assignment of the human and mouse prion genes to homologous chromosomes. *Proc. Natl. Acad. Sci. USA* **83** 7358–7362.

Stahl, N., Borchelt, D.R., Hsiao, K. and Prusiner, S.B. (1987) Glycolipid modification of the scrapie prion protein. *Cell* **51** 229–240.

Stahl, N., Borchelt, D.R. and Prusiner, S.B. (1990) Differential release of cellular and scrapie prion proteins from cellular membranes by phosphatidylinositol-specific phospholipase C. *Biochemistry* **29** 5405–5412. Tomlinson, B.E. and Corsellis, J.A.N. (1984) Ageing and dementias. In: Adams, J.H., Corsellis, J.A.N. and Duchen, L.W. (eds) *Greenfield's Neuropathology,* 4th edn. Wiley, New York, pp. 951–1025.

Tsuji, S. and Kuroiwa, Y. (1983) Creutzfeldt–Jakob disease in Japan. *Neurology* **33** 1503–1506.

Weller, R.O. (1989) Iatrogenic transmission of Creutzfeldt–Jakob disease. *Psychiatry Med.* **19** 1–4.

Westaway, D., Goodman, P.A., Mirenda, C.A., McKinley, M.P., Carlson, G.A. and Prusiner, S.B. (1987) Distinct prion proteins in short and long scrapie incubation period mice. *Cell* **51** 651–662. DWestaway, D., Carlson, G.A. and Prusiner, S.B. (1989) Unravelling prion diseases through molecular genetics. *Tins.* **12** 221–227.

Wilesmith, J.W., Wells, G.A.H., Cranwell, M.P. and Ryan, J.B.M. (1988) Bovine spongiform encephalopathy. *Vet. Rec.* **123** 638.

Wilesmith, J.W., Ryan, J.B.M. and Atkinson, M.J. (1991) Bovine spongiform encephalopathy: epidemiology studies on the origin. *Vet. Rec.* **128** 199.

Will, R.G. and Mathews, W.B. (1984) A retrospective study of Creutzfeldt–Jakob disease in England and Wales 1970–79 I: clinical features. *J. Neurol. Neurosurg. Psychiatry* **47** 134–140.

Wyatt, J.M., Pearson, G.R., Smerdon, T.N., Gruffydd-Jones, T.J., Wells, G.A.H. and Wilesmith, J.W. (1991) Naturally occurring scrapie-like spongiform encephalopathy in five domestic cats. *Vet. Rec.* **129** 233–236.

Yost, C.S., Lopez, C.D., Prusiner, S.B., Meyers, R.M. and Lingappa, V.R. (1990) Non-hydrophobic extracytoplasmic determinant of stop transfer in the prion protein. *Nature (London)* **343** 669–672.

Part IV Animal prion diseases

22

Bovine spongiform encephalopathy: a brief epidemiography, 1985–1991

J.W. Wilesmith

ABSTRACT

The unanticipated occurrence of a transmissible spongiform encephalopathy in the cattle population in Great Britain obviously required an epidemiological study. This chapter describes a brief review of this aspect of the research on bovine spongiform encephalopathy. The main findings discussed relate to the source of infection, the reasons for the occurrence of the disease, the descriptive epidemiological features and the prospects for the future incidence.

INTRODUCTION

The archetypical member of the group of diseases now generally referred to as transmissible spongiform encephalopathies (TSEs) is scrapie of sheep. This is a disease for which there is convincing documentary evidence of its presence in Great Britain for the last two centuries. More recently, earlier this century, cases of naturally occurring scrapie in goats have been recorded in a number of countries including Great Britain. Co-grazing of sheep and cattle was not uncommon in Great Britain in the past without any untoward effects. The recognition in November 1986 of an apparently novel neurological disease of domestic cattle in which the histopathological lesions in the central nervous system were reminiscent of those in sheep scrapie was therefore somewhat unexpected and of some considerable interest. Although only four cases had been confirmed histopathologically by April 1987 their occurrence was regarded with due seriousness. A vital step was to disseminate information on the clinical signs and to request the voluntary notification of cases to facilitate more detailed pathological studies and to obtain an initial epidemiological assessment. As a result, although the number of cases accumulated was still then small, the disease could not be regarded as a sporadic event and epidemiological studies were commenced in June 1987.

The occurrence of BSE has considerably raised the profile of the TSEs and brought to the attention of both the scientific world and the general public the uncertainties about the nature of the transmissible agent involved. It has also naturally generated a large multidisciplinary research programme to study BSE and provided a stimulus for basic research on these pathogens. However, this chapter is confined to providing a very brief view of the findings of the epidemiological studies to date (January 1992).

RESULTS AND DISCUSSION

Aetiological studies

At the start of the epidemiological studies there was no definite evidence that BSE was a member of the TSEs although the pathological findings were highly suggestive of a scrapie-like disease (Wells et al., 1987). A cautious, catholic approach to considering potential aetiological hypotheses was therefore made in the first-stage epidemiological study. The list of such hypotheses was not confined merely to possible vehicles of a scrapie-like agent such as vaccines. It also included the possibilities that BSE was entirely a genetically based disease or that the observed histopathological lesions in the CNS were due to an intoxication by a novel chemical compound incorporated in agricultural chemicals or veterinary therapeutic products.

The complete list of potential aetiological hypotheses (Wilesmith et al., 1988; Wilesmith, 1991) was investigated in a detailed case study of 200 histopathologically confirmed cases and their herds. This was achieved by December 1987 by means of the usual questioning of herd owners and herdsmen using a standard questionnaire and forms. Although by this time there was evidence that BSE was a transmissible disease (Fraser et al., 1988; Wells et al., 1987) this investigation confirmed that BSE was neither merely a genetic disease nor due to a chemical intoxication. The examination of possible vehicles for a scrapie-like agent ruled out all possible sources, including contact with sheep on the affected farms, except for the feeding of concentrate rations potentially containing meat and bone meal derived from the rendering of animal products. Further detailed studies on a sample of affected animals revealed that, where complete feeding histories were available and details of the rate of inclusion of meat and bone meal in the feedstuffs were available from the manufacturers, all animals had received meat and bone meal (Wilesmith et al., 1988; Wilesmith, 1991).

In addition to the collection of information for the evaluation of the aetiological hypotheses, this initial case study involved the collection of detailed descriptive epidemiological data on the affected animals and their herds. A more complete description of the descriptive epidemiological features is given below, but the initial findings were consistent with the meat and bone meal hypothesis. The features of note were that all cases appeared to be index cases and the form of the epidemic was typical of that with an extended common source. Important in this respect was that the disease occurred contemporaneously throughout Great Britain (Wilesmith 1992). The greater incidence of affected dairy herds compared with beef suckler herds also provided sustenance to the hypothesis. Concentrate feeding is much less common in the latter type of herd and consequently the risk of animals born in beef suckler herds becoming infected is remarkably small (Wilesmith *et al.*, 1992b). Other aspects of the descriptive epidemiological features of BSE in the British Isles which were consistent with the meat and bone meal hypothesis have been discussed previously (Wilesmith, 1991).

Further analytical epidemiological studies of the hypothesis by means of a cohort study of herds not being fed concentrates containing meat and bone meal and herds in which animals had been fed rations containing meat and bone meal were not possible. However, a case-control study in which the inclusion of meat and bone meal in concentrate rations fed to calves was compared in affected and unaffected herds was possible. The results of this study supported the hypothesis that the source of exposure for cattle was meat and bone meal (Wilesmith *et al.*, 1992a).

The start of the epidemic and reasons for exposure

One of the objectives of the initial epidemiological study was to determine whether BSE was truly a new disease and, if so, to establish when the first case occurred. Questioning of herd owners and veterinary surgeons revealed that BSE was novel and the first cases, based on clinical histories, occurred in April 1985 (Wilesmith *et al.*, 1988).

An initial estimate of the time when effective exposure of the cattle population commenced was obtained using computer-bases simulation studies. These revealed that exposure must have commenced abruptly in the winter of 1981–1982 (Wilesmith *et al.*, 1988). In addition, there was a predictive element to the results of the simulation studies. This was that the age-specific incidences in animals 5 years old and greater would increase from 1988 to 1990, but not significantly affect the total incidence. This change has occurred (Fig. 1) and is consistent with exposure commencing in 1981–1982. The explanation for this is that, given the protracted incubation period and the finding that the majority of affected animals must have been exposed in calfhood, then the full extent of the incubation period was not apparent at the beginning of 1988 when the initial simulation studies were conducted. However, with the passage of time a fuller appreciation of the incubation period distribution has been possible (Wilesmith *et al.*, 1992b).

The next logical stage of the investigation was to determine the reasons for this apparently sudden exposure of cattle to a scrapie-like agent via meat and bone meal. The inclusion of meat and bone meal as a protein supplement in cattle feedstuffs was not a recent introduction and there was no evidence for a major change in its rate of inclusion. It was also possible to rule out the emergence of a mutant strain of the sheep scrapie agent which was pathogenic for cattle as a reason for this sudden onset of exposure. This was evident form the finding that the disease occurred simultaneously throughout Great Britain (Wilesmith, 1992) and the form of the epidemic curve (Wilesmith *et al.*, 1988). A number of possible factors which could have played a role

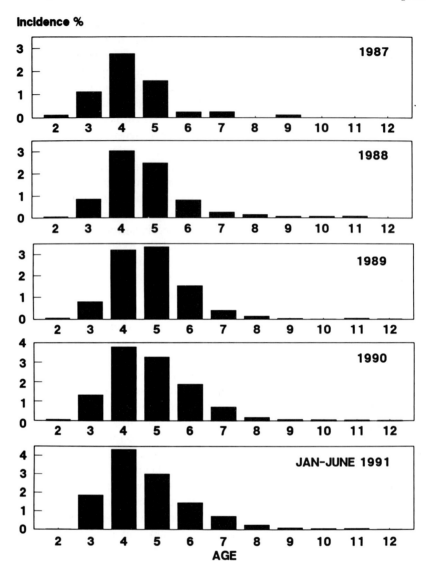

Fig. 1.   Age-specific incidences of BSE in affected herds in each year 1987 to 1990 and for affected herds January to June 1991.

in the alteration of the exposure of cattle to the sheep scrapie agent were, however, identified (Wilesmith *et al.*, 1988). A survey of rendering plants in Great Britain in 1988 revealed that two major changes in the processes used had occurred. The first was a change from batch rendering to the more energy-efficient continuous processes. The first continuous processing plant was commissioned in 1972 and other plants changes to this type of process gradually over a period of years. No difference in the mean maximum temperature achieved in the two types of process was observed, but the particle size of the material subjected to the continuous rendering processes was smaller than that in batch processes. This change could not therefore be associated with the alternation in exposure of cattle. The second change was the reduction in the number of plants using hydrocarbon solvent extraction of meat and bone meal, a process used to maximize the yield to tallow. This change occurred over a short period of time which was coincidental with the putative onset of effective exposure in 1981–1982. An examination of the details of the solvent extraction process revealed that the reduction in its use removed two treatments which could have potentially reduced the titre of the scrapie agent in the final product. These were the direct action of the solvent and the additional heat treatment, including the application of superheated steam, necessary to remove the solvent. In addition the solvent had an indirect effect by reducing the lipid content, which would enhance the effect of the heat treatment. The conclusion from this study of rendering practices was that the cessation of solvent extraction in all but two plants, both in Scotland, was the most probable reason for the commencement of effective exposure sufficient to result in the expression of clinical disease (Wilesmith *et al.*, 1991).

Descriptive features of epidemic

From November 1986 until BSE became a statutorily notifiable disease in June 1988 (Order, 1988) cases were voluntarily notified, but detailed epidemiological data were obtained from all such cases. Data collection has continued since June 1988 and the brains of all suspect cases which are reported and compulsorily slaughtered under the legislation, or which die, are examined histopathologically. This has enabled a relatively detailed analyses of the descriptive epidemiology.

No predisposition for the development of BSE has been observed with breed or sex. The disease has been confined essentially to adult animals with the youngest case being 22 months of age at clinical onset. The modal age at onset of clinical signs is 4 to 5 years (Fig.1) and this undoubtedly reflects the incubation period. The initial and continuing simulation studies have indicated that the majority of affected animals have been exposed in calfhood i.e. in their first six months of life. This does not imply a difference in susceptibility to infection with age. Meat and bone meal may be included in feedstuffs for all age classes of cattle and cases have occurred in which potential exposure only occurred in adulthood. However, as the average lifespan of a dairy cow in Great Britain is just in excess of 5 years and animals enter the adult herd at 2.5 years of age many animals exposed in adulthood will not survive the incubation period.

Animals in dairy herds are considerably more at risk than those in beef suckler herds. This is simply explained by the difference in the nutritional requirements and feeding regimes between these two types of herds, concentrate feeding being more common in dairy herds than in beef herds. A detailed examination of the cases occurring in beef suckler herds has revealed that the majority were cross-bred animals originating from dairy herds where they were more likely to have been exposed (Wilesmith *et al.*, 1992a). Such beef suckler herds, where the only animals affected have been purchased, rather than the whole herd having been potentially exposed from the common feed source, continue to be closely monitored as they could provide a valuable indication of the occurrence of direct cattle-to-cattle transmission, if this were to occur, in addition to the other studies concerned with this important aspect.

Within affected herds the mean annual incidence is approximately 2% of adult animals (Table 1) and currently some 50% of affected herds have only experienced one case. As discussed previously (Wilesmith, 1992) this within-herd incidence underestimates the true attack rate in exposed animals. Also, estimates of the effective exposure rates from the clinical incidence are biased because of the protracted incubation period and the fact that cows do not complete their natural lifespan. However, although the potential for exposure to meat and bone meal for animals born and reared in dairy herds before 18 July 1988 was relatively great the risk of exposure to infection and becoming infected was considerably lower. This is because it is unlikely that the infectious agent was homogenously distributed in meat and bone meal, and therefore in the finished feedstuffs, and the observation that the oral route appears to be a relatively inefficient means of infecting a number of species (Taylor, 1989).

Table 1.   Mean within herd incidence of BSE in affected herds in Great Britain

	1988		1989		1990		1991	
	January–June	July–December	January–June	July–December	January–June	July–December	January–June	July–December[a]
Within herd incidence (%)	1.9	1.8	2.0	1.9	2.1	2.2	2.3	2.1

[a] Provisional incidence.

The marked geographical variation in the incidence of affected herds evident from the beginning of the epidemic has remained unaltered (Wilesmith *et al.*, 1992b). The greatest proportion of herds affected is in the south of England. Taking account of the confounding factor of herd size by standardization has revealed that there is a true geographical variation in the risk of animals contracting BSE. This variation in risk is simply due to a variation in the risk of exposure from infected meat and bone meal. This variation comprises three elements (Wilesmith *et al.*, 1992b). The first is that there was not a constant use and inclusion rate of meat and bone meal across all manufacturers and all proprietary cattle feedstuffs and each manufacturer does not have an equable market geographically. The second is that there was a geographical variation in the proportion of meat and bone meal produced by further processing of the intermediate product of rendering, greaves, which involves a double heat treatment. The third element was the continued use of the hydrocarbon solvent extraction process in only two plants, both in Scotland, which produced the majority of meat and bone meal in Scotland. The individual contribution of each of these is impossible to quantify, but it is likely that they were acting in concert.

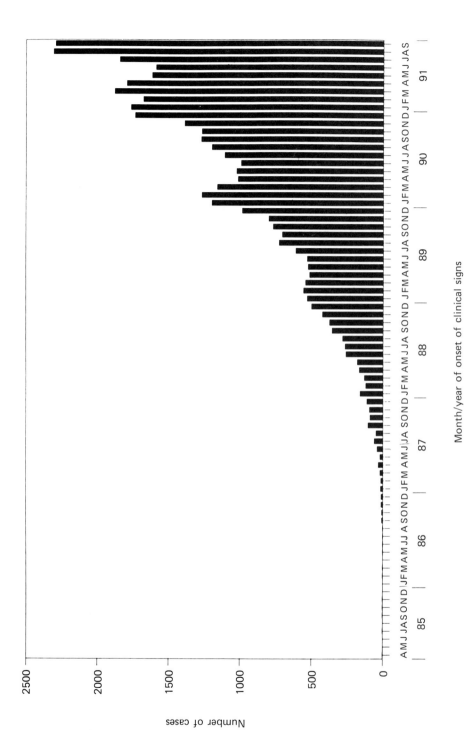

Month/year of onset of clinical signs

Fig. 2. The epidemic curve of cases of BSE reported from April 1985 to September 1991 in Great Britain.

NUMBER OF CASES

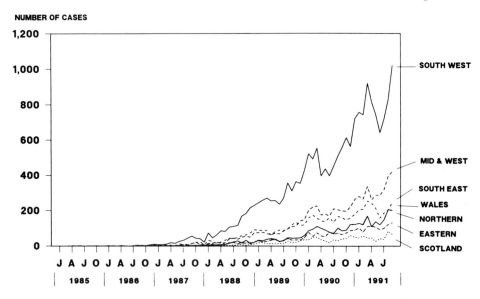

Fig. 3.   The epidemic curves of cases of BSE reported from 1985 to September 1991 in each region of Great Britain.

The current annual incidence, based on the month with the greatest incidence thus far, is 6 cases per 1 000 adult animals in Great Britain. The epidemic curve of confirmed cases of BSE up to September 1991 is shown in Fig. 2. Because of the inconstant ascertainment rate over time this represents an incomplete picture of the epidemic and therefore presents some difficulties in interpretation until June 1988 (Wilesmith, 1992b). Since this time, when the disease was made statutorily notifiable, there has been a complete and constant ascertainment of cases. During the first six months of 1989 the monthly incidence was remarkably constant, at 500 cases per month, and was as expected from the initial common source. However, from July 1989 the observed incidence increased. A detailed investigation confirmed that this was a true increase (Wilesmith 1991, 1992). Further analyses were directed at determining whether this increase could be attributed to the occurrence of cattle-to-cattle transmission, otherwise undetected from the detailed monitoring of the epidemic and specific studies and analyses. It was found that the within-herd incidence had remained constant (Table 1) and therefore the increase in the national incidence could not be attributed to horizontal transmission. Also, the increase was not localized geographically and had occurred uniformly throughout Great Britain (Fig. 3). The epidemiological picture was therefore of an increase in the number of herds affected rather than an increase in the attack rate within exposed herds.

This increase in incidence is consistent with the hypothesis that infection was due to the inclusion of infected cattle tissues in material rendered to produce meat and bone meal used in cattle feedstuffs, which commenced in 1984–1985 before the occurrence of the first cases of BSE. It would have continued until the introduction in July 1988 of the legislation prohibiting the feeding of ruminant-derived protein to ruminants (Order, 1988), but has produced a major effect on the incidence. This recycling is akin to 'passage' of scrapie in rodent models, in which a reduction in the incubation period has been observed (Kimberlin *et al.*, 1987, 1989). However, at present there is no evidence that the incubation period of BSE is getting shorter, although this is the subject of continued epidemiological analyses.

Hypotheses as to the source of infection

The primary hypothesis for the original source of infection for cattle is scrapie from sheep. This would have been from the inclusion of infected tissues from preclinical, scrapie-infected sheep rather than solely from tissues from clinically affected sheep (Wilesmith, 1992). As discussed above, the epidemiological findings indicate that such exposure would have been to a strain of the sheep scrapie agent already prevalent throughout the national sheep flock and not a novel mutant strain.

An important epidemiological feature of BSE is the apparent sudden onset of effective exposure. One hypothesis is that cattle did not present an insurmountable species barrier to the naturally occurring sheep scrapie agent. Therefore cattle may be naturally susceptible to sheep scrapie, but had not previously experienced a sufficient exposure to result in a detectable incidence of BSE.

A secondary hypothesis for the source of infection has been identified (Wilesmith *et al.*, 1991). As the inclusion of meat and bone meal in cattle feed is not a recent event and because it is very unlikely that any of the rendering processes used would totally disinfect meat and bone meal of the transmissible agent, there has been the possibility of exposure to cattle previously. If scrapie had been transmitted from sheep to cattle or cattle had been subclinically infected for some time, there has been considerable scope for increasing the prevalence of infection in the cattle population. This is also true in the USA, as rendered ruminant tissues are not excluded from animal feedstuffs and rendering processes are essentially now the same in the two countries. BSE has not, however, been diagnosed to date in the USA (USDA, 1991).

There are obvious difficulties in determining the reasons for the occurrence of BSE in Great Britain in 1985. This appears to have resulted from the presence of a combination of a number of basic risk factors. The most important of these are a large sheep population in relation to the cattle population and a sheep population infected with scrapie and the feeding of meat and bone meal derived from ruminants to cattle. These risk factors have only been enumerated in and comparisons made between Great Britain, The United States of America and Spain, and there are considerable national differences suggesting an exceedingly low risk for cattle in both of the latter countries (Walker *et al.*, 1991; M. Perote and J.W. Wilesmith, unpublished).

The future course of the epidemic

Not unexpectedly there has been considerable interest in both the short-term and long-term prospects for the future incidence of BSE in Great Britain. The basic scenarios for the future course of the epidemic have been discussed previously (Wilesmith and Wells, 1991). At the current stage of the epidemic, when the earliest time is approaching when the effect of the statutory intervention, introduced in July 1988, could be observed, a number of aspects of the epidemiology are under continuous review.

An obvious aspect is a decline in the incidence in the two year old age class, during 1991 and subsequently, due to the diminished number of these animals exposed from the common food source. The age-specific incidences depicted in Fig. 1 for affected herds in 1991 are liable to change with the accumulation of all data for cases with a clinical onset in 1991. However, the incidence in two year old animals in 1991 was lower than in 1990. A more specific aspect is the date of birth of the more recently affected two year old animals. At the time of writing it is possible to consider cases with a clinical onset before 1 July 1991 as all data for these have been obtained. A comparison of the distribution of two year old cases by month and year of birth with a clinical onset in the six months January to June 1990 (i.e. born between January 1987 and June 1988) with those with an onset in the comparable period for 1991 (i.e. born between January 1988 and June 1989) is one possible analysis. This indicates that the number and percentage of cases with a date of birth in the latter five months of 1988 and with an onset in the first six months of 1991 are lower than expected compared with the previous year. No two year old cases with an onset from January to June 1991 had been born in the five month period August to December 1988, or subsequently, compared with 65.4% of two year old cases with an onset from January to June 1990 born in the period August 1987 to June 1988.

In addition to such analyses, specific, detailed epidemiological investigations are being conducted on confirmed cases of BSE born after 18 July 1988. To date only one animal, born in November 1988, has any suspicion of being a case of maternal transmission in that the dam had also developed BSE and the risk of an accidental exposure to ruminant-derived protein was low. The consequences for the future course of the epidemic should BSE be transmitted only by maternal transmission, i.e. only from affected dams to their progeny and not to unrelated individuals, are minimal. The effective contact rate of this means transmission is insufficient to maintain the disease in the cattle population (Wilesmith and Wells, 1991). However, the occurrence of maternal transmission of BSE, which is considered to occur to sheep scrapie probably as a result of exposure to infected placentae, which also present a risk for other members of the flock, could be a harbinger of an analogous situation in cattle. Consequently other epidemiological analyses and studies are in progress in addition to the assays of tissues, such as placenta, from BSE-affected animals in laboratory rodent models.

A cohort study comprising a comparison of the incidence of BSE in the progeny of confirmed dams and of contemporary born progeny of unaffected dams in the same herds is in progress. It will be some time before definitive results will be available. In the meantime, the incidence of BSE in the progeny of confirmed cases is being monitored in the field and compared with the expected incidence from the food borne source. Again, no excess incidence has been observed in the progeny of confirmed cases (J.W. Wilesmith, J.B.M. Ryan and L.J. Hoinville, unpublished).

Other aspects of the epidemiology which are subject to continuous monitoring and review include the epidemic curve of newly affected herds. This analysis is most appropriately confined to those herds in which the affected animal, or animals, was homebred. The epidemic curve of these herds, up to September 1991, is depicted in Fig. 4 and there is now evidence of a decline in the rate of occurrence of such herds.

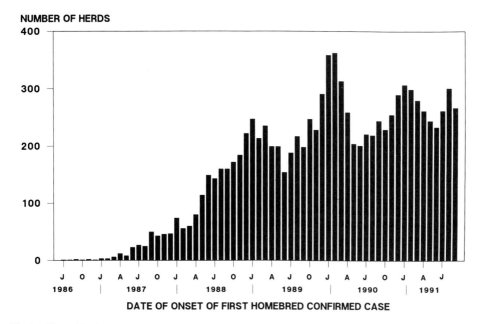

Fig. 4.   The epidemic curve of newly affected herds in Great Britain in which BSE occurred in a homebred animal from June 1986 to September 1991.

This epidemic curve is of interest as, in the absence of cattle-to-cattle transmission, it is likely to exhibit a decline before the epidemic curve of cases. The observed decline provides some optimism as, given the time of the start of the epidemic, the incubation period distribution and the movement of cattle between herds, a sufficient period has elapsed for cattle-to-cattle transmission to have influenced the rate of occurrence of newly affected herds. The observed decline has been described as analogous to 'the exhaustion of susceptibles in the population' (Wilesmith et al., 1992b). Alternatively, it can be regarded as an exhaustion of the exposed population as it is unlikely that the cattle in all herds were exposed to infected meat and bone meal in their feedstuffs. Also, the legislation would have limited the proportion of herds exposed.

A full appreciation of what the future holds for the BSE epidemic will, however, only be possible with the passage of time and continuous re-analysis of these and other epidemiological aspects together with the acquisition of the results from the relevant laboratory-based studies which are in progress.

CONCLUSIONS

The occurrence of transmissible spongiform encephalopathy in cattle in Great Britain was undoubtedly unexpected. It has had considerable economic consequences for the country as a result of the costs of control and for the agricultural community because of the loss of export markets, and temporary effects on the consumption on beef domestically.

The epidemiological evidence provides strong evidence that the effective exposure of cattle to a scrapie-like agent occurred via the feeding of meat and bone meal. This exposure commenced in 1981–1982 and was related to the reduction in the use of hydrocarbon solvent extraction of fat from meat and bone meal during the rendering process. The exposure from the food source was prevented by the introduction of legislation in July 1988. A considerable proportion of the effort in understanding the epidemiology of BSE is directed at providing an insight into the future course of the epidemic. The main concern is obviously whether horizontal transmission can occur between cattle at a rate sufficient to maintain the epidemic. Specific studies have been instigated to investigate this, in addition to the detailed monitoring of the epidemic. However, at present, there is no indication that horizontal transmission is a significant factor.

ACKNOWLEDGEMENTS

The author is indebted to a large number of people for their assistance in conducting these epidemiological studies. Thanks are especially due to colleagues in the Veterinary Field Service for the collection of much of the data. I am particularly grateful of Judi Ryan and the whole BSE Team in the Epidemiology Department for their unstinting and excellent work in maintaining the epidemiological database and conducting the associated computer programming and to Dawn Cockle for typing the original manuscript.

REFERENCES

Fraser, H., McConnell, I., Wells, G.A.H. and Dawson, M. (1988) Transmission of bovine spongiform encephalopathy to mice. *Vet. Rec.* **123** 472.

Kimberlin, R.H., Cole, S. and Walker, C.A. (1987) Temporary and permanent modifications to a single strain of mouse scrapie on transmission to rats and hamsters. *J. Gen. Virol.* **68** 1875–1881.

Kimberlin, R.H., Walker, C.A. and Fraser, H. (1989) The genomic identity of different strains of mouse scrapie is expressed in hamsters and preserved on re-isolation in mice. *J. Gen. Virol.* **70** 2017–2025.

Order (1988) *The Bovine Spongiform Encephalopathy Order* 1988. *Statutory Instrument* 1988 *no* 1345, *HMSO*, London.

Taylor, D.M. (1989) Bovine spongiform encephalopathy and human health. *Vet. Rec.* **125** 413–415.

USDA (1991) Retrospective surveillance for bovine spongiform encephalopathy (BSE) in the United States. *Animal Health Insight*, Winter 1991. USDA:APHIS:VS, Fort Collins, USA pp 11–16.

Walker, K.D., Hueston, W.D., Hurd, H.S. and Wilesmith, J.W. (1991) Comparison of bovine spongiform encephalopathy risk factors in the United States. *J. Am. Vet. Med. Assoc.* **11** 1554–1561.

Wells, G.A.H., Scott, A.C., Johnson, C.T., Gunning, R.F., Hancock, R.D., Jeffrey, M., Dawson, M. and Bradley, R. (1987) A novel progressive spongiform encephalopathy in cattle. *Vet Rec* **121** 419–420.

Wilesmith, J.W. (1991) Bovine spongiform encephalopathy: epidemiological approaches, trials and tribulations. In: *Proc. 6th International Symposium on Veterinary Epidemiology and Economics*, pp 32–43.

Wilesmith, J.W. (1992) Epidemiology of bovine spongiform encephalopathy. *Semin. Virol.* **2** 239–245.

Wilesmith, J.W. and Wells, G.A.H. (1991) Bovine spongiform encephalopathy. In: Chesebro, B.W. and Oldstone, M. (eds) *Scrapie, Creutzfeldt–Jakob Disease and Other Spongiform Encephalophathies*. Current Topics in Microbiology and Immunology, Vol. 172. Springer, pp. 21–38.

Wilesmith, J.W., Wells, G.A.H., Cranwell, M.P. and Ryan, J.B.M. (1988) Bovine spongiform encephalopathy: epidemiological studies. *Vet. Rec.* **123** 638–644.

Wilesmith, J.W., Ryan, J.B.M. and Atkinson, M.J. (1991) Bovine spongiform encephalopathy: epidemiological studies on the origin. *Vet. Rec.* **128** 199–203.

Wilesmith, J.W., Ryan, J.B.M. and Hueston, W.D. (1992a) Bovine spongiform encephalopathy: case-control studies of calf feeding practices and meat and bone meal inclusion in proprietary concentrates. *Res. Vet. Sci.* **52** 325–351.

Wilesmith, J.W., Ryan, J.B.M., Hueston, W.D. and Hoinville, L.J. (1992b) Bovine spongiform encephalopathy: descriptive epidemiological features 1985– 1990. *Vet. Rec.* **130** 90–94.

23

The discovery of bovine spongiform encephalopathy and observations on the vacuolar changes

Gerald A. H. Wells, Stephen A. C. Hawkins, William J. Hadlow[†] and Yvonne I. Spencer

ABSTRACT

Bovine spongiform encephalopathy (BSE) was recognized in November 1986 from the microscopic examination of the brains of adult domestic dairy cows that presented with a novel progressive neurological syndrome. Records indicate that the syndrome was first observed in April 1985.

On the subsequent demonstration of molecular pathological criteria and transmissibility, the nosologic identity of BSE was confirmed as a neurodegenerative disease caused by an unconventional pathogen.

Studies of the neuropathology of the natural disease have shown consistency in the pattern of distribution and severity of vacuolar changes in the brain. Constant vacuolar changes in the medulla have made it possible to simplify the histological examination for statutory diagnosis. Close similarity of the profiles of vacuolar changes in the naturally occurring disease and experimental BSE of cattle is in contrast to the findings in sheep scrapie and suggests a uniform pathogenesis of BSE. The possible significance of the observations in relation to host–agent properties and origins of the disease are discussed.

[†] Dr W. J. Hadlow was a visiting research officer at the Central Veterinary Laboratory – Weybridge, Addlestone, Surrey, UK, June–August 1990.

INTRODUCTION

The characteristic encephalopathic changes recognized in the brains of domestic cattle in November 1986 and called bovine spongiform encephalopathy (BSE) (Wells *et al.*, 1987) provided the first clear indication of the occurrence of a scrapie-like disease in a major food animal species other than sheep and goats. The circumstances of this discovery are, given the consequences, of considerable interest both in relation to such diseases and from a wider perspective of detecting novel diseases in domestic animal populations.

The nationwide epidemic of BSE in the United Kingdom (UK) (Wilesmith *et al.*, 1988, 1991, 1992a) posed a further precedent among scrapie-like diseases in providing an unequivocal example of dietary transmissibility. The constancy of route and method of exposure established for BSE in cattle (Wilesmith and Wells, 1991) has refocused interest in ruminant species as natural hosts of such diseases (Kimberlin, 1990).

Neuropathology has had a pivotal role in the initial recognition, case definition and diagnosis of BSE. Indeed, as reflected in the unifying term transmissible spongiform encephalopathies (TSEs), the discipline has been important in establishing the nosologic position of all scrapie-like diseases. Dependence upon rodent bioassay systems for detection and characterization of the unconventional causal pathogens of the TSEs also has required the extensive use of neurohistological methods in experimental approaches to understand better such diseases.

Relatively uninterpretable variation in the neuropathology of natural scrapie was a major incentive to the development of independent experimental rodent models in which the expression of lesions proved to be a basis for understanding host–agent interactions (Fraser, 1976). BSE, on the other hand, has not presented such variability of the distribution of brain lesions (Wells and Wilesmith, 1989). This may indicate important biological differences between BSE and natural sheep scrapie, as already suggested by their contrasting epidemiologic features.

This paper first recounts the process of the initial recognition and classification of BSE. Second, observations on the topographic distribution and pattern of severity of vacuolar changes in the brain in the natural and the experimental disease of cattle are described and discussed in the context of the BSE epidemic in the UK.

THE RECOGNITION AND NOSOLOGY OF BSE

Chronology of events leading to recognition

The initial recognition of BSE in the UK as a bovine equivalent of scrapie of sheep and goats was not a sudden revelation comparable with that which evoked Archimedes' ejaculation of 'heureka' on finding the principle of specific gravity. Instead, it was a realization resulting from cumulative observations made by farmers and stockmen, field veterinarians in private practice, and veterinary investigation officers and pathologists of the State Veterinary Service, over a period of some 18 months (Table 1). In retrospect, the process was largely facilitated by many factors vested in the infrastructure of the veterinary services to the UK livestock industries, which included provision for referrals of pathological material from unusual incidents

Table 1.   Chronology of key events in the identification of BSE in the UK[a]

Date	Event	Reference
April 1985	Clinical onset of a novel neurological syndrome in an adult Friesian–Holstein cow (herd A) ('chronic hypersensitivity and incoordination syndrome')	Whitaker and Johnson (1988)
November 1985– February 1987	Clinical onset of nine additional cases in herd A (November 85 (2), February 86, June 86, October 86 (3), January 87, February 87)	Whitaker and Johnson (1988), Wells *et al.* (1987)
June 1986	Spongiform encephalopathy (SE) diagnosed in a captive African antelope[b] (nyala (*Tragelaphus angasi*)) with a scrapie-like syndrome	Ministry of Agriculture, Fisheries and Food (1987), Jeffrey and Wells (1988)
November 1986	SE diagnosed in brains of two domestic cows from a single herd in South-East England (herd A).	Wells *et al.* (1987)
December 1986– May 1987	SE diagnosed in four cows, one from herd A and three from a further three herds throughout Southern England.	Wells *et al.* (1987)

[a]Data for the period reviewed which were acquired after the histopathological diagnosis of the initial six cases (May 1987) are not included.
[b]The family Bovidae comprises ten subfamilies, including Caprinae (sheep, goats) and Bovinae (cattle and several tribes of antelope indigenous to Africa and Asia; includes nyala).

of disease. The first observation, in November 1986, of brain changes in domestic cattle that closely resembled those of scrapie (Wells *et al.*, 1987) probably provided the most substantial component in directing investigations toward confirming the nature of the disorder and signalling its potential significance.

By November 1986, however, case records had already established the apparent novelty and progressive course of the associated clinical syndrome. The experience within herd A (see Table 1) from which the first histopathological diagnosis was made was particularly important in this respect. Between April 1985, the reported time of onset of the first clinically suspected cases (Wells *et al.*, 1987; Wilesmith *et al.*, 1988), and February 1987, 10 cattle in this herd were recorded with a similar range of clinical signs that progressed over 1 to 5 months (Whitaker and Johnson, 1988). Each of the brains from three cows in this series had the characteristic pathological changes. The first cow to present in this herd was described as having 'a change in character', including aggressive behaviour (Whitaker and Johnson, 1988). A diagnosis of cystic ovaries, a relatively common cause of intraspecific aggression and nymphomania in dairy cows, was made. Although therapy for the cysts was successful, behavioural changes persisted. Additional signs of gait ataxia and hyperaesthesia ensued. Further treatments for differential diagnoses of such signs were not successful. The clinical signs progressed over a 3 month period, and the cow was finally slaughtered after becoming recumbent at pasture. The degree of awareness of the syndrome gained from the experience was fundamental to the subsequent identification of other affected cattle in this herd.

Confidence in appreciating the novelty and in establishing the definition of the clinical disorder in cattle before epidemiological studies were started in June 1987 (Wilesmith et al., 1988) was gained also from three other herds that had multiple cases of the syndrome (Wells et al., 1987). In these, as in herd A, the stockmen were the first to recognize and to be intrigued by the similarities in presenting signs among cases. The basis for recognizing future cases was thus established.

Given the low within-herd incidence of BSE throughout the epidemic to date (December 1991) (Wilesmith et al., 1988, 1992a; Wilesmith, personal communication), it is significant that herds with multiple cases were the first to be identified. Indeed, if the disease had occurred at a lower within-herd incidence or sporadically, with a long time between cases, it probably would have remained undetected. A major variable in this detection process is the threshold at which veterinary investigation of sporadically occurring diseases is considered financially viable or sufficiently important. Stockmen and veterinarians, generally, later became more aware of the signs of BSE and an increased reporting rate in the UK occurred from mid-1987 (Wilesmith et al., 1988). Also, two cases were reported from Oman (Carolan et al., 1990) and one from the Falkland Islands (Wells and Wilesmith, unpublished data) in 1989. In both of these latter incidents, the affected cattle had been exported from the UK as young adults from herds in which cases of BSE subsequently occurred (Wilesmith, personal communication).

The clinical signs of apprehension, hyperaesthesia, and incoordination of gait observed initially, have, despite clinical observation of several tens of thousands of cases during the epidemic, remained the cardinal indicators of BSE (Wilesmith et al., 1988, 1992b). The absence of consistent clinical biochemical abnormalities and failure of sustained responses to therapy in cases from the first four herds investigated (Wells et al., 1987; Johnson, personal communication) were also important in ruling out hypomagnesaemia and nervous ketosis, the two metabolic disorders of dairy cattle common in the UK that have presenting signs most like those of BSE.

The definition of clinical and histopathological diagnostic criteria for BSE (Wells et al., 1987, 1989; Wilesmith et al., 1988, 1992b) has made provision for BSE to be included in animal disease surveillance programmes in other countries. As a consequence, detection of the disease has been facilitated in the Republic of Ireland, France, and Switzerland, countries where the incidence is low relative to that in the UK (Veterinary Record, 1991; Gafner, 1991).

Fig. 1.   Spongiform change of grey matter neuropil in the nucleus tractus solitarii in BSE. Haematoxylin and eosin (HE), × 160.

Homology with diseases caused by unconventional transmissible pathogens

Histopathological observations of BSE have established the striking light microscopic parallels between scrapie and other TSEs (Wells *et al.*, 1987, 1991). Like scrapie, the bilaterally symmetrical vacuolar lesions are largely confined to the brain stem and are accompanied by neuronal degeneration and an astrocytic reaction. Cerebral amyloidosis, observed as light microscopic plaques, is recorded to a variable extent in the TSEs, except transmissible mink encephalopathy. It is present in BSE, though in only a small proportion of cases and then only as sparse focal deposits.

Contrasting somewhat with findings in scrapie, the principal form of vacuolar change in BSE is spongiform change of grey matter neuropil (Fig. 1) (Wells *et al.*, 1989), as has been defined (Masters and Richardson, 1978). Even so, vacuolation of neuronal perikarya, often with marked distention of the cell membrane (Fig. 2), is prominent in certain grey matter areas.

Homology of BSE with scrapie and other TSEs was further confirmed by the demonstration in brain tissue of the protease-resistant core (PrP27-30) of a disease-specific isoform of the host membrane protein (PrP^C), a molecular marker of these diseases (Hope *et al.*, 1988; Scott *et al.*, 1990). Also, PrP27-30 has been shown to be the major protein of the characteristic fibrils, equivalent to scrapie-associated fibrils (SAFs), visualized by electron microscopy in proteinase K treated, detergent extracts of BSE-affected brains (Hope *et al.*, 1988; Scott *et al.*, 1990; Wells *et al.*, 1987). These studies have confirmed the specificity of fibril detection in BSE, but the frequency of

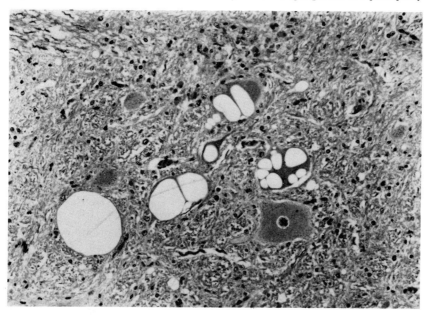

Fig. 2.   Vacuolation of neuronal perikarya with distention of the cell membrane in BSE. HE, × 160.

detection is dependent on the brain region examined. Fibrils are detected most often in the basal nuclei, diencephalon, mesencephalon, and medulla; their prevalence approximates the severity of vacuolar changes (Scott *et al.*, 1990).

Transmissibility of BSE was established in mice inoculated parenterally with brain homogenates (Fraser *et al.*, 1988). Subsequently, successful transmissions were made also by parenteral inoculations of BSE brain to cattle (Dawson *et al.*, 1990a) and a pig (Dawson *et al.*, 1990b). Experimental dietary transmission with brain material has been reported only in mice (Barlow and Middleton, 1990).

THE VACUOLAR PATHOLOGY OF BSE

Distribution of vacuolar changes
Among the first six cases of BSE to be examined the neuroanatomical distribution of vacuolar changes was closely similar (Wells *et al.*, 1987). Later, histopathologic examinations of a large number of brains revealed the consistency of the predominantly brain stem distribution of vacuolar changes and proved the effectiveness of establishing the diagnosis on a single standard section of the medulla oblongata (Wells *et al.*, 1989). Each of the two forms of vacuolar change has consistent localizations, with vacuolation of neuronal perikarya principally in the nuclei vestibulares, formatio reticularis, and nucleus ruber. Irrespective of the form of the vacuolation, the main

sites affected are the nucleus tractus solitarii, nucleus tractus spinalis N. trigemini, nuclei vestibulares and formatio reticularis in the medulla oblongata, substantia grisea centralis in the mesencephalon, paraventricular area in the diencephalon and the nuclei septi in the telencephalon. Vacuolar changes are most intense in the medulla, mesencephalon, and diencephalon (Scott *et al.*, 1990; Wells *et al.*, 1989, 1991). In cerebellum, hippocampus, cortex cerebri, and basal nuclei, they are usually correspondingly mild (Hope *et al.*, 1988; Scott *et al.*, 1990).

Minimal vacuolar change in the presence of advanced clinical disease well recognized in natural and experimental sheep scrapie (Fraser, 1976) and occurs also in a small proportion of clinically suspect cases of BSE (Wells *et al.*, 1989).

Vacuolation profiles
In mouse models of scrapie a system of scoring the intensity of vacuolation in specific areas of the brain has been used to provide highly reproducible results in distinguishing strains of the scrapie pathogen (Bruce and Fraser, 1991; Fraser, 1976; Fraser and Dickinson, 1968). Provided that other variables are strictly controlled, the distribution of the lesions and the shape of the profile are characteristic of the agent strain.

Because of the many uncontrolled variables impinging on natural scrapie, variations in lesion distribution patterns in affected sheep have been considered biologically uninterpretable (Fraser, 1976). The constancy of the distribution of vacuolar changes in BSE (Wells and Wilesmith, 1989) led to further study of the relationship between severity and distribution of vacuolation by using methods based on the lesion profile system.

A series of 100 brains from affected cattle killed at arbitrary but recorded times through the clinical course, and sampled before July 1989, were examined (Wells and Wilesmith, 1989; Wells *et al.*, 1990). Excluded from the study were any clinically suspected cases of BSE in which the histopathological diagnosis was inconclusive because of minimal vacuolar changes. The sample had a breed distribution that approximated the breed frequency in the national cattle population and consisted of 95 Friesian–Holstein and 5 of other breeds. Brains were removed, fixed by immersion in 10% formol saline, and processed by routine histological methods. Six standard transverse sections stained with haematoxylin and eosin represented medulla oblongata (cut at the obex), pons, mesencephalon and diencephalon. The severity of vacuolation was scored on a 0–4 scale (Fig. 3), based on the lesion profile method. In contrast to this method, in which 9 grey matter areas of the brain are scored, 64 named neuroanatomic groups, identified from a stereotaxic atlas of the Friesian cow brain (Lignereux, 1986), were scored individually.

The frequency of vacuolation and the mean vacuolation score were calculated for each neuronal group and the results tabulated according to descending mean scores for each region of the brain (Table 2). Results were excluded where fewer than 10 observations of a given nuclear group were made in the 100 brains.

Vacuolar change was constantly present in 18 of the 64 identified neuronal groups. No identified neuronal group was invariably unaffected. In the medulla three nuclei were consistently affected: nucleus tractus spinalis N. trigemini, nucleus tractus solitarii and nucleus olivaris accessorius medialis. Seven other neuronal groups in the medulla

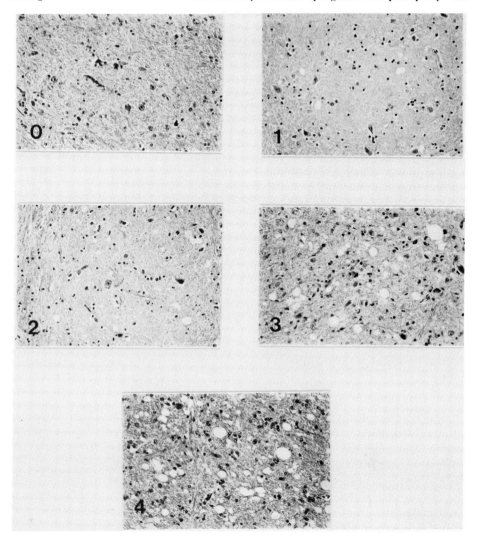

Fig. 3.   Range of severities of vacuolar change corresponding to the scores assigned (0–4), based on the lesion profile method. Examples from nucleus tractus solitarii. HE, × 125.

were affected at frequencies ranging from 80% to 98%. In the pons, the rostral projection of the nucleus tractus spinalis N. trigemini was invariably affected, as were the nuclei vestibulares and the formatio reticularis. In the mesencephalon, 9 of the 14 neuronal groups scored were always involved: nucleus ruber, formatio reticularis, substantia nigra, colliculus rostralis, substantia grisea centralis, nucleus pretectalis caudalis, nucleus pretectalis lateralis, nucleus motorius oculomotorii, and nucleus tegmenti laterodorsalis. In the diencephalon, no neuronal group was always affected, but the hypothalamus was most frequently involved (in 90% of cases).

Table 2. Frequency of occurrence and mean score of vacuolar change according to neuron group in medulla, pons, mesencephalon and diencephalon of field cases ($n = 100$)

Neuron group	Occurrence (%)	Mean vacuolar score
Medulla		
Nucleus tractus spinalis N. trigemini	100	3.0
Nucleus tractus solitarii	100	2.9
Nucleus olivaris accessorius medialis	100	2.0
Nucleus parasympathicus N. vagi	95	2.0
Nucleus olivaris accessorius dorsalis	98	1.5
Nucleus olivaris	96	1.5
Nucleus cuneatus medialis	85	1.2
Formatio reticularis	89	1.1
Nucleus gracilis	88	1.1
Nucleus cuneatus lateralis	80	1.1
Nucleus funiculi lateralis	70	1.0
Nucleus ambiguus	64	0.7
Nucleus motorius N. hypoglossi	43	0.4
Raphe	31	0.3
Pons		
Nucleus tractus spinalis N. trigemini	100	2.4
Formatio reticularis	100	1.9
Nucleus raphe magnus	95	1.8
Nucleus vestibularis caudalis	100	1.7
Nucleus vestibularis medialis	100	1.6
Nucleus vestibularis lateralis	100	1.6
Nucleus dorsalis corporis trapezoidei	94	1.5
Nucleus parasympathicus N. facialis	97	1.4
Raphe	94	1.3
Nucleus vestibularis rostralis	90	1.2
Nucleus motorius N. facialis	97	1.1
Nucleus motorius N. abducentis	88	1.1
Nucleus cochlearis	81	1.0
Nucleus fasciculi teres	51	0.7
Mesencephalon		
Substantia grisea centralis	100	3.2
Formatio reticularis	100	1.8
Nucleus pretectalis caudalis	100	1.8
Colliculus rostralis	100	1.7
Nucleus pretectalis lateralis	100	1.7

Nucleus tegmenti laterodorsalis	100	1.6
Nucleus tractus mesencephalici N. trigemini	99	1.6
Substantia nigra	100	1.5
Nucleus ruber	100	1.2
Nucleus motorius N. oculomotorii	100	1.1
Nucleus geniculatus medialis	84	1.0
Nucleus terminalis dorsalis	81	0.9
Nucleus interpeduncularis	67	0.7
Nucleus tractus optici	47	0.5
Diencephalon		
Regio hypothalamica intermedia/caudalis	90	1.9
Nucleus ventralis medialis	64	1.2
Nucleus centralis	53	1.1
Nucleus ventralis caudalis pars lateralis	54	1.0
Nucleus ventralis rostralis	46	0.9
Nucleus ventralis lateralis	41	0.7
Nucleus geniculatus medialis	40	0.7
Nucleus lateralis caudalis	34	0.6
Nucleus dorsomedialis thalami	32	0.5
Nucleus rhomboideus	26	0.5
Nucleus interventralis	14	0.3
Nucleus parafascicularis	12	0.2
Substantia nigra	12	0.1
Nucleus habenularis medialis	10	0.1
Nucleus geniculatus lateralis pars ventralis	9	0.1
Nucleus rostralis dorsalis	9	0.1
Nuclei paraventriculares thalami	7	0.1
Substantia grisea centralis	7	0.1
Nucleus paratenialis	7	0.1
Nucleus interstitialis	6	0.1
Nucleus rostralis ventralis	2	<0.1
Nucleus reticulatus thalami	2	<0.1

For each of the brain regions, those neuronal groups always affected had the greatest mean vacuolation scores. Also, in general, the hierarchies of mean vacuolation scores and the frequency of involvement of neuronal groups corresponded. Mean scores were generally lower in the diencephalon than in the more caudal regions of the brain.

A similar study of vacuolar profiles was carried out on the brains of a random subset of ten cattle from sixteen in which BSE was experimentally transmitted by simultaneous intracerebral and intravenous inoculations (Dawson *et al.*, 1990a, 1991). The Friesian–Holstein and Jersey breeds of cattle were equally represented in this study. In the brains of these cattle, the major distribution pattern of vacuolation was

Table 3.   Frequency of occurrence and mean score of vacuolar change according to neuron group in medulla, pons and mesencephalon of experimental cases ($n = 10$)[a]

Neuron group	Occurrence (%)	Mean vacuolar score
Medulla		
Nucleus tractus spinalis N. trigemini	100	2.8
Nucleus tractus solitarii	100	2.2
Nucleus olivaris accessorius medialis	90	2.0
Nucleus olivaris	100	1.7
Nucleus olivaris accessorius dorsalis	100	1.7
Nucleus funiculi lateralis	88	1.6
Nucleus cuneatus medialis	90	1.4
Nucleus gracilis	90	1.4
Formatio reticularis	100	1.2
Nucleus ambiguus	90	1.2
Nucleus cuneatus lateralis	80	1.1
Nucleus parasympathicus N. vagi	40	0.4
Nucleus motorius N. hypoglossi	40	0.4
Raphe	0	0.0
Pons		
Nucleus tractus spinalis N. trigemini	100	2.6
Nucleus motorius N. abducentis	89	1.7
Nucleus vestibularis caudalis	100	1.6
Nucleus vestibularis medialis	100	1.6
Nucleus vestibularis lateralis	100	1.6
Nucleus raphe magnus	90	1.5
Formatio reticularis	100	1.4
Nucleus dorsalis corporis trapezoidei	100	1.4
Nucleus cochlearis	90	1.4
Nucleus parasympathicus N. facialis	88	1.1
Nucleus motorius N. facialis	90	0.9
Raphe	70	0.8
Nucleus vestibularis rostralis	67	0.7
Nucleus fasciculi teres	0	0.0
Mesencephalon		
Substantia grisea centralis	100	3.2
Nucleus pretectalis caudalis	100	2.5
Nucleus pretectalis lateralis	100	2.5
Formatio reticularis	100	2.3
Colliculus rostralis	100	2.0
Nucleus tractus mesencephalici N. trigemini	100	2.0

Nucleus tegmenti laterodorsalis	100	1.8
Substantia nigra	100	1.8
Nucleus geniculatus medialis	80	1.4
Nucleus motorius N. oculomotorii	100	1.1
Nucleus ruber	90	0.9
Nucleus interpeduncularis	90	0.9

ᵃVacuolar profiling was conducted on 10 of 16 cattle to which BSE was transmitted. Inocula were prepared from the brain stem of four BSE affected Friesian–Holstein cows from different parts of Great Britain. Recipients were of principally two different breeds, Friesian–Holstein and Jersey, and included four sets of dizygous twins. Test and control groups were assembled to give equitable distribution of animals according to breed, sex, and sibling status. Inoculations were carried out at 4–5 months of age. The 16 test animals were each inoculated simultaneously with 1 ml intracerebrally and intravenously of a 10% brain homogenate. Eight control calves were inoculated similarly with saline. Transmissions were successful in all challenged animals with overt clinical signs of BSE evident 74–90 weeks post-inoculation. Control cattle remain healthy to date (January 1992).

similar to that of field cases. The hierarchy of mean vacuolar scores according to neuronal group, scored only for brain stem regions (Table 3), was, with minor specific exceptions, also similar to that of the field cases. The most notable difference in the frequency and severity of vacuolar change between the two series was in the nucleus parasympathicus N. vagi in the medulla oblongata. In the field cases this nucleus was affected in 95 of the 100 brains examined and had a mean score of 2.0, whereas in the 10 experimental cases it was affected only mildly, and in only 4, with a mean score of 0.4. In the mesencephalon, marginally greater mean scores for many neuronal groups were recorded for the experimental disease than for field cases. Results of studies of the diencephalon in the experimental disease are as yet incomplete.

From these data and additional preliminary observations of other major parts of the brain a mean maximum vacuolar score for each brain region was calculated for 10 of the field cases, selected as clinically terminal, and for the experimental set. The mean maximum score for any given brain region was determined from the greatest intensity of vacuolation within that region for each case, irrespective of the individual neuron groups involved. The result for each of the brain regions was placed in a hierarchy according to score (Table 4). A caudal–rostral diminution in severity was apparent in the field cases with the highest scores in the medulla and mesencephalon and the lowest in the cortex cerebri. With two exceptions, the field and experimental cases were similar. In the experimental disease, vacuolation in the thalamus and hypothalamus was relatively more severe and that in the medulla less so.

Two precedents for the transmission of unconventional pathogen-induced diseases, other than BSE, to cattle were recently reported: the transmission of transmissible mink encephalopathy (TME) (Marsh et al., 1991) and scrapie (Gibbs et al., 1990). While there are no detailed vacuolar profile data from these studies, a brief comparison of the neurohistologic findings with our observations in BSE is pertinent.

As part of an investigation of an incident of TME in Stetsonville, WI, USA, in which a mink rancher reportedly had never fed sheep products but did feed a large proportion of products from fallen or diseased dairy cattle, two Holstein steer calves were inoculated intracerebrally with the affected mink brain (Marsh et al., 1991).

Table 4. Hierarchy of descending mean maximum vacuolar scores according to brain region ($n = 10$)

Field case sample		Experimental transmission case sample	
Mesencephalon	3.7	Diencephalon	3.6
Medulla	3.2	Mesencephalon	3.2
Pons	2.8	Medulla	2.8
Diencephalon	2.5	Pons	2.6
Hippocampus	1.5	Hippocampus	2.3
Corpus striatum/nuclei septi	1.3	Corpus striatum/nuclei septi	1.4
Cortex frontalis	0.4	Cortex frontalis	0.7
Cortex parietalis	0.3	Cortex parietalis	0.1
Cortex occipitalis	0.2	Cortex occipitalis	0.1

Spongiform encephalopathy occurred in both animals, 18 and 19 months after inoculation. Spongiform changes were present in the grey matter of the brain stem but not in the cortex cerebri. Additional observations in one of these steers (Wells and Marsh, unpublished) indicated a distribution of vacuolar changes and a hierarchy of lesion severities not unlike those in cattle affected with experimental BSE, but with relatively greater involvement of the nuclei septi. The biological properties of the Stetsonville TME pathogen (Marsh *et al.*, 1991) were more consistent with the epidemiology of naturally occurring TME than were isolates of the scrapie pathogen from several different sources when tested in mink (Hanson *et al.*, 1971; Marsh and Hanson, 1979). That cattle may be a source of exposure for mink to the TME pathogen has been considered for many years (Hartsough and Burger, 1965; Marsh *et al.*, 1969) and, with the Stetsonville incident, has led to the suggestion that a naturally occurring scrapie-like infection may exist in cattle in the USA, albeit with a very low incidence of disease (Marsh *et al.*, 1991; Taylor, 1988).

Scrapie from sheep after passage in sheep, and from goat after passage in goats, has also been transmitted to cattle in the USA (Gibbs *et al.*, 1990). Three of ten cattle, challenged at 8-11 months of age by intracerebral, intramuscular, subcutaneous and oral routes, developed neurological signs 27–48 months after inoculation. Although PrP27-30 was detected in the brains of all three animals and therefore confirms the transmissibility of scrapie to cattle, the only reported histopathologic changes comprised a gliosis and sparse vacuolation considered insufficient for a histopathological diagnosis of scrapie-like encephalopathy.

Why profile uniformity in BSE?

In experimental mouse models of scrapie the distribution and intensity of neuroparenchymal vacuolation, the lesion profile, has been shown to be influenced by several factors; the most important are pathogen strain and host genotype (Dickinson and Fraser, 1979; Fraser, 1976; Fraser and Dickinson, 1973; Outram, 1976). In sheep scrapie and in the other naturally occurring TSEs, interpretation of variation in the neuropathological profile is confounded by the variability of these factors (Fraser, 1976).

Constancy of the pathology in BSE, in both natural cases and on primary parenteral transmission to cattle with brain tissue of four donors each from a different geographic area, indicates a regularization in the morphologic expression, mimicking that seen under controlled conditions in experimental rodent models. It may also suggest that factors controlling expression of the lesions, at least in the circumstances of the UK epidemic, are invariable.

In the mouse models of scrapie *Sinc* is the major gene controlling incubation period characteristics and has a close linkage with the PrP gene. Recent studies of mouse lines congenic for *Sinc* have further confirmed that to a lesser extent it also influences the lesion profile (Bruce *et al.*, 1991).

As yet there is little information on any effects of differences in host genotype on the occurrence and morphologic expression of BSE. However, 93% of affected cattle have been Friesian–Holstein types, with a relatively small male gene pool, and throughout the epidemic there has been no indication of a breed-associated suscep-tibility (Wilesmith *et al.*, 1992a). In the present study, no major differences were found in the vacuolar profiles among the different affected breeds either in the field case series or the experimental cases. Although different alleles of the bovine PrP gene have been found in the Holstein–Friesian breed, no linkage to the incidence of BSE has yet been shown (Goldmann *et al.*, 1991).

As well as defined effects of host genes on the lesion profile in mouse models of scrapie, effects are also related to route of exposure. In the BSE epidemic, epidemiologi-cal data to date support a common foodborne source of exposure via meat and bone meal to a scrapie-like agent (Wilesmith *et al.*, 1992a). The minor variation in relative severity of vacuolation, but not its distribution, between the field and experimental cases of BSE may well be attributable to differences of routes of exposure. Intracerebral inoculation, which was directed in a semistereotaxic way into the midbrain, may have been particularly important in establishing the more severe changes in the dien-cephalon of experimentally infected animals. These observations are consistent with the effects of different inoculation routes mainly on the height (lesion severity) rather than the shape (lesion distribution) of lesion profiles for scrapie in mice (Fraser, 1976; Kimberlin *et al.*, 1987a).

There were other variables, notably those of age and sex, among cattle in these studies but no effects were apparent on the vacuolar profiles. The range of ages at onset of clinical signs in the field case study (36–120 months) broadly reflected the age-specific incidence in the epidemic (Wilesmith *et al.*, 1992a). The overwhelming predominance of females in the cattle population militated against the occurrence of bulls in the field case series. In the experimental series, however, seven of the ten cattle examined were castrate males.

In mouse models of scrapie, the major factor influencing the lesion profile is, however, the strain of pathogen. The genomic basis for strain differences remains unclear, but strains can nevertheless be defined by the characteristic and stable properties they show after serial passage in mice (see Bruce and Fraser, 1991, for review).

If, in cattle with BSE, as in the mouse models, the strain of pathogen is the major determinant of the neuropathology, uniformity of the vacuolar changes in the natural and experimental disease in cattle would suggest that properties of the pathogen are stable. Whether this represents singularity of the strain of the pathogen, which as previously suggested (Wilesmith and Wells, 1991) could arise either from earlier selection in cattle from a pool of scrapie agents or from the effective exposure of cattle to a single scrapie strain, probably cannot now be determined. Recycling of infected cattle tissues, before the ruminant protein ban, provided the means by which the pathogen was passaged in cattle, even before the disease was recognized. This may therefore have contributed to the stability of the causal pathogen.

It is not known to what extent strains isolated in mice are representative of extant field strains of the scrapie pathogen, but the argument for a stable BSE pathogen is further supported by the uniformity of primary transmissions of BSE to four genotypes of mice inoculated with the same four sources of BSE pathogen used in the primary parenteral transmission of the disease to cattle (Fraser *et al.*, 1988). Furthermore, the properties of the BSE pathogen on primary transmission and subpassage in mice are unlike those of previous characterized sheep-derived isolates in mice (Fraser *et al.*, 1990). This may have analogies with the experimental phenomenon of mutation of the pathogen which can occur when infection is established in a new host species inoculated with a defined strain of the scrapie pathogen (Kimberlin *et al.*, 1987b, 1989). The precise relationship between sheep scrapie and BSE remains unclear. The only source of contamination of meat and bone meal with a scrapie-like unconventional pathogen, before BSE, is presumed to be that of natural scrapie, and risk factors for the potential exposure of cattle in Great Britain are considered to have been greater than elsewhere (Walker *et al.*, 1991; Wilesmith and Wells, 1991).

Because the putative onset of effective exposure in cattle is consistent only with a change in exposure to a scrapie-like pathogen (Wilesmith *et al.*, 1992a), it follows that the BSE pathogen would necessarily have been nationally distributed in meat and bone meal before events which produced the change in exposure. The process which, in experimental animal models at least, can give rise to a novel stable strain in a new host species may have therefore occurred in cattle a considerable period of time before the change in exposure took place. The possibility that in cattle this process resulted in subclinical disease or a very low incidence of clinical disease before April 1985 has been suggested, but there is no evidence to support such occurrences (Wilesmith and Wells, 1991).

It might be proposed, however, that, irrespective of the source or variability of the scrapie-like pathogen or the route of exposure, cattle have a species-related stereotypic neurohistological response controlling selective vulnerability or targetting within the nervous system. Although the evidence is limited, the Stetsonville TME transmission to cattle (Marsh *et al.*, 1991) would support this notion. However, the apparent absence of convincing vacuolar changes in cattle affected with experimental scrapie (Gibbs *et al.*, 1990) suggests that the hypothesis may not be true when the pathogen is derived from sheep- or goat-passaged natural scrapie, at least from inocula obtained in North America.

Whatever the reasons for uniformity of the pathological response in BSE, the conclusion that it reflects a relatively constant neural pathogenesis offers a degree of predictability for using cattle to study temporal and spatial progression of the BSE pathogen in the natural host.

CONCLUSIONS

The clinical observations of BSE from April 1985, before its microscopic recognition in November 1986, had an important role in establishing the nature and novelty of the disease in the UK domestic cattle population. When the homology of BSE with diseases caused by unconventional pathogens was confirmed the criteria for defining the disease in cattle were established to facilitate case finding for epidemiological studies in the UK and for surveillance in many other countries.

Invariability of the pattern of vacuolar changes in the brains of affected cattle in a sample of cases from the epidemic before July 1989 and in experimentally induced cases indicates a stereotypic pathogenetic response in cattle to the BSE pathogen. We suggest that this indicates that a single strain of scrapie-like pathogen causes BSE, but its precise relationship to strains causing natural scrapie in sheep remains obscure.

REFERENCES

Barlow, R.M. and Middleton, D.J. (1990) Dietary transmission of bovine spongiform encephalopathy to mice. *Vet. Rec.* **126** 111–112.

Bruce, M.E. and Fraser, H. (1991). Scrapie strain variation and its implications. In: Chesebro, B.W. (ed.) *Transmissible Spongiform Encephalopathies. Scrapie, BSE and Related Human Disorders.* Current Topics in Microbiology and Immunology, Vol. 172, Springer, Berlin, pp. 125–138.

Bruce, M.E., McConnell, I., Fraser, H. and Dickinson, A.G. (1991) The disease characteristics of different strains of scrapie in *Sinc* congenic mouse lines: implications for the nature of the agent and host control of pathogenesis. *J. Gen. Virol.* **72** 595–603.

Carolan, D.J.P., Wells, G.A.H. and Wilesmith, J.W. (1990) BSE in Oman. *Vet. Rec.* **126** 92.

Dawson, M., Wells, G.A.H. and Parker, B.N.J. (1990a) Preliminary evidence of the experimental transmissibility of bovine spongiform encephalopathy to cattle. *Vet. Rec.* **126** 112–113.

Dawson, M., Wells, G.A.H., Parker, B.N.J. and Scott, A.C. (1990b) Primary parenteral transmission of bovine spongiform encephalopathy to the pig. *Vet. Rec.* **127** 338.

Dawson, M., Wells, G.A.H., Parker, B.N.J. and Scott, A.C. (1991) Transmission studies of BSE in cattle, hamsters, pigs and domestic fowl. In: Bradley, R., Savey, M. and Marchant, B.A. (eds) *Subacute Spongiform Encephalopathies. Proc. Seminar in the CEC Agricultural Research Programme, Brussels, November 12–14, 1990.* Current Topics in Veterinary Medicine and Animal Science, Vol. 55, Kluwer, Dordrecht, pp. 25–32.

Dickinson, A.G. and Fraser, H. (1979). An assessment of the genetics of scrapie in sheep and mice. In: Prusiner, S.B. and Hadlow, W.J. (eds) *Slow Transmissible Diseases of the Nervous System* Vol. 1. Academic Press, New York, pp. 367–385.

Fraser, H. (1976) The pathology of natural and experimental scrapie. In: Kimberlin, R.H. (ed.) *Slow Virus Diseases of Animals and Man.* North-Holland, Amsterdam, pp. 267–305.

Fraser, H. and Dickinson, A.G. (1968) The sequential development of the brain lesions of scrapie in three strains of mice. *J. Comp. Pathol.* **78** 301–311.

Fraser, H. and Dickinson, A.G. (1973) Scrapie in mice. Agent-strain differences in the distribution and intensity of grey matter vacuolation. *J. Comp. Pathol.* **83** 29–40.

Fraser, H., McConnell, I., Wells, G.A.H. and Dawson, M. (1988) Transmission of bovine spongiform encephalopathy to mice. *Vet. Rec.* **123** 472.

Fraser, H., McConnell, I., Bruce, M.E. and Wells, G.A.H. (1990) Transmission of bovine spongiform encephalopathy (BSE) to mice. In: *Proc. XIth International Congress of Neuropathology, September 2–8, 1990, Kyoto, Japan.* Abstract IV-C-16.

Gafner, P. (1991) Bovine spongiform encephalopathy in Switzerland. *OIE Dis. Inf.* **4** 41.

Gibbs, C.J., Jr, Safar, J., Ceroni, M., Di Martino, A., Clark, W.W. and Hourrigan, J.L. (1990) Experimental transmission of scrapie to cattle. *Lancet* **335** 1275.

Goldmann, W., Hunter, N., Martin, T., Dawson, M. and Hope, J. (1991) Different forms of the bovine PrP gene have five or six copies of a short G–C-rich element within the protein-coding exon. *J. Gen. Virol.* **72** 201–204.

Hanson, R.P., Eckroade, R.J., Marsh, R.F., Zu Rhein, G.M., Kanitz, C.L. and Gustafson, D.P. (1971) Susceptibility of mink to sheep scrapie. *Science* **172** 859–861.

Hartsough, G.R. and Burger, D. (1965) Encephalopathy of mink. I. Epizootiologic and clinical observations. *J. Infect. Dis.* **115** 387–392.

Hope, J., Reekie, L.J.D., Hunter, N., Multhaup, G., Beyreuther, K., White, H., Scott, A.C., Stack, M.J., Dawson, M. and Wells, G.A.H. (1988) Fibrils from brains of cows with new cattle disease contain scrapie-associated protein. *Nature (London)* **336** 390–392.

Jeffrey, M. and Wells, G.A.H. (1988) Spongiform encephalopathy in a nyala (*Tragelaphus angasi*). *Vet. Pathol.* **25** 398–399.

Kimberlin, R.H. (1990) Unconventional slow viruses. In: Collier, L.H. and Timbury, M.C. (eds) *Topley and Wilson's Principles of Bacteriology, Virology and Immunity,* 8th edn. Arnold, London, pp. 671–693.

Kimberlin, R.H., Cole, S. and Walker, C.A. (1987a) Pathogenesis of scrapie is faster when infection is intraspinal instead of intracerebral. *Microb. Pathog.* **2** 405–415.

Kimberlin, R.H., Cole, S. and Walker, C.A. (1987b) Temporary and permanent modifications to a single strain of mouse scrapie on transmission to rats and hamsters. *J. Gen. Virol.* **68** 1875–1881.

Kimberlin, R.H., Cole, S. and Walker, C.A. (1989) The genomic identity of different strains of mouse scrapie is expressed in hamsters and preserved on reisolation in mice. *J. Gen. Virol.* **70** 2017–2025.

Lignereux, Y. (1986) Atlas stéréotaxique de l'encephale de la vache frisonne. *Doctoral Thesis.* L'Université Paul Sabatier de Toulouse, France.

Marsh, R.F. and Hanson, R.P. (1979) On the origin of transmissible mink encephalopathy. In: Prusiner, S.B. and Hadlow, W.J. (eds) *Slow Transmissible Diseases of the Nervous System*, Vol. 1. Academic Press, New York, pp. 451–460.

Marsh, R.F., Burger, D., Eckroade, R., Zu Rhein, G.M. and Hanson, R.P. (1969) A preliminary report on the experimental host range of the transmissible mink encephalopathy agent. *J. Infect. Dis.* **120** 713–719.

Marsh, R.F., Bessen, R.A., Lehmann, S. and Hartsough, G.R. (1991) Epidemiological and experimental studies on a new incident of transmissible mink encephalopathy. *J. Gen. Virol.* **72** 589–594.

Masters, C.L. and Richardson, E.P., Jr (1978) Subacute spongiform encephalopathy (Creutzfeldt—Jakob disease). The nature and progression of spongiform change. *Brain* **101** 333–344.

Ministry of Agriculture, Fisheries and Food (1987) Scrapie-like disease in a captive nyala. In: *Animal Health, 1986. Report of the Chief Veterinary Officer.* HMSO, London, p. 69.

Outram, G.W. (1976) The pathogenesis of scrapie in mice. In: Kimberlin, R.H. (ed.) *Slow Virus Diseases of Animals and Man.* North-Holland, Amsterdam, pp. 325–357.

Scott, A.C., Wells, G.A.H., Stack, M.J., White, H. and Dawson, M. (1990) Bovine spongiform encephalopathy: detection and quantitation of fibrils, fibril protein (PrP) and vacuolation in brain. *Vet. Microbiol.* **23** 295–304.

Taylor, D.M. (1988) Occult BSE in the USA? *Vet. Rec.* **123** 138.

Veterinary Record (1991). International disease surveillance. Quarterly report April to June 1991 *Vet. Rec.* **129** 108.

Walker, K.D., Hueston, W.D., Hurd, S. and Wilesmith, J.W. (1991) Comparison of bovine spongiform encephalopathy risk factors in the United States and Great Britain. *J. Am. Vet. Med. Assoc.* **199** 1554–1561.

Wells, G.A.H. and Scott, A.C. (1988) Neuronal vacuolation and spongiosus: a novel encephalopathy of adult cattle. *Neuropathol. Appl. Neurobiol.* **14** 247.

Wells, G.A.H. and Wilesmith, J.W. (1989) The distribution pattern of neuronal vacuolation in bovine spongiform encephalopathy (BSE) is constant. *Neuropathol. Appl. Neurobiol.* **15** 591.

Wells, G.A.H., Scott, A.C., Johnson, C. T., Gunning, R.F., Hancock, R.D., Jeffrey, M., Dawson, M. and Bradley, R. (1987) A novel progressive spongiform encephalopathy in cattle. *Vet. Rec.* **121** 419–420.

Wells, G.A.H., Hancock, R.D., Cooley, W.A., Richards, M.S., Higgins, R.J. and David, G.P. (1989) Bovine spongiform encephalopathy: diagnostic significance of vacuolar changes in selected nuclei of the medulla oblongata. *Vet. Rec.* **125** 521–524.

Wells, G.A.H., Wilesmith, J.W. and Dawson, M. (1990) Bovine spongiform encephalopathy (BSE) is a common source epidemic with remarkable pathogenetic uniformity. In: *Proc. of the XIth International Congress of Neuropathology, September 2-8, 1990, Kyoto, Japan.* Abstract IV-C-15.

Wells, G.A.H., Wilesmith, J.W. and McGill, I.S. (1991) Bovine spongiform encephalopathy: a neuropathological perspective. *Brain Pathol.* **1** 69–78.

Whitaker, C.J. and Johnson, C. (1988). A neurological syndrome. In: Gibbs, H.A. (ed.) *British Cattle Veterinary Association Proceedings for 1987–1988.* Schering-Plough Animal Health, New Jersey, pp. 64–71.

Wilesmith, J.W. and Wells, G.A.H. (1991) Bovine spongiform encephalopathy. In: Chesebro, B.W. (ed.) *Transmissible Spongiform Encephalopathies. Scrapie, BSE and Related Human Disorders.* Current Topics in Microbiology and Immunology, Vol. 172, Springer, Berlin, pp. 21–38.

Wilesmith, J.W., Wells, G.A.H., Cranwell, M.P. and Ryan, J.B.M. (1988) Bovine spongiform encephalopathy: epidemiological studies. *Vet. Rec.* **123** 638–644.

Wilesmith, J.W., Ryan, J.B.M. and Atkinson, M.J. (1991) Bovine spongiform encephalopathy: epidemiological studies on the origin. *Vet. Rec.* **128** 199–203.

Wilesmith, J.W., Ryan, J.B.M., Hueston, W.D. and Hoinville, L.J. (1992a) Bovine spongiform encephalopathy: epidemiological features 1985–1990. *Vet. Rec.* **130** 90–94.

Wilesmith, J.W., Hoinville, L.J., Ryan, J.B.M. and Sayers, A.R. (1992b) Bovine spongiform encephalopathy: aspects of the clinical picture and analyses of possible changes 1986–1990. *Vet. Rec.* **130** 197–201.

24

Association of ovine PrP protein variants with the *Sip* gene and their similarity to bovine PrP protein

Wilfred Goldmann

INTRODUCTION

Transmissible, degenerative encephalopathies are characterized by impairment of brain function after a long preclinical incubation period and accumulation of fibrillar protein aggregates. A unique feature is the modification of a host protein, PrP protein, into a partially protease-resistant form as a result of either covalent modifications or conformational changes (Hope and Hunter, 1988, for review). Since the first scrapie transmission from sheep to mice many studies were centred on experimental scrapie in hamster and mice, so that aspects of disease such as pathogenesis, scrapie strain variation and host genetic control have mostly been analysed in these rodent models (Kimberlin, 1990, for review). With the report of first full sequence of PrP protein from hamster and its confirmation as a host gene (Oesch *et al.*, 1985) a new molecular approach to understanding the role of this protein in scrapie became possible. When the subsequent molecular studies revealed that many if not all mammals encode a PrP protein and confirmed a high degree of sequence homology between PrP proteins of species such as mouse (Locht *et al.*, 1986), man (Kretzschmar *et al.*, 1986) and sheep (Goldmann *et al.*, 1990), rodents became also the model to investigate molecular processes in scrapie and scrapie-like diseases. It was therefore not surprising that an association between PrP protein and the host gene *Sinc*, which controls the incubation period of experimental scrapie in mice, was first reported for this species (Carlson *et al.*, 1986; Hunter *et al.*, 1987). Eventually, the two amino acid substitutions found in the PrP protein of inbred mice which differed in their *Sinc* genotypes (Westaway *et al.*, 1987) could be interpreted as an indication that PrP gene (*Prn-p*) and *Sinc* gene might be congruent. Strong support for PrP association with disease came also from studies of Creutzfeldt–Jakob disease (CJD) and Gerstmann–Sträussler–Scheinker syndrome (GSS), in which the incidence of disease could be linked to several different single amino acid changes in human PrP protein as well as to insertions of octapeptide repeats into the *N*-terminal protein region (Hsiao and Prusiner, 1990, for review). Although these linkage data in the human disease are further evidence for the involvement of PrP in pathogenesis, they can reveal neither mechanisms nor the stage in disease development at which PrP is active. These limitations might be overcome by transgenic animal models (Scott *et al.*, 1989; Prusiner *et al.*, 1990; Hsiao *et al.*, 1990), but their significance for unravelling processes of the natural disease has to be carefully assessed.

An alternative means to investigate a natural encephalopathy is provided by scrapie in sheep and goats, and steady progress in our knowledge of experimental scrapie in sheep since the 1950s might contribute to the understanding of the way these natural diseases are transmitted and develop (Dickinson, 1976). Research in sheep scrapie is also of considerable economic importance for reduction of scrapie in affected flocks as well as maintaining scrapie-free areas. A further reason to refocus attention on ruminants is given by the outbreak of BSE in cattle. BSE has been confirmed as a scrapie-like disease (Wells *et al.*, 1987, Hope *et al.*, 1988) and was apparently caused by feeding scrapie contaminated bone-meal (Wilesmith *et al.*, 1991). Here we might have encountered an example of transmission between related species and in this respect it seems to be essential to know the similarity between the PrP genes of cattle and sheep in order to allow predictions on the development of BSE. Two variants of bovine PrP protein have been presented (Goldmann *et al.*, 1991a), raising the possibility of investigating their linkage with BSE incidence.

Host genetic control in experimental sheep scrapie, similar to that in mouse scrapie, was demonstrated with the creation of selected Cheviot flocks (positive and negative line Neuropathogenesis Unit (NPU) Cheviot sheep (Dickinson *et al.*, 1968)) with predictable SSBP/1 scrapie susceptibility. This led to the definition of the *Sip* gene, the equivalent to the murine *Sinc* gene (Dickinson and Outram, 1988; Hunter, 1992). Recently, Hunter *et al.*, (1989) confirmed an association between PrP and *Sip* gene in experimental scrapie based on the incubation period data produced in these positive and negative NPU Cheviot lines. As with mice, RFLPs (*Eco*RI and *Hind* III) could be linked to the *Sip* alleles sA and pA (haplotype e1h2 associated with sA, haplotype e3h1 associated with pA; for details see also Hunter (1992)). Two ovine PrP alleles have already been described for a Suffolk sheep (Goldmann *et al.*, 1990), but this animal was not available for scrapie challenge to verify its *Sip* genotype. Therefore we analysed the PrP coding region of NPU Cheviot sheep with different *Sip* genotypes for mutations linked to *Sip* phenotypes (Goldmann *et al.*, 1991b).

RESULTS AND DISCUSSION

Ovine PrP gene polymorphisms

Hunter *et al.*, (1989, 1991) and Chapter 28 have highlighted the polymorphic nature of the PrP gene outside of its coding region. Not only were three polymorphic restriction sites (two for *Eco*RI and one for *Hind* III) discovered, but also most of the combinations in which these RFLP sites might occur were found in genotype analysis of different sheep breeds. Two of these RFLPs could unambiguously be associated with the alleles of the *Sip* gene in NPU Cheviot sheep (haplotypes e1h2 and e3h1), but the incidence of natural scrapie was correlated only up to 90% with the *Eco*RI RFLP and seemed not associated with the *Hind* III RFLP. We have established sequence information for the entire PrP coding regions of three of these haplotypes (e1h2, positive line NPU-Cheviot; e3h1, negative line Cheviot; e1h1, Suffolk) (Goldmann *et al.*, 1990, 1991b). Three nucleotide differences (codons 136, 154 and 171) were found, all of which have led to single amino acid changes in the predicted PrP proteins. On the other hand, no variation of the number of octapeptide repeat elements, as reported for bovine PrP (see below), has been found so far in sheep. (Codons and amino acids are numbered differently. The amino terminus of mature ovine and bovine PrP after cleavage of the signal peptide (codons 1 to 24) has been described as Lys–Lys–Arg (Hope *et al.*, 1988). We therefore refer to codon 25 (Lys) as PrP amino acid 1 (Lys1).)

The A–G transition in codon 171 conveying a glutamine$_{147}$ to arginine$_{147}$ substitution had been described previously for a Suffolk sheep (Goldmann *et al.*, 1990) and is now also confirmed for Cheviot sheep, whereas not only the alanine$_{112}$ to valine$_{112}$ substitution (C–T transition in codon 136) but also the arginine$_{130}$ to histidine$_{130}$ change (A–G transition at codon 154) were unprecedented in PrP proteins. The three polymorphic amino acids have been found in four different combinations, defining four PrP protein variants (using the amino acid single letter code): (1) PrP^{VRQ} (Val$_{112}$, Arg$_{130}$, Gln$_{147}$), (2) PrP^{AHQ} (Ala$_{112}$, His$_{130}$, Gln$_{147}$), (3) PrP^{ARR} (Ala$_{112}$, Arg$_{130}$, Arg$_{147}$) and (4) PrP^{ARQ} (Ala$_{112}$, Arg$_{130}$, Gln$_{147}$).

Association of ovine PrP protein variants with the *Sip* alleles
The mutations in codons 136 and 154 created polymorphic restriction sites for *Rsp*XI (or its isoschizomer *Bsp*HI), a restriction endonuclease that recognizes TCATGA (therefore the codons for valine$_{112}$ or histidine$_{130}$) but not C̲CATGA (alanine$_{112}$) or TC̲GTGA (arginine$_{130}$). To evaluate linkage of the protein polymorphisms to *Sip* phenotypes, we exploited the *Rsp*XI restriction site polymorphisms as markers for the changes in codon 136 (designated *Rsp*XI-112) and in codon 154 (designated *Rsp*XI-130). Additionally, some of the sheep were sequenced at codon 147, as no RFLP exists to distinguish between arg$_{147}$ and gln$_{147}$. The analysis was performed on *in vitro* amplified (PCR) DNA fragments and an update of results presented in Goldmann *et al.*, (1991b) is summarized in Table 1. All 29 positive line Cheviots encoded variant PrP^{VRQ} on at least one allele, whereas none of the 17 negative line animals encoded the variant PrP^{VRQ}. A frequency of 40% for the PrP^{AHQ} variant in the negative line was observed, but might not yet reflect the accurate allele frequency as the sample number of sheep tested is still small. A few positive line sheep with a homozygous genotype (e1h2/e1h2) were found to be heterozygous for *Rsp*XI-112 encoding PrP^{VRQ} and PrP^{ARQ}.

Table 1. PrP coding region *Rsp*XI analysis in NPU Cheviot sheep

	Restriction sites present (one or both alleles)				
	*Rsp*XI–112		*Rsp*XI–130		*Pst* I[a]
	Both	One	Both	One	Both
Positive line (*Sip* sAsA or sApA) n = 29	9	20	0	1	29
Negative line (*Sip* pApA) n = 17	0	0	0	7	17

n, number of sheep tested; presence of *Rsp*XI–112 indicates PrP variant with Val_{112}, not Ala_{112}; presence of *Rsp*XI–130 indicates PrP variant with His_{130} not Arg_{130}.
[a] Control digestion.

Although after more than 25 years of selective breeding both NPU Cheviot lines are highly divergent in their response to scrapie, the positive line (*Sip* sAsA and *Sip* sApA) might accidentally have a small number of *Sip* pApA amongst them. By incorporating incubation time data (subcutaneous injection of SSBP/1 scrapie) which were available for 23 sheep, association of PrP^{VRQ} with SSBP/1 scrapie susceptibility could be demonstrated (Table 2). From all data available for the three polymorphic amino acids we conclude that the amino acid 112 substitution appears to be associated with survival time differences following experimental challenge with scrapie source SSBP/1. The results suggest that $valine_{112}$ is associated with high susceptibility and with *Sip* sA, the dominant allele, and $alanine_{112}$ with low susceptibility and with *Sip* pA, the recessive allele. A slightly longer incubation period in putative *Sip* sApA than in putative *Sip* sAsA has been noted recently (Foster and Hunter, 1991).

We have reported three amino acid substitutions in ovine PrP, but at present it seems likely that only one of them conveys the difference between the previously characterized *Sip* sA, pA phenotype groups (as defined by SSBP/1 scrapie challenge). Similarly, Westaway *et al.*, (1987) have reported two amino acid substitutions in the murine PrP protein, which correlate with one or other of the *Sinc* alleles (the homologue of *Sip*) at codons 108 and 189. However, both mouse substitutions have so far always occurred together, and this lack of natural variation meant that it has as yet not been possible to decide whether one or both might be necessary to induce *Sinc* phenotype differences.

The SSBP/1 isolate might differentiate sheep carrying PrP^{VRQ} from PrP^{AHQ}, PrP^{ARR} and PrP^{ARQ}, resulting in three PrP protein variants associated with negative line sheep. Dickinson *et al.*, (1986) pointed out the variability of incubation times in NPU Cheviot negative line *Sip* pApA sheep after intracerebal injection with SSBP/1 and

Table 2. Ovine PrP polymorphism in amino acid 112 (valine/alanine) and its association with scrapie incubation period in NPU Cheviot sheep

	Positive line		Negative line
Number of sheep	6	13	4
*Rsp*XI–112^a digestion of PrP coding region	++	+−	−−
SSBP/1 incubation period (subcutaneous route) (days ± se)	167 ± 13	317 ± 83	Survivors for 1095, 1825
			>1100^b, >1130^b

^{a} Presence of *Rsp*XI–112 site (++ or +−) indicates PrP variant with Val_{112} on both or one allele, and thus absence of *Rsp*XI–112 (−+ or −−) indicates AlA_{112} on one or both alleles.
^{b} Sheep are still alive at date of writing.

raised the possibility of subgroups in sheep with low susceptibility. Other strain isolates such as CH1641, which induces short incubation periods in negative line Cheviots (Foster and Dickinson, 1988a), may allow us to associate phenotypical differences with other PrP variants, i.e. PrP^{AHQ} and PrP^{ARQ}.

The bovine PrP gene

BSE in British cattle has reinforced interest in molecular structure of the bovine PrP gene, polymorphisms within this gene and host genes, homologous to *Sip* in sheep, controlling disease incidence and incubation period. The data available at present for the bovine PrP gene are confirming the close relationship between cattle and sheep. In each case the genome contains a single copy PrP gene (Hunter *et al.*, 1989) which in brain is transcribed into a 4.5 kb mRNA (Goldmann *et al.*, 1990). However, although the restriction map of bovine PrP gene with the enzymes *Eco*RI and *Hind* III looked similar to sheep PrP haplotypes (Hunter *et al.*, 1989), an association of RFLPs similar to the ovine PrP gene with scrapie incubation period (see above) has not been detected in the bovine genome. Only one RFLP had so far been determined: within the protein-coding region a silent mutation (C-to-T transition) created a polymorphism for *Hind* II with fragments of either 7.2 kb and 0.8 kb or 8 kb (Goldmann *et al.*, 1991a). This polymorphism was independently found for *Hinc* II (isochizomer of *Hind* II) as the only RFLP (out of 16 enzymes) in the genome of various cattle breeds (Ryan and Womack, 1991), but any association with the incidence of BSE needs to be investigated.

Only polymerase chain reaction (PCR) (Erlich *et al.*, 1991) and sequence analysis of bovine DNA revealed the presence of two variants of the bovine PrP protein. Both PrP proteins differ in the number of glycine-rich peptides encoded by GC-rich elements

Table 3. Single amino acid differences between the PrP proteins of sheep and cattle

Codon	98	100	136	146	154	158	171	189	208
Amino acid	74	76	112	122	130	134	147	165	184
Sheep									
Cheviot	Ser	Ser	Ala	Asn	Arg	Tyr	Arg	Gln	Ile
			Val		His		Gln		
Suffolk	Ser	Ser	Ala	Asn	Arg	Tyr	Arg	Gln	Ile
							Gln		
Cattle									
Holstein–Friesian	Thr	Gly	Ala	Ser	Arg	His	Gln	Glu	Met

Amino acids are shown in three-letter code. Codon numbers refer to the open reading frame, whereas amino acid numbering starts with Lys (codon 25) as the first residue of mature PrP protein (Lys_1).

(24 or 27 nucleotides in length) in this gene. One allele has six copies (PrP_6) of these elements whereas another has five (PrP_5). Within a gene, each 24 nucleotide element has a unique DNA sequence but codes for identical octapeptides: Pro–His–Gly–Gly–Gly–Trp–Gly–Gln. Apart from rare human alleles in pedigrees of CJD (Owen *et al.*, 1990; Collinge *et al.*, 1989), PrP genes from human and all other mammals sequenced to date encode proteins containing five copies of this repeat (Oesch *et al.*, 1985; Locht *et al.*, 1986; Kretzschmar *et al.*, 1986; Lowenstein *et al.*, 1990; Goldmann *et al.*, 1990). Thus, the PrP_5 variant might be regarded as producing the original form of bovine PrP and the mutant allele arose by insertion of element number three in PrP_6. Out of 12 cattle, we found 8 animals to be homozygous for genes with six copies of the repeat peptide ($PrP_6 : PrP_6$), while 4 were heterozygous ($PrP_6 : PrP_5$). Two confirmed cases of BSE occurred in ($PrP_6 : PrP_6$) homozygous animals, which were also homozygous for the coding region *Hind* II site. Unfortunately, the *Hind* II RFLP is not informative with respect to the described allelism as the PrP_6-encoding allele exists with or without the *Hind* II site (out of eight $PrP_6 : PrP_6$ homozygous cattle four contained the *Hind* II site on both alleles and four only on one).

The bovine PrP_5 protein showed only 18 nucleotide differences when compared with the Suffolk variant of ovine PrP, six of them leading to amino acid differences. However, the divergency is higher (eight amino acids) in comparison with the different Cheviot PrP alleles (Table 3). In contrast, murine and bovine PrP differ in more than 25 amino acids, though, not surprisingly, many of them were confined to the signal peptide. Like the intraspecies PrP variation, which is linked to disease, interspecies PrP protein variation might influence the transmissibility between species. Candidate differences between sheep and bovine PrP might be the single amino acid differences in positions 74, 76, 122, 134, 165, 184 (Table 1), whereas important differences between ruminants and rodents might be at residues 76 (Ser/Gly or Asn), 117 (Leu or Ile), 124 (Tyr or Trp) and 163 (Val or Ile).

PrP structure and disease associated polymorphisms

There are now a large number of PrP protein variants in various species and most of them have been associated with the onset or timing of disease. Therefore it might well be that the as-yet unlinked PrP variants, i.e. the two bovine PrP forms, will be associated in one way or the other with distinctive disease phenotypes. Remarkably these polymorphisms are spread over almost the entire protein, with the restriction that they all occur in the mature, that is post-translational trimmed, PrP protein. Hence it is becoming increasingly more difficult to define only one region which is associated with phenotype differences. Nevertheless, at least one correlation between structure and disease associated polymorphisms has been suggested: *in vitro* translation–translocation studies indicated that a domain between amino acid 70–100 may be crucial for the correct folding, integration and targeting of the nascent PrP protein in the cell membrane (Hay *et al.*, 1987; Lopez *et al.*, 1990; Yost *et al.*, 1990). Five PrP variants with substitutions between amino acids 80 and 95 (two human, two hamster and one murine PrP) support the idea that defects in the intracellular processing of PrP may underly the molecular pathology and timing of scrapie-like diseases (Taraboulos *et al.*, 1990). However, our data linking *Sip* alleles in Cheviot sheep to amino acid polymorphisms at amino acid 112 (and to a lesser extent at amino acids 130 and 147) fill the gap between this region and another polymorphic domain further towards the carboxyl terminus of the protein (amino acids 155–180), in which at present another five substitutions (four human, one murine PrP) have been found, all associated with disease. If some of the single amino acid mutations can be linked to alterations in membrane transport, does the third polymorphic domain in PrP, the *N*-terminal octapeptide repeat region, cause alterations in the same or a similar mechanism? Analysing bovine or human PrP variants with six or more repeats might provide clues to answer this question. If the mechanisms linking this *N*-terminal PrP domain or the other substitutions with disease are different, it is conceivable that their effects are additive (Baker *et al.*, 1991; Palmer *et al.*, 1991).

CONCLUSION

Linkage of PrP protein mutations to disease phenotypes might be the first step to understanding the underlying mechanisms of incubation period control. The natural variation of PrP provided in these sheep can be used as a source for experimental analysis of how molecular prediposition (the PrP variant) leads to precise incubation period phenotype. It is as yet only possible to distiguish three phenotypes, short or medium incubation period and survivor (*Sip*: sAsA, sApA and pApA) in sheep. In the future it may be possible to describe more subgroups, as there is the potential for ten different PrP protein combinations in NPU Cheviots: four sAsA or sApA genotypes and six pApA genotypes. How many of these potential combinations act in an *in vivo* situation cannot be predicted at present, but it is conceivable that it will lead to a better understanding of field strains and incubation periods of natural scrapie, as certain PrP haplotypes are predominant in the natural disease (Hunter *et al.*, 1991; Hunter 1992) and the *Sip* gene has been demonstrated to act in natural scrapie as well (Foster and Dickinson, 1988b).

ACKNOWLEDGEMENT

I would like to thank Dr J. Hope, Dr N. Hunter and J. Foster for invaluable discussions and support, and G. Benson for excellent technical assistance.

REFERENCES

Baker, H.F., Poulter, M., Crow, T.J., Frith, C.D., Lofthouse, R., Ridley, R.M. and Collinge, J. (1991) Amino-acid polymorphism in human prion protein and age at death in inherited prion disease. *Lancet* **337** 1286.

Carlson, G.A., Kingsbury, D.T., Goodman, P.A., Coleman, S., Marshall, S.T., DeArmond, S., Westaway, D. and Prusiner, S.B. (1986) Linkage of prion protein and scrapie incubation time genes. *Cell* **46** 503–511.

Collinge, J., Owen, F., Lofthouse, R., Shah, T., Harding, A.E., Poulter, M., Boughey, A.M. and Crow, T.J. (1989) Diagnosis of Gerstmann–Sträussler syndrome in familial dementia with prion protein gene analysis. *Lancet* **II** 15–17.

Dickinson, A.G. (1976) Scrapie in sheep and goats. In: Kimberlin, R. (ed.) *Slow Virus Diseases of Animals and Man* North Holland, Amsterdam.

Dickinson, A.G. and Outram, G.W. (1988) Genetic aspects of unconventional virus infections: the basis of the virino hypothesis. In: *Novel Infectious Agents and the Central Nervous System.* Ciba Foundation Symposium 135, Wiley-Interscience, London, pp. 63–83.

Dickinson, A.G., Stamp, J.T., Renwick, C.C. and Rennie, J.C. (1968) Some factors controlling the incidence of scrapie in Cheviot sheep injected with a Cheviot-passaged scrapie agent. *J. Comp. Pathol.* **78** 313–321.

Erlich, H.A., Gelfand, D. and Sninsky, J.J. (1991) Recent advances in the polymerase chain reaction. *Science* **252** 1643–1651.

Foster, J.D. and Dickinson, A.G. (1988a) The unusual properties of CH1641, a sheep-passages isolate of scrapie. *Vet. Rec.* **123** 5–8.

Foster, J.D. and Dickinson, A.G. (1988b) Genetic control of scrapie in Cheviot and Suffolk sheep. *Vet. Rec.* **123** 159.

Foster, J.D. and Hunter, N. (1991) Partial dominance of the Sip gene in the control of experimental scrapie in Cheviot sheep. *Vet. Rec.* **128** 548–549.

Goldmann, W., Hunter, N., Foster, J.D., Salbaum, J.M., Beyreuther, K. and Hope, J. (1990) Two alleles of a neural protein gene linked to scrapie in sheep. *Proc. Natl. Acad. Sci. USA* **87** 2476–2480.

Goldmann, W., Hunter, N., Martin, T., Dawson, M. and Hope, J. (1991a) Different forms of the bovine PrP gene have five or six copies of a short, G–C-rich element within the protein-coding exon. *J. Gen. Virol.* **72** 201–204.

Goldmann, W., Hunter, N., Benson, G., Foster, J.D. and Hope, J. (1991b) Different scrapie associated fibril proteins (PrP) are encoded by lines of sheep selected for different alleles of the *Sip* gene. *J. Gen. Virol.* **72** 2411–2417

Hay, B., Barry, R.A., Lieberburg, I., Prusiner, S.B. and Lingappa, V.R. (1987) Biogenesis and transmembrane orientation of the cellular isoform of the scrapie prion protein. *Mol. Cell. Biol.* **7** 914–920.

Hope, J. and Hunter, N. (1988) Scrapie-associated fibrils, PrP protein and the *Sinc* gene. In: *Novel Infectious Agents and the Central Nervous System*. Ciba Foundation Symposium 135, Wiley-Interscience, (London,) pp. 146–158.

Hope, J., Reekie, L.D., Hunter, N., Multhaup, G., Beyreuther, K., White, H., Scott, A.C., Stack, M.J., Dawson, M. and Wells, G.A.H. (1988) Fibrils from brains of cows with new cattle disease contain scrapie-associate protein *Nature (London)* **336** 390–392.

Hsiao, K. and Prusiner, S.B. (1990) Inherited human prion diseases. *Neurology* **40** 1820–1827.

Hsiao, K., Scott, M., Foster, D., Groth, D.F., DeArmond, S.J. and Prusiner, S.B. (1990) Spontaneous neurodegeneration in transgenic mice with mutant prion protein *Science* **250** 1587–1590.

Hunter, N. (1992) Association of PrP gene polymorphisms with the incidence of natural scrapie in British sheep. In: Prusiner, S.B., Collinge, J., Powell, J. and Anderson, B. (eds) *Prion Diseases of Humans and Animals*. Ellis Horwood, Chichester, Chap. 28.

Hunter, N., Hope, J., McConnell, I. and Dickinson, A.G. (1987) Linkage of the scrapie-associated fibril protein (PrP) gene and *Sinc* using congenic mice and restriction fragment length polymorphism analysis. *J. Gen. Virol* **68** 2711–2716.

Hunter, N., Foster, J.D., Dickinson, A.G. and Hope, J. (1989) Linkage of the gene for the scrapie-associated fibril protein (PrP) to the *Sip* gene in Cheviot sheep. *Vet. Rec.* **124** 364–366.

Hunter, N., Foster, J.D., Benson, G. and Hope J. (1991) Restriction fragment length polymorphisms of the scrapie-associated fibril protein (PrP) gene and their association with susceptibility to natural scrapie in British sheep. *J. Gen. Virol.* **72** 1287–1292.

Kimberlin, R. (1990) Unconventional 'slow' viruses. In: Collier, L.H. and Timbury, M.C. (eds) *Principles of Bacteriology, Virology and Immunity*, Vol. 4, *Virology*. Topley and Wilson's, London, pp. 671–693.

Kretzschmar, H.A., Stowring, L.E., Westaway, D., Stubblebine, W.H., Prusiner, S.B. and DeArmond, S.J. (1986) Molecular cloning of a human prion protein cDNA. *DNA* **5** 315–324.

Locht, C., Chesebro, B., Race, R. and Keith, J.M. (1986) Molecular cloning and complete sequence of prion protein cDNA from mouse brain infected with the scrapie agent. *Proc. Natl. Acad. Sci. USA* **83** 6372–6376.

Lopez, C.D., Yost, C.S., Prusiner, S.B., Myers, R.M. and Lingappa, V.R. (1990) Unusual topogenic sequence directs prion protein biogenesis. *Science* **24** 226–229.

Lowenstein, D.H., Butler, D.A., Westaway, D., McKinley, M.P., DeArmond, S.J. and Prusiner, S.B. (1990) Three hamster species with different scrapie incubation times and neuropathological features encode distinct prion proteins. *Mol. Cell. Biol.* **10** 1153–1163.

Oesch, B., Westaway, D., Wälchli, M., McKinley, M.P., Kent, S.B.H., Aebersold, R., Barry, R.A., Tempest, P., Teplow, D.B., Hood, L.E., Prusiner, S.B. and Weissmann, C. (1985) A cellular gene encodes scrapie PrP 27–30 protein. *Cell* **40** 735–746.

Owen, F., Poulter, M., Shah, T., Collinge, J., Lofthouse, R., Baker, H., Ridley, R.,

McVey, J. and Crow, T.J. (1990) An in-frame insertion in the prion protein gene in familial Creutzfeldt–Jakob disease. *Mol. Brain Res.* **7** 273–276.

Palmer, M.S., Dryden, A.J., Hughes, J.T. and Collinge, J. (1991) Homozygous prion protein genotype predisposes to sporadic Creutzfeldt–Jakob disease. *Nature (London)* **352** 240–342.

Prusiner, S.B., Scott, M., Foster, D., Pan, K.M., Groth, D., Mirenda, C., Torchia, M., Yang, S.L., Serban, D., Carlson, G.A., Hoppe, P.C., Westaway, D. and DeArmond, S.J. (1990) Transgenetic studies implicate interaction between homologous PrP isoforms in scrapie prion replication. *Cell* **63** 673–686.

Ryan, A.M. and Womack, J.E. (1991) Somatic cell mapping and RFLP studies of the prion protein gene in cattle. In: *Proc. 7th North American Colloquium on Domestic Animal Cytogenetics and Gene Mapping.* pp. 117–121.

Scott, M., Foster, D., Mirenda, C., Serban, D., Coufal, F., Walchli, M., Torchia, M., Groth, D., Carlson, G., DeArmond, S.J., Westaway, D. and Prusiner, S.B. (1989) Transgenic mice expressing hamster prion protein produce species-specific scrapie infectivity and amyloid plaques. *Cell* **59** 847–857.

Taraboulos, A., Serban, D. and Prusiner, S.B. (1990) Scrapie prion proteins accumulate in the cytoplasm of persistently infected cultured cells. *J. Cell Biol.* **110** 2117–2132.

Wells, G.A., Scott, A.C., Johnson, C,T., Gunning, R.F., Hancock, R.D., Jeffrey, M., Dawson, M. and Bradley, R. (1987) A novel progressive spongiform encephalopathy in cattle. *Vet. Rec.* **121** 419–420.

Westaway, D., Goodman, P.A., Mirenda, C.A., McKinley, M.P., Carlson, G.A. and Prusiner, S.B. (1987) Distinct prion proteins in short and long scrapie incubation period mice. *Cell* **51** 651–662.

Wilesmith, J.W., Ryan, J.B.M. and Atkinson, M.J. (1991) Bovine spongiform encephalopathy: epidemiological studies on the origin. *Vet. Rec.* **128** 199–203.

Yost, C.S., Lopez, C.D., Prusiner, S.B., Myers, R.M. and Lingappa, V.R. (1990) Non-hydrophobic extracytoplasmic determinant of stop transfer in the prion protein. *Nature (London)* **343** 669–672.

25

Bovine spongiform encephalopathy: the history, scientific, political and social issues

R. Bradley and R. C. Lowson

ABSTRACT

Like scrapie of sheep, a disease known for centuries, bovine spongiform encephalopathy (BSE), discovered in 1986, has not been implicated in any human disease. Nevertheless, legislation has been introduced to eliminate any remote risk that the agent of BSE could affect the human population. Unprecedented media attention attracted by the mystery and controversy surrounding the epidemic in Britain and unilateral action by Germany, France and Italy in prohibiting imports of beef from Britain resulted in a significant social impact on the British Agricultural Industry quelled finally by the report of the House of Commons Agriculture Committee. Britain has been the world leader in public and scientific education on BSE leading to a less severe reaction elsewhere despite confirmation of disease in France and Switzerland.

INTRODUCTION

Scrapie, a natural disease of sheep, has been known in Europe, sometimes in epidemic form, since the 18th century. Though McGowan (1922) reported 'The earliest definite record of the occurrence of scrapie in Britain that one could find was in 1732', it was Leopoldt (1759) in Germany who is often quoted as being the first to describe the disease. According to Parry (1983) scrapie has occurred in many European countries since 1700, sometimes sporadically and associated with importation and sometimes as epidemics or severe outbreaks such as occurred in Germany, France, England and Wales in the period 1780–1820. Curiously the disease was virtually unknown in

England and Wales between 1880 and 1910, is now rare in Germany and is reported to have a low incidence in France.

Experimental studies on the enigmatic disease gained momentum in the late 19th and in the 20th century. Besnoit and Morel (1898) at Toulouse, for example, were the first to describe vacuoles of the perikaryon of ventral horn cells. On 11 June 1898, Besnoit transfused a ewe with 1.825 kg of blood from an animal with scrapie but no disease had resulted in the recipient after 9 months. He also, conducted negative cohabitation experiments. In 1918 Sir John M'Fadyean at the Royal Veterinary College in London 'noticed that *post mortem* examination does not reveal any visceral lesions which can be regarded as part of the disease' (M'Fadyean, 1918) and transmission experiments with blood, cerebrospinal fluid, skin, allantois, foetal cotyledons, intestinal contents or contact were all negative after 18 months. He concluded 'The long period of incubation is a very serious obstacle to the experimental investigation of scrapie.'

Gaiger (1924) was among the first in more recent times to put a specific economic cost to scrapie, reporting one sheep farmer who had lost £600 in 1 year from scrapie, a considerable sum in those days. He also stated the main research tasks were to find the agent of scrapie and to develop a test for its detection ('in rams').

Cuillé and Chelle (1936), two excellent professors and research workers from Toulouse, were the first to demonstrate conclusively that scrapie was a transmissible disease following intraocular exposure of two ewes to an emulsion of either spinal cord or cerebral hemispheres on 6th July 1934. Symptoms first appeared after 15 months in the first ewe and 22 months in the second. Controls did not develop disease. Chelle (1942) reported for the first time natural scrapie in a goat and referred to descriptions of 'scrapie' in horses, cattle and zoo animals as something very different. Pattison *et al.* (1959) recognized goats as universally susceptible to scrapie, and transmitted scrapie orally via placenta of affected sheep to both sheep and goats.

This brief historical introduction recounting some of the major events preceding the advent of BSE would be incomplete without reference to the major contributions of Dr W. J. Hadlow and Dr R. L. Chandler. Hadlow, a distinguished American veterinary pathologist, made a singular, pertinent observation on the similarities between scrapie, a disease he had worked with, and kuru, a human disease of mystifying origin that affected the Fore tribe in the eastern highlands of New Guinea and which tribe practised endocannibalism. Kuru had first been reported by Gajdusek and Zigas (1957). In the *Lancet* Hadlow (1959) wrote 'The overall resemblance between kuru and scrapie is too impressive to be ignored.' Hadlow suggested experimental induction of kuru in a laboratory primate, which was later achieved by Carleton Gajdusek and his coworkers (Gajdusek *et al.*, 1966) and contributed to the award in 1976 of the Nobel prize for Medicine to Dr Gajdusek.

Hadlow and coworkers also conducted exemplary studies on tissue infectivity during the pre-clinical and clinical phases of natural scrapie in sheep (Hadlow *et al.*, 1982) and similar studies in the clinical phase in goats (Hadlow *et al.*, 1980). These were to become of immense importance later in assessing the probability of infectivity in tissues from cattle with BSE.

At the AFRC Institute for Research in Animal Diseases at Compton, as a result of his work with three different strains of mice which were differentially susceptible to Johne's disease, Chandler (1962) made the suggestion to Pattison that mice might also be susceptible to scrapie. Subsequently Chandler was the first to transmit the disease successfully to this now valuable model. Dickinson and colleagues developed specific in-bred strains of mice with allelic variations of the major scrapie incubation period (*Sinc*) controlling gene. These mice are now used extensively in research and form the basis for infectivity assays and strain typing of field isolates of scrapie and BSE. They have also stimulated research into the genetic control of natural disease.

This introduction serves to show that the objectives of research have changed little over the centuries and much has been done. However, the work has been laborious and slow to yield practical results that are immediately useful for the control or elimination of natural scrapie and scrapie-like diseases.

Controversy still rages over whether scrapie-like diseases have a hereditary or environmental cause. The evidence from kuru and BSE is strongly in favour of an infectious origin, whereas Gerstmann–Sträussler–Scheinker (GSS) syndrome and familial Creutzfeldt–Jakob disease of man provide support for a hereditary basis for the diseases—or is it only increasing susceptibility to a ubiquitous agent? It is well known that all disease has an environmental and a genetic component. What is now clear is that GSS syndrome is both hereditary and infectious. How the agent genome (if it has one—and that is another disputed point) interacts with the host genome is crucial in determining whether infection and/or disease will result in the host. The various theories of agent structure (prion, virino or virus) each strongly supported by their protagonists and their ingenious experiments are still not finally resolved.

The remainder of this paper will be devoted to the advent of BSE and the effect this has had on the British agricultural industry, the social impact on its people and the stimulus it has given to research aimed at control and eventual elimination of the disease.

RESULTS AND DISCUSSION

There have been many vague references in the literature to natural scrapie in other species of animals. One notable exception was that described by Sir Stewart Stockman, Chief Veterinary Officer and Director of the Central Veterinary Laboratory, in a lecture given to the Yarrow and Ettrick Pastoral Society on 23rd October 1913 (Stockman, 1913). He stated 'I have seen a description of what purported to be a case of scrapie in an ox in an old French writing and although the symptoms described were very like those of scrapie it can hardly be accepted from this one case that the ox is susceptible.' The search for a more authoritative account of this and other reports of scrapie outside the ovine species continues. Retrospectively, the first indication of a new twist to the scrapie story came in June 1986. The *Report of the Chief Veterinary Officer* (Anon, 1986) referred to a pathologically confirmed case of a scrapie-like disease in a captive nyala in southern England. To date, four other exotic species in captivity in Britain have been affected similarly. Curiously the nyala case caused no stir in the media until several years later when BSE was very much in the limelight.

Fig. 1.   Holstein Friesian cow with BSE showing apprehension, arched back, abnormal head posture, straight abducted hind limbs and staring eyes. Crown Copyright. (Courtesy Dr A.E. Wrathall – Pathology Department, Central Veterinary Laboratory, Weybridge.)

As a result of the combined efforts of the Veterinary Investigation Service and the Pathology Department at Weybridge, BSE (Fig. 1) was first diagnosed and confirmed by brain histopathology in two adult Holstein Friesian cows in November 1986 (Wells *et al.*, 1987). At first, with few reports of suspects or confirmations in the months following its discovery and because of its immediate recognition as a scrapie-like disease, the main concern was with animal health. Were these cases just scientific curiosities, was BSE truly related to the group of diseases known as the subacute, transmissible spongiform encephalopathies (STSEs) caused by unconventional agents, and was there any possibility that BSE would have a different host range to scrapie and thus possibly present a risk to public health? These were some of the important questions to answer at that time. As time progressed with the advent of more cases it was clear that BSE was not just a scientific curiosity but a problem of importance for the cattle industry and the Government to tackle. In the meantime, scientists at Weybridge confirmed the presence of scrapie-associated fibrils (SAFs) in detergent-treated brain extracts treated with proteinase K, negatively stained and examined by electron microscopy (Wells *et al.*, 1987). The major protein of these fibrils (PrP) was identified as the bovine homologue of that found in scrapie (Hope *et al.*, 1988). At the Institute for Animal Health, Neuropathogenesis Unit, in Edinburgh, successful transmission of BSE from four cows from different geographical areas of the country was accomplished in mice (Fraser *et al.*, 1988). These features collectively confirmed that BSE was a new member of the STSE group of diseases (Table 1).

Table 1.   Naturally occurring transmissible spongiform encephalopathies

Host	Disease	Reported distribution
Man	Kuru	Papua New Guinea
	Creutzfeldt–Jakob disease (CJD) (Iatrogenic, Sporadic Familial) Gerstmann–Sträussler (–Scheinker) (GSS) syndrome	Worldwide
Sheep, goats	Scrapie	Widely distributed, but not reported in Argentina, Australia, New Zealand, Uruguay and some other countries
Mule deer, elk	Chronic wasting disease (CWD)	North America
Farmed mink	Transmissible mink encephalopathy (TME)	North America, Europe
Cattle	Bovine spongiform encephalopathy (BSE)	UK/Republic of Ireland, France, Switzerland, Oman, Falkland Islands

Southwood Working Party

To deal with the third major question the Secretary of State for Health and Minister for Agriculture, Fisheries and Food set up a Working Party in May 1988 chaired by Sir Richard Southwood to advise on the implications of BSE in relation to both animal health and any possible human health hazards and to advise on any necessary measures. Interim recommendations were made during 1988 and a report was published in February 1989 (Working Party on Bovine Spongiform Encephalopathy, 1989). All the recommendations were, so far as was possible, promptly implemented along with other actions deemed necessary either at that time or later when the results of research suggested modifications to be desirable.

Animal health controls

The animal health controls were implemented in 1988 and in June of that year the disease was made notifiable and movement restrictions were placed upon suspect animals (not the herd because research had shown that each case was an index case resulting from exposure to a scrapie-like agent in the food and there was no evidence of cow-to-cow transmission). Isolation of pregnant suspect cows close to calving was required whilst calving and for 72 hours afterwards and placenta, uterine discharges and soiled bedding were required to be safely destroyed by burning or burial.

In July 1988, the feeding of ruminant protein to ruminant animals was prohibited. In September 1990 on advice from the Spongiform Encephalopathy Advisory Committee (see below), following experimental transmission of BSE from affected cow brain to one pig after massive parenteral challenge, the feeding of specified bovine offals (brain, spinal cord, tonsil, thymus, spleen and intestine) or protein derived from them to any species of animal or bird was prohibited.

Public health controls

From August 1988, cases of BSE were slaughtered and compensation paid and then carcases were destroyed so that they could not enter any food chain. Virtually all carcases are now incinerated. Though there is no evidence at all for infectivity being present in udder or milk (Hadlow *et al.*, 1982; Working Party on Bovine Spongiform Encephalopathy, 1989), the Southwood Working Party recommended that, as a precautionary measure, milk from suspects should also be destroyed; this was effective from December 1988. Finally, but from November 1989, as with animals, no specified bovine offals or protein derived from them were permitted to enter the human food chain.

The Southwood Working Party also recommended that an expert consultative committee on research be set up. This was established in February 1989. The interim report on the research required, proposed and in progress was presented to Ministers in June 1989 and subsequently published (Consultative Committee on Research into Spongiform Encephalopathies, 1989). This Committee chaired by Dr D. A. J. Tyrrell was subsequently reconstituted as the Spongiform Encephalopathy Advisory Committee. A number of other recommendations of the Southwood Working Party were also implemented. These included monitoring the offspring of affected animals, monitoring cases of Creutzfeldt–Jakob disease, and informing the Health and Safety Executive, Committee on Safety of Medicines, Committee on Dental and Surgical Materials and Veterinary Products Committee of the emergence of BSE. In the event actions were taken to protect human and animal patients and occupationally exposed workers (veterinarians, herdsmen, abattoir and research personnel) with the aid of the responsible controlling authorities, appropriate industries and professions.

Epidemiology

The epidemiological study of the BSE epidemic, now the most extensive applied to an animal disease, was initiated in June 1987 and will be reported elsewhere. Apart from the professionalism and skill with which it was conducted two points of clarification are worthy of mention.

First, although BSE was first confirmed in November 1986, it was as a result of retrospective epidemiological studies, farmer interrogation and questionnaire completion that, beyond reasonable doubt, BSE was shown to have existed as a clinical entity at a very low incidence as early as April 1985 but not before. Clearly these early cases were not confirmed by brain histology.

Second, on the 23 August 1991, the number of confirmed BSE cases was 35 100, which some have claimed is well beyond the predictions made in the Southwood Working Party Report, paragraph 6.1 (Working Party on Bovine Spongiform Encephalopathy, 1989)—namely a presentation of 350–400 cases per month and a total expectation of 17 000–20 000 cases from cows currently alive and subclinically infected. Careful reading of that paragraph indicates the conditions which qualified the predictions that were made. The Working Party assumed minimal and undetectable effects of either an increase of ovine material or the inclusion of recycled infected bovine material in the cattle food chain before the ruminant feed ban was implemented in July 1988. However, it is now apparent that the latter and possibly the former factors were actually significant and probably responsible for both the increased presentation rate and increased total number of cases beyond the prediction. Nevertheless, the other predictions of a fall in incidence from 1992 and, if there is no cattle-to-cattle transmission, a virtual elimination of disease soon after 1996 should hold good.

Human health risks

The main point to make is that, although we do not know yet that the BSE agent is or is not a human pathogen, the likelihood of transmission of disease to man, taking account of all the evidence from scrapie and that which we have from BSE, is remote (Working Party on Bovine Spongiform Encephalopathy, 1989). If BSE is not a human pathogen there is nothing to worry about. If it is, then protection can only be afforded by eliminating or reducing exposure of man to a level of infectivity that could not produce infection or disease in the new host. The main factors determining risk are derived from scrapie research.

First, the agent is localized in significant quantities only in the lymphoreticular system (LRS) and central nervous system (CNS). In the early stages after infection (up to 8 months in lambs (Hadlow, 1982)) there is no detectable infectivity in any tissue. Infection is then detected initially in the LRS and finally the CNS, particularly in the second half of the incubation period and clinical phases of disease.

Second, the oral route of exposure is inefficient compared with parenteral (especially brain-to-brain) exposure and, third, the species barrier is an efficient form of protection.

The nature of this last-mentioned factor forms part of the current research programme and includes that concerning the species comparison of the PrP gene sequence and molecular structure of the PrP^C for which it codes. The British Government's actions to protect public and animal health, therefore, have been based upon scientific evidence and aimed at reducing or eliminating exposure of animals or people by removing affected animals, their milk and specified offals (from all cattle over six months of age) from all food chains. Furthermore, a very large research programme costing many millions of pounds has been established as recommended by the Consultative Committee on Research.

Public concern and the media

So what has all the fuss been about? Public concern reached a peak in 1990, three and a half years after the first case was discovered. Concern was generated on several counts, not least perhaps the fact that, once BSE was recognized as having a food-scare flavour, it became a prime candidate for media interest. A significant factor was the inappropriate, but eye- and ear-catching, use of the colloquial appellation or misnomer 'mad cow disease', that rapidly substituted for the scientific name ascribed to the disease (Wells *et al.,* 1987). Cows were not mad and the title was misleading; nevertheless, it caught public attention and conjured up one of the ingredients needed for a good story—fear. An association was made between BSE and the human counterpart—CJD, a rare disease affecting about one person per 1–2 million worldwide—a disease which ended in death after a distressing and rapid decline in mental health and for which there was no cure. Some more industrious reporters delved superficially into the literature and found accounts of an incidence of CJD which were 30 times the normal level in a group of Libyan Jews living in Israel who were purported to consume sheep brains and eyeballs as part of their diet. The press failed to report that the scientific reason for this increased incidence of CJD was unconnected with diet but related to a codon 200 mutation in the PrP gene (Goldfarb *et al.,* 1990) which was found in all patients examined (and indeed in other familial CJD patients elsewhere who had more standard dietary habits). Such incomplete stories added fuel to the fire and convinced at least some of a gullible public, lacking knowledge of the whole subject area, that perhaps there really was something in the falsely promoted food scare.

During the course of the media build-up various aspects of the agricultural industry were highlighted. Most important was the fact that cattle and particularly calves on dairy farms did not spend all their time munching grass or hay but were additionally fed concentrate rations containing meat and bone meal derived from their own species. The Southwood Working Party (Working Party on Bovine Spongiform Encephalopathy, 1989) noted this and questioned the wisdom of methods which may expose susceptible species of animals to pathogens and asked for this general issue to be addressed, which indeed it was by the Expert Group in animal feeding stuffs (also recommended by the House of Commons Agriculture Committee (Bovine Spongiform Encephalopathy (BSE) Agriculture Committee, 1990)) chaired by Professor G. E. Lamming.

In fact the feeding of meat and bone meal to farm animals had been going on without concern or public comment for decades. Some reporters and readers of their reports were deceived by the newly acquired information in the unfamiliar field of rendering and actually believed it was the raw, uncooked waste offals from abattoirs that were fed to cattle instead of the not unpleasant, heat-treated, meat and bone meal derived from them.

Given the predisposition on the part of the media and public opinion to take an interest in food scares, there is no doubt at all that BSE represented a very good one. Importantly, whatever the risk from BSE was, or might be construed to be, it was a new risk. Most of the population accept the everyday risks of death or injury, such as those from road accidents. The risks can be calculated and are reduced by two means: first, by government actions—introduction of speed limits, enforced wearing of seat belts and breath analysing; second, by providing enough information so that the individual can reduce even further his or her risk by selection of safer options, such as travelling by alternative means of transport. Both methods of risk reduction were adopted in BSE but there were initial difficulties because a new risk is different. It takes time to understand the risks and to quantify them. The mystery of this group of diseases adds further to the problem. The causal agent is not known and in human sporadic CJD the origin is also unknown. There is no immune response and so there are no tests that can be applied in the live animal to detect infection. The agent, whatever it is, is resistant to heat, radiation and chemical disinfection— though there are effective and practical ways of destroying it which are used regularly in hospitals, laboratories, farms and so on.

The final ingredient for a good story is controversy, and there is no shortage of that. Even in the scientific community, workers with decades of experience of working with these disease have opposing views, for example, on the nature of the agent. That is a healthy state of affairs and provides the stimulus for research to reach the truth for the benefit of all. What was different with BSE was that, whenever a new piece of information came to light, certain sections of the media presented unbalanced views to the public, some at the extreme end of the spectrum of rationality. At a critical time in the first half of 1990 when Germany, France and Italy unlawfully, and without scientific evidence of risk, ceased importation of British beef and live cattle, media attention was especially prominent and was coincident with a temporary loss of public confidence in British beef. In Parliament also there was fierce discussion on the subject leading finally to a debate and investigation of the BSE problem by the House of Commons Agriculture Committee, particularly to address the question— is beef safe? The Agriculture Committee reported comprehensively on 10 July 1990 stating that it was safe (Bovine Spongiform Encephalopathy (BSE) Agriculture Committee, 1990), and this did much to dispel earlier fears.

Consumer groups

Consumer groups and others prepared written information which was sometimes inaccurate and caused further public concern, though the intentions were otherwise. Doubt was particularly cast on the use of mechanically recovered meat in human food, on the safety of sausages and meat pies and the advisability of permitting bovine heads to be split in abattoirs. The last-mentioned practice, though only operated in a few abattoirs, was soon banned but there was little or no substance in the other criticisms. The Ministry of Agriculture, Fisheries and Food maintained contact with trade organizations and consumer groups and meetings were convened specifically to discuss relevant BSE issues and to correct errors of fact. Slowly but surely accuracy improved, the public were better informed and fears subsided. Some reporters soon got the message that the subject was very complex and that accuracy of fact was a top priority as the public became more educated and discerning.

Exports

In time the media made greater efforts to achieve this objective of accuracy. Some extended television programmes dealt sensibly with the subject and contributed to public understanding. There was still a tendency, however, for the use of eye-catching sensational headlines to otherwise well-written articles, especially when a new piece of information was released. This and previous non-factual reporting assisted in damaging Britain's export trade in cattle and cattle products but in truth there were genuine reasons why some restrictions on trade were necessary. Some countries nevertheless over-reacted and banned everything remotely connected with cattle, and sometimes sheep too, including milk and hides and skins. As with the internal situation this was dealt with by discussion with the appropriate authorities in the countries concerned by visits to them, by inviting and receiving foreign delegations and by discussion within appropriate international organizations such as the Office International des Epizooties (OIE). These meetings have resulted in dissemination of knowledge, guidance on risk assessment and restoration of a good deal of trade subject to various conditions.

School Beef

Two final points about beef should be made. First, certain local authorities banned the use of British beef in school meals. This was inconsistent with the risks involved since no infectivity had ever been detected in muscle from cases of natural scrapie or, in so-far incomplete experiments, from BSE-affected cows. Decisions were made under pressure from parents and others who were concerned for the safety of their children. Some of the concern resulted, not from an inherent fear of infection in beef itself, but rather from concern from cross-contamination with either brain or spinal cord during the slaughter process and before the production of beefburgers which formed a significant part of the beef used in school meals. This problem was dealt with by indicating that a very high proportion of British beef at the time was derived from animals not exposed to BSE via the food, ensuring that heads were no longer permitted to be split in the abattoir and that the spinal cord was removed after carcase splitting as a discrete anatomical entity. The risks of cross-contamination were therefore miniscule. Slowly but surely bans were lifted and local authorities now permit pupils in their schools to eat beef precisely as they can at home.

Pets and petfood

The second point concerns the unilateral action taken by Germany, France and Italy preventing the importation of British beef and live animals during the early part of 1990, contrary to the rules governing trade laid down by the European Community. This action coincided with the discovery of a scrapie-like spongiform encephalopathy in a domestic cat (Wyatt et al., 1990) which further exacerbated public fear. Members of the Pet Food Manufacturers' Association had already taken voluntary steps to eliminate bovine offals that might contain the BSE agent from their products from before the time of imposition of the bovine-specified offals ban for man (November 1989) and which, since September 1990, applies to all species. Thus, companion animals fed proprietary food had been and are protected equally with humans and for at least as long.

The European Community (EC)
BSE has been regularly discussed under the auspices of the Scientific Veterinary Committee from July 1988 onwards. From 28 July 1989, a ban was imposed on the export from the United Kingdom of live cattle born prior to the ruminant feed ban, or which were the offspring of affected cattle. On 1 March 1990 live cattle exports were restricted to calves under six months of age and from the 30th of that month a ban was imposed upon the export of specified bovine offals and certain organs for pharmaceutical purposes. From 1 April 1990 Member States were required to notify all cases of BSE to the Commission and other Member States.

From November 1989 until January 1990 several important meetings of scientific experts and the Scientific Veterinary Committee were held which culminated in the publication of a document giving the Committee's final opinions on a set of relevant questions which included those concerned with human and animal health.

On 6 June 1990, a joint meeting of the EC Animal and Public Health Sections of the Scientific Veterinary Committee met in Brussels to discuss the new occurrence of feline spongiform encephalopathy and inaccurate reports of confirmation of BSE in animals born after 18 July 1988. No further animal or public health risks were identified and with minor textual adjustment the earlier opinions were endorsed unanimously. They were similarly endorsed by the Standing Veterinary Committee on the same day. These meetings were followed by a meeting of the Agriculture Council on 6–7 June attended by Ministers from Member States. The conclusions of the Council on its considerations of BSE enabled immediate re-establishment of trading in beef, subject to certain conditions, and of live calves certified to be under six months of age, which were not the offspring of cows in which BSE was suspected or confirmed.

At the time of writing and since the Agriculture Committee's Report in July 1990 some significant developments have taken place. A European Community funded research programme has been initiated to evaluate scientifically the effectiveness of current rendering practices in destroying the BSE–scrapie agent, to monitor CJD in man and negative rabies suspects in cattle in other European countries and to develop a test for infection in the live animal using current and future molecular biological technology. BSE has been detected in France (four confirmed cases by September 1991) and in Switzerland (seven confirmed cases by September 1991). The number of feline spongiform encephalopathy cases in the United Kingdom has risen to only 19 by the same date and the media interest in BSE in Britain has declined to an unobtrusive level.

Meat sales
Significantly, in Britain, the percentage change in household purchases of beef in four-week periods compared with the same periods in the previous year, as determined by the Meat and Livestock Commission–AGB for the months of April–July in 1990 and 1991, show that, beef consumption in Britain has returned to its former level (Table 2). Not unexpectedly on past evidence, this important fact receives very little media attention and it is left to the industry itself to promote its own image.

Table 2.   Percentage changes in household purchases of beef in four-week periods compared
with the same periods in the previous year

	1990	1991
April	−5.3	+4.6
May	−24.7	+26.3
June	−8.6	+10.9
July	−16.7	+9.0

Source: MLC/AGB.

Social effects

So what has all this meant in the social context? There has been a significant economic impact involving Government expenditure for compensation and carcase disposal. The agricultural industry, including specifically farmers, abattoir owners, wholesale and retail butchers, renderers, cattle feed compounders and suppliers and not least knackers, has borne the brunt of the losses. A number of bankruptcies have been reported and more are feared (Bovine Spongiform Encephalopathy (BSE) Agriculture Committee, 1990). A contribution has been made to these from a changed financial structure wherein renderers, unable to sell meat and bone meal, have had to charge to remove and process offals instead of paying for them. Likewise, knackers have had to charge for removal of carcases instead of paying for them. Restriction on exports, particularly in pedigree cattle, has also had a significant effect on those farmers and companies specializing in this trade. In general, because the mean within-herd incidence of BSE is just below two per one hundred adult cows, BSE is not an economic problem to individual farmers especially as 100% compensation (subject to a ceiling) has been paid since 14 February 1990. However, a few individual farmers have suffered financial hardship especially if they have had a high within-herd incidence of disease—particularly if this occurred in successive age cohorts and even more so in a pedigree herd which supplied breeding animals for sale or export. However, the British Agricultural Industry is very resilient; it has to a large extent come to terms with the situation and overcome the temporary setbacks of 1990. One redeeming feature has been the judgement of the Bovine Spongiform Encephalopathy (BSE) Agriculture Committee (1990) that, after hearing evidence from scientific experts (including those who expressed doubts), administrators, the Minister of Agriculture, Fisheries and Food, the Chief Veterinary Officer and the Chief Medical Officer (who had previously stated that British beef can be safely eaten by everyone, both adults and children and including patients in hospital), declared that the measures introduced by the Government should reassure people that eating beef is safe.

Have there been any beneficiaries in this saga? The answer must be a qualified yes. The public have been protected at an early stage from any significant risks to their health by the prompt recognition of the disease by the animal health surveillance activities of the State Veterinary Service, the knowledge and skills of the expert scientific advisers and the control measures that were instituted as a result. These measures also contributed significantly to the economic performance of an important national industry which, though dented by the effects of BSE, would have been even more seriously damaged without Government intervention. The Government also demonstrated its ability to control a difficult and complex new situation in a capable manner. There has been a welcome injection of funding for research into BSE and related diseases. Importantly, new teams previously outside the established specialist field of spongiform encephalopathy research have been stimulated to join the programme. Undoubtedly they will contribute new knowledge and ideas for the successful investigation of this enigmatic group of diseases which may lead to the development of a much needed test to detect infection in the live animal. Finally, the world at large has benefitted and will benefit from the British experience, from the results of the intensive research programme that has been initiated and from the information communicated directly or under the auspices of international organizations such as the Office International des Epizooties (OIE) on the intricacies of the epidemic, what brought it about, how it has been controlled and the risks for other countries from external or internal sources of infection. As a result, in other European countries in which BSE has now been confirmed, the effects, whilst not insignificant, have not been as severe as in Britain. Britain can be justly proud of its response to the advent of BSE despite some unwelcome aspects of the epidemic and its effects on its people. We now enter a phase of optimism with the rapid decline of the epidemic in Britain in sight.

CONCLUSIONS

BSE in Britain, a new disease first observed clinically in 1985, is an extended common source epidemic with an origin in infected meat and bone meal.

Animal health and public health have been protected by several legislative actions preventing access of susceptible or potentially susceptible species including man to infective material in sufficient quantity to cause disease.

British beef as a commodity is regarded by the British Government, the European Community and the Office International des Epizooties as safe for human consumption and can be freely traded subject to certain conditions.

The BSE epidemic in Britain is predicted to decline from 1992 onwards and will be virtually eliminated by 1996 provided that there is no cow-to-cow transmission.

A large research programme on BSE and other spongiform encephalopathies has been instituted, some of it seeking to answer similar questions posed by scrapie researchers over sixty years ago.

Public concern about BSE was stimulated by unwarranted, persistent media attention aided by the actions of some individuals. After a period financially damaging to the British beef industry, concern has now diminished following the Report by the House of Commons Agriculture Committee but not without social consequences for some individuals. Sales of British beef have now returned to their former level.

There are lessons for all to learn from the epidemic and how it has been handled. We now enter a phase of optimism with a genuine prospect of being able to eliminate the disease in the foreseeable future and, despite the occurrence of a few cases in mainland Europe, to beat this unwelcome visitor into submission.

ACKNOWLEDGEMENTS

The authors thank many colleagues, too numerous to mention, for access to data, the Meat and Livestock Commission and AGB for statistics on beef purchases, library staff at the Central Veterinary Laboratory, Weybridge, the Royal College of Veterinary Surgeons, London, the Institute of Agricultural History, Reading, and Professor J. Brugere Picoux, Drs H. Baron and E. Barrairon from France all for supply of historical books, papers, figures or information, Dr D. Tarry and Miss M. Fimmel for translations and Mrs Y. Spencer for preparation of visual aids.

REFERENCES

Anon (1986) *Report of the Chief Veterinary Officer.* Her Majesty's Stationery Office, London, p. 69.

Besnoit, C. and Morel, C. (1898) Note sur les lésions nerveuses de la tremblante du mouton. *Rev. Vét. (Toulouse)* **23** 397–400.

Bovine Spongiform Encephalopathy (BSE) Agriculture Committee (1990) *Fifth Report.* HMSO, London.

Chandler, R.L. (1962) Encephalopathy in mice. *Lancet* **1** 107–108 (Letter).

Chelle, P.-L. (1942) Un cas de tremblante chez la Chévre. *Bull. Acad. Vét. Fr.* **15** 294–295.

Consultative Committee on Research into Spongiform Encephalopathies (1989) *Report.* Department of Health, Ministry of Agriculture, Fisheries and Food, London.

Cuillé, J. and Chelle, P.-L. (1936) La maladie dite tremblante du mouton est-elle inoculable? *C.R. Hebd. Séanc. Acad. Sci. Paris* **203** 1552–1554.

Fraser, H., McConnell, I., Wells, G.A.H. and Dawson, M. (1988) Transmission of bovine spongiform encephalopathy to mice. *Vet. Rec.* **123** 472.

Gaiger, S.H. (1924) Scrapie. *J. Comp. Pathol.* **37** 259–277.

Gajdusek, D.C. and Zigas, V. (1957) Degenerative disease of the central nervous system in New Guinea. The endemic occurrence of "kuru" in the native population. *N. Engl. J. Med.* **257** 974–978.

Gajdusek, D.C., Gibbs, C.J., Jr, and Alpers, M. (1966). Experimental transmission of a kuru-like syndrome to chimpanzees. *Nature (London)* **209** 794–796.

Goldfarb, L.B., Korczyn, A.D., Brown, P., Chapman, J. and Gajdusek, D.C. (1990) Mutation in codon 200 of scrapie amyloid precursor gene linked to Creutzfeldt Jakob disease in Sephardic Jews of Libyan and non-Libyan origin. *Lancet* **336** 637–638.

Hadlow, W.J. (1959) Scrapie and kuru. *Lancet* **2** 289–290.

Hadlow, W.J., Kennedy, R.C., Race, R.E. and Eklund, C.M. (1980) Virologic and neurohistologic findings in dairy goats affected with natural scrapie. *Vet. Pathol.* **17** 187–199.

Hadlow, W.J., Kennedy, R.C. and Race, R.E. (1982) Natural infection of suffolk sheep with scrapie virus. *J. Infect. Dis.* **146** 657–664.

Hope, J., Reekie, L.J.D., Hunter, N., Multhaup, G., Beyreuther, K., White, H., Scott, A.C., Stack, M.J., Dawson, M. and Wells, G.A.H. (1988) Fibrils from brains of cows with new cattle disease contain scrapie-associated protein. *Nature (London)* **336** 390–392.

Leopoldt, J. G. (1759) *Nützliche und auf die Erfahrung gegründete Einleitung zu der Landwirtschaft* **5** 344–360.

M'Fadyean, J. (1918) Scrapie. *J. Comp. Pathol. Ther.* **31** 102–131.

McGowan, J.P. (1922) Scrapie in sheep. *Scot. J. Agric.* **5** 365–375.

Parry, H.B. (1983) In: Oppenheimer, D.R. (ed.) *Scrapie Disease in Sheep.* Academic Press, London.

Pattison, I.H., Gordon, W.S. and Millson, G.C. (1959) Experimental production of scrapie in goats. *J. Comp. Pathol.* **69** 300–312.

Stockman, S. (1913) Scrapie: an obscure disease in sheep. *J. Comp. Pathol.* **26** 317–327.

Wells, G.A.H., Scott, A.C., Johnson, C.T., Gunning, R.F., Hancock, R.D., Jeffrey, M., Dawson, M. and Bradley, R. (1987) A novel progressive spongiform encephalopathy in cattle. *Vet. Rec.* **121** 419–420.

Working Party on Bovine Spongiform Encephalopathy (1989) *Report.* Department of Health, Ministry of Agriculture, Fisheries and Food, London.

Wyatt, J.M., Pearson, G.R., Smerdon, T., Gruffydd-Jones, T.J. and Wells, G.A.H. (1990) Spongiform encephalopathy in a cat. *Vet. Rec.* **126** 513.

26

Transmissible mink encephalopathy

R.F. Marsh

ABSTRACT

Transmissible mink encephalopathy (TME) is a rare food-bone disease of ranch-raised mink caused by an as-yet unidentified contaminant. Because TME is clinicopathologically similar to scrapie, the most probable origin of TME is feeding mink scrapie-infected sheep. However, epidemiologic observations and experimental testing of mink susceptibility to sheep scrapie have been unable to confirm this association. Investigation of a new incident of TME in Stetsonville, Wisconsin, in 1985 revealed that the mink rancher was a 'dead stock' feeder using mostly downer dairy cows and a few horses. He had never knowingly fed sheep to his mink.

To examine whether this incident of TME may have occurred by feeding infected cattle to mink, two Holstein steers were inoculated intracerebrally (i.c.) with mink brain from the Stetsonville ranch. They developed a fatal spongiform encephalopathy in 18 and 19 months. More importantly, both bovine brains remained highly pathogenic for mink, producing disease 4 months after i.c. inoculation and only 7 months after oral exposure. These findings indicate that there is little species barrier effect between mink and cattle and they are compatible with the Stetsonville incident of TME being produced by feeding mink infected cattle. They also suggest that there exists an unrecognized scrapie-like infection of cattle in the United States.

INTRODUCTION

Transmissible mink encephalopathy (TME) was initially reported by Hartsough and Burger in 1965 (Hartsough and Burger, 1965; Burger and Hartsough, 1965).

Retrospectively, TME first occurred on two commercial mink ranches in Wisconsin and Minnesota in 1947. All adult animals on the Wisconsin ranch became affected with a progressive neurologic disease featuring locomotor incoordination, somnolence, debilitation, and death. Mink on the Minnesota ranch developed an identical disease at the same time but the morbidity in this instance was limited to only 125 animals received seven months earlier from the Wisconsin mink ranch.

TME did not occur again until 1961 when five ranches in Wisconsin, all sharing a common source of mink feed, were observed to have a similar debilitating illness in 10% to 30% of their adult animals, all of which died. In 1963, TME was seen on one mink ranch in Idaho (Hadlow, 1965) and on two ranches in Wisconsin using the same food source. Outside of the United States, TME has occurred on mink ranches in Canada (Hadlow and Karstad, 1968), Finland (unpublished, histopathologic diagnosis confirmed by R. F Marsh from brain slides sent from A. Kangas), East Germany (Hartung et al., 1970; Johannsen and Hartung, 1970), and Russia (Danilov et al., 1974; Duker et al., 1986).

This article will review previous findings on TME emphasizing studies that may contribute unique insights into our understanding of the transmissible spongiform encephalopathies.

CHARACTERIZATION OF THE TRANSMISSIBLE AGENT

Physicochemical properties
The original studies of Burger and Hartsough demonstrated that the TME agent was filterable through 0.5 ψm Seitz filters, resistant to heating in a boiling water bath for 15 minutes, and resistant to treatment with 0.3% formalin for 12 hours at 37°C (Burger and Hartsough, 1965). Additional studies later showed that the transmissible agent was less than 50 nm, sensitive to diethylether, resistant to ultraviolet irradiation, relatively resistant to 10% formalin when in minced brain tissue, and sensitive to proteolytic digestion with Pronase (Marsh and Hanson, 1969).

Biological properties
TME has been experimentally transmitted to European ferrets (Marsh et al., 1969a; Eckroade et al., 1973), striped skunks and raccoons (Eckroade et al., 1973), American sable (pine marten) and beech marten (Hartung et al., 1975), Syrian golden (Marsh et al., 1969a) and Chinese (Kimberlin et al., 1986) hamsters, rhesus monkeys (Marsh et al., 1969a; Eckroade et al., 1970), stumptail macaques (Eckroade et al., 1970), squirrel monkeys (Eckroade et al., 1970), sheep and goats (Hadlow et al., 1987), and cattle (Marsh et al., 1991). Attempts to transmit TME to mice have as yet been unsuccessful (Marsh et al., 1969a; Taylor et al., 1986).

The clinicopathologic syndromes in these recipient species have been unremarkable in the sense that all developed progressive neurologic illnesses accompanied by spongiform encephalopathy. However, differences have been observed in the behaviour of TME on backpassage to mink. TME produced in the skunk raccoon (Eckroade et al., 1973), rhesus and squirrel monkey and stumptail macaque and (Eckroade,

1972), sheep (Hadlow *et al.*, 1987) and goat (Hanson and Marsh, 1973; Hadlow *et al.*, 1987), bovine (Marsh *et al.*, 1991), and early passages in Syrian hamsters (Hanson and Marsh, 1973; Marsh and Hanson, 1979; Bessen and Marsh, 1992) all passage relatively easily back into mink. On the other hand, high passage Syrian hamster TME (Marsh and Hanson, 1979; Bessen and Marsh, 1992) and ferret TME (unpublished) are not transmissible back to mink.

These findings suggest that the mink may provide and important model to study species barrier effects and those factors that influence the host range of these unconventional neuropathic agents. Previous studies in transgenic mice have shown that the prion protein (PrP) gene is a major determinant of species susceptibility and length of incubation period in experimental mouse and hamster scrapie (Scott *et al.*, 1989). Therefore, current studies comparing PrP genes in mink and ferrets (J. Bartz *et al.*, in preparation) may disclose important information on why these two closely related mustelids have such different susceptibilities to TME. Alternatively, host restriction in the mink–ferret system may not be due to differences in their PrP genes but rather to host selection of agent subpopulations as is seen in TME infection in outbred hamsters (Bessen and Marsh, 1992). Here a minor component in the original TME mink brain inoculum is more pathogenic for hamsters than the mink pathogen that produces a long incubation disease in hamsters that has distinct clinical signs and histopathologic lesions.

PATHOLOGY

Natural (Burger and Hartsough, 1965; Johannsen and Hartung, 1970; Marsh *et al.*, 1991) and experimental (Marsh *et al.*, 1969b; Eckroade *et al.*, 1979) TME are characterized by widespread microvacuolation (spongiform degeneration) of the gray matter in the cerebral cortex, corpus striatum, and midbrain. The one exception to this is the response of mink homozygous recessive for the Aleutian gene (aa) to experimental intracerebral inoculation of TME. These animals have very little spongiform degeneration if inoculated after a year and a half of age (Marsh *et al.*, 1976), yet the length of incubation, clinical signs, astrocytic response, and infectivity titre of the TME agent are the same as in other mink (Aa or AA). Since all aa mink exhibit the Chediak–Higashi (CH) syndrome, a hereditary disorder of lysosomal structure and function, it was proposed that the lack of spongiform degeneration in aged CH mink may be due to the failure of lysosomes in reactive astrocytes to release their hydrolytic enzymes (Marsh *et al.*, 1976).

Today, the CH mink model of TME may provide important new information on the role of lysosomes in the modification of the normal cellular form of PrP to the disease-specific form (PrP^{TME}; see Fig. 1). PrP^{TME} from non-CH and aged CH mink should be compared for possible biochemical and/or physical differences that may indicate specific means of post-translational or conformational modification. These studies become even more compelling in light of recent findings associating lysosomal proteases with amyloid deposits in Alzheimer brain (Cataldo and Nixon, 1990).

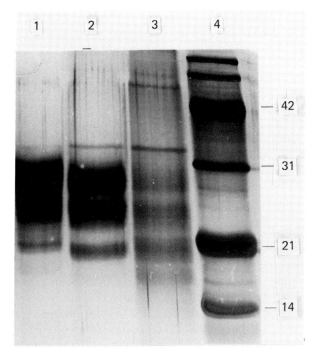

Fig. 1.   Prion protein purified from hamster (lane 1), mink (lane 2), and bovine (lane 3) TME brains using a modification of the Hilmert and Diringer (1984) method. Protein (1 μg) from each sample was stained with silver after separation by SDS–PAGE. Molecular weight markers in lane 4 are indicated on right.

ORIGIN OF TME

Epidemiology

Natural TME is a food-borne disease having minimum incubation periods of 7 to 12 months. Since the disease is not transmitted to offspring unless these animals cannibalize their affected mothers (Hartsough and Burger, 1965), TME is a dead-end infection with the mink serving as an aberrant host because of its misfortunate exposure to an unidentified feed ingredient.

The similarities between TME and scrapie suggest an obvious source of the contaminated ingredient, scrapie-infected sheep. However, there has been no clear epidemiologic finding linking the feeding of sheep to mink with TME. Conversely, there is one incident of TME where the ranch owner was confident he had never fed sheep (Marsh et al., 1991). However, this information becomes difficult to assess because of the long periods between exposure and the onset of disease, and the fact that some by-product mixtures of uncertain composition are fed. It would seem that epidemiologic observations per se do not provide sufficient evidence to identify scrapie-infected sheep as the source of TME.

Experimental studies

To test the experimental susceptibility of mink to scrapie, six sources of sheep brain from the United Kingdom, one drowsy goat brain, and 14 mouse adapted 'strains' (all gifts from Alan Dickinson) were injected intracerebrally into a total of 65 mink. Only one of these animals developed a TME-like disease after an incubation period of 22 months (Marsh and Hanson, 1979). Other experiments testing American sources of sheep scrapie have resulted in all mink injected intracerebrally with either of two scrapie-infected Suffolk sheep brains developing TME-like disease in 11–12 months (Hanson *et al.*, 1971) and 16–24 months (Marsh and Hanson, 1979) post-infection. In these same studies, a brain from an American Cheviot sheep with scrapie produced no disease in mink and no scrapie sheep or goat brain was infectious for mink by oral exposure. Although these findings do not seem to support the premise that TME results from feeding mink scrapie-infected sheep tissues, they clearly show that different sources of sheep scrapie can vary in their pathogenicity for mink and, therefore, it is possible that there may exist a sheep scrapie agent capable of producing disease in mink 7 to 12 months after ingestion.

A new incident of TME in Stetsonville, Wisconsin, in 1985 has expanded the search for the origin of TME beyond sheep and suggests that cattle may have been the source of infection (Marsh and Hartsough, 1985, 1988; Marsh *et al.*, 1991). This observation is based on the finding that the mink rancher involved was a 'dead stock' feeder using mostly downer dairy cows and a few horses. To investigate the possibility that this incident of TME may have been caused by feeding mink infected cattle, two Holstein steers were intracerebrally inoculated with mink brain. Each developed a fatal spongiform encephalopathy 18 and 19 months post-infection. Brain tissue from the affected steers produced disease in mink 4 months after intracerebral inoculation and only 7 months after oral exposure (Marsh *et al.*, 1991).

DISCUSSION

Each of the diseases that comprise the transmissible spongiform encephalopathies have unique epidemiologic and experimental aspects. Scrapie is the oldest, but in several ways the least understood. How is the agent transmitted in sheep? What is the genetic influence on sheep susceptibilty? Creutzfeldt–Jakob disease (CJD) provides both a familial form that can be traced to specific changes in the PrP gene (Goldgaber *et al.*, 1989) and a sporadic form for which there is no evidence for natural transmission. If this implies that CJD can occur spontaneously in the human population at an incidence rate of about one per one million population, the origin of kuru can be simply explained as the misfortunate inclusion of a spontaneous case of CJD in the diet of a tribe of New Guinea natives practicing the unusual act of cannibalism. A man is a mink is a cow is a cat, etc.

However, the origin of TME is not as obvious. Epidemiologic and experimental findings on the Stetsonville incident of TME suggest that cattle may have been the source of infection. If this is true, there must exist an unrecognized scrapie-like disease of cattle in the United States (Marsh and Hartsough, 1985). This possibility is especially significant when considering the epidemiology of bovine spongiform encephalopathy (BSE), a disease initiated by feeding rendered animal protein from scrapie-infected sheep to cattle, then perpetuated by feeding cows to cows.

The United States did not previously feed large amounts of animal protein to ruminants. Then, about 5–6 years ago, a trend developed to feed 'by-pass' protein that encouraged the addition of less digestible protein sources, such as meat and bone meal, to cattle rations. With this change in feeding practice, we have now closed the circle of transmission for any putative unconventional slow virus infection of cattle. Hopefully, this will not happen. BSE is a disease that many investigators believe almost did not occur in Great Britain because of so many events being dependent on one another. However, is it wise to ignore the lessons of BSE and to trust only luck to avoid the economic impact of a BSE-like disease in the United States?

ACKNOWLEDGEMENTS

I would like to dedicate this paper to the memory of Dr Robert P. Hanson who directed the early studies on TME as graduate student advisor to myself, Dr Dieter Burger, and Dr Robert Eckroade. Thanks for your patience and the long insightful conversations. It would not have been the same without you.

REFERENCES

Bessen, R.A. and Marsh, R.F. (1992) Identification of two biologically distinct strains of transmissible mink encephalopathy in hamsters. *J. gen. Virol.* **73** 329–334.

Burger, D. and Hartsough, G.R. (1965) Encephalopathy of mink. II. Experimental and natural transmission. *J. Infect. Dis.* **115** 393–399.

Cataldo, A.M., and Nixon, R.A. (1990) Enzymatically active lysosomal proteases are associated with amyloid deposits in Alzheimer brain. *Proc. Natl. Acad. Sci. USA* **87** 3861–3865.

Danilov, E.P., Bukina, N.S. and Akulova, B.P. (1974) Encephalopathy in mink. *Krolikovod. Zverovod.* **17** 34.

Duker, I.I., Geller, V.I., Chizhov, V.A., Roikhel, V.M., Pogodina, V.V., Fokina, G.I., Sobolev, S.G. and Korolev, M.B. (1986) Clinical and morphological investigation of transmissible mink encephalopathy. *Vopr. Virusol* **31** 220–225.

Eckroade, R.J. (1972) Neuropathology and experimental transmission to other species of transmissible mink encephalopathy. *PhD Thesis.* University of Wisconsin-Madison.

Eckroade, R.J., Zu Rhein, G.M., Marsh, R.F. and Hanson, R.P. (1970) Transmissible mink encephalopathy: experimental transmission to the squirrel monkey. *Science* **169** 1088–1090.

Eckroade, R.J., Zu Rhein, G.M. and Hanson, R.P. (1973) Transmissible mink encephalopathy in carnivores: clinical, light and electron microscopic studies in raccoons, skunks and ferrets. J. Wildlife Dis. **9** 229–240.

Eckroade, R.J., Zu Rhein, G.M. and Hanson, R.P. (1979) Experimental transmissible mink encephalopathy: brain lesions and their sequential development in mink. In: Prusiner, S.B. and Hadlow. W.J. (eds) *Slow Transmissible Diseases of the Nervous System*, Vol. 1. Academic Press, New York, pp. 409–449.

Goldgaber, D., Goldfarb, L.G., Brown, P., Asher, D.M., Brown, W.T., Lin, S., Teener, J.W., Feinstone, S.M., Rubenstein, R., Kascsak, R.J., Boellaard, J.W. and Gajdusek, D.C. (1989) Mutations in familial Creutzfeldt–Jakob disease and Gerstmann–Sträussler–Scheinker's syndrome. *J. Exp. Neurol.* **106** 204–212.

Hadlow, W.J. (1965) Discussion of paper by D. Burger and G.R. Hartsough. In: Gajdusek, D.C., Gibbs C.J., Jr. and Alpers, M. (eds) *Slow, Latent, and Temperate Virus Infections.* NINDB Monograph 2, US. Government Printing Office, Washington DC. pp. 303–305.

Hadlow, W.J. and Karstad, L. (1968) Transmissible encephalopathy of mink in Ontario. *Can. Vet. J.* **9** 193–195.

Hadlow, W.J., Race, R.E. and Kennedy, R.C. (1987) Experimental infection of sheep and goats with transmissible mink encephalopathy virus. *Can. J. Vet. Res.* **51** 135–144.

Hanson, R.P. and Marsh, R.F. (1973) Biology of transmissible mink encephalopathy and scrapie. In: Zeman, W. and Lennette, E.H. (eds) *Slow Virus Diseases.* Baltimore, M.D., Williams and Wilkins, pp. 10–15.

Hanson, R.P., Eckroade, R.J., Marsh, R.F., Zu Rhein, G.M., Kanitz, C.L. and Gustafson, D.P. (1971) Susceptibility of mink to sheep scrapie. *Science* **172** 859–861.

Hartsough, G.R. and Burger, D. (1965) Encephalopathy of mink. I. Epizootiologic and clinical observations. *J. Infect. Dis.* **115** 387–392.

Hartung, J., Zimmermann, H. and Johannsen, U. (1970) Infectious encephalopathy in mink. I Clinico-epidemiological and experimental studies. *Monatsh. Veterinaermed* **25** 385–388.

Hartung, J., Johannsen, U. and Zimmermann, H. (1975) Infectious encephalopathy in mink. III. Results of field surveys and experimental transmission. *Monatsh. Veterinaermed* **30** 23–27.

Hilmert, H. and Diringer, H. (1984) A rapid and efficient method to enrich SAF-protein from scrapie brains of hamsters. *Biosci. Rep.* **4** 165–170.

Johannsen, H. and Hartung, J. (1970) Infectious encephalopathy in mink. II Pathological studies. *Monatsh. Veterinaermed* **25** 389–395.

Kimberlin, R.H., Cole, S. and Walker, C.A. (1986) Transmissible mink encephalopathy (TME) in Chinese hamsters: identification of two strains of TME and comparisons with scrapie. *Neuropathol. Appl. Neurobiol* **12** 197–206.

Marsh, R.F., and Hanson, R.P. (1969) Physical and chemical properties of the transmissible mink encephalopathy agent. *J. Virol.* **3** 176–181.

Marsh, R.F. and Hanson, R.P. (1979) On the origin of transmissible mink encephalopathy. In: Prusiner, S.B. and Hadlow, W.J. (eds) *Slow Transmissible Diseases of the Nervous System*, Vol. 1. Academic Press, New York, pp. 451–460.

Marsh, R.F. and Hartsough, G.R. (1985) Is there a scrapie-like disease in cattle? In: *Proc. 89th Annual Meeting of the US Animal Health Association.* pp. 8–9.

Marsh. R.F. and Hartsough, G.R. (1988) Evidence that transmissible mink encephalopathy results from feeding infected cattle. In: Murphy, B.D. and Hunter D.B. (eds) *Proc. IV International Congress on Fur Animal Production.* Canada Mink Breeders Association, Toronto, pp. 204–207.

Marsh, R.F., Burger, D., Eckroade, R., Zu Rhein, G.M. and Hanson, R.P. (1969a) A preliminary report on the experimental host range of the transmissible mink encephalopathy agent. *J. Infect. Dis.* **120** 713–719.

Marsh, R.F., Burger, D. and Hanson, R.P. (1969b) Transmissible mink encephalopathy: behavior of the disease agent in mink. *Am. J. Vet. Res.* **30** 1637–1642.

Marsh, R.F., Sipe, J.C., Morse, S.S. and Hanson, R.P. (1976) Transmissible mink encephalopathy: reduced spongiform degeneration in aged mink of the Chediak–Higashi genotype. *Lab. Invest.* **34** 381–386.

Marsh, R.F., Bessen, R.A., Lehmann, S. and Hartsough, G.R. (1991) Epidemiological and experimental studies on a new incident of transmissible mink encephalopathy. *J. Gen. Virol.* **72** 589–594.

Scott, M., Foster, D., Mirenda, C., Serban, D., Coufal, F., Wälchi, M., Torchia, M., Groth, D., Carlson. G., DeArmond, S.J., Westaway, D. and Prusiner, S.B. (1989) Transgenic mice expressing hamster prion protein produce specific scrapie infectivity and amyloid plaques. *Cell* **59** 847–857.

Taylor, D.M., Dickinson, A.G., Fraser, H. and Marsh, R.F. (1986) Evidence that transmissible mink encephalopathy agent is biologically inactive in mice. *Neuropathol. Appl. Neurobiol.* **12** 207–215.

27

The lymphoreticular system in the pathogenesis of scrapie

H. Fraser, M.E. Bruce, D. Davies, C.F. Farquhar and P.A. McBride

ABSTRACT

Although scrapie and its homologues are diseases of the central nervous system, the infectious agencies which cause them occur or replicate in many non-CNS tissues such as spleen and lymph nodes during the preclinical phase. There are no immune responses in these infections, and no agent or 'virus' specific antigens. Genetic or surgical asplenia prolongs scrapie incubation periods with high doses of ME7 scrapie, indicating that replication occurs in spleen. This loss of replicative capability does not regenerate if a delay of many weeks after splenectomy lapses before infection is introduced. The complete neutrality for any aspect of scrapie pathogenesis of whole-body radiation suggests that the cells in the lymphoreticular system (LRS) where replication occurs do not proliferate.

INTRODUCTION

The complex tissue interactions in pathogenesis experiments of a disease process which can only be studied in the whole animal and is caused by a pathogen which cannot be recognized poses ambiguities in interpretation. Scrapie-like infections present particular difficulties as their detection depends on indirect methods of identification using bioassay and the detection of protease-resistant protein (Hope *et al.*, 1986, 1988) whose relationship with the infectious agent is uncertain, and because there are no agent-specific antigens or immunity in these diseases (see Brown, 1990). A knowledge of the early events will be important if these types of diseases are to be controlled in the future. Many questions arise concerning the early steps in establishment of the infection, such as those preceeding, those coincident with and those subsequent to infection of peripheral tissues, particularly the lymphoreticular system (LRS). Most experimental studies depend on systemic routes of infection, which may bypass early stages of the natural disease. It is likely that post-natal infection in natural scrapie in sheep includes some oral exposure, and placenta may be a source of infection both pre- and post-natally, although this is by no means certain. The extent and means of foetal infection are unclear, and a true vertical or even a germline infection cannot be excluded on present evidence. A knowledge of the early steps in the establishment of infection in peripheral organs, including LRS, in natural scrapie will be important in understanding the epidemiology of the disease. Many approaches have been made in attempts to understand the lymphoreticular phase of scrapie infection and mouse-passaged Creutzfeldt–Jakob disease, using various routes of infection, sequential bioassay of peripheral organs, genetic mutants, splenectomy, thymectomy, whole-body irradiation, spleen cell separation, immunomodulation, polyanions and various other compounds which act on immune cells (Eklund et al., 1967; Dickinson and Fraser, 1972; Lavelle et al., 1972; Outram et al., 1974; Fraser and Dickinson, 1970, 1978; Clarke and Kimberlin, 1984; Ehlers and Diringer, 1984; Bruce, 1985; Farquhar and Dickinson, 1986; Kimberlin and Walker, 1986a; Fraser and Farquhar, 1987; Mohri et al., 1987; Kimberlin and Walker, 1989; Robinson and Gorham, 1990). The identification of PrP and the disease-specific protease-resistant form of this protein in lymphoid and other tissues and cells provides an important approach to the recognition of infection in addition to, and hopefully replacing, bioassay (Doi et al., 1988; Ikegami et al., 1991).

There is good evidence to suggest that the target cells in the central nervous system in which the primary events in all scrapie-type infections occur are the neurones. The object of this presentation is to review and summarize some work aimed at investigating the peripheral pathogenesis of scrapie and to suggest a tentative identity of cells in the lymphoreticular system in which replication of scrapie occurs.

RESULTS AND DISCUSSION

Splenectomy
It has been known for some time that splenectomy before or shortly after scrapie infection can prolong the incubation period following peripheral infection (Fraser and Dickinson, 1970, 1978; Kimberlin and Walker, 1989). Spleen removal before intraperitoneal infection consistently extends incubation period and this effect is not reversed if the interval is prolonged for many weeks (Fraser and Dickinson, 1978). With subcutaneous infection, splenectomized mice infected with ME7 scrapie do not show a consistently longer incubation period (Table 1), and splenectomy had no effect with this route in mice infected with the 139A scrapie strain (Kimberlin and Walker, 1989). Splenectomy has no significant effect on the efficiency of subsequent intraperitoneal infection with ME7 or 139A scrapie strains. Incubation periods were consistently prolonged with high but sometimes not with low doses of infection with

Table 1. Effect of splenectomy on ME7 scrapie injected intraperitoneally or subcutaneously 4 days later into C57BL mice

Incubation period, mean ± SE (days		Increase in incubation period (%)	Probability of significant increase in incubation period due to splenectomy
Sham	Splenectomy		
Intraperitoneal injection			
270 ± 11	320 ± 11	19	<0.01
Subcutaneous injection			
307 ± 17	352 ± 14	15	<0.05
317 ± 6	321 ± 1	1	NS
304 ± 10	324 ± 4	7	NS

NS, not significant.

ME7 in C57BL mice and 139A in BSC mice (Table 2), although in C57BL mice the 139A strain produced a prolongation of incubation period throughout the dilution series. Similar results were obtained in earlier studies with 139A scrapie (Kimberlin and Walker, 1989), and here also the prolongation in incubation period was present throughout the dilution range. In the later study splenectomy prior to intravenous infection with 139A significantly reduced the efficiency of infection by about a hundredfold, although other work with ME7 showed that whereas perivenously injected mice, splenectomized 1 day before or after injection, had incubation periods nearly 25% longer than controls, this difference was absent in intravenously injected animals (Fraser and Dickinson, 1978). There are therefore differences in the detail of the effect of spleen removal between strains of agent or mouse and with routes of infection; spleen may be a resevoir of intravenously administered 139A scrapie when splenectomy reduces the efficiency of infection to that of the intraperitoneal route (Kimberlin and Walker, 1978, 1989).

Spleen removal soon after intraperitoneal injection with serial dilutions of ME7 scrapie considerably reduces the efficiency of infection by removing a major source of peripheral replication (Table 3). The approximate half log loss of effective titre following this spleen removal suggests that about 50% of peripheral replication is retained elsewhere. By between about 3 and 4 weeks after infection with ME7 or 139A scrapie, when neuroinvasion of the spinal cord from agent which has replicated in the spleen has occurred (Kimberlin and Walker, 1989), splenectomy has no further delay on incubation period (Table 4).

With some models of the spongiform encephalopathies the role of the spleen may be less than that confirmed in mice with 139A and ME7 scrapie, and splenectomy in hamsters has no effect on the incubation period after intraperitoneal infection of 263K scrapie (Kimberlin and Walker, 1986b). A surprising result is our (Fraser and McConnell, unpublished) failure to identify the BSE agent in spleen from terminally

Table 2. Effect of splenectomy on infectibility with serial scrapie dilutions. ME7 or 139A scrapie injected intraperitoneally into C57BL or BSC mice 7–10 days after splenectomy

Strain	Dilution (tenfold)	Incubation period, mean ± SE (days)		Prolongation due to splenectomy (%)	Titre	
		Sham	SX[a]		Sham operation	Splenectomy
ME7	− 1	234 ± 2	310 ± 4	32		
	− 4	336 ± 14	361 ± 10	7		
	− 5	(351)	(358)[b]	− 2	˷− 4.17	− 4.03
ME7	− 2 × 5	266 ± 12	305 ± 12	17		
	− 3 × 5	302 ± 7	318 ± 15	5		
	− 4 × 5	332 ± 4	365 ± 7	10		
	− 5 × 5	–	(344)	–	− 3.68	− 3.76
ME7	− 2	284 ± 9	371 ± 13	31		
	− 3	324 ± 6	367 ± 17	13		
	− 4 × 5	331 ± 22	(365)	10	− 3.12	− 3.53
ME7	− 2	286 ± 12	332 ± 9	16		
	− 3	328 ± 17	386 ± 7	18		
	− 4 × 5	331 ± 13	358 ± 21	8		
	− 4	354 ± 10	372 ± 7	5		
	− 5 × 5	346 ± 8	345 ± 17	0	− 3.34	− 3.27
ME7	− 1	258 ± 4	298 ± 11	15		
	− 2	278 ± 4	325 ± 6	17		
	− 3	316 ± 2	341 ± 4	8		
	− 4	324 ± 5	355 ± 3	10		
	− 5	(350)	366 ± 15	5	− 4.44	− 4.73
139A[c]	− 2	217 ± 10	258 ± 10	19		
	− 3	230 ± 4	291 ± 5	26		
	− 4	250 ± 8	320 ± 6	28		
	− 5	(260)	(316)	21	− 4.3	− 4.3
139A[d]	− 2	191 ± 0	204 ± 7	7		
	− 3	198 ± 7	227 ± 7	15		
	− 4	203 ± 3	222 ± 8	9		
	− 5	–	(219)	–	− 4.5	− 4.83

[a] Splenectomy.
[b] Single values.
[c] C57BL mice (ME7 experiments were all in C57BL mice).
[d] BSC mice.

affected cattle from which infection has been isolated from brain (Fraser *et al.*, 1988) and spleen may not be important in the pathogenesis of CJD infection in mice (Mohri *et al.*, 1987). Splenectomy had only a marginal 5% prolongation on the incubation period in C57BL mice subsequently infected intraperitoneally with hundredfold dilutions of 87A scrapie brain (splenectomy 545 ± 13, sham operated 518 ± 9 days, mean ± SE).

Table 3. Effect of splenectomy 7–8 days after intraperitoneal infection with serial dilution of ME7 scrapie

Mouse strain	Dilution (tenfold)	Incubation period, mean ± SE days		Prolongation due to splenectomy (%)	Titre	
		Sham	SX[a]		Sham operation	Splenectomy
BRVR	− 2	282 ± 5	327 ± 14	15		
	− 3	310 ± 6	359 ± 10	16	−	−
VL	− 2	303 ± 7	335 ± 4	11		
	− 3	325 ± 6	356 ± 6	9		
	− 4	338 ± 11	−	−	− 3.5	− 3.16
VL	− 1	252 ± 7	281 ± 14	11		
	− 2	286 ± 5	315 ± 6	10		
	− 3	311 ± 3	339 ± 4	9		
	− 4	308 ± 0	347 ± 5	12	− 3.9	− 3.17

[a] Splenectomy.

Whole-body irradiation

A considerable amount of data has accumulated to confirm the surprising failure of low or high doses of whole-body ionizing irradiation, administered as single or fractionated doses, at various times before or after infection with different strains of scrapie, to have any effect on scrapie pathology and pathogenesis, incubation period, infectibility with low doses of agent, or levels of agent in spleen from irradiated infected mice (Fraser and Farquhar, 1987; Fraser et al., 1989a, b). The level of ME7 scrapie in whole spleens was assayed 15 weeks after intracerebral infection of C3H mice in 5 experiments, and, despite the (3–6)-fold loss of spleen weight due to irradiation-induced loss of dividing populations of spleen cells, there was no loss of infection from the whole organs by bioassay (Table 5).

Cell identification

The overwhelming suggestion from the thymectomy, splenectomy and ionizing radiation studies is that scrapie replication outside the CNS is independent of lymphoid cells, and depends on non-dividing, long-lived cells which are not replaced from a stem-cell population in adulthood (Fraser et al., 1989a); their possible identity as follicular dendritic cells (FDCs) has previously been made (Clarke and Kimberlin, 1984; Fraser and Farquhar, 1987). FDCs are radiation resistant (Jaroslov and Nossal, 1966; Brown et al., 1973; Tew et al., 1982), whereas dendritic cells (interdigitating cells, IDCs) are radiosensitive (Eikelenboom, 1978) and proliferate in vivo (Steinman et al., 1974). The presence of cells with the morphology and location of FDCs in the spleen and lymphoid tissue of mice, which immunostain with anti-PrP, and which trap immune complexes, but remain unstained with other cell-identity markers (Thy-1,

Table 4. Effect of splenectomy 21–168 days after intraperitoneal infection of C3H mice (and see Table 3) with 10–50 i.p. ID50 infectious units of ME7 scrapie

Splenectomy days post-infection	Prolongation (%)	Probability of significant change
[7	9–16	<0.001 (see Table 3)]
21	5.9	<0.001
42	2.3	NS
56	2.3	NS
70	−3.3	NS
84	−2.3	NS
98	0.0	NS
112	−3.6	NS
126	−2.62	NS
168	0.0	NS

F4/80, NLDC-145, MIDC-8), provides an important confirmation of the suggestion that the target cells for scrapie replication in the LRS are FDCs (McBride *et al.*, in tpress), although it obviously should not be assumed that cells which immunolabel with anti-PrP necessarily participate in scrapie replication. The presence of PrP, and its mRNA, in human lymphoid cells and its increase in lymphocyte activation suggests that it is a lymphocyte surface molecule and might participate in immunoreactivity (Cashman *et al.*, 1990). Unidentified PrP-positive cells are also present in thymus, although it has been known for many years that neither neonatal nor adult thymectomy had any effect on the subsequent disease in scrapie-infected mice

Table 5. Levels of scrapie in whole spleen after gamma-irradiation. C3H mice were infected intracerebrally with 10^5 i.c. ID50s of ME7 scrapie, and irradiated with 10 Gy at 105 dpi. Spleens taken 4 days post-irradiation were bioassayed by intracerebral injection of VL mice

Experiment	0 Gy		10 Gy	
	Spleen weight (mg)	incubation period, mean ± SE (days)	Spleen weight (mg)	Incubation period, mean ± SE (days)
1	106	166 ± 2	22	171 ± 2
2	116	171 ± 3	19	171 ± 2
3	97	167 ± 2	30	169 ± 2
4	140	168 ± 3	60	168 ± 3
5	88	181 ± 2	16	183 ± 1

(McFarlin *et al.*, 1971; Fraser and Dickinson, 1978). FDCs do not have a bone marrow origin (Humphrey *et al.*, 1984; Kraal *et al.*, 1986), in contrast to IDCs (Steinman, 1991) and , in irradiated recipients, neither isogeneic (Table 6) nor homogeneic bone marrow grafting between mice of differing *Sinc* genotype has any effect on scrapie incubation period (Fraser and Farquhar, 1987; Fraser *et al.*, 1989a). The reduction in susceptibility to scrapie after steroid administration (Outram *et al.*, 1974) to which FDCs are sensitive is consistent with their proposed role in the replication of the agencies of scrapie. Although these data generally support the candidacy of FDCs as subserving scrapie replication in the LRS, the finding of unidentified PrP-positive cells in thymus, which contains IDCs but not FDCs, and in other non-lymphoid tissues which contain neither, but where moderate–high scrapie titres occur in mice (Eklund *et al.*, 1967), is not wholly consistent with this suggestion. Furthermore, the failure of scrapie infections to disturb immune responsiveness (see Brown, 1990) is difficult to reconcile with the replication of their causal agencies in cells which subserve a central function in immunity, unless, unlike in the CNS, replication can occur in cells without causing their dysfunction! Therefore further work is clearly needed to identify the lymphoreticular cell substrate system which permits agent replication in the LRS, and to establish the role of FDCs within this system.

Table 6. Effect of 15 Gy gamma-irradiation on incubation period of scrapie in VMs7 mice. Mice were infected intraperitoneally with 10^5 i.c. ID50s of 22A scrapie, irradiated with 2.5 Gy at 50 day intervals (100–350 dpi), and supplemented with isogeneic bone marrow one day post-irradiation (design equivalent to Fig. 1 (Fraser *et al.*, 1989a) but in which *Sinc* p7, VM, bone marrow was injected into *Sinc* s7p7, C57BL × VM, recipients)

Treatment	Incubation period, mean $\pm$ SE (days)	Survival range in days[a]
Unirradiated controls	613 ± 12	84–682
Unirradiated controls with bone marrow	618 ± 11	94–676
Irradiated	593 ± 23	146–683
Irradiation lethality	—	127–763[b]

[a] From scrapie injection date.
[b] Not scrapie injected.

REFERENCES

Brown, J.C., Harris, G., Papamichail, M., Sljivic, V.S. and Holborow, E.J. (1973) The localisation of aggregated human gamma-globulin in the spleens of normal mice. *Immunology* **24** 955–968.

Brown, P. (1990) The phantasmagoric immunology of transmissible spongiform encephalopathy. In: Waksman, B.H. (ed.) *Immunologic Mechanisms in Neurologic and Psychiatric Disease*. Raven Press, New York, pp. 305–313.

Bruce, M.E. (1985) Agent replication dynamics in a long incubation period model of mouse scrapie. *J. Gen. Virol.* **66** 2517–2522.

Cashman, N.R., Loertscher, R., Nalbantoglu, J., Shaw, I., Kascsak, R.J., Bolton, D.C. and Bendheim, P.E. (1990) Cellular isoform of the scrapie agent protein participates in lymphocyte activation. *Cell* **61** 185–192.

Clarke, M.C. and Kimberlin, R.H. (1984) Multiplication of scrapie agent in mouse spleen. *Res. Vet. Sci.* **9** 215–225.

Dickinson, A.G. and Fraser, H. (1972) Scrapie: effect of Dh gene on incubation period of extraneurally injected agent. *Heredity* **29** 91–93.

Doi, S., Ito, M., Shinagawa, M., Sato, G., Isomura, H. and Goto, H. (1988) Western blot detection of scrapie-associated fibril protein in tissues outside the central nervous system from preclinical scrapie-infected mice. *J. Gen. Virol.* **69** 955–960.

Ehlers, B. and Diringer, H. (1984) Dextran sulphate 500 delays and prevents mouse-scrapie by impairment of agent replication in spleen. *J. Gen. Virol.* **65** 1325–1330.

Eikelenboom, P. (1978) Characterisation of non-lymphoid cells in the white pulp of the mouse spleen: an *in vivo* and *in vitro* study. *Cell Tissue Res.* **195** 445–460.

Eklund, C.M., Kennedy, R.C. and Hadlow, W.J. (1967) Pathogenesis of scrapie infection in the mouse. *J. Infect. Dis.* **117** 15–22.

Farquhar, C.F. and Dickinson, A.G. (1986) Prolongation of scrapie incubation period by an injection of dextran sulphate 500 within the month before or after infection. *J. Gen. Virol.* **67** 463–473.

Fraser, H. and Dickinson, A.G. (1970) Pathogenesis of scrapie in the mouse: the role of the spleen. *Nature (London)* **226** 462–463.

Fraser, H. and Dickinson, A.G. (1978) Studies of the lymphoreticular system in the pathogenesis of scrapie: the role of the spleen and thymus. *J. Comp. Pathol.* **88** 563–573.

Fraser, H. and Farquhar, C.F. (1987) Ionising radiation has no influence on scrapie incubation period in mice. *Vet. Microbiol.* **13** 32–41.

Fraser, H., McConnell, I., Wells, G.A.H. and Dawson, M. (1988) Transmission of bovine spongiform encephalopathy to mice. *Vet. Rec.* **123** 472.

Fraser, H., Davies, D., McConnell, I. and Farquhar, C.F. (1989a) Are radiation-resistant, post-mitotic, long-lived (RRPMLL) cells involved in scrapie replication? In: Court, L.A., Dormont, D., Bown, P. and Kingsbury, D.T. (eds) *Unconventional Virus Diseases of the Central Nervous System*. Commissariat à l'Énergie Atomique, Fontenay-aux-Roses, pp. 563–574.

Fraser, H., Farquhar, C.F., McConnell, I. and Davies, D. (1989b) The scrapie disease process is unaffected by ionising radiation. In: Iqbal, K., Wisniewski, H.M. and Winblad B. (eds) *Alzheimer's Disease and Related Disorders*. Alan, R. Liss, New York, pp. 653–658.

Hope, J., Morton, L.J.D., Farquhar, C.F., Multhaup, G., Beyreuther, K. and Kimberlin, R.H. (1986) The major protein of scrapie-associated fibrils (SAF) has the same size, charge distribution and N-terminal protein sequence as predicted for the normal brain protein (PrP). *EMBO J.* **5** 2591–2597.

Hope, J., Multhaup, G., Reekie, L.J.D., Kimberlin, R.H. and Beyreuther, K. (1988) Molecular pathology of scrapie-associated fibril protein (PrP) in mouse brain affected by the ME7 strain of scrapie. *Eur. J. Biochem.* **172** 271–277.

Humphrey, J.H., Grennan, D. and Sundaram, V. (1984) The origin of follicular dendritic cells in the mouse and the mechanism for trapping immune complexes on them. *Eur. J. Immunol.* **140** 859–865.

Ikegami, Y., Ito, M., Isomura, H., Momotani, E., Sasaki, K., Muramatsu, Y., Ishiguro, N. and Shinagawa, M. (1991) Preclinical and clinical diagnosis of scrapie by detection of PrP protein in tissues of sheep. *Vet. Rec.* **128** 271–275.

Jaroslov, B.N. and Nossal, G.J.V. (1968) Effects of X-irradiation on antigen localisation in lymphoid follicles. *Aust. J. Exp. Biol. Med.* **44** 609–628.

Kimberlin, R.H. and Walker, C.A. (1978) Pathogenesis of mouse scrapie: effect of route of inoculation on infectivity titres and dose–response curves. *J. Comp. Pathol.* **88** 39–47.

Kimberlin, R.H. and Walker, C.A. (1986a) Suppression of scrapie infection in mice by heteropolyanion 23, dextran sulphate, and some other polyanions. *Antimicrob. Agents Chemother.* **30** 409–413.

Kimberlin, R.H. and Walker, C.A. (1986b) Pathogenesis of scrapie (strain 263K) in hamsters infected intracerebrally, intraperitoneally, or intraocularly. *J. Gen. Virol.* **67** 255–263.

Kimberlin, R.H. and Walker, C.A. (1989) The role of the spleen in the neuroinvasion of scrapie in mice. *Virus Res.* **12** 201–212.

Kraal, G., Breel, B., Janse, M. and Bruin, G. (1986) Langerhans cells, veiled cells, and interdigitating cells in the mouse recognised by a monoclonal antibody. *J. Exp. Med.* **163** 981–997.

Lavelle, G.C., Sturman, L. and Hadlow, W.J. (1972) Isolation from mouse spleen of cell populations with high specific infectivity for scrapie virus. *Infect. Immunol.* **5** 319–323.

McBride, P.A., Eikelenboom, P., Kraal, G., Fraser, H. and Bruce, M.E. (in press) PrP protein is associated with follicular dendritic cells of spleens and lymph nodes in uninfected and scrapie-infected mice. *J. Pathol.*

McFarlin, D.F., Raff, M.C., Simpson, E. and Nehlsen, S.H. (1971) Scrapie in immunologically deficient mice. *Nature (London)* **233** 336.

Mohri, S., Handa, S. and Tateishi, J. (1987) Lack of effect of thymus and spleen on the incubation period of Creutzfeldt–Jakob disease in mice. *J. Gen. Virol.* **68** 1187–1189.

Outram, G.W., Dickinson, A.G. and Fraser, H. (1974) Reduced susceptibility to scrapie in mice after steroid administration. *Nature (London)* **249** 855–856.

Robinson, M.M. and Gorham, J.R. (1990) Pathogenesis of hamster scrapie. Adherent splenocytes are associated with relatively high levels of infectivity. *Arch. Virol.* **112** 283–289.

Steinman, R.M. (1991) The dendritic cell system and its role in immunogenicity. *Annu. Rev. Immunol.* **9** 271–296.

Steinman, R.M., Lustig, D.S. and Cohn, Z.A. (1974) Identification of a novel cell type in peripheral lymphoid organs of mice. III. Functional properties *in vivo. J. Exp. Med.* **139** 1431–1445.

Tew, J.G., Thorbecke, J. and Steinman, R.M. (1982) Dendritic cells in the immune response: characteristics and recommended nomenclature (a report from the Reticuloendothelial Society Committee on Nomenclature). *J. Reticuloendothel. Soc.* **31** 317–380.

28

Association of PrP gene polymorphisms with the incidence of natural scrapie in British sheep

Nora Hunter

ABSTRACT

In NPU Cheviot sheep there are several PrP gene restriction fragment length polymorphisms (RFLPs) which can be used to predict an animal's response to experimental infection with SSBP/1 scrapie. To determine whether these RFLPs would be equally useful dealing with other breeds and with the natural disease, 167 sheep of 32 breeds and crossbreeds affected by natural scrapie throughout Britain were analysed for two PrP gene RFLPs–*Eco*RI and *Hind*III. An *Eco*RI polymorphic fragment (e1, 6.8 kb) appears to have an association with the incidence of disease as 88% of the sheep tested carried this fragment, 59% as e1e1 homozygotes. Such homozygotes (PrP e1e1) are less common in several samples of unaffected sheep. The *Hind*III RFLP appears to be less generally informative.

INTRODUCTION

Scrapie is usually studied in laboratory rodents and great progress has been made in our understanding of the disease by using these well-controlled model systems. The natural disease in sheep has been less extensively investigated in recent years as it is difficult to study in the field: its appearance is unpredictable, there is no *in vitro* detection system for the infectious agent, there is no pre-clinical, non-invasive diagnostic test and many sheep owners are reluctant to admit having scrapie in their flocks. There is, therefore, no way to tell whether a healthy flock is at risk from infection and no means other than direct experimental challenge to predict with any certainty how a sheep will respond to scrapie.

The average age of sheep at appearance of scrapie symptoms is 3.5 years (Stamp, 1962; Dickinson *et al.*, 1964) with most animals affected between 2.5 and 4.5 years. From the time of the first recording of scrapie in Britain, it has been noticed to run in families. This has been subsequently interpreted in different ways—for example Parry (1960, 1962, 1984) believed that scrapie was a genetic disease caused by a defective and recessive gene. The demonstration of the transmissibility of scrapie both experimentally (Chandler, 1963; and numerous others) and by contagion (Brotherston *et al.*, 1968; Dickinson, *et al.*, 1974; Hourringan *et al.*, 1979) has led others to suggest that the familial appearance of scrapie is the result of inherited susceptibility to an infectious and transmissible agent (Asher *et al.*, 1983).

Studies on the genetic control in sheep of the incidence of scrapie have been greatly facilitated by the work of Alan Dickinson who selected lines of Cheviot sheep (Neuropathogenesis Unit (NPU) Cheviot sheep) for differing incubation periods following injection with SSBP/1 scrapie (Dickinson, 1976). The gene which controls experimental scrapie incubation period was discovered in NPU Cheviots, is called *Sip* (for scrapie incubation period) and has two alleles, sA and pA (Dickinson and Outram, 1988). Dickinson described negative line Cheviots as *Sip* pApA: they survive subcutaneous (s.c.) injection of SSBP/1. Using the intracerebral (i.c.) route, most negative line sheep succumb with long incubation periods (800 to 1000 days) but a proportion survive. The genotype of positive line sheep was described as *Sip* sAsA or sApA (as sA was thought to be fully dominant) and these develop SSBP/1 scrapie in 150 to 400 days depending on the route of infection.

Sip is also believed to control the incidence of natural scrapie (Foster and Dickinson, 1988b) but in this case the 'susceptible' allele (sA) was said to be recessive. Other workers have also said that natural scrapie affects only those animals homozygous for a recessive allele of a single gene (e.g. Gordon, 1966; Parry, 1984; Millot *et al.*, 1988)—if this proves to be *Sip* (as seems likely from the work of Foster and Dickinson (1988b) it remains to be explained why a gene has an allele apparently dominant in the control of incidence of experimental scrapie and recessive with the natural disease.

A search for biochemical markers for the *Sip* gene or for differences in scrapie incidence has greatly benefitted from the work in mice which demonstrated a genetic linkage between *Sinc* (the mouse homologue of *Sip*) and the PrP gene (Carlson *et al.*, 1986; Hunter *et al.*, 1987; Westaway *et al.*, 1987). PrP is a host glycoprotein found in abnormal fibrillar form in brain extracts of mammals affected by scrapie-like diseases. An association was found between restriction fragment length polymorphisms (RFLPs) of the sheep PrP gene and the alleles of the *Sip* gene in NPU Cheviot sheep (Hunter *et al.*, 1989; Foster and Hunter, 1991). Using the enzyme *Eco*RI, a 6.8 kb fragment (e1) was associated with *Sip* sA (short SSBP/1 incubation period) and a fragment of 4.0 kb (e3) with *Sip* pA. A third fragment of 5.2 kb (e2) is relatively rare in the Cheviots and has not yet been unambiguously associated with a *Sip* allele. A second enzyme, *Hind*III, also produced polymorphic fragments—fragments of 5.0 kb (h1) and 3.4 kb (h2) have been associated *Sip* pA and sA respectively. The positions of e1, e3, h1 and h2 are indicated in Fig. 1.

Using RFLP analysis, it was possible to show a partial separation of PrP e1e1/h2h2 (putative *Sip* sAsA) from PrP e1e3/h1h2 (putative *Sip* sApA) animals in terms of their

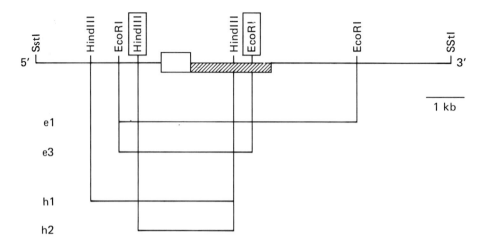

Fig. 1.   Restriction map of sheep PrP gene with the positions of the polymorphic *Eco*RI and *Hind*III sites (in black boxes) which give rise to fragments e1, e3, h1 and h2. The PrP protein coding exon is indicated by an open box and untranslated 3′ transcribed region by a hatched box.

mean incubation period with SSBP/1 scrapie (Foster and Hunter, 1991) and to suggest that the dominance of the sA allele of *Sip* in the control of SSBP/1 scrapie is not full but partial. These studies led to a survey of sheep affected by natural scrapie from all over Britain to establish whether the PrP gene RFLPs were as useful in predicting susceptibility to natural as well as experimental scrapie.

RESULTS AND DISCUSSION

Scrapie sheep survey

During the period between July 1988 and February 1991 sheep throughout Britain affected by natural scrapie were tested for their PrP gene RFLP genotypes. Samples were obtained from Veterinary Investigation (VI) centres from sheep suspected of having scrapie and sent in for diagnosis by the owner. VI centre staff carried out histopathological analysis on the brains of these animals and only those confirmed as scrapie positive in this manner were included in the study. Approximately 10% of total samples were 'path negative' and excluded—it is therefore important to examine brain sections in order to be sure that the clinical diagnosis is correct.

RFLP typing with the restriction enzymes *Eco*RI and *Hind*III was carried out on genomic DNA from 167 scrapie sheep using as probe pNPU42 a Suffolk sheep PrP genomic clone (Goldmann *et al.*, 1990; Hunter *et al.*, 1991). The sheep were of 32 breeds and crossbreeds and represented a good range of British and continental breeds. The average age of the sheep was 3.3 years ranging from 1.2 to 8 years (the age was not known for nine animals). This agrees well with previous studies of ages of scrapie sheep (Dickinson *et al.*, 1964). No breed contracted scrapie markedly earlier

Table 1.   PrP gene *Eco*RI RFLP genotypes in sheep affected by natural scrapie

PrP genotype[a]	Frequency (%)
e1e1	59
e1e3	28
e3e3	10
e1e2	1
e2e2	1
e2e3	1

[a] Fragment sizes on Southern analysis of genomic DNA from 167 sheep probed with a sheep PrP gene coding region genomic clone: e1 is 6.8 kb, e2 is 5.2 kb and e3 is 4.0 kb.

than any other but the numbers of each breed tested were probably too small to be conclusive on this point.

*Eco*RI *analysis*

The results from PrP gene *Eco*RI RFLP analysis from all 167 sheep are presented in Table 1. Approximately 59% of the natural scrapie sheep were PrP e1e1, 28% were PrP e1e3 and 10% were PrP e3e3. The remaining 3% carried e2 (with 1% e1e2). Therefore in this study, 88% of scrapie sheep carried PrP e1, the fragment associated with short incubation period of experimental scrapie in NPU Cheviot sheep. The approximate RFLP type frequencies were as follows: e1, 74%; e2, 2%; e3, 24%.

*Hind*III *analysis*

Most, but not all, of the natural scrapie animals were also analysed with *Hind*III and the results are presented in Table 2. Of 113 sheep , 29% were PrP h1h1, 45% were h1h2 and 26% were h2h2. These frequencies are not significantly different from a 1:2:1 ratio and seem to indicate that the *Hind*III polymorphism, although still successful at marking differences in susceptibility to experimental scrapie in NPU Cheviots, is not of general application.

Table 2.   PrP gene *Hind*III RFLP genotypes in sheep affected by natural scrapie

PrP genotype[a]	Frequency (%)
h1h2	42
h1h1	29
h2h2	26

[a] Fragment sizes on Southern analysis of genomic DNA from 113 sheep probed with a sheep PrP gene coding region genomic clone: h1 is 5.0 kb and h2 is 3.4 kb.

Table 3.   *Eco*RI RFLP genotype analysis of the PrP gene in sheep affected by natural scrapie compared with unaffected flockmates

	Total	Number of each genotype						χ^2	P
		e1e1	e1e2	e1e3	e2e2	e2e3	e3e3		
Group II[a]									
Scrapie	14	11	0	3	0	0	0	15.52	<0.005
Unaffected	58	11	2	31	0	1	13		
Group IV[a]									
Scrapie	21	0	0	10	0	2	9	2.46	>0.6
Unaffected	35	2	0	13	1	1	18		

	Total number of sheep	Frequency (%) of each RFLP type			χ^2	P
		e1	e2	e3		
Group II						
Scrapie	14	89	0	11	12.63	<0.005
Unaffected	58	47	3	50		
Group IV						
Scrapie	21	24	5	71	0.007	>0.995
Unaffected	35	24	4	71		

[a] Group II, Shetland sheep; group IV, Scottish Halfbred sheep.

Group studies

In order to assess the significance of these results comparisons were made between the PrP genotypes of scrapie cases and age-matched flockmates. The results from two of these sets of analyses are presented in Table 3 (*Eco*RI) and Table 4 (*Hind*III) representing group II and group IV sheep form Hunter *et al.*, (1991). Group II consists of Shetland sheep from Shetland and was selected for further analysis because the scrapie cases were all PrP e1 carriers (88% of the main survey). Four flocks provided sheep for this group but genotype analysis suggested little difference amongst the flocks and the results were pooled. Fourteen scrapie cases were compared with 58 age-matched unaffected flockmates. Group IV were sampled from a flock of Scottish Halfbred sheep; 21 scrapie cases are compared with 35 unaffected age-matched flockmates. This flock was chosen for study because it had provided some scrapie cases which were e3e3 homozygotes, relatively uncommon (10%) in the main survey of Table 1.

Table 4. *Hind*III RFLP genotypes analysis of the PrP gene in sheep affected by natural scrapie compared with unaffected flockmates

	Total[a]	Number of each genotype			χ^2	P
		h1h1	h1h2	h2h2		
Group II[b]						
Scrapie	13	1	8	4	1.14	>0.5
Unaffected	57	25	28	4		
Group IV[b]						
Scrapie	21	9	12	0	0.65	>0.7
Unaffected	35	20	11	4		

Total number of sheep	Frequency (%) of each RFLP type		χ^2	P
	h1	h2		
Group II				
Scrapie	13	38	62	0.41	>0.5
Unaffected	57	68	32		
Group IV				
Scrapie	21	71	29	0.002	>0.9
Unaffected	35	73	27		

[a] Numbers of sheep tested with *Hind*III are not the same as tested with *Eco*RI in Table 3.
[b] Group II, Shetland sheep; group IV, Scottish Halfbred sheep.

EcoRI analysis

In Group II there was an excess of PrP e1e1 genotypes in the scrapie cases ($\chi^2 = 15.52$; $P < 0.005$) compared with their unaffected flockmates and there were no e3e3 scrapie cases despite this genotype being found at quite high frequency in the flock sample (Table 3). PrP e1 is much more common in group II scrapie cases than in the healthy sample ($\chi^2 = 12.63$; $P < 0.005$).

Group IV appeared to be different. The 21 scrapie cases had very similar PrP RFLP genotypes to the 25 scrapie cases ($\chi^2 = 2.46$; $P > 0.6$) and all of the scrapie cases carried e3, approximately 50% as homozygotes (Table 3). The frequencies of e1, e2 and e3 were almost exactly the same in the scrapie and healthy samples. Group IV scrapie sheep have a higher frequency of e3 (71%) than those in group II (11%) and a lower frequency of e1 (24%) compared with 89% in group II. The unaffected sheep from the two groups have similar PrP RFLP type frequencies with higher numbers of e3 than e1. Other studies (Hunter *et al.*, 1991, unpublished observations) have suggested that scrapie-affected flocks usually resemble group II, with PrP e1 carriers most affected by scrapie. The scrapie sheep in group IV could be affected by a different strain of scrapie. There is experimental evidence in NPU Cheviots for such different scrapie strains—negative line (*Sip* pApA) sheep have a higher incidence of disease following i.c. injection of CH1641 scrapie than do positive line *Sip* sA carriers. However, alternative explanations for the group IV PrP e3 scrapie sheep are that there is a breed difference in the RFLP appearance and/or breakdown in the linkage between PrP e1 and disease incidence.

HindIII analysis
Again the *Hind*III analysis of these flocks suggests that this enzyme is less informative than *Eco*RI; the results are presented in Table 4. The scrapie sheep from neither group II nor group IV have significantly different *Hind*III PrP RFLP genotypes from their healthy flockmates.

Combined **Eco***RI and* **Hind***III RFLP genotypes analysis*
It was not possible to combine the RFLP genotypes into haplotype analysis with any certainty; as e1, e2 and e3 can each occur with either h1 or h2 (see genotype analysis in Table 5). Although some PrP e and h combinations are more common than others and the genotypes found in 113 scrapie sheep are not in the same relative frequencies in 187 healthy sheep, this is probably due only to the different ratios of the *Eco*RI RFLP types. The two polymorphic restriction sites are not fixed with respect to each other in all groups of sheep and although the *Eco*RI polymorphism continues to be useful (with nearly 90% of scrapie sheep carrying PrP e1) it is now necessary to evaluate PrP protein coding region polymorphisms for linkage to disease incidence. Several of these have now been found in Suffolk (Goldmann *et al.*, 1990) and NPU Cheviot sheep (Goldmann *et al.*, 1991; and Chapter 24) and are currently under investigation in naturally infected scrapie sheep.

The *Sip* gene and control of incidence of scrapie
How do the results presented in this paper affect the debate about linkage between PrP and *Sip* and the difference in relative dominance of the alleles of the *Sip* gene depending on whether the sheep are challenged by experimental or natural scrapie? *Sip* was described originally (and is still defined) as the gene which controls the incubation period of SSBP/1 scrapie in NPU Cheviot sheep following experimental challenge. In this case the sA allele is dominant over the pA allele. In other lines of sheep bred for differing susceptibility to experimental challenge, the alleles of the gene (probably *Sip*) controlling the response of the sheep are at least co-dominant with heterozygous (putative *Sip* sApA) and homozygous (putative *Sip* sAsA) 'susceptible' sheep having clearly distinguishable incubation periods (Nussbaum *et al.*, 1975; Hoare *et al.*, 1977; Davies and Kimberlin, 1985).

Sip is also thought to act in control of the incidence of natural scrapie but sA was said to be recessive in the crossing experiments of Foster and Dickinson (1988b). Other studies on sheep flocks with high incidence of natural scrapie have also concluded that scrapie incidence is under the control of a single gene and that only homozygous susceptible sheep develop scrapie (Parry, 1960, 1962, 1984; Millot *et al.*,

Table 5. Comparison of PrP RFLP genotypes in healthy and naturally infected scrapie sheep

PrP genotypes	Frequencies (%)	
	Natural scrapie sheep	Unaffected sheep[a]
e1e1 h1h2	26	13
e1e1 h2h2	25	3
e1e3 h1h2	16	24
e3e3 h1h1	11	27
e1e1 h1h1	8	7
e1e3 h1h1	8	20
e2e2 h2h2	2	0.6
e1e2 h1h2	1	0.4
e2e3 h1h2	1	0.4
e3e3 h2h2	1	2
e3e3 h1h2	0	1
e2e3 h1h1	0	0.6
e1e2 h1h1	0	0.4
e1e3 h2h2	0	0.4
Total number of sheep	113	183

[a] Unaffected sheep are of several breeds—Suffolk, Shetland and Scottish Halfbreds. These include the unaffected animals from Tables 3 and 4.

1988). However, all these studies (including Foster and Dickinson, 1988b) have had to make assumptions about genotype based on the presence or absence of scrapie symptoms in a sheep or its parents or progeny. Even in an environment where there is high exposure to natural scrapie infection, it is not possible to be absolutely certain that every individual animal has had equal exposure to infection, It is therefore dangerous to assume that a healthy animal is 'resistant'—it may be susceptible and simply not have received a high enough dose of scrapie to develop disease symptoms within its lifetime. It is equally dangerous to assume that a scrapie animal is homozygous 'susceptible'. Dickinson and Outram (1988) have suggested that the small anomalies sometimes found in expected scrapie incidence figures may mean that heterozygotes do in some circumstances develop natural scrapie. Indeed, the calculations in Foster and Dickinson (1988b) could include a small percentage of heterozygotes amongst the scrapie victims and still arrive at the same incidence values. Millot et al., (1988) found that not all offspring of presumed homozygous scrapie susceptible animals developed scrapie and therefore the parents could have included some heterozygotes. A similar phenomenon was also noted by Gordon (1966).

The apparently opposing genetic observations (experimental versus natural) can be rationalized if the incidence of scrapie is controlled by *Sip* in such a way that the product of each allele (sA and pA) has an effect on scrapie replication with differing relative dominance depending on the breed of sheep, dose and route of infection and, perhaps, strain of scrapie. The argument is really about whether or not heterozygotes develop scrapie. So this would mean that most *Sip* sApA sheep would develop experimental scrapie at times closer to *Sip* sAsA sheep and natural scrapie at times closer to *Sip* pApA sheep. The wide spread in timing of scrapie appearance in both the experimental and natural disease would then be due to other factors operating on outbred animals kept in non-uniform field conditions.

Using polymorphisms of the PrP gene it may become possible to describe more exactly the genetic control of scrapie susceptibility in sheep. PrP gene non-coding region RFLPs have already suggested that *Sip* sAsA and sApA Cheviot sheep can be distinguished by incubation period differences (Foster and Hunter, 1991). The present study has found that PrP gene heterozygotes develop natural scrapie—whether these animals are really *Sip* gene heterozygotes also remains to be established.

CONCLUSIONS

Nearly 90% of naturally infected scrapie sheep of more than 30 breeds and crossbreeds carry the PrP gene *Eco*RI fragment e1, almost 60% as homozygotes. This fragment is associated with short incubation period of experimental scrapie in NPU Cheviot sheep. A PrP gene RFLP with *Hind*III is less informative.

In most studies where scrapie sheep are compared with age-matched unaffected flockmates, scrapie sheep are more likely to carry e1 than healthy sheep. These scrapie sheep included a high number e1e1 homozygotes. In one study, however, scrapie sheep genotypes frequencies were very similar to unaffected flockmates and none of the scrapie sheep were e1e1. Reasons for this unusual flock include breed differences in, or breakdown of, the association of PrP e1 with incidence of disease but the flock could also be affected by a different strain of scrapie.

The complex sheep PrP genotype information produced by this study provides a baseline for further investigations into the association of PrP gene coding region polymorphisms with natural scrapie.

ACKNOWLEDGEMENTS

The author gratefully acknowledges the following: Veterinary Investigation Service staff from various centres in Scotland, Wales and England for providing samples, histopathology results and information; The Shetland Flock Health Association Ltd and all farmers, shepherds and vets who supplied sheep material; Grace Benson and James D. Foster for techinical assistance and information gathering and Professor W. G. Hill FRS for assistance with data analysis.

REFERENCES

Asher, D.M., Masters, C.L., Gajdusek, D.C. and Gibbs, C.J. (1983) Familial spongiform encephalopathies. In: Kety, S.S., Rowland, L.P., Sidman, R.L. and Matthysse, S.W. (eds) *Genetics of Neurological and Psychiatric Disorders*. Raven Press, New York, pp. 273–291.

Brotherston, J.G., Renwick, C.C., Stamp, J.T. and Zlotnik, I. (1968) Spread of scrapie by contact to goats and sheep. *J. Comp. Pathol.* **78** 9–17.

Carlson, G.A., Kingsbury, D.T., Goodman, P.A., Coleman, S., Marshall, S.T., DeArmond, S., Westaway, D. and Prusiner, S.B. (1986) Linkage of prion protein and scrapie incubation time genes. *Cell* **46** 503–511.

Chandler, R.L. (1963) Experimental scrapie in the mouse. *Res. Vet. Sci.* **4** 276–285.

Davies, D.C. and Kimberlin, R.H. (1985) Selection of Swaledale of reduced susceptibility to experimental scrapie. *Vet. Rec.* **116** 211–214.

Dickinson, A.G. (1976) Scrapie in sheep and goats. In:Kimberlin, R.H. (ed) *Slow Virus Diseases of Animals and Man* North-Holland, Amsterdam, pp. 209–241.

Dickinson, A.G. and Outram, G.W. (1988) Genetic aspects of unconventional virus infections: the basis of the virino hypothesis. In:Bock, G. and Marsh J. (eds) *Novel Infectious Agents and the Central Nervous System.* Ciba Foundation Symposium 135, Wiley-Interscience, pp. 63–83.

Dickinson, A.G., Young, G.B., Stamp, J.T. and Renwick, C.C. (1964) A note on the distribution of scrapie in sheep of different ages. *Anim. Prod.* **6** 375–377.

Dickinson, A.G., Stamp, J.T., Renwick, C.C. and Rennie, J.C. (1968) Some factors controlling the incidence of scrapie in Cheviot sheep injected with a Cheviot-passaged scrapie agent. *J. Comp. Pathol.* **78** 313–321.

Dickinson, A.G., Stamp, J.T. and Renwick, C.C. (1974) Maternal and lateral transmission of scrapie in sheep. *J. Comp. Pathol.* **84** 19–25.

Foster, J.D. and Dickinson, A.G. (1988a) The unusual properties of CH1641, a sheep-passaged isolate of scrapie. *Vet. Rec.* **123** 5–8.

Foster, J.D. and Dickinson, A.G. (1988b) Genetic control of scrapie in Cheviot and Suffolk sheep. *Vet. Rec.* **123** 159.

Foster, J.D. and Hunter, N. (1991) Partial dominance of the *Sip* gene in the control of experimental scrapie in Cheviot sheep. *Vet. Rec.* **128** 548–549.

Goldmann, W. (1992) Association of ovine PrP protein variants with the *Sip* gene and their similarity to bovine PrP protein. In: Prusiner, S.B., Collinge, J. Powell, J. and Anderton, B. (eds) Prion Diseases of Humans and Animals. Ellis Horwood, Chichester.

Goldmann, W., Hunter, N., Foster, J.D., Salbaum, J.M., Beyreuther, K. and Hope, J. (1990) Two alleles of a neural protein gene linked to scrapie in sheep. *Proc. Natl. Acad. Sci. USA* **87** 2476–2480.

Goldmann, W., Hunter, N., Benson, G., Foster, J.D. and Hope, J. (1991) Different scrapie-associated fibril proteins (PrP) are encoded by lines of sheep selected for different alleles of the *Sip* gene. *J. Gen. Virol* **72** 2411–2417.

Gordon, W.S. (1966) Variation in susceptibility of sheep to scrapie and genetic implications. In: *Report of Scrapie Seminar.* 1964 ARS 91–53 US Department of Agriculture, pp. 53–67.

Hoare, M., Davies, D.C. and Pattison, I.H. (1977) Experimental production of scrapie-resistant Swaledale sheep. *Vet. Rec.* **101** 482–484.

Hourrigan, J., Klingsporm, A., Clark, W.W. and deCamp, M. (1979) Epidemiology of scrapie in the United States. In: Prusiner, S.B. and Hadlow, W.J. (eds) *Slow Transmissible Diseases of the Nervous System*, Vol. 1. Academic Press, New York, pp. 331–356.

Hunter, N., Hope, J., McConnell, I. and Dickinson, A.G. (1987) Linkage of the scrapie-associated fibril protein (PrP) gene and *Sinc* using congenic mice and restriction fragment length polymorphism analysis. *J. Gen. Virol.* **68** 2711–2716.

Hunter, N., Foster, J.D., Dickinson, A.G. and Hope, J. (1989) Linkage of the gene for the scrapie-associated fibril protein (PrP) to the *Sip* gene in Cheviot sheep. *Vet. Rec.* **124** 364–366.

Hunter, N., Foster, J.D., Benson, G. and Hope, J. (1991) Restriction fragment length polymorphisms of the scrapie-associated fibril protein (PrP) gene and their association with susceptibility to natural scrapie in British sheep. *J. Gen. Virol.* **72** 1287–1292.

Millot, P., Chatelain, J., Dautheville, C., Salmon, D. and Cathala, F. (1988) Sheep major histocompatibility (OLA) complex: linkage between a scrapie susceptibility/resistance locus and the OLA complex in lle-de-France sheep progenies. *Immunogenetics* **27** 1–11.

Nussbaum, R.E., Henderson, W.M., Pattison. I.H., Elcock, N.V. and Davies, D.C. (1975) The establishment of sheep flocks of predictable susceptibility to experimental scrapie. *Res. Vet. Sci.* **18** 49–58.

Parry, H.B. (1960) Scrapie: a transmissible hereditary disease of sheep. *Nature (London)* **185** 441–443.

Parry, H.B. (1962) Scrapie: a transmissible and hereditary disease of sheep. *Heredity* **17** 75–105.

Parry, H.B. (1984) *Scrapie* Academic Press, London.

Stamp, J.T. (1962) Scrapie: a transmissible disease of sheep, *Vet. Rec.* **74** 357–362.

Westaway, D., Goodman, P.A., Mirenda, C.A., McKinley, M.P., Carlson, G.A. and Prusiner, S.B. (1987) Distinct prion proteins in short and long scrapie incubation period mice. *Cell* **51** 651–662.

29

PrP gene allelic variants and natural scrapie in French Ile-de-France and Romanov sheep

J.L. Laplanche, J. Chatelain, S. Thomas, M. Dussaucy, J. Brugere-Picoux and J.M. Launay

ABSTRACT

DNA samples from 153 sheep from two breeds, Ile-de-France and Romanov, including 29 natural scrapie cases, were screened for mutations in the PrP gene coding sequence by using polymerase chain reaction and denaturing gradient gel electrophoresis. Four predicted aminoacid substitutions in PrP were found: 171 Gln → Arg, 154 Arg → His, 136 Ala → Val and 112 Met → Thr. The 136Ala allele appeared to be associated with resistance to the disease whereas 136Val correlated with susceptibility to natural scrapie but with partial dominance. These data reassert the pivotal role of PrP in natural scrapie and allow the identification of susceptible and resistant genotypes which could be of importance in natural disease control.

INTRODUCTION

Scrapie is a fatal, progressive degenerative disorder of the central nervous system which occurs as a natural infection in sheep and goats. Its transmissibility was first demonstrated by Cuille and Chelle (1938). The disease is characterized by a post-translational modification of a host-encoded protein called prion protein (PrP) in brain, resulting in an abnormal isoform which is consistently associated with disease and infectivity. Studies of PrP genes in mice and hamsters with experimental scrapie, in humans with genetic prion encephalopathies, and in transgenic mice carrying foreign PrP genes showed that PrP plays a major role in pathogenesis, including control of incubation times and neuropathological features (see Prusiner 1991 for review). The sheep PrP gene, which is closely linked—if not identical—to the *Sip* gene, also appears to be involved in control of incubation time and incidence of sheep scrapie, both natural and experimental (Hunter *et al.*, 1989, 1991). We therefore decided to look for specific disease-associated mutations within the PrP coding sequence in sheep.

To search efficiently for mutations, our strategy has been (i) to collect DNA samples from a large number of healthy and naturally scrapie-affected sheep, (ii) to amplify the sheep PrP gene coding sequence by using polymerase chain reaction (PCR) and (iii) to assay the amplification products for sequence variations by using denaturing gradient gel electrophoresis (DGGE). We studied 153 sheep from two breeds, Ile-de-France and Romanov, including 29 neuropathologically verified scrapie cases from three endemically affected flocks (Table 1).

Table 1.   Sheep population characteristics

Breed (flock)	Total sheep number by flock	Studied sheep	
		Scrapied	Healthy
Romanov (R)	500	14	27
Ile-de-France (IdF1)	100	7	35
Ile-de-France (IdF2)	1500	8	29
Ile-de-France (BN)	150	0	33

Romanov flock (R) provided cases from Oct. 1990 to July 1991. IdF1 is a previously described Ile-de-France flock (Millot *et al.*, 1985; Chatelain *et al.*, 1986; Chatelain and Dautheville-Guibal, 1989) from which cases were collected from April 1989 to May 1991. IdF2, a second Ile-de-France flock geographically distinct from IdF1, provided cases from October 1990 to June 1991. The third Ile-de-France flock from the 'Bergerie Nationale' (BN) is considered as free from scrapie. Healthy animals were sheep without clinical symptoms of scrapie randomly selected from each flock.

RESULTS AND DISCUSSION

DGGE reveals several alleles of the sheep PrP gene

DGGE allows the separation of DNA molecules differing by as little as a single base change. The change in sequence causes the fragment to melt at different denaturant concentrations and thus to be retarded at different positions in the gel (Myers *et al.*, 1987). Based upon the sheep PrP gene sequence (Goldmann *et al.*, 1990), three pairs of primers were selected using the Meltmap program (Lerman and Silverstein, 1987) to produce suitable fragments for analysis by DGGE (Table 2). The inclusion of a GC clamp has been shown to increase the sensitivity of DGGE (Sheffield *et al.*, 1989). Screening of DNA samples by DGGE displayed about 20 different patterns of bands reflecting existence of several alleles of the sheep PrP gene (Fig. 1). Preliminary examination of the PCR products on a neutral polyacrylamide gel allowed us to ascertain that shifted bands observed on the denaturing gel were not due to allele size differences. It is noteworthy that, in all studied sheep, neither deletion nor insertion in the octapeptide repeat region was detected in spite of their occurrence in cattle (Goldmann *et al.*, 1991a) and in humans (Laplanche *et al.*, 1990; Owen *et al.*, 1991).

Table 2.   Primers

Primer	Position	Sequence
P2	645–626	5'GTGTGTGTTGCTTGACTGTG3'
P3	217–236	5'GCAACCGCTATCCACCTCAG3'
P4	872–849	5'AAGAAGATAATGAAAACAGGAAGG3'
P5	GC clamp + 217 + 236	5'GCCCGCCGTCCCGGCCCGACCCCCGGGC-GTCCGGCGCCCGGCAACGCTATCCACCTCAG3'
P6	522–503	5'GGTCCTCATAGTCATTGCCA3'

Sheep PrP protein variants

Double-strand sequencing of PCR products showing shifted bands on denaturing gradient gel revealed at least four point mutations leading to aminoacid differences in the predicted PrP protein.

The first of these was a G-to-A substitution at codon 171, initially described in Suffolk sheep (Goldmann *et al.*, 1990), leading to a glutamine-to-arginine change. A dot blot analysis performed on all samples indicated that this mutation was present in both studied breeds, but at a low frequency (7.4%, $n = 54$) in the Romanov sample. Whether this allele is naturally rare in this breed or is a result of selection in the flock remains to be established.

The second mutation, a G-to-A substitution at codon 154, led to an arginine-to-histidine change.

The third, a C-to-T substitution at the second base of codon 136, introduced a valine instead of an alanine.

Both substitutions at codons 154 and 136 occurred in a highly conserved region among species. The creation of a Bsp H1 restriction site at both variant codons confirmed the nucleotidic change and was used to determine codon genotypes in both breeds (see below).

The fourth mutation, a C-to-T substitution at codon 112 resulting in a methionine-to-threonine change, was found in only three Ile-de-France sheep from the IdF2 flock, two of which were genetically linked and scrapie affected.

Ile-de-France and Romanov sheep affected by natural scrapie carried a 136Val allele

In order to determine codon 154 and 136 genotypes, PCR products of all sheep were digested by Bsp H1 (Fig. 2). The 154His allele was found in only nine unaffected Romanov, one of which was homozygous. None of the Ile-de-France sheep carried this allele. By contrast, the 136Val allele was constantly found in scrapie-affected sheep from both breeds, whatever the geographic origin of the flock. The 136Val

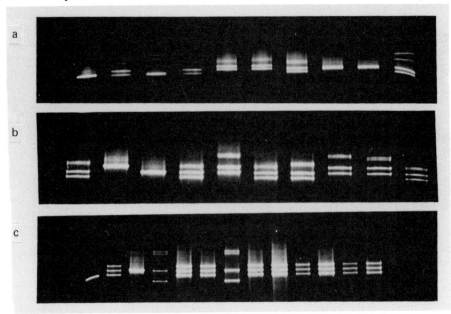

Fig. 1.   Examples of denaturing gradient gel patterns of PCR products. PCR were performed as described elsewhere (Saiki *et al.*, 1988) using (a) P3–P4, (b) P2–P3, and (c) P5–P6 as primers (see Table 2). Gel apparatus and conditions were as described elsewhere (Myers *et al.*, 1987; Sheffield *et al.*, 1989). From 10 to 15 μl of each PCR product were loaded on a 6.5% polyacrylamide gel containing a linearly increasing gradient from 20% to 80% denaturant (100% denaturant = 7 M urea – 40% formamide) and electrophoresed for 6 hours at 160 V. Gels were then ethidium bromide stained andexamined under an UV light source. Lower bands corresponded to homoduplexes and upper bands to heteroduplexes formed in heterozygous individuals between the two alleles during the reassorting of strands that occurs during PCR. In some cases, the two heteroduplex DNA fragments did notseparate from each other.

allelic frequencies were significantly higher in scrapied sheep than in their apparently healthy flockmates and the unaffected BN flock (Table 3). As shown in Table 4, the difference between distribution of codon 136 genotypes in scrapied and healthy animals was highly significant ($\chi^2 = 64.33$, P $\ll$ 0.001). All scrapie-affected sheep were 136Ala/Val or 136Val/Val. It is noteworthy that the 136Val/Val genotype was observed only in the scrapie samples. Interestingly, the 136Ala/Ala genotype was never found within scrapie-affected sheep despite its significant occurrence in their healthy flockmates and in the unaffected BN flock. These results strongly suggest that 136Ala is a resistance allele and 136Val a susceptibility allele to natural scrapie in Ile-de-France and Romanov sheep. Concomitantly with this study, Goldmann *et al.* (1991b) also described codon 171, 154 and 136 mutations in experimentally scrapied Cheviot sheep and linked the 136Val and 136Ala alleles to *Sip* sA and *Sip* pA, which respectively confer increased and decreased susceptibility to scrapie after injection of SSBP/1.

The 136Val susceptibility allele is partially dominant in natural scrapie.

Several apparently healthy sheep from both breeds appeared heterozygous at codon 136 (Table 4), suggesting that the 136Val susceptibility allele was not fully dominant

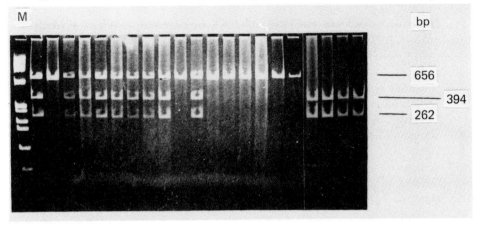

Fig. 2.   Codon 136 genotypes of 21 samples assessed by Bsp H1 digestion of P3–P4 PCR products. Bsp H1 cuts codon 136 when it encodes Val. The 656 bp fragment of amplified Val allele is digested to give two bands of 394 and 262 bp. Lane M: Phi X 174/HaeIII.

Table 3.   Frequency of 136 codon alleles in sheep affected by natural scrapie compared with unaffected flockmates

Samples		n	136 codon alleles (%)		χ^2 (Yates)	P
			136Ala	136Val		
R	Scrapie	14	21	79	20.21	<0.001
	Healthy	27	76	24		
IdF1	Scrapie	7	43	57	8.21	<0.01
	Healthy	35	83	17		
IdF2	Scrapie	8	37	63	4.61	<0.05
	Healthy	29	71	29		
BN	Healthy	33	79	21		

Flocks designated as in Table 1. n = total number of sheep. χ^2 and P were determined by comparison of scrapie samples with healthy samples using Yates' correction.

in natural scrapie. Accordingly, the partial dominance of the sA allele of the *Sip* gene was recently reported in experimental scrapie (Foster and Hunter, 1991). In order to investigate whether specific PrP proteins distinguished these unaffected sheep from scrapie-affected ones, data obtained from sheep heterozygous for codon 136 which were older than the maximal age at onset of clinical disease were compiled. In our samples, the mean age at clinical onset of scrapie was 2.4 years (range 1.5–3.5) in Romanov and 4.5 years (range 2.0–7.0) in Ile-de-France sheep. Only genotypes from animals older than 3.5 years (Romanov) and 7 years (Ile-de-France) were considered. As indicated in Table 5, 12/15 healthy sheep were double heterozygous, i.e. 136Ala/Val and either 154 Arg/His or 171Arg/Gln. By contrast, 15/18 codon 136

Table 4. Codon 136 genotypes in sheep affected by natural scrapie compared with unaffected flockmates

Samples		n	136Ala/Ala	136Ala/Val	136Val/Val	χ^2	P
R	Scrapie	14	0	6	8	22.70	<0.001
	Healthy	27	14	13	0		
IdF1	Scrapie	7	0	6	1	13.20	<0.01
	Healthy	35	23	12	0		
IdF2	Scrapie	8	0	6	2	10.83	<0.01
	Healthy	29	12	17	0		
BN	Healthy	33	19	14	0		
All samples							
	Scrapie	29	0	18	11	64.33	<0.001
	Healthy	124	68	56	0		

Flocks designated as in Table 1. n = total number of sheep. χ^2 and P were determined by comparison of scrapie samples with healthy samples.

heterozygous scrapied sheep were homozygous 154Arg/Arg and 171Gln/Gln. Interestingly, the three diseased Ile-de-France sheep carrying the 171Arg/Gln genotype (one from the IdF1 flock and two from the IdF2 flock) were the latest to came down with scrapie, at 7 years of age. Also unusual were the three Romanov sheep still healthy at the age of 5.5 years, despite having the same genotypes as affected animals. These results suggest that (i) 136Val is a partial dominant allele in natural scrapie, (ii) 171Arg and perhaps 154His alleles were associated with a prolonged incubation period in sheep carrying the susceptibility 136Val allele, and (iii) other genetic or environmental factors modulated individual susceptibility to scrapie. The two genetically linked (same unaffected sire) scrapie-affected sheep from flock IdF2 carrying the variant 112Thr allele were also heterozygous for codon 136. The sheep codon 112 corresponds to the mouse polymorphic codon 108, encoding Phe or Leu, associated with the incubation time in mice (Westaway *et al.*, 1987). However, the codon 112 polymorphism does not seem to influence the onset of the disease in these two sheep since they developed scrapie at 4.5 and 6 years respectively, though they carried the same genotypes at the PrP locus.

Recent experimental data have shown that the primary structure of PrP modulates the pathogenesis of the disease (Scott *et al.*, 1989; Lowenstein *et al.*, 1990; Prusiner *et al.*, 1990). One proposed mechanism for the initiation and extension of the disease is that an abnormal isoform PrP^{Sc} might combine with the cellular isoform PrP^c and catalyse its conversion into PrP^{Sc}. The efficiency of this conversion should depend on the sequence identity of the two isoforms. Whether the PrP^{Sc} form initiating the process is naturally introduced into sheep from exogenous sources or results from a spontaneous conversion of a variant PrP remains unknown. Whatever the mechanism, PrP^c carrying a valine at position 136 should be more susceptible to pathological

Table 5. Codon 154 and 171 genotypes in scrapie-affected and unaffected sheep heterozygous at codon 136

Sheep		136 Ala/Val 145 Arg/Arg 171 Gln/Gln	136 Ala/Val 154 Arg/Arg 171 Arg/Gln	136 Ala/Val 154 Arg/His 171 Gln/Gln
R	Scrapie ($n = 6$)	6	–	–
	Healthy[a] ($n = 6$)	3	2	1
IdF1 and 2	Scrapie ($n = 12$)	9	3	–
	Healthy[a] ($n = 11$)	–	11	–

Flocks designated as in Table 1. Data from IdF1 and IdF2 were pooled (n = number of sheep).
[a] Healthy sheep were over the maximal age of clinical onset ($R > 3.5$, IdF < 7 years).

conversion into PrP^{Sc}, leading to its aggregation and deposition in brain. According to this hypothesis, the disease process would extend more rapidly in homozygous 136Val/Val sheep, because of sequence identity of proteins, than in sheep encoding two PrPs differing by two amino acids, at crucial positions 136 and either 171 or 154.

CONCLUSION

Until now, no preclinical marker of natural scrapie has been available. We report here that all scrapie-affected sheep from two French breeds, Ile-de-France and Romanov, encode a PrP with a valine at position 136. Whether sheep from other endemically affected breeds carry the same or distinct mutation(s) remains to be established. Evidence for distinct prion proteins encoded by scrapied sheep might offer the possibility to control natural scrapie by selection of genitors carrying only the resistance genotype.

ACKNOWLEDGMENTS

We thank G. Durand and J. Lebreton for their help in the collection of scrapie-affected sheep, Professors M. Goossens and M. Vidaud for initiation to DGGE, Dr L. Lerman for the generous gift of the Meltmap program, Dr D. Westaway for helpful discussions, and Dr H. Baron for his critical reading and improvement of the manuscript.

REFERENCES

Chatelain, J. and Dautheville-Guibal (1989) Ovine scrapie: follow-up of sheep belonging to an endemic scrapie-infected flock. *Eur. J. Epidemiol.* **5** 113–116.
Chatelain, J., Baron, H., Baille, V., Bourdonnais, A., Delasnerie-Laupretre, N. and Cathala, F. (1986) Study of endemic scrapie in a flock of Ile-de-France sheep. *Eur. J. Epidemiol.* **2** 31–35.

Cuille, J. and Chelle, P.L. (1938) La tremblante du mouton est bien inoculable. *C.R. Acad. Sci. Paris* **206** 78–79.

Foster, J.D. and Dickinson, A.G. (1988) Genetic control of scrapie in Cheviot and Suffolk sheep. *Vet. Rec.* **123** 159.

Foster, J.D. and Hunter, N. (1991) Partial dominance of the sA allele of the Sip gene for controlling experimental scrapie. *Vet. Rec.* **128** 548–549.

Goldmann, W., Hunter, N., Foster, J.D., Salbaum, J.M., Beyreuther, K. and Hope, J. (1990) Two alleles of a neural protein linked to scrapie in sheep. *Proc. Natl. Acad. Sci. USA* **87** 2476–2480.

Goldmann, W., Hunter, N., Martin, T., Dawson, M. and Hope, J. (1991a) Different forms of the bovine PrP gene have five or six copies of a short GC-rich element within the protein-coding exon. *J. Gen. Virol.* **72** 201–204.

Goldmann, W., Hunter, N., Benson, G., Foster, J.D. and Hope, J. (1991b) Different scrapie-associated fibril proteins (PrP) are encoded by lines of sheep selected for different alleles of the Sip gene. *J. Gen. Virol.* **72** 2411–2417.

Hunter, N., Foster, J.D., Dickinson, A.G. and Hope, J. (1989) Linkage of the gene for the scrapie-associated fibril protein (PrP) to the Sip gene in Cheviot sheep. *Vet. Rec.* **124** 364–366.

Hunter, N., Foster, J.D., Benson, G. and Hope, J. (1991) Restriction fragment length polymorphisms of the scrapie-associated fibril protein (PrP) gene and their association with susceptibility to natural scrapie in British sheep. *J. Gen. Virol.* **72** 1287–1292.

Laplanche, J.L., Chatelain, J., Launay, J.M., Gazengel, C. and Vidaud, M. (1990) Deletion in prion protein gene in a Moroccan family. *Nucleic Acids Res.* **18** 6745.

Lerman, L.S. and Silverstein, K. (1987) Computational simulation of DNA melting and its application to denaturing gradient gel electrophoresis. *Methods Enzymol.* **155** 482–501.

Lowenstein, D., Butler, D., Westaway, D., McKinley, M, DeArmond, S. and Prusiner, S. (1990) Three hamster species with different scrapie incubation times and neuropathological features encode distinct prion proteins. *Mol. Cell. Biol.* **10** 1153–1163.

Millot, P., Chatelain, J. and Cathala, F. (1985) Sheep major histocompatibility complex OLA: gene frequencies in two french breeds with scrapie. *Immunogenetics* **21** 117–123.

Myers, R.M., Maniatis, T. and Lerman, L.S. (1987) Detection and localization of single bases changes by denaturing gradient gel electrophoresis. *Methods Enzymol.* **155** 501–527.

Owen, F., Poulter, M., Collinge, J., Leach, M., Shah, T., Lofthouse, R., Chen, Y., Crow, T.J., Harding, A.E., Hardy, J. and Rossor, M.N. (1991) Insertions in the prion protein gene in atypical dementias. *Exp. Neurol.* **112** 240–242.

Prusiner, S. (1991) Molecular biology of prion disease. *Science* **252** 1515–1522.

Prusiner, S., Scott, M., Foster, D., Pan, K., Groth, D., Mirenda, C., Torchia, M., Yang, S., Serban, D., Carlson, G., Hoppe, P., Westaway, D. and DeArmond, S. (1990) Transgenetic studies implicate interactions between homologous PrP iso-forms in scrapie prion replication. *Cell* **63** 673–686.

Saiki, R.K., Gelfand, D.H., Stoffel, S., Scharf, S.J., Higuchi, R., Horn, G.T., Mullis, K.B. and Erlich, H.A. (1988) Primer-directed enzymatic amplification of DNA with a thermostable DNA polymerase. *Science* **239** 487–494.

Scott, M., Foster, D., Mirenda, C., Serban, F., Coufal, M., Walchli, M., Torchia, M., Groth, D., Carlson, G., DeArmond, D., Westaway, D. and Prusiner, S. (1989) Transgenic mice expressing hamster prion protein produce species-specific scrapie infectivity and amyloid plaques. *Cell* **59** 847–857.

Sheffield, V.C., Cox, D.R., Lerman, L.S. and Myers, R.M. (1989) Attachment of a 40-base-pair G + C-rich sequence (GC-clamp) to genomic DNA fragments by the polymerase chain reaction results in improved method of single-base changes. *Proc. Natl. Acad. Sci. USA* **86** 232–236.

Westaway, D., Goodman, P., Mirenda, C., McKinley, M., Carlson, G. and Prusiner, S. (1987) Distinct prion proteins in short and long scrapie incubation period mice. *Cell* **51** 651–662.

Part V   Prions and prion proteins

30

Nucleic acids and scrapie prions

Detlev Riesner, Klaus Kellings, Norbert Meyer, Carol Mirenda and Stanley B. Prusiner

ABSTRACT

Different approaches are described in the literature to search for a putative nucleic acid component of scrapie or CJD infectious prion particle. The influence of physical, chemical and enzymatic treatment on the infectivity is regarded as an argument against a nucleic acid component. Occurrence of particular nucleic acids in infectious fractions was studied by gel electrophoresis, fractionation by ultracentrifugation, hybridization with cDNA libraries and with specific hybridization probes, and searching for non-host sequences by molecular cloning and PCR. Applying the newly developed method of return refocusing gel electrophoresis, nucleic acids including those which are heterogeneous in size were analysed. A paucity of heterogeneous nucleic acids was detected. Molecules with a chain length above 100 nucleotides were present with a particle-to-infectivity ratio below unity. Thus, nucleic acids larger than 100 nucleotides cannot be essential components of scrapie infectivity. A critical comparison of all data available shows that, in spite of multiple studies, the issue of a scrapie-specific nucleic acid remains unresolved even though it has not been possible to isolate a nucleic acid which is an essential component of the infectious prion.

INTRODUCTION

The issue of whether or not the scrapie agent contains a putative nucleic acid component has been problematic for more than two decades. Early experiments with ionizing and UV radiation by Alper and colleagues raised the possibility that the infectious scrapie prion might not contain a nucleic acid (Alper *et al.*, 1966, 1967).

We describe here the application of gel electrophoretic techniques to search directly for nucleic acids in fractions highly enriched for scrapie infectivity.

The search for a scrapie-specific nucleic acid presents a significant challenge. No convincing identification of such a nucleic acid has been reported to date. On the other hand, exclusion of its existence is a difficult task. It can always be argued that the putative nucleic acid has unusual features and thus has eluded detection.

Three approaches have been used to search for a scrapie-specific nucleic acid. These include (1) inactivation of prions by procedures that modify or hydrolyse nucleic acids, (2) characterization of nucleic acids in fractions enriched for prion infectivity, and (3) search for a nucleic acid unique to scrapie-infected preparations.

RESULTS AND DISCUSSION

Resistance of prions to procedures that hydrolyse or modify nucleic acids

The first suggestion for an unusually small size of the scrapie prion came from studies using ionizing and UV radiation (Alper et al., 1966). Infectious prions in highly purified fractions were subjected to procedures that hydrolyze, modify, denature or shear biopolymers (McKinley et al., 1981; Diener et al., 1982; Bellinger-Kawahara et al., 1987). Procedures were chosen that target either protein or nucleic acids but not both. These procedures were tested on viroids which are small infectious pathogens composed exclusively of RNA. In similar studies viruses and bacteriophages served as controls. The results of studies comparing viroids with prions are summarized in Table 1. Prions resisted inactivation to procedures that hydrolyse, modify or shear nucleic acids but were inactivated by treatments which denatured or modify proteins. The same results were obtained with crude brain homogenates, purified prion preparations (Bellinger-Kawahara et al., 1987; Brown et al., 1990), prion liposomes (Gabizon et al., 1988a), and in prions from cultured neuroblastoma cells (Neary et al., 1991).

This approach depends upon extrapolating the properties of known viruses, bacteriophages, viroids, nucleic acids and proteins to the enigmatic prion. Hypothetical nucleic acids within infectious scrapie prion particles are assumed to have properties similar to those exhibited by nucleic acids; (1) free in solution, (2) in contact with viral proteins or (3) present in cellular extracts. It is unclear whether this assumption is justified since the major protein component of infectious prions, PrP^{Sc}, differs from most viral and cellular proteins with respect to solubility, resistance to proteolysis and propensity for aggregation. Studies described below demonstrate that small nucleic acids up to several hundred nucleotides in length withstand harsh procedures that hydrolyse nucleic acids in prion preparations. Although the wide variety of procedures used in these studies argue that a nucleic acid component within the prion particle is unlikely, they do not exclude such a molecule.

Nucleic acids in fractions enriched for prion infectivity

In some studies, nucleic acids were reported to be found only in fractions prepared from scrapie-infected animals. In other studies, copurification of nucleic acids and

Table 1. Stabilities of the scrapie agent and viroids after chemical and enzymatic treatment

Chemical treatment	Concentration	PSTV	Scrapie agent
Et$_2$PC	10–20 mM	$(-)$	$+$
NH$_2$OH	0.1–0.5 M	$+$	$-$
Psoralen (AMT)	10–500 μg/ml	$+$	$-$
Phenol	Saturated	$-$	$+$
SDS	1–10%	$-$	$+$
Zn^{2+}	2 mM	$+$	$-$
Urea	3–8 M	$-$	$+$
Alkali	pH 10	$(-)$	$+$
KSCN	1 M	$-$	$+$
RNase A	0.1–100 μg/ml	$+$	$-$
DNase	100 μg/ml	$-$	$-$
Proteinase K	100 μg/ml	$-$	$+$
Trypsin	100 μg/ml	$-$	$+$

$+$, inactivation; $-$, no change; $(-)$, small change.
Reproduced in modified form from Diener *et al.* (1982) and Prusiner and McKinley (1987).

scrapie infectivity was described. These results were interpreted as evidence indicative of a scrapie-specific nucleic acid. Of course, the detection of a nucleic acid under these circumstances does not mean that it is an essential component of the infectious prion particle. In no case reported to date was a convincing correlation with scrapie infectivity demonstrated.

In Table 2, the different methodological approaches used in the search for a scrapie-specific nucleic acid are listed. Some authors interpreted their results as indicative for a scrapie-specific nucleic acid, some as not contradictory to such a nucleic acid, and others stated that they could not find an indication for a scrapie-specific nucleic acid.

Search for a nucleic acid unique to scrapie-infected preparations
Because of the many unsuccessful attempts to identify a nucleic acid component required for scrapie infectivity, we set out to develop an experimental paradigm that would offer the possibility of definitely excluding a nucleic acid if prions are devoid of such molecules (Meyer *et al.*, 1991). We chose silver staining of nucleic acids after polyacrylamide gel electrophoresis (PAGE) since this method is independent of the structure of the polynucleotide. Furthermore, we assumed that one infectious unit must contain at least one nucleic acid molecule if it is to be essential for infectivity. In other words, the particle (nucleic acid molecule)-to-infectivity ratio (P/I) has to be at least unity. Because the properties of the hypothetical scrapie-specific nucleic acid

Table 2. Correlation of the presence of nucleic acids and infectivity

---

Differences in the patterns of nucleic acids from infected and non-infected tissue

(1) End-labelling and gel electrophoresis (German *et al.*, 1985; Dees *et al.*, 1985; Castle *et al.*, 1987)
(2) Differential hybridization of a cDNA library (Duguid *et al.*, 1988; Duguid and Dinauer, 1989; Aiken *et al.*, 1989)

Nucleic acid analysis of purified infectious material

(1) Copurification of nucleic acids and infectivity in ultracentrifugation (Sklaviadis *et al.*, 1989)
(2) Detection of retroviral and polyadenylated RNA by PCR technique: (Murdoch *et al.*, 1990; Akowitz *et al.*, 1990)
(3) Detection of mitrochondrial DNA by hybridization (Aiken *et al.*, 1990)
(4) Detection of ssDNA in electron microscopy (Narang *et al.*, 1988, 1991)
(5) Molecular cloning of residual nucleic acids in highly purified prions (Oesch *et al.*, 1988)

---

are unknown, we considered that such a molecule might be either DNA or RNA, single or double stranded, circular or linear, as well as capped, chemically modified or covalently bound to proteins. In addition, it might be heterogeneous in size. Since high levels of scrapie infectivity have been found only in animals, it was not possible to incorporate radiolabelled nucleotides into the putative scrapie nucleic acid. Cell culture systems producing high levels of prions would obviate this problem but such cultures are not available. A general method for the analysis of all the kinds of nucleic acids listed above is PAGE combined with silver staining. Depending on the size of the nucleic acid 20 to 200 pg of nucleic acid per gel band can be detected. For example, 170 pg would be contained in 3×10^9 ID_{50} units of infectious prions if the hypothetical nucleic acid were 100 nt in length and the particle-to-infectivity ratio were unity. Obviously, studies on a smaller scrapie-specific nucleic acid would require higher titres while studies on a larger nucleic acid would need lower titres to yield the same mass of nucleic acid. The sensitivity of silver staining after conventional PAGE decreases substantially if the nucleic acid is heterogeneous in size. To be able to detect heterogeneous nucleic acids, the method of return refocusing gel electrophoresis (RRGE) was developed.

Search for homogeneous nucleic acids by PAGE
Scrapie prions were purified from brains of Syrian hamsters using a discontinuous sucrose gradient (Prusiner *et al.*, 1983). The infectious prions consisted mainly of rod-shaped aggregates composed largely of PrP 27-30. The prion rods were precipitated from the sucrose gradient fractions with ethanol and subjected to a nucleic acid degradation protocol using DNase and Zn^{2+}. No significant change in prion titre was detected after digestion with DNase and exposure to Zn^{2+} ions in

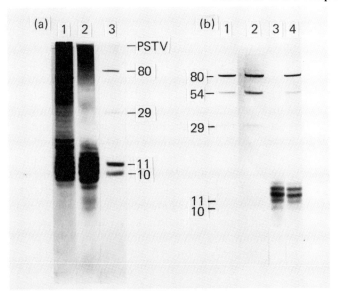

Fig. 1.   Analysis of nucleic acids in prion rod fractions by 20% PAGE before and after DNase and Zn^{21+} treatment. (A) Nucleic acids without and with DNase and Zn^{2+} treatment. Lane 1, prion rods not treated with DNase and Zn^{2+}; lane 2, prion rods treated with DNase and Zn^{2+}. Lane 3, marker nucleic acids: circular potato spindle tuber viroid (PSTVd, 300 pg), tRNA (80 nt, 1 ng), oligo DNA (29 nt, 200 pg), oligo RNA (11 nt, 3 ng), oligo RNA (10 nt, 2 ng). (B) Comparable amounts of control nucleic acids (number of molecules) and prions (ID_{50} units). 3×10^{10} molecules of control nucleic acids and 1.2×10^{10} ID_{50} units of prions recovered from sucrose gradients were analysed. Lane 1, control nucleic acids: tFNA (80 nt, 1.4 ng), oligo RNA (11 nt, 0, 19 ng), oligo RNA (10 nt, 0.17 ng); lane 2, control nucleic acids as in lane 1 but treated by the deproteinization procedure; lane 3, prions treated with DNase and Zn^{2+} and by deproteinization;lane 4, control nucleic acids as in lane 1 added to prions after DNase and Zn^{2+} treatment but before deproteinization. The bands of 10 and 11 nt are weak in the photographic reproduction but were clearly seen in the original gel. Reproduced from Meyer *et al.* (1991).

agreement with earlier observations (cf. Table 1). The samples were deproteinized by boiling in 2% SDS, digesting with proteinase K and extracting once with phenol and once with phenol:chloroform (1:1) or in later experiments two times with phenol: chloroform:isoamylalcohol (50:48:2), and precipitated with ethanol.

    PAGE analysis of the DNase-digested and Zn^{2+}-treated prions showed some background smearing as well as distinct bands migrating near the dye front (Fig. 1(A), lane 2). Omission of DNase digestion and Zn^{2+} hydrolysis resulted in a prion fraction with a large number of stained bands throughout the lane in addition to the rapidly migrating molecules (Fig. 1(A), lane 1). The size of the rapidly migrating molecules was estimated to range between 8 and 15 bases as judged by the migration of control nucleic acids (Fig. 1(A), lane 3). From the analysis of the sample after additional treatments with DNase, Zn^{2+} ions and NaOH prior to electrophoresis as well as phenol–sulphuric acid measurements, we concluded that the rapidly migrating bands observed on PAGE might be either complex Asn-linked oligosaccharides released from PrP 27-30 during proteinase K digestion or non-covalently bound sugar polymers which purified with the prion rods. Control nucleic acids were analysed in the same way as infectious scrapie prions. In Fig. 1(B), lane 1 demonstrates that

3×10^{10} molecules of each of the control nucleic acids were readily detected and lane 2 shows that this amount could be recovered using the protocol described (Meyer *et al.*, 1991). If the control nucleic acids were hydrolysed with DNase I and Zn^{2+}, no silver-stained bands were detected (data not shown). Silver staining of prions after PAGE showed no signal above 20 nt (Fig. 1(B), lane 3). If the control nucleic acid molecules (see also lanes 1 and 2) were added to the scrapie sample after DNase and Zn^{2+} treatment but before destroying the infectivity with SDS and boiling (Fig. 1(B), lane 4), all of them were visible except those of 10 nt and 11 nt in length because these were hidden by the non-nucleic acid molecules purifying with the prions.

The prion infectivity was determined to be 1.2×10^{10} ID_{50} unit immediately prior to deproteinization. Since 170 pg nucleic acid in a single band was assumed to be the limit of detection, 1.2×10^{10} nucleic acid molecules of at least 25 nt in length would have been detected. The bioassay requires us to quantify this conclusion since imprecision in scrapie prion titre determinations by a factor of 10 attends these measurements. Because the analysis utilized 20% polyacrylamide gels, only nucleic acids < 300 nt in length could be analysed. Larger nucleic acids were assessed using a modified procedure.

The method of RRGE for the analysis of heterogeneous nucleic acids

Although remote and unprecedented, we considered the possibility that prions contain nucleic acid molecules of non-uniform length. In such a case, the nucleic acids would migrate during PAGE as many bands and each band might either be below the threshold for detection or not resolved from neighbouring bands, resulting in a smear of staining.

With the method of RRGE, nucleic acids were separated from other molecules staining with silver and focused into one sharp band. Using RRGE (Fig. 2), the heterogeneous nucleic acids were detected with a sensitivity close to that attained with a homogeneous nucleic acid.

In an initial set of experiments, a 15% polyacrylamide gel matrix was used for RRGE but this restricted our analyses to polynucleotides smaller than 200 bases. In a recent study (Kellings *et al.*, 1992), the size range of the polynucleotides analysed was extended to 1100 bases by using a 9% polyacrylamide gel matrix. In order to obtain a quantitative estimate of the amount of nucleic acid present in a band, known amounts of reference nucleic acids were analysed simultaneously in adjacent slots (cf. lanes 1–3 in Fig. 2).

Detection of heterogeneous nucleic acids in purified prion preparations

Prion samples which were prepared by a protocol similar to that used for the studies (cf. Fig. 1) were evaluated by RRGE (Fig. 3). After the first electrophoresis only a faintly stained smear in gel sections a and b as well as a ladder of silver-stained bands comigrating with oligonucleotides ranging from 4 to 12 bases were visible. After RRGE, distinct silver-stained bands were visualized from pieces a and b and weaker bands from pieces c to e. The total amount of heterogeneous nucleic acid in the prion sample was estimated to be about 20 ng.

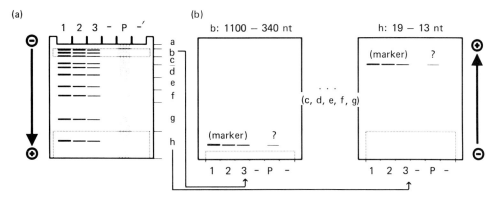

Fig. 2.   Scheme of return refocusing gel electrophoresis (RRGE). After conventional PAGE (e.g. 100 min, 150 V) as a first step, heterogeneous nucleic acids are dispersed over the whole length of the lane (lane P in A). The lane is cut into a few segments (a–h), each corresponding to a well-defined range of M_r. The segments are repolymerized into the bottom of new gel matrices (B) and second electrophoresis (250 V) is performed with reversed polarity so that the nucleic acids migrate into the new gel matrix. Because all nucleic acids in a gel segment begin migration from the same position at the beginning of the first PAGE, they meet again after reversal of the polarity of the second electrophoresis if the second run is stopped at a definite time. By adding SDS to the second PAGE, the focusing effect still works for nucleic acids while other substances such as proteins and polysaccharides remain dispersed. This is a significant advantage since proteins, like nucleic acids, stain with silver. The times of refocusing of different gel segments are chosen optimal for the different segments (between 42 and 48 min). The unknown nucleic acid of the prion sample is determined by comparison with the nucleic acid markers of known concentrations (markers 1, 2, 3). Only the two gel segments b and h are given as an example for the refocusing step; gel segment a is not used for refocusing. Reproduced from Kellings *et al.* (1992).

Nuclease digestion studies were carried out prior to RRGE in order to confirm the nucleic acid nature of the bands and to differentiate between RNA and DNA. The RNA and DNA detected differed in size distribution: nucleic acids below 50 nt were mainly RNA while longer molecules were primarily DNA. The ladder in the range of 4 to 12 bases was not composed of nucleic acids (cf. Meyer *et al.*, 1991).

During the course of investigations by RRGE, procedures were developed for dispersing prion rods into detergent–lipid–protein complexes (DLPCs) and liposomes with retention of infectivity (Gabizon *et al.*, 1987, 1988b). We thought that nucleic acids which were possibly protected from degradation by inclusion within the rod-shaped aggregates might become accessible to nucleases upon formation of the DLPCs.

Purified prion samples were transformed into DLPCs composed of sodium cholate and egg lecitin as described (Gabizon *et al.*, 1988a; Meyer *et al.*, 1991). The DLPCs were digested either with a mixture of DNase, Bal31 and RNaseA or a mixture of micrococcus nuclease, alkaline phosphatase, RNaseA and phophodiesterase. Control nucleic acids, which were added to the DLPC fraction and treated as described above, were degraded to mono- and di-nucleotides. As shown in Fig. 4, a ten-fold reduction in nucleic acid content of the prion preparations was observed after DLPC formation followed by nuclease digestion.

In Fig. 5, the nucleic acids analysed ranged in size from 13 to 1100 nucleotides.

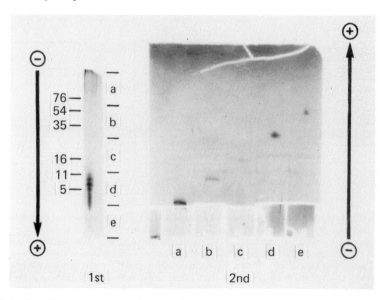

Fig. 3. Detection of heterogeneous nucleic acids in prion rods by RRGE. Prion rods were treated with DNase and Zn^{2+} and deproteinized prior to RRGE. In this RRGE, the sample contained $10^{7.3}$ ID_{50} units as measured by bioassay prior to boiling in SDS. The separation run is indicated as '1st'. Positions of marker oligonucleotides are indicated at the left. An identical gel lane but without staining was cut into gel segments (a to e) and analysed in the refocusing run, indicated '2nd'. Marker nucleic acids as depicted in the scheme of Fig. 2 were not analysed in this experiment. The total amount of nucleic acid was estimated as 20 ng. Reproduced from Meyer *et al.* (1991).

The amount of nucleic acid present in the faint bands of the prion sample was estimated by comparison with the markers designated (1), (2) and (3).

Ratio of nucleic acid molecules per infectious unit

The scrapie infectivity was monitored by incubation time interval bioassays (Prusiner *et al.*, 1982) at each step in the preparation. Separation of the prion rods from the sucrose used for discontinuous gradient centrifugation resulted in a loss of infectivity of 1 to 3 orders of magnitude; this was probably due to aggregation as well as denaturation (Meyer *et al.*, 1991). It is noteworthy that the dispersion of ethanol-precipitated prion rods into DLPCs frequently increased the titre more than ten-fold.

Based on the amount of nucleic acid estimated from RRGE and the titres of the prion fractions prior to boiling in SDS, the P/I ratio of nucleic acid molecules to ID_{50} units was calculated. If the nucleic acids that were detected are related to scrapie infectivity, then one of two alternative paradigms must attend. First, a putative scrapie-specific nucleic acid of uniform length might be hidden amongst an ensemble of background nucleic acids. Such a scrapie-specific nucleic acid would not have been detected by PAGE (see Fig. 1) even if it were present in sufficient amounts. Second, a scrapie-specific polynucleotide might be heterogeneous in length. In Fig. 6 the numbers of nucleic acid molecules per ID_{50} unit are plotted as a function of their length as estimated from the individual gel sections. In this plot the calculation was based upon the first paradigm, i.e. a well-defined scrapie-specific nucleic acid among

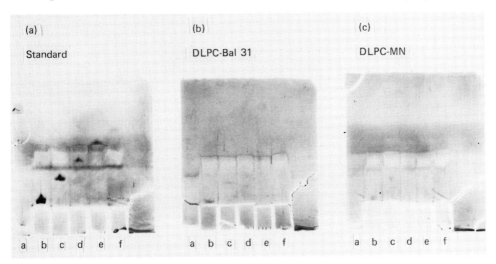

Fig. 4.　RRGE of the nucleic acids in prion preparations before and after DLPC formation with subsequent nuclease digestions. A sample of prion rods was DNase and Zn^{2+}-treated and divided into three aliquots. Aliquot A was directly deproteinized, aliquots B and C were transformed into DLPC, aliquot B was digested by Bal 31, aliquot C by MN, and both aliquots were deproteinized. Obviously, the nucleic acid amount was drastically reduced to the 'Bal 31' DLPC as well as the 'MN'-DLPC treatment. Following sizes and amounts of nucleic acid were estimated. Total sample A: 43 ng (gel segment 6 (65–200 nt), 20 ng; c (64–35 nt), 13 ng; d (34–17 nt), 3 ng; e (16–6 nt), 5 ng; f (5–3 nt), 2 ng). Total sample B 4 ng (b 200–64 nt), 2.5 ng; c (63–27 nt), 1 ng;d (26–12 nt), 0.5 ng). Total sample C: 3 ng (b (200–64 nt), 2 ng; c (63–29 nt), 0.5 ng; d (28–14 nt), 0.5 ng). The corresponding infectivities are for sample A $10^{7.7}$ ID_{50}, B $10^{8.7}$ ID_{50}, and C $10^{8.3}$ ID_{50}. Reproduced from Meyer et al. (1991).

the heterogeneous background nucleic acids. Data from all experiments published (Meyer et al., 1991; Kellings et al., 1992) are shown in the plot.

A significant decrease in the number of nucleic acid molecules per ID_{50} unit was found as the size of the polynucleotide increased. For small nucleic acid molecules (20 nt) about 10 to several hundred molecules per ID_{50} unit were estimated. If the scrapie-specific nucleic acid were longer (>100 nt), the particle-to-infectivity ratio would fall below unity. In comparison with our initial report (Meyer et al., 1991) more nucleic acids in the range of smaller sizes (<60 nt) were found in a second series of experiments (Kellings et al., 1992). This was probably due to the more effective extraction procedure, the omission of DNase digestion of prions in the rod state and better quantification. This recent study demonstrated that for nucleic acids above 200 nt in length the ratio P/I fell below unity by several orders of magnitude.

Any conclusions about a hypothetical, scrapie-specific nucleic acid must be made with respect to a particular model. The first model envisions a hypothetical, scrapie-specific nucleic acid that is a well-defined molecular species hidden in the smear of heterogeneous nucleic acids which represent preparative impurities of the sample. In this case, molecules longer than about 100 nt were not present in concentrations above one molecule per infectious unit. If we assume an order of magnitude error in the bioassay, the limit would be 160 instead of 100 nucleotides chain length. We concluded that larger nucleic acids can be excluded as scrapie-specific candidates but smaller ones cannot. Nonetheless, no functional significance can be

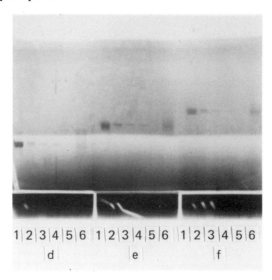

Fig. 5.   RRGE of a prion sample after DLPC formation and nuclease digestion. The refocused gel segments d to f (cf. Fig. 2) are shown. Each gel segment consists of six lanes: (1) sonicated calf thymus DNA, 300 ng (total); (2), (3), (4) nucleic acid markets 1, 2, 3 (cf. Fig. 2); (5) empty; (6) prion sample. The comparison of band intensities had to be carried out on the original gel, not on the photographic reproduction. The amounts of nucleic acids in the prion sample were estimated as 300 pg (d), 450 pg (e), and 600 pg (f). Reproduced from Kellings *et al.* (992).

assigned to any of the smaller molecules. Further studies are needed to determine whether integrity of any of these putative , scrapie-specific nucleic acids is required for infectivity.

The second model describes a hypothetical, scrapie-specific nucleic acid that is heterogeneous in length; in this case, more of the detected nucleic acid molecules would serve as candidates. All of the molecules detected or at least all those within a particular size range must be considered. The sum of all molecules with a chain length longer than 240 nucleotides is less than one molecule per infectious unit. We wish to emphasize that a hypothetical, heterogeneous scrapie polynucleotide is an extreme assumption, because there is no precedent for this model in biology. Whatever the structure of a hypothetical, scrapie genome may be, it would have to be constructed of molecules smaller than 140 nt. While this statement applies to all known types of nucleic acids, nucleic acids of an unknown chemistry cannot be excluded.

CONCLUSIONS

Since nucleic acid molecules smaller than about 100 nucleotides have been detected in highly purified prion preparations with a particle-to-infectivity ratio larger than unity, these molecules cannot be excluded as components to infectivity. We might list a few speculations about the functional relevance of the small nucleic acids detected. Possibly, the relationship shown in Fig. 6 simply reflects the decreasing efficiency of nuclease degradation with decreasing size of the nucleic acid. In this case, the nucleic

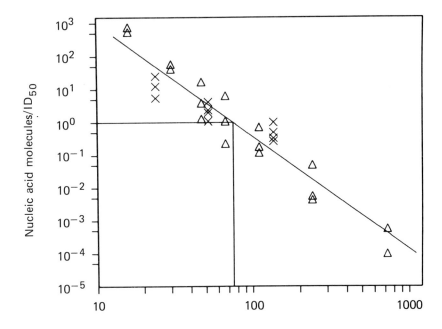

Fig. 6.   Ratio P/l from five independent prion samples. Only for small nucleic acids (< 80 nt) is the ratio P/l is above unity. The calculation is based on the assumption that the hypothetical scrapie-specific nucleic acid is a well defined molecular species among the heterogeneous nucleic acids. Data ($\times$) were taken from Meyer *et al.* (1991), ($\triangle$), from Kellings *et al.* (1992). The straight line is an interpolation in order to determine the average of the length of the nucleic acids at P/l of unity.

acids would be mere impurities in the highly purified preparations without any functional relevance for scrapie infectivity. It is worth noting that prion preparations tend to protect nucleic acids against degradation. If, however, the nucleic acid would be essential for infectivity it is highly improbable that it acts as an mRNA for a peptide. Although small peptide products from ribosomal translation are well documented, they are not known to be translated from an mRNA as small as 100 nucleotides. Instead, the peptides are derived from a larger propeptide by limited proteolysis. Thus, we doubt that such a small polynucleotide would function to encode a protein but rather it might possess some regulatory function. Small regulatory nucleic acids, in most cases RNAs, might act as ribozymes, primers for replication or transcription, antisense RNA for inhibition of transcription, guide RNA for editing or as RNA which forms splicing sites. In any of these scenarios, the genetic information would exert a regulatory influence on the host cell. Such a function might be the molecular basis for different scrapie strains. However, we hasten to state that no experimental evidence exists to support any of the speculations offered above.

While the issue of scrapie strains remains unresolved, it seems most appropriate to apply additional nucleic acid degradation procedures to determine whether any small nucleic acids survive in highly infectious prion preparations.

ACKNOWLEDGMENTS

This work was supported by grants from the National Institute of Health (Washington), the Fonds der Chemischen Industrie and the Minister für Wissenschaft und Forschung von NRW.

REFERENCES

Aiken, J.M., Williamson, J.L. and Marsh, R.F. (1989) Evidence of mitochondrial involvement in scrapie infection. *J. Virol.* **63** 1686–1694.

Aiken, J.M., Williamson, J.L., Borchardt, L.M. and Marsch, R.F. (1990) Presence of mitochondrial D-loop DNA in scrapie-infected brain preparation enriched for the prion protein. *J. Virol.* **64** 3265–3268.

Akowitz, A., Sklaviadis, T., Manuelidis, E.E. and Manuelidis, L. (1990) Nuclease-resistant polyadenylated RNAs of significant size are detected by PCR in highly purified Creutzfeldt–Jakob disease preparations. *Microbiol. Pathog.* **9** 33–45.

Alper, T., Haig, D.A. and Clarke, M.C. (1966) The exceptionally small size of the scrapie agent. *Biochem. Biophys. Res. Commun.* **22** 278–284.

Bellinger-Kawahara, C., Diener, T.O., McKinley, P., Groth, D.F., Smith, D.R. and Prusiner, S.B. (1987) Purified scrapie prions resist inactivation by procedures that hydrolyze, modify, or shear nucleic acids. *Virology* **160** 271–274.

Brown, P., Liberski, P.P., Wolff, A. and Gajdusek, D.C. (1990) Conservation of infectivity in purified fibrillary extracts of scrapie-infected hamster brain after sequential enzymatic digestion and polyacrylamide gel electrophoresis. *Proc. Natl. Acad. Sci. USA* **87** 7240–7244.

Brown, P., Goldfarb, L.G., Brown, W.T., Goldgraber, O., Rubenstein, R., Jascsak, R. J., Guiroy, D.C., Piccardo, P., Boellaard, J.W. and Gajdusek, D.C. (1991) Chemical and molecular genetic study of a large German kindred with Gerstmann-Sträusler–Scheinker syndrome. *Neurology* **41** 375–379.

Castle, B.E., Dees, C., German, T.L. and Marsh, R.F. (1987) Effects of different methods of purification of aggregation of scrapie infectivity. *J. Gen. Virol.* **68** 225–231.

Dees, C., McMillan, B.C., Wade, W.F., German, T.L. and Marsh, R.F. (1985) Characterization of nucleic acids in membrane vesicles from scrapie-infected hamster brain. *J. Virol.* **55** 126–132.

Diener, T.O., McKinley, M.P. and Prusiner, S.B. (1982) Viroids and prions. *Proc. Natl. Acad. Sci. USA* **79** 5220–5224.

Duguid, J.R. and Dinauer, M.C. (1989) Library subtraction of *in vitro* cDNA libraries to identify differentially expressed genes in scrapie infection. *Nucleic Acids Res.* **18** 2789–2792.

Duguid, J.R., Rohwer, R.G. and Seed, B. (1988) Isolation of cDNAs of scrapie modulated RNAs by subtractive hybridization of a cDNA library. *Proc. Natl. Acad. Sci. USA* **85** 5738–5742.

Gabizon, R., McKinley, M.P. and Prusiner, S.B. (1987) Purified prion proteins and scrapie infectivity copartition into liposomes. *Proc. Natl. Acad. Sci. USA* **84** 4017–4021.

Gabizon, R., McKinley, M.P., Groth, D.F., Kenaga, L. and Prusiner, S.B. (1988a) Properties of scrapie prion protein liposomes. *J. Biol. Chem.* **263** 4950–4955.

Gabizon, R., Groth, D.F., McKinley, M.P. and Prusiner, S.B. (1988b) Immunoaffinity purification and neutralization of scrapie prion infectivity. *Proc. Natl. Acad. Sci. USA* **85** 6617–6612.

German, T.L., McMillan, B.C., Castle, B.E., Dees, C., Wade, W.F. and Marsch, R.F. (1985) Comparison of RNA from healthy and scrapie-infected hamster brain. *J. Gen. Virol.* **66** 839–844.

Kellings, K., Meyer, N., Mirenda, C., Prusiner, S.B. and Riesner, D. (1992) Further analysis of nucleic acids in purified scrapie prion preparations by an improved return refocusing gel electrophoresis (RRGE). *J. Gen. Virol.* **73** 1025–1029.

McKinley, M.P., Masiarz, F.R., Prusiner, S.B. (1981) Reversible chemical modification of the scrapie agent. *Science* **214** 1259–1261.

Meyer, N., Rosenbaum, V., Schmidt, B., Gilles, K., Mirenda, C., Groth, D., Prusiner, S.B. and Riesner, D. (1991) Search for a putative scrapie genome in purified prion fractions reveals a paucity of nucleic acids. *J. Gen. Virol.* **72** 37–49.

Murdoch, G.H., Sklaviadis, T., Manuelidis, E.E. and Manuelidis, L. (1990) Potential retroviral RNAs in Creutzfeldt–Jakob disease. *J. Virol.* **64** 1477–1486.

Narang, H.K., Asher, D.M. and Gajdusek, D.C. (1988) Evidence that DNA is present in abnormal tubulofilamentous structures found in scrapie. *Proc. Natl. Acad. Sci. USA* **85** 3575-3579.

Narang, H.K., Caughey, B., Ernst, D., Race, R.E. and Chesebro, B. (1991) Protease sensitivity and nuclease resistance of the scrapie agent propagated *in vitro* in neuroblastoma cells. *J. Virol.* **65** 1031–1034.

Oesch, B., Groth, D.F., Prusiner, S.B. and Weissman, C. (1988) Search for a scrapie-specific nucleic acid: a progress report. In: Bock, G. and Marsh, J. (eds) *Novel Infectious Agents and the Central Nervous System.* Ciba Foundation Symposium, Vol. 135, Wiley, New York, pp. 209–223.

Prusiner, S.B. and McKinley, M.P. (eds) (1987) *Prions: Novel Infectious Pathogens Causing Scrapie and Creutzfeldt–Jakob Disease.* Academic Press, New York.

Prusiner, S.B., Cochran, S.P., Groth, D.F., Downey, D.E., Bowman, K.A. and Martinez, H.M. (1982) Measurement of the scrapie agent using an incubation time interval assay. *Ann. Neurol.* **11** 353–358.

Prusiner, S.B., McKinley, M.P., Bowman, K.A., Bolton, D.C., Bendheim, P.E., Groth, D.F. and Glenner, G.G. (1983) Scrapie prions aggregate to form amyloid-like birefringent rods. *Cell* **35** 349–358.

Sklaviadis, T.K., Manuelidis, L. and Manuelidis, E.E. (1989) Physical properties of the Creutzfeldt–Jakob disease agent. *J. Virol.* **63** 1212–1222.

31

Modified host nucleic acids: a role in scrapie infection?

Judd M. Aiken

ABSTRACT

The search for the identity of the scrapie agent has produced experimental data that are wrought with contradiction. Many lines of evidence argue that a scrapie-specific nucleic acid is a necessary component of the agent. Experimental evidence, however, places a number of constraints on the characteristics of such a scrapie-specific nucleic acid. Irradiation data indicate it is either very small or unusually well protected, while cDNA hybridization studies have failed to identify nucleic acids unique to scrapie-infected tissue. Our efforts to characterize the scrapie agent have led us to propose a mitochondrial involvement in the disease. Of particular interest is the mitochondrial D-loop fragment, a small single-stranded DNA that binds to the control region of the mitochondrial genome. Such a modified host nucleic acid is compatible with the restrictions experimental evidence imposes on the putative scrapie agent nucleic acid.

INTRODUCTION

The supposition that nucleic acid is the heritable material is central to biological dogma. In the field of scrapie research, this is, however, an issue of considerable controversy. Because of the scrapie agent's ability to replicate and mutate, it would seem unreasonable for a nucleic acid not to be part of the aetiologic agent. Experimental evidence, to date, suggests an essential role for a host encoded protein, the prion protein (PrP), and imposes a number of restrictions on the characteristics of a putative nucleic acid component of the scrapie agent. In this paper, the nature of these restrictions will be discussed, a model that is consistent with the experimental evidence will be proposed and an example given of one host encoded nucleic acid that is compatible with that model.

NUCLEIC ACID STUDIES

Both direct and indirect analyses have been used to ascertain whether nucleic acid is a vital component of the scrapie agent. The direct methods have involved either differential hybridization methodologies such as subtraction hybridization (Duguid *et al.*, 1988, 1989; Duguid and Dinauer, 1990) to identify a scrapie-specific nucleic acid or quantitation of nucleic acids in nuclease treated infectious preparations (Meyer *et al.*, 1991). Indirect methods have involved experiments such as nuclease treatment of highly infectious samples (Prusiner, 1982) as well as UV and gamma irradiation experiments (Alper *et al.*, 1967, 1978; Latarjet *et al.*, 1970).

Subtractive hybridization is the most sensitive method available for comparing nucleic acid populations. It has recently been combined with the polymerase chain reaction to provide a very powerful tool for the identification of rare nucleic acids (Duguid and Dinauer, 1990). The analysis of poly A+ RNA isolated from scrapie-infected tissue identified a number of cellular RNAs that were preferentially expressed but no scrapie-specific nucleic acids were identified. These findings suggest that it is unlikely that the scrapie agent is a unique poly A+ RNA species greater than 150 nt in length (the minimum size of the poly A+ RNA used in these studies).   Agent inactivation studies using UV irradiation as well as ionizing radiation made it clear that any essential nucleic acid component of the scrapie agent must be an unusual one. The scrapie agent is much more resistant to UV and ionizing irradiation than are conventional viruses (Alper *et al.*, 1967, 1978; Latarjet *et al.*, 1970; Bellinger-Kawahara *et al.*, 1987). Some researchers have argued that these data support their contention that nucleic acid is not involved in scrapie infectivity (Bellinger-Kawahara *et al.*, 1987) while others have concluded that the scrapie genome responds in a manner consistent with that of small viruses (Rohwer, 1984, 1986).

One final factor that merits consideration is the data that links heritable forms of Creutzfeldt–Jakob disease (CJD) and Gerstmann–Sträussler–Scheinker (GSS) syndrome with mutations in the PrP allele of these individuals (Hsiao *et al.*, 1989; Doh-ura *et al.*, 1989; Owen *et al.*, 1990). The data suggest that individuals with specific PrP mutations will eventually be afflicted with the disease. This would indicate that any putative nucleic acid component of the agent cannot originate from an exogenous viral infection but instead may be provided by the host itself.

MODIFIED HOST NUCLEIC ACID

I would like to propose a concept that involves a nucleic acid and yet is compatible with previous experimental findings. First, I would like to suggest that the scrapie agent nucleic acid is not of exogenous origin (viral or viroid) but rather has sequence identity with some component of the host genome. Given that the putative scrapie agent nucleic acid has strong sequence identity with the host genome, either an RNA or DNA molecule would be compatible with the experimental evidence. Second, since the size of the nucleic acid must be small, we are very likely dealing with a nucleic acid that does not encode a functional polypeptide but rather affects a control region for cellular DNA replication and/or transcription.   It is well established that protein is a necessary component of the scrapie agent. The obvious candidate is PrP. We are, therefore, suggesting that a host encoded protein, probably PrP, in combination with a modified host nucleic acid constitutes the scrapie agent. This model is similar in many aspects with the virino hypothesis (Dickinson and Outram, 1979; Kimberlin, 1982). Common to both theories is the concept of a nucleic acid being protected by a host encoded protein. What is unique to this present proposal is that the nucleic acid component is either an RNA or a DNA molecule having significant sequence identity with some component of the host genome and that this nucleic acid affects cellular function probably at the DNA replication and/or transcription level. This proposal is compatible with nucleic acid data and provides an explanation for the many enigmatic characteristics of the scrapie agent. It raises the possibility, however, that the search for the putative scrapie agent nucleic acid may be exceedingly difficult since the difference between host and agent nucleic acid could be as small as a single nucleotide.   Does such a nucleic acid species exist? Data from our laboratory have suggested a candidate molecule that is consistent with the experimental evidence just described. These studies originated when we identified mitochondrial nucleic acids as preferentially occurring in cytoskeletal preparations purified from scrapie-infected tissue compared with similar preparations isolated from uninfected animals (Aiken et al., 1989).

These results led us to investigate the infectivity levels of mitochondria purified from scrapie-infected tissue. Fractionation and subfractionation of mitochondria was performed by traditional mitochondrial methodologies. Mitochondria were purified from scrapie—infected brain homogenates by banding on sucrose gradients. The outer membrane of mitochondria was removed either by osmotic shock or by digitonin treatment to produce an inner membrane vesicle referred to as the mitoplast. Purified mitochondria were found to contain high infectivity (10^8 LD_{50}) while the mitoplasts showed no loss of infectivity ($10^{8.5}$ LD_{50}; Aiken et al., 1990).

Scrapie infectivity, however, is known to separate into fractions much smaller and less dense than intact mitochondria. Fractions as small as 40S have been shown to contain significant infectivity (Malone et al., 1979; Prusiner et al., 1978, 1979). A standard method of subfractionating intact mitochondria or mitoplasts is to create inner membrane vesicles (submitochondrial particles) by detergent or sonic disruption. These particles retain respiratory chain and energy transducing processes while enzymes located in the mitochondrial matrix and intermembrane space are lost.

Submitochondrial particles were produced by the sonication of gradient purified mitochondria. After centrifugation to remove the mitochondrial outer membranes as well as unruptured mitochondria, submitochondrial particles were pelleted by high speed centrifugation. These particles were then resuspended and infectivity determined by intracerebral inoculation. Similar to the mitoplasts, no loss of infectivity was observed when preparing submitochondrial particles from the mitochondrial fraction (Table 1).

Table 1.   Infectivity levels of mitochondrial fractions

Fraction	LD_{50}
Mitochondria	10^8
Mitoplast	$10^{8.5}$
Submitochondrial particles	10^9

If mitochondrial nucleic acid is involved in scrapie infection, it should be present in nuclease-treated PrP protein enriched preparations having high infectivity. We have tested numerous samples from two such preparations: (i) the prion protein preparation and (ii) the scrapie associated fibril (SAF) preparation. Both preparations contain abnormal rod or fibril-like structures that are composed primarily of a single glycosylated protein, the prion protein (PrP). Both preparations are highly infectious and resistant to mild proteinase K treatment. PrP is not protease resistant in uninfected animals. In addition to partial resistance to protease treatment, infectivity in these preparations is also resistant to micrococcal nuclease, Zn^{2+} hydrolysis and DNase I digestion (Prusiner, 1982).

Standard nucleic acid extraction procedures were used on the PrP-enriched samples including proteinase K digestion followed by phenol:chloroform extraction. Nuclease treated samples were originally analysed by slot blot analysis. These experiments indicated the samples reacted strongly to a mitochondrial DNA probe from the D-loop region of the genome (Aiken et al., 1990). We then investigated whether the D-loop molecules present in the sample were intact by analysis of polymerase chain reaction products. The primers used in the amplification were chosen internal to the D-loop fragment and the expected 550 nt fragment was produced upon amplification. Quantitation using PCR indicated approximately 10^{10} D-loop molecules were present in the 10^9 LD_{50} sample. PrP-enriched fractions prepared by the Hilmert and Diringer method (Hilmert and Diringer, 1984) or by purifying scrapie associated fibrils (Merz et al., 1981) were analysed by Southern blot hybridization. Only the small 550 nt D-loop fragment reacted with the mitochondrial D-loop probe. A signal was not identifiable in the 16 kb region of the gel indicating an absence or limited quantity of mitochondrial genome in the PrP-enriched fractions. These studies, therefore, indicate an enrichment for a specific component of mitochondrial sequences, the D-loop region, in PrP-enriched preparations.

We have found infectious, nuclease treated PrP enriched preparations to contain a specific component of the mitochondrial genome, the D-loop fragment. The D-loop fragment is a single-stranded, heterologously sized DNA molecule. It interacts with a region of the mitochondrial genome (the D-loop region) displacing the parental heavy strand to produce a novel triple-stranded structure. The D-loop region is the control region of the mitochondrial genome containing an initiation site for DNA replication as well as initiation sites for transcription. This may facilitate the introduction and integration of small complementary nucleic acids into the mitochondrial genome. An association between mitochondrial DNA sequences and scrapie infection has also been shown by Narang and associates. Analysis of abnormal tubulofilamentous structures consistently found in spongiform encephalopathies by electron microscopy (Narang *et al.*, 1987) identifies multimeric mitochondrial DNA and single-stranded DNA (approximately 0.49×10^6 daltons) in the scrapie-infected preparations (Narang, 1990). Further analysis confirms the identification of the multimeric DNA as mitochondrial in origin and that there is an increase in multimeric mitochondrial DNA in nucleic acid purified from scrapie-infected hamsters compared with uninfected controls (Narang *et al.*, 1991).

If the scrapie agent nucleic acid resides in the mitochondria, two criteria must be satisfied: (1) infectivity should purify with mitochondria and (2) the nucleic acid should be present in the prion protein enriched preparations. Our data support these predictions; however, the most critical piece of evidence, the discovery of a scrapie-unique nucleic acid, remains elusive.

SUMMARY

Although there is a substantial body of data to support the contention that PrP plays a critical role in scrapie infection, it is premature to conclude that a nucleic acid is not part of the infectious agent. Experimental evidence to date, however, certainly limits the characteristics such a nucleic acid can have. Based upon this experimental evidence, I have proposed a model that hypothesizes the existence of a modified host nucleic acid being a component of the etiologic agent responsible for scrapie.

ACKNOWLEDGMENTS

This research was supported by a National Institutes of Health grant (1 R29 AI29487-01) and by an Alzheimer's Disease Research Grant, a program of the American Health Assistance Foundation, Rockville, Maryland.

REFERENCES

Aiken, J.M., Williamson, J.L. and Marsh, R.F. (1989) Evidence of mitochondrial involvement in scrapie infection. *J. Virol.* **63** 1689–1694.

Aiken, J.M., Williamson, J.L., Borchardt, L.M. and Marsh, R.F. (1990) Presence of mitochondrial D-loop DNA in scrapie-infected brain preparations enriched for the prion protein. *J. Virol.* **64** 3265–3268.

Alper, T., Cramp, W.A., Haig, D.A., and Clarke, M.C..(1967) Does the agent of scrapie replicate without nucleic acid? *Nature (London)* **214** 764–766.

Alper, T., Haig, D.A., and Clarke, M.C. (1978) The scrapie agent: evidence against its dependence for replication on intrinsic nucleic acid. *J. Gen. Virol.* **41** 503–516.

Bellinger-Kawahara, C., Cleaver, J.E., Diener, T.O. and Prusiner, S.B. (1987) Purified scrapie prions resist inactivation by UV irradiation. *J. Virol.* **61** 159–166.

Dickinson, A.G. and Outram, G.W. (1979) The scrapie replication-site hypothesis and its implications for pathogenesis. In: S.B. Prusiner and W.J. Hadlow (eds) *Slow*

Transmissible Diseases of the Nervous System, Vol. 2. Academic Press, New York, pp. 387–406.

Doh-ura, K., Tateishi, J., Sasaki, H., Kitamoto, T. and Sakaki, Y. (1989) Pro→ Leu change at position 102 of prion protein is the most common but not the sole mutation related to Gerstmann–Sträussler syndrome. *Biochem. Biophys. Res. Commun.* **163** 974–979.

Duguid, J.R., and Dinauer, M.C. (1990) Library subtraction of *in vitro* cDNA libraries to identify differentially expressed genes in scrapie infection. *Nucl. Acids Res.* **18** 2789–2792.

Duguid, J.R., Rohwer, R.G. and Seed, B. (1988) Isolation of cDNAs of scrapie-modulated RNAs by subtractive hybridization of a cDNA library. *Proc. Natl. Acad. Sci. USA* **85** 5738–5742.

Duguid, J.R., Bohmont, C.W., Liu, N. and Tourtellotte, W.W. (1989) Changes in brain expression shared by scrapie and Alzheimer disease. *Proc. Natl. Acad. Sci. USA.* **86** 7260–7264.

Hilmert, H., and Diringer, H. (1984) A rapid and efficient method to enrich SAF-protein from scrapie brains of hamsters. *Biosci. Rep.* **4** 165–170.

Hsiao, K., Baker, H.F., Crow, T.J., Poulter, M., Owen, F., Terwilliger, J.D., Westaway, D., Ott, J. and Prusiner, S.B. (1989) Linkage of a prion protein missense variant of Gerstmann–Sträussler syndrome. *Nature* **338** 342–345.

Kimberlin, R.H. (1982) Reflections on the nature of the scrapie agent. *Trends in Biochemical Sciences* **7** 392–394.

Latarjet, R., Muel, B., Haig, D.A., Clarke, M.C. and Alper, T. (1970) Inactivation of the scrapie agent by near monochromatic ultraviolet light. *Nature (London)* **227** 1341–1343.

Malone, T.G., Marsh, R.F., Hanson, R.P. and Semancik, J.S. (1979) Evidence for the low molecular weight nature of scrapie agent. *Nature (London)* **278** 575–576.

Merz, P.A., Somerville, R.A., Wisniewski, H.M. and Iqbal, K. (1981) Abnormal fibrils in scrapie-infected brain. *Acta Neuropathol. (Berlin)* **54** 63–74.

Meyer, N., Rosenbaum, V., Schmidt, B., Gilles, K., Mirenda, C., Groth, D., Prusiner, S.B. and Riesner, D. (1991) Search for a putative scrapie genome in purified prion fractions reveals a paucity of nucleic acids. *J. Gen Virol.* **72** 37–49.

Narang, H.K. (1990) Detection of single-stranded DNA in scrapie-infected brain by electron microscopy. *J. Mol. Biol.* **216** 469–473.

Narang, H.K., Asher, D.M. and Gajdusek, D.C. (1987) Tubulofilaments in negatively stained scrapie-infected brains: relationship to scrapie-associated fibrils. *Proc. Natl. Acad. Sci. USA* **84** 7730–7734.

Narang, H.K., Millar, N.S., Asher, D.M. and Gajdusek, D.C. (1991) Increased multimeric mitochondrial DNA in the brain of scrapie-infected hamsters. *Intervirology* **32** 316–324.

Owen, F., Poulter, M., Shah, T., Collinge, J., Lofthouse, R., Baker, H., Ridley, R., McVey, J. and Crow, T.J. (1990) An in-frame insertion in the prion protein gene in familial Creutzfeldt–Jakob disease. *Mol. Brain Res.* **7** 273–276.

Prusiner, S.B. (1982) Novel proteinaceous infectious particles cause scrapie. *Science* **216** 136–144.

Prusiner, S.B., Hadlow, W.J., Eklund, C.M., Race, R.E., and Cochran, S.P. (1978) Sedimentation characteristics of the scrapie agent from murine spleen and brain. *Biochemistry* **17** 4987–4992.

Rohwer, R.G. (1984) Scrapie infectious agent is virus-like in size and susceptibility to inactivation. *Nature (London)* **308** 658–662.

Prusiner, S.B., Garfin, D.E., Baringer, J.R. and Cochran, S.P. (1979) On the partial purification and apparent hydrophobicity of the scrapie agent. In: S.B. Prusiner and H.J.Hadlow (eds) *Slow Transmissible Diseases of the Nervous System*, Vol. 2. Academic Press, New York, pp. 425–466.

Rohwer, R.G. (1986) Estimation of scrapie nucleic acid MW from standard curves for virus sensitivity to ionizing radiation. *Nature (London)* **320** 381.

32

Cataloguing post-translational modifications of the scrapie prion protein by mass spectrometry

Neil Stahl, Michael A. Baldwin, David Teplow, Leroy Hood, Ron Beavis, Brian Chait, Bradford W. Gibson, Alma L. Burlingame and Stanley B. Prusiner.

ABSTRACT

The only identified component of the scrapie prion is PrP^{Sc}, a glycosylinositol phospholipid (GPI)-linked protein with disease-specific physical properties that distinguish it from the cellular isoform (PrP^C) found in uninfected animals. The result of many studies indicate that PrP^{Sc} is derived from PrP^C by a post-translational event that may reflect a chemical modification, a conformational change, or tight association with other cellular components. We have undertaken the structural determination of Syrian hamster PrP^{Sc} to uncover all post-translational modifications and to resolve whether a chemical difference might distinguish it from the Prp^C. Purified PrP^{Sc} or PrP 27-30 was solubilized, digested with endoproteinase Lys-C and phosphatidylinositol-specific phospholipase C, and the peptides were purified by reverse phase HPLC. Every peptide expected to arise from PrP^{Sc} based on the gene sequence has been identified, and the mass measured by liquid secondary ion, electrospray, or laser desorption mass spectrometry. Additionally, all of PrP 27-30 has been verified by Edman sequencing. We have found that the mass of every peptide matches that predicted from translation of the gene sequence. Quantitative amino acid analysis of the peptides indicates recoveries ranging from 65–100%, asserting that the majority of the protein molecules have been examined. Analysis of the N-terminus of PrP^{Sc} showed no evidence for the uncharacterized arginine modifications that have been reported at positions 25 and 37, suggesting that they are either variably present or labile under acidic conditions. Thus the complete catalogue of post-translational modifications identified to date includes the two N-linked oligosaccharides, removal of the amino-terminal signal sequence and C-terminal GPI signal peptide, addition of the GPI at serine 231 in Syrian hamster PrP^{Sc}, formation of a disulphide between the only two cysteines in the protein, and the presence of a truncated C-terminus ending at glycine 228 in $\sim$15% of the purified molecules. We have also found that the N-terminus of PrP 27-30 is heterogenous; molecules have been identified starting after every histidine or tryptophan between residues 73 and 90, consistent with the preference of proteinase K for cleavage after aromatic residues. We have confirmed this catalogue of post-translational modifications by laser desorption mass spectrometry of the entire PrP^{Sc} molecule after peptide N-glycosidase F digestion to remove the heterogenous N-linked oligosaccharides. This analysis gave a measured peak top of 25 460 Da, which is a close match to masses of 25 185–25 638 predicted for the amino acid sequence and the structures of the GPI anchors that have also been determined. Although detailed comparison with the PrP^C structure remains to be completed, there is no obvious way that the structure of the PrP^{Sc} post-translational modifications could give rise to the unusual physical properties and allow PrP^{Sc} to become a prion component. While the possibility remains that only a small fraction of the PrP^{Sc} molecules contain a scrapie-specific chemical modification, it seems more likely that PrP^{Sc} differs from PrP^C by conformation or the presence of other tightly bound molecules.

INTRODUCTION

Scrapie prions are extraordinary in that there is no evidence suggesting that a nucleic acid component is required for transmission of the disease (Alper *et al.*, 1967; Oesch *et al.*, 1988; Meyer *et al.*, 1991; Prusiner, 1991). The only identified constituent of the prion is PrP^{Sc}, a host-encoded protein with disease-specific physical properties that distinguish it from a normal isoform denoted PrP^C (Oesch *et al.*, 1985; Basler *et al.*, 1986). While many suppositions have been made as to the nature of the prion and its mode of replication, many investigators have settled on the hypothesis that PrP^C and PrP^{Sc} differ by a post-translational event (Basler *et al.*, 1986; Borchelt *et al.*, 1990; Borchelt *et al.*, 1990; Caughey *et al.*, 1991; Palmer *et al.*, 1991). Pulse-chase radiolabelling experiments in scrapie-infected neuroblastoma cells have revealed that PrP^{Sc} acquires the cardinal property of protease resistance post-translationally with a half-time of 2 hours after the beginning of the chase period (Borchelt *et al.*, 1990). It is currently unknown whether this post-translational event represents a chemical modification, a stable conformational change, or tight association with other cellular components (Stahl and Prusiner, 1991). In order to determine whether a chemical modification might distinguish PrP^{Sc} from PrP^C, we have endeavoured to identify and structurally characterize all of the post-translational modifications of PrP^{Sc} to unveil candidates that might play a role in properties or replication of prions.

Numerous post-translational modifications of both PrP^{Sc} and PrP^C are known.

(1) There are two N-linked carbohydrate acceptor sites, both of which are occupied for the majority of PrP^{Sc} and PrP^C molecules (Bolton *et al.*, 1985; Haraguchi *et*

Fig. 1. PrP^{Sc} peptides observed by mass spectrometry. The predicted amino acid sequence of PrP is shown with the two N-linked oligosaccharide sites marked. The largest start site of PrP 27–30 begins on the second line. Inverted triangles indicate potential sites for cleavage by endoproteinase Lys-C, and the resultant peptides are referred to as K1 through K12. Underlined portions signify peptides that have been observed by either liquid secondary ion (LSIMS, unbroken line) or electrospray (ESP, broken line) mass spectrometry. The asterisk marks the amino acid to which the GPI is attached as Ser$_{231}$ (Stahl et al., 1990a). N-terminal and C-terminal regions that are not found in the mature protein embody signal sequences that direct targetting to the endoplasmic reticulum and addition of the GPI respectively.

al., 1989). Structures of the complex N-linked oligosaccharides have been determined for PrP^{Sc} (Endo *et al.*, 1989). However, ablation of the N-linked sites by site-directed mutagenesis does not prevent the formation of protease-resistant PrP^{Sc} upon expression of the mutant protein in scrapie-infected neuroblastoma cells, asserting that the N-linked carbohydrates are not essential (Taraboulos *et al.*, 1990).

(2) Both PrP^{Sc} and PrP^{C} are modified by glycosylinositol phospholipid (GPI) anchors (Stahl *et al.*, 1987). Although both the PrP^{C} and PrP^{Sc} GPI anchors are sensitive to cleavage by phosphatidylinositol-specific phospholipase C (PIPLC) after denaturation (Stahl *et al.*, 1990b), only PrP^{C} and not PrP^{Sc} can be released from cellular membranes by PIPLC (Stahl *et al.*, 1990b; Caughey *et al.*, 1990). Characterization of the PrP^{Sc} GPI revealed novel structures (Baldwin *et al.*, 1990a,b), including the presence of sialic acid (Stahl *et al.*, 1991). Although sialic acid has not previously been identified in GPI anchors, this is not a unique feature of PrP^{Sc} as it was also shown to be a component of the PrP^{C} GPI (Stahl *et al.*, 1992).

(3) A C-terminal peptide is removed from Syrian hamster PrP^{Sc} upon addition of the GPI anchor to Ser$_{231}$ (Stahl *et al.*, 1990a). Although not rigorously proven, the identical migration of the PrP^{C} C-terminal peptide by reverse phase HPLC and capillary electrophoresis suggests that the attachment point of the GPI to PrP^{C} will be the same (Stahl *et al.*, 1992)

(4) Approximately 15% of purified Syrian hamster PrP^{Sc} and PrP 27-30 molecules are truncated to end at Gly$_{228}$, and are thus missing the final three acids and the GPI anchor (Stahl *et al.*, 1990a). Although unproven, it is possible that this product results from cleavage by a dibasic-specific protease and subsequent trimming of the C-terminal arginines by a carboxypeptidase B-like activity. Many cells apparently contain enzyme systems that are capable of processing precursors at these sites (Misumi *et al.*, 1991; Hatsuzawa *et al.*, 1990).

(5) A 22 amino acid N-terminal signal sequence is removed from both PrP^{Sc} (Hope *et al.*, 1988; Turk *et al.*, 1988) and PrP^{C} (Turk *et al.*, 1988) upon targetting of the proteins to the endoplasmic reticulum.

(6) Both PrP^{C} and PrP^{Sc} contain only two cysteines that form a single disulphide (Turk *et al.*, 1988).

(7) Two arginine residues at the Nterminus of PrP^{Sc} may bear unknown modifications that give either new peaks (Hope *et al.*, 1988) or missing cycles (Turk *et al.*, 1988) during Edman degradation. Similar analysis of PrP^{C} also showed a missing residue at the cycle corresponding to the first of these arginines (Turk *et al.*, 1988). The arginine modification of PrP^{Sc} is either variable or labile since arginine is often observed at reasonable recoveries during Edman sequencing (Hope *et al.*, 1986; Bolton *et al.*, 19887: Safar *et al.*, 1990).

Identification and characterization of every post-translational modification on a protein has been simplified by recent advances in mass spectrometry (Burlingame and McCloskey, 1990). We have used electrospray mass spectrometry (ESP) to measure the masses of large peptides and the glycolipid anchor with accuracies approaching 1 in 10 000. Furthermore, the mass of intact proteins can be ascertained

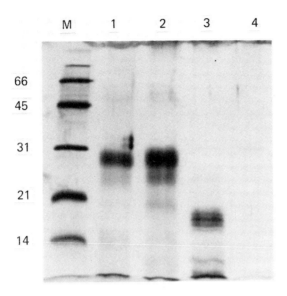

Fig. 2.   Endeoproteinase Lys-C digestion of PrP 27–30. Preparations of infections prion were denatured, solubilized, reduced and carboxymethylated, and digested with endoproteinase Lys-C as described (Stahl *et al.*, 1990a), then analysed by silver staining (Turk *et al.*, 1988) after electrophoresis on 12% acrylamide gels (Laemmli, 1970). 'M' signifies molecular weight markers of the indicated size (BioRad). Lane 1: infectious prions boiled in sample buffer. Lane 1: infectious prions boiled in sample buffer. Lane 2: sample after denaturation in guanidine hydrochloride and carboxymethylation. Lane 3: sample following overnight digestion with endoproteinase Lys-C. Lane 4: endoproteinase Lys-C alone. The prominent triplet in lane 3 represents the glycosylated K8 peptide.

with an accuracy of 1 in 1000 at the 1 pmol level by UV matrix-assisted laser desorption mass spectrometry (Beavis and Chait, 1990). We have used LDMS to measure the mass of PrP^{Sc} as corroboration of the structures determined for individual and the GPI anchor, as well as to exclude the possibility that other modifications might exist that had escaped detection in our analysis of purified peptides.

In this report, we describe the results of our search for post-translational modifications of Syrian hamster PrP^{Sc} and its protease-resistant core PrP 27–30. Mass spectrometry and Edman sequencing were carried out on peptides purified by reverse phase HPLC after digestion of PrP^{Sc} by endoproteinase Lys-C. Quantitative amino acid analysis of the purified fractions indicated a recovery of 65–90% for most peptides. We conclude from this analysis that the majority of PrP^{Sc} protein molecules possess an amino acid sequence that exactly matches that predicted from the gene or cDNA. Furthermore, LDMS analysis of PrP^{Sc} following digestion with peptide N-glycosidase F (PNGase), which removes the heterogenous N-Linked oligosaccharides, revealed masses that matched those expected from the peptide backbone and the known structures of the GPI anchor. Although it is conceivable that only a small fraction of the PrP that purifies with scrapie prion infectivity is chemically modified in a unique fashion, it seems to us more likely that PrP^{Sc} and PrP^{C} differ by conformation and/or the presence of another molecule which is tightly bound.

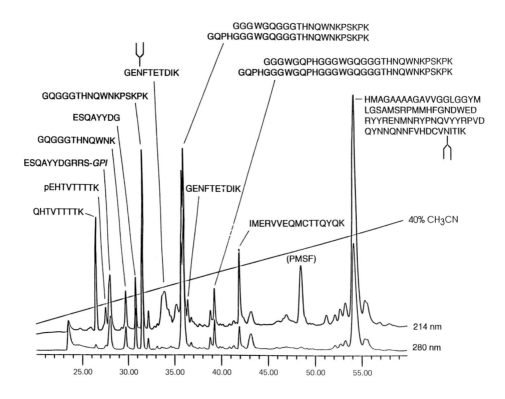

Fig. 3.   Reverse phase HPLC of endo Lys-C digested prion proteins. Preparations of PrP 27–30 were denatured, digested with endo Lys-C, and the resultant peptides purified by reverse phase HPLC as described (Stahl *et al.*, 1990a). The diagonal line indicates the fraction of buffer B (80% acetonitrile) vs. time.

RESULTS

Endoproteinase Lys-C Digestion of PrP 27-30

We used endoproteinase Lys-C to generate peptides from PrP 27-30 and PrP^{Sc} for analysis. The nucleotide sequence for PrP predicts 11 lysines resulting in 12 peptides, which we refer to as K1 through K12 (Fig. 1). Numerous procedures were evaluated for rendering PrP 27-30 soluble and completely susceptible to proteolysis, as detailed elsewhere (Stahl *et al.*, 1992). The best procedure involved denaturation (with concomitant loss of scrapie infectivity) in 6 M guanidine hydrochloride (GdnHCl), cleavage of the disulphide bond with dithiothreitol, alkylation of cysteines with iodoacetate, precipitation from the GdnHCl with 20 volumes of ethanol, and solubilization and digestion of the pellet in 0.1% SDS (Baldwin *et al.*, 1990b). Endoproteinase Lys-C digestion in 0.1% SDS after this procedure routinely gave complete disappearance of PrP 27-30 upon analysis by SDS PAGE with the accompanying appearance of the glycosylated triplet of the K8 peptide migrating with $M_r \sim 16$ kDa (Fig. 2). The sample was then incubated with PIPLC, which proved to be completely effective at cleaving the diradylglycerol from the GPI anchor in a solution containing 0.1% SDS (Stahl *et al.*, 1990a).

Peptides generated by endo Lys-C were purified by reverse phase HPLC following removal of the SDS by precipitation with excess GdnHCl (Shively, 1986) to give chromatograms like that shown in Fig. 3. Although the preparations of PrP^{Sc} were less pure than those for PrP 27-30, disparities in the chromatograms should reveal the elution positions of the corresponding *N*-terminal peptides than differ as the result of the digest with proteinase K.

Edman sequencing of PrP 27-30 and PrP^{Sc} peptides

Each peptide purified by reverse phase HPLC was subjected to gas phase Edman degradation. The large K8 peptide was further subdigested to allow sequencing of the *C*-terminal half. Details of the residues recovered at each cycle are presented elsewhere (Stahl *et al.*, 1992). Every peptide sequenced matched that predicted from the gene or cDNA sequences. As reported previously, mature PrP^{Sc} begins with Lys$_{23}$ after removal of the *N*-terminal signal sequence (Hope *et al.*, 1986; Turk *et al.*, 1988), and ends at Ser$_{231}$ upon addition of the GPI (Stahl *et al.*, 1990a). As discussed below, the tetrapeptide K7 was not retained on the C18 column, and was verified by mass spectrometry following derivatization. Furthermore, the large peptide from the *N*-terminus of PrP^{Sc} containing the octarepeats has not yet been sequenced by the Edman procedure, but has been verified by ESP (see below).

Mass spectrometry of PrP 27-30 and PrP^{Sc} peptides

Table 1 summarizes the recoveries and mass spectrometry of some of the peptides purified by HPLC following digestion of PrP 27-30 and PrP^{Sc} with endo Lys-C. Each of these peptides are discussed below.

Table 1. Peptide HPLC elution times, recoveries and molecular masses

Peptide	Residues	Time	Recovery (%)		Mass[c]	Predicted mass[d]
			GP[a]	Rods[b]		
K4a[e]	90–101	34.0	92	92	1283.5	1283.6
K5–K6[f]	90–106	39.2	~50	~50	1820.0	1821.9
K7	107–110	nr[g]	nd[h]		625.6	652.3[i]
K8	111–185	59.2	79	10–80	8607.8[j]	8608.6
K9	186–194	28.8	92	67	1016.6	1016.5
K10 (CHO)	195–204	36.7	95	101	1154.5[j]	1154.5
K11	205–220	45.7	95	64	2044.9	2045.0
K12	221–231	31.0	92[k]	98[k]	1374.4[l]	1374.6
K12a	221–228	39.2	~15[l]	~15	932.3	932.4

[a] Recovery from gel-purified PrP 27-30.
[b] Recovery from PrP 27-30 in prion rods.
[c] Masses measured by LSIMS, expect for K8 which was measured by ESP. In most cases the quoted masses are the means of several measurements made on different occasions.
[d] Based on monisotopic masses except for K8 which is based on average atomic masses.
[e] Quantitation takes account of all N-terminal species.
[f] Observed only as K4–K5–K6.
[g] Not retained.
[h] Not done.
[i] Mass of derivative (Stultz et al., 1989; Stahl et al., 1992).
[j] After digestion with PNGase, which converts a glycosylated asparagine into an aspartic acid.
[k] Quantitation includes K12a which varied from 10–20%.
[l] After incubation in aqueous hydrofluoric acid, which leaves an ethanolamine attached to the peptide.

K1 and K2

The first two residues of mature PrP[Sc] and PrP[C] are lysines, and thus have been designated K1 and K2. Although this dipeptide was not recovered in a retained fraction by reverse phase HPLC, lysine is always observed in both of the first two cycles by Edman degradation (Hope et al., 1988; Turk et al., 1988). Furthermore, recovery of K3 (see below) would require that endo Lys-C accurately cleave after the preceding lysine, consistent with no modification at this location.

K3–K4–K5–K6

Cleavage by endo Lys-C between lysine and proline is variable and does not always proceed to completion. Although we did not recover K3 alone, we did find peptides corresponding to uncleaved K3–K4. Furthermore, both K5 and K6 begin with prolines, and are also observed attached to K4 as partially uncleaved products. Thus the large N-terminal peptide of PrP[Sc] is recovered after endo Lys-C digestion as a heterogenous product with partial cleavages at both the amino and carboxy termini. The prominent peak eluting near 45 minutes in the chromatogram of PrP[Sc] contains

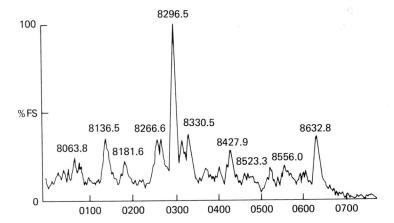

Fig. 4.   Electrospray mass spectrometry of PrP^{Sc} N-terminal peptide. The large peptide corresponding to K3–K4–K5–K6, which elutes near 47 minutes in the HPLC was subjected to electrospray. The electrospray data were transformed to show a scale of molecular weight instead of mass/charge. Both the electrospray spectrum and the electropherogram show one major peak, the mass of which matches that predicted for K3–K4–K5–K6 to within one mass unit.

these peptides. Fig. 4 shows an electrospray mass spectrum of a portion of this peak, and reveals a peptide with a mass predicted for the sum of K3–K4–K5–K6.

In contrast to previous reports (Hope *et al.*, 1988; Turk *et al.*, 1988), we found no evidence for modifications on either R_{25} or R_{37}. Known modifications of arginine include addition of one to three methyl groups, phosphorylation, deamidation, and ADP-ribosylation (Wold, 1981). The occurrence of any of these on the K4 peptide would have been easily detected by the ESP analysis. Furthermore, a tryptic subdigest of these peptides gave a smaller peptide (Fig. 1) containing these arginines whose mass measured by high resolution LSIMS also gave no indication of modifications. Since it is conceivable that amino acid sequencing artifacts gave rise to the missing cycles or novel peaks, the status of these putative modifications remains uncertain.

K5 and K6 were never found as independent peptides; they were only observed when an incomplete endo Lys-C cleavage left them attached to K4. K5 contains Pro_{102}, which when substituted as a Leu in ataxic GSS gives rise to neurodegeneration (Hsiao *et al.*, 1989, 1990). We estimate that $\sim 50\%$ of the K5–K6 peptides was recovered attached to K4. We did not find peptides that were cleaved between K5 and K6. Whether this may be due to the sequence or secondary structure of the peptide, or the presence of a *cis* proline at position 105, is unknown.

PrP 27-30 N-termini

Heterogeneity at the N-terminus was reported in the initial sequencing of PrP 27-30 following purification in the presence of proteinase K (Prusiner *et al.*, 1984). Analysis of various HPLC fractions from the endo Lys-C digest of PrP 27-30 revealed the presence of a ragged N-terminus with five different start sites. We observed peptides starting at Gly_{90}; Gly_{86}, Gly_{82}, Gly_{78}, and Gly_{74}. This can be rationalized by the preference for proteinase K to cleave after aromatic residues; this array of N-termini would arise by cleavage after every tryptophan and histidine in this portion of the protein. Like the K4 peptide discussed above, there are also peptides containing the K5–K6 extension at their C-termini resulting from the resistance of the lysine–proline bonds to endo Lys-C. In fact, only 50% of the PrP 27-30 N-terminal peptides are recovered without K5–K6 attached. It is unlikely that this N-terminal diversity plays a role in scrapie infectivity or amyloid rod formation since full-length PrP^{Sc} is associated with infectivity (Hope *et al.*, 1988; McKinley *et al.*, 1991), and treatment of PrP^{Sc} with other proteases such as pronase or trypsin also gives rise to the formation of rods (McKinley *et al.*, 1991).

K7

K7 was not recovered in any HPLC fraction that was retained on the column, and a synthetic peptide of the predicted sequence eluted in the unretained fractions when coinjected with 1 M GdnHCI. We thus employed a derivitization procedure (Stultz *et al.*, 1989) that could be applied to the unretained fraction in the presence of GdnHCI that would increase the retention of the peptide on the C18 column, as reported elsewhere (Stahl *et al.*, 1992). Reinjection of the derivatized fraction gave a retained peak that showed the correct amino acid composition and matched the predicted mass when measured by LSIMS. Although the recovery of the peptide was ∼50%, it is likely that this results from incomplete derivatization, or the addition of a second derivatization group on the lysine ε-amino group, which would limit detection by LSIMS.

K8

K8 is the largest endo Lys-C product with 75 amino acids, and contains the highly conserved hydrophobic region, the first N-linked glycosylation site, and one of the cysteine residues involved in the disulfide bond. Analysis of HPLC-purified fraction by SDS PAGE reveals that >90% of K8 is glycosylated resulting in a large degree of heterogeneity that gives a difuse triplet upon staining with silver (Fig. 5, insert). Incubation of the fraction with PNGase removes the carbohydrate to give a single band (Fig. 5, insert). ESP analysis of the PNGase-treated peptides shows a peak of the predicted mass (Fig. 5).

To confirm the structure of K8, we performed proteolytic subdigests, followed by HPLC repurification and mass determination by LSIMS (Stahl *et al.*, 1992). Although K8 contains five arginine residues, preferential cleavage at Arg$_{156}$ was observed upon incubation of the peptide with 3 μg/ml trypsin for 3 hours at 37°C. The repurified C-terminal half of this peptide was then further digested with endoproteinase Asp-N. Overnight incubation of K8 in trypsin gave more thorough cleavages and release of the remaining peptide fragments. Peptides of the expected mass and amino acid sequence were recovered from the digest and provided complete coverage of K8 (Stahl *et al.*, 1992). None of the data showed any indication of either post-translational modifications or amino acid substitutions in this region of PrP 27-30.

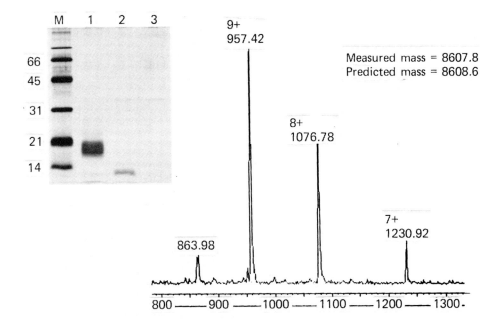

Fig. 5. Electrospray mass spectrometry of K8 following digestion with PNGase. HPLC purified K8 peptide was digested overnight with PNGase in 50 mM Tris hydrochloride (pH 8.5) and 1% β-octyglucoside, then repurified by HPLC and analysed by electrospray mass spectrometry as described (Stahl *et al.*, 1992). The insert shows purified K8 without PNGase (lane 1), K8 after PNGase (lane 2), an equal volume of PNGase alone (lane 3), and molecular weight markers (M). The electrospray spectrum shows four peaks with the indicated mass to charge ratio, which give an overall molecular mass of 8607.8.

K9

Although the K9 peptide contains a run of threonines that form a potential site for *O*-glycosylation, the unmodified peptide was recovered in good yield. We routinely found that endo Lys-C cleavage between K9 and K10 was incomplete, but the purified K9–K10 fragment could be digested to completion after removal of the carbohydrate with PNGase (not shown).

K10

The K10 peptide is the site of the second *N*-linked oligosaccharide acceptor site. Approximately 20% of the peptide was unglycosylated, while the other 80% contained heterogenous *N*-linked carbohydrates. Analysis of the glycosylated peptide by LSIMS gave no peaks unless the sample was first deglycosylated with PNGase (not shown). As predicted, hydrolysis by PNGase converted the glycosylated asparagine to an aspartic acid, which increased the mass of the peptide by 1 unit and allowed confirmation of the exact location of glycosylation by tandem mass spectrometry (Kaur *et al.*, 1990).

K11

The K11 peptide contains the second cysteine involved in the disulfide bond. Most of K11 was recovered in a single peak with the predicted mass, but we occasionally found a portion with an extra 16 or 32 mass units resulting from oxidation of one or both methionines to form methionine sulphoxide. It is unclear whether PrP^{Sc} in the brain is oxidized at these sited, or whether they become oxidized upon purification. Also, it is possible that the treatment of PrP with DTT preceding carboxymethylation could partially reverse the oxidation to give methionine. We also observed that a significant fraction of K11 tended to bleed slowly off the column following the main peak, which could explain the relatively low recovery of the peptide observed in some experiments. The structure of the tailing K11 was identical to that eluting in the main peak as judged by Edman sequencing, migration on CE, and LSIMS analysis.

K12

The characterization of the *C*-terminal peptide K12, which revealed attachment of ethanolamine to the α-carboxyl group of Ser$_{231}$, has been published (Stahl *et al.*, 1990a). Electrospray mass spectrometry of K12-GPI gave masses corresponding to those for the unmodified peptide linked to the six structures of the GPI glycan that were independently determined (Baldwin *et al.*, 1990a; Stahl *et al.*, 1991, 1992). This indicates that there are no other modifications of the peptide.

We previously found that $\sim 15\%$ of the *C*-terminus was recovered in a truncated form ending at Gly$_{228}$ (Stahl *et al.*, 1990a), and is thus missing the final three amino acids and the GPI. Although unproven, it is possible that this post-translational modification might occur through cleavage following the two arginines at position 230 by a dibasic-specific protease, followed by removal of the *C*-terminal arginines by a carboxypeptidase B-like activity (Hatsuzawa *et al.*, 1990; Stahl *et al.*, 1990; Misumi *et al.*, 1991). Soluble forms of PrP^{C} exist (Caughey *et al.*, 1988; Stahl *et al.*, 1990c), and are missing most or all of the GPI anchor (D. Borchelt, M. Rogers, N. Stahl and S. B. Prusiner, submitted for publication).

LDMS of PrP^{Sc}

Laser desorption mass spectrometry of PrP^{Sc} was performed to verify the peptide data and to ensure that no significant chemical modifcations were undetected. The technique consists of dissolving the protein at ~ 1 pmol/μl in a detergent-free buffer containing the UV absorbing compound sinapinic acid (Beavis and Chait, 1990). The hydrophobic nature of PrP^{Sc} required the use of 70% formic acid for solubilization; attempts with other solvent systems or additional solutes were unsuccessful. Reduced and carboxymethylated PrP^{Sc} was digested with PNGase to remove the *N*-linked oligosaccharides, and then precipitated with 20 volumes of ethanol and resuspended in 70% formic acid. The LDMS spectrum shows a somewhat broad peak centred at 25 526 mass units (Fig. 6). The theoretical mass of PrP^{Sc} was calculated from the mass of the amino acid backbone from Lys$_{23}$ to Ser$_{231}$ (23090), and the known structures of four predominant glycoforms of the GPI anchor (Stahl *et al.*, 1992). The latter values required an assumption as to the nature of the second lipid moiety that is attached to the GPI along with stearic acid (Stahl *et al.*, 1987), which we assumed

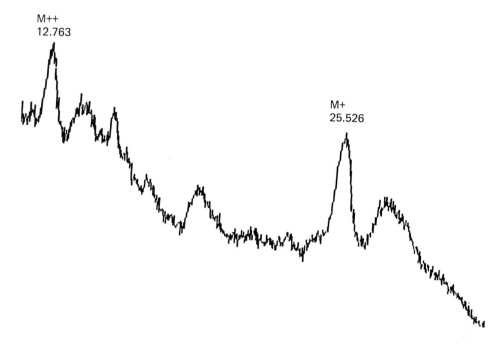

Fig. 6.   Laser desorption mass spectrometry of PrP^{Sc} following digestion with PNGase. The protein was precipitated with cold ethanol and separated by centrifugation, then dissolved in 70% formic acid containing 10 mg/μl sinapinic acid. An aliquot of 0.5 μl was applied to the laser target and air dried. The sample was irradiated with pulsed laser radiation of wavelength 355 nm, giving ions which were analyzed in a 2 m linear flight tube. Multiple single-shot spectra (50–100) were accumulated to produce the final spectrum. The y-axis of the spectrum shown is linear with respect to time, which is proportional to the square root of mass. Mass calibration was carried out in a separate analysis using myoglobin as an internal standard (data not shown).

was an 18 carbon ether-linked lipid. This calculation gave overall values of 25 185, 25 347, 25 476, and 25 638, which closely match the observed value. The mass estimate derived by LDMS represents an average of the different GPI glycoforms, which cannot be resolved by this method. Allowing for peak tailing due to the formation of matrix adducts, the weighted average of the theoretical masses gives a broad peak with a maximum in the region 25 350–25 450. We therefore conclude that there is at most 50–150 mass units difference between the actual and theoretical values.

It should be emphasized that we might not detect a substantial modification that occurs on only 10% of the protein molecules, which might not be clearly resolved from the main peak. The broad peak centred near mass 28 000 in the spectrum most likely results from the singly N-glycosylated protein following incomplete digest of PrP^{Sc} by PNGase, which can be observed by silver staining of the preparation after SDS PAGE.

DISCUSSION

Much data argue persuasively that PrP^{Sc}, or its protease-resistant core PrP 27-30, constitutes an important component of the scrapie prion (Prusiner, 1991). Furthermore, there is excellent evidence that PrP^{Sc} acquires its unusual properties of protease resistance and aggregation as the result of a post-translational event (Borchelt et al., 1990; Taraboulos et al., 1990). This event could represent a chemical modification to PrP, a conformational change, or association with other cellular components. It is likely that determining the difference between PrP^{Sc} and PrP^{C}, the normal isoform found in uninfected brain tissue, will lead to a better understanding of how prions replicate. For this reason we have determined the chemical structure of PrP^{Sc} to reveal any unusual features that might confer upon PrP^{Sc} its unique physical properties and the capacity to become a prion component. We summarize here the results of the Edman sequencing and mass spectrometric mapping of all peptides and post-translational chemical modification of PrP^{Sc}.

The results of this investigation confirm that the amino acid sequence of PrP 27-30 exactly matches that predicted from the gene (Basler et al., 1986) and cDNA (Oesch et al., 1985) sequences. Furthermore, the mass measured for every peptide derived from PrP^{Sc} also matches that predicted from the gene sequence once the known post-translational modifications are taken into account (Fig. 1, Table 1). High resolution LSIMS analysis was carried out on the peptides encompassing PrP 27-30, which gives a mass accuracy within 0.3 mass units. Thus any post-translational modification or amino acid substitution that resulted in a difference of a single mass unit would have been detected. The long N-terminal endo Lys-C peptide of PrP^{Sc} containing the octapeptide repeats (K3–K4–K5–K6) was analyzed by ESP and gave a measured mass that differed from the theoretical mass by only 1 unit, which is within the error of the technique at this mass range.

The mass of intact PrP^{Sc} measured by laser desorption mass spectrometry verifies the results of the peptide mass spectrometry. After removal of the N-linked carbohydrate with PNGase, the measured mass of PrP^{Sc} matches that predicted from the amino acid sequence and the known structures of the GPI anchor within 50–150 mass units (Fig. 6). The exact identity of the lipid attached at the GPI is not known, and therefore we assumed that an 18 carbon alkyl group was ether linked at the glycerol along with stearic acid (Stahl et al., 1987). The presence of a 24 carbon alkyl chain would eliminate the difference between the observed and calculated values. This result indicates that there is no chemical modification greater than 100 mass units that is present on the majority of PrP^{Sc} molecules and has gone undetected in the mass spectrometric survey of the endo Lys-C peptides.

One caveat of this analysis is that it is possible that a chemical modification existing on only 10% of the protein molecules could go undetected. Since the particle:infectivity ratio of scrapie prions is 10^4–10^5 molecules of PrP^{Sc} per ID$_{50}$, it is formally possible that only a small percentage of PrP 27-30 actually exhibits a chemical modification that causes the disease. This seems unlikely, however, since the physical properties of all PrP molecules that copurify with scrapie infectivity are altered to give both protease resistance and aggregation. A second caveat is that the putative chemical modifications that distinguish PrP^C from PrP^{Sc} might be labile under the conditions used to purify PrP^{Sc} or PrP 27-30. Although the extreme resistance of scrapie prion infectivity to inactivation argues against this proposition, it remains as a formal possibility.

The absence of any detectable chemical group that distinguishes PrP^{Sc} from PrP^C has profound implications for the mechanism of prion replication. Familial cases of CJD and GSS in humans are associated with point mutations or insertions of the PrP open reading frame (Hsiao *et al.*, 1989, 1991a, b; Goldgaber *et al.*, 1989; Owen *et al.*, 1989). Furthermore, transgenic mice expressing MoPrP with a mutation corresponding to the Leu_{102} GSS mutation become spontaneously ill with a neurodegenerative disease, and exhibit *de novo* synthesis of transmissible prions in their brain tissue (Hsiao *et al.*, 1990). As a corollary to these findings, it has been hypothesized that cases of sporadic CJD might result from somatic mutations in the PrP gene (Prusiner *et al.*, 1991; Palmer *et al.*, 1991). The results of our study argue that experimental scrapie does not entail an analogous mechanism involving RNA editing or other phenomena that might alter the PrP amino acid sequence.

We also find no evidence in favour of a post-translational mechanism that features chemical modification in the synthesis of PrP^{Sc}. Although the exact chemical structure of PrP^C is not yet known, the types and structures of the PrP^{Sc} post-translational modifications (see Introduction) do not offer any obvious candidates that could account for the altered physical properties of PrP^{Sc}. Experiments in scrapie-infected cell cultures indicate that PrP can acquire protease resistance in the absence of the N-linked carbohydrate (Taraboulos *et al.*, 1990). Although the PrP^{Sc} GPI anchor has a novel structure containing sialic acid, this monosaccharide is also found on the PrP^C GPI (Stahl *et al.*, 1992). Unless subtle differences in the GPI glycosidic linkages or lipid distinguish PrP^C and PrP^{Sc}, it is likely that the distinct properties of PrP^{Sc} are not the result of a chemical modification found on a majority of the protein molecules. Analysis of PrP^C by laser desorption mass spectrometry should reveal whether there really are any mass differences. If the conclusion is upheld that PrP^C and PrP^{Sc} do not differ chemically, then it is likely that the two isoforms must differ only in conformation or by tight binding to other cellular components (Prusiner, 1991). A major unanswered question is whether other molecules are required in addition to PrP^{Sc} to form a prion. Although analysis of infectious prions by SDS PAGE reveals only PrP 27-30 upon silver staining, we have identified several other molecules that copurify in every preparation of prions that have been examined (Stahl *et al.*, 1992). Although none of these consists of nucleic acid, one is an aggregating compound that contains covalently bound fatty acids. It remains to be determined whether these molecules represent contaminants or additional prion components that might bond to PrP 27-30 and play an essential role in prion replication or pathogenesis.

An important fact that any viable prion model must explain is the existence of prion 'strains' or distinct isolates that have unique heritable characteristics with respect to incubation time and neuropathology (Bruce and Dickinson, 1987; Kimberlin *et al.*, 1987; Carp and Callahan, 1991; Weissmann, 1991). The extent of prion diversity giving rise to these distinct isolates is unknown. Whether these isolates result from distinct conformations of PrP^{Sc} or result from the binding of a second putative

prion component remains to be established. A recent proposal involves the participation of a small, non-essential cellular nucleic acid that influences the phenotype of
the scrapie symptoms (Weissman, 1991); this theory is currently being experimentally
tested. Whether the other molecules that copurify with PrP 27-30 could also impart
heritable characteristics to a prion is an intriguing possibility.

REFERENCES

Alper, T., Cramp, W.A., Haig, D.A. and Clarke, M.C. (1967) Does the agent of scrapie
  replicate without nucleic acid? *Nature (London)* **214** 764–766.
Baldwin, M.A., Stahl, N., Reinders, L.G., Gibson, B.W., Prusiner, S.B. and Burlingame,
  A.L. (1990a) Permethylation and tandem mass spectrometry of oligosaccharides
  having free hexosamine: analysis of the glycoinositol phospholipid anchor glycan
  from the scrapie prion protein. *Anal. Biochem.* **191** 174–182.
Baldwin, M.A., Stahl, N., Burlingame, A.L. and Prusiner, S.B. (1990b) Structure
  determination of glycoinositol phospholipid anchors by permethylation and tandem
  mass spectrometry. *Methods: Companion Methods Enzymol.* **1** 306–314.
Basler, K., Oesch, B., Scott, M., Westaway, D., Walchli, M., Groth, D.F.,McKinley,
  M.P., Prusiner, S.B. and Weissmann, C. (1986) Scrapie and cellular PrP isoforms
  are encoded by the same chromosomal gene. *Cell* **46** 417–428.
Beavis, R.C. and Chait, B. (1990) High accuracy molecular mass determinations of
  proteins using matrix assisted laser desorption mass spectrometry. *Anal. Chem.* **62**
  1836–1840.
Bolton, D.C., Mayer, R.K. and Prusiner, S.B. (1985) Scrapie 27-30 is a sialoglycoprotein. *J. Virol* **53** 596–606.
Bolton, D.C., Bendheim, P.E., Marmorstein, A.D. and Potempska, A. (1987) Isolation
  and structure studies of the intact scrapie agent protein. *Arch. Biochem. Biophys.*
  258 579–590.
Borchelt, D.R., Scott, M., Taraboulos, A., Stahl, N. and Prusiner, S.B. (1990) Scrapie
  and cellular prion protein differ in their kinetics of synthesis and topology in
  cultured cells. *J. Cell Biol.* **110** 743–752.
Bruce, M.E. and Dickinson, A.G. (1987) Biological evidence that the scrapie agent
  has an independent genome. *J. Gen. Virol.* **68** 79–89.
Burlingame, A.L. and McCloskey, J.A. (eds) (1990) *Biological Mass Spectrometry.*
  Elsevier, New York.
Carp, R.I. and Callahan, S.M. (1991) Variation in the characteristics of 10 mouse-
  passaged scrapie lines derived from five scrapie-positive sheep. *J. Gen. Virol.* **72**
  293–298.
Caughey, B.W., Race, R.E., Vogel, M., Buchmeir, M.J. and Chesebro, B. (1988) *In
  vitro* expression in eukaryotic cells of a prion protein gene cloned from scrapie-
  infected mouse brain. *Pro. Natl. Acad. Sci. USA* **85** 4657–4661.
Caughey, B., Neary, K., Butler, R., Ernst, D., Perry, L., Chesebro, B. and Race, R.
  (1990) Normal and scrapie-associated forms of prion protein differ in their
  sensitivities to phospholipase and proteases in intact neuroblastoma cells. *J. Virol.*
  64 1093–1101.

Caughey, B.W., Dong, A., Bhat, K.S., Ernst, D., Hayes, S.F. and Caughey, W.S. (1991) Secondary structure analysis of the scrapie-associated protein PrP 27-30 in water by infrared spectroscopy. *Biochemistry* **30** 7672–7680.

Endo, T., Groth, D., Prusiner, S.B. and Kobata, A. (1989) Diversity of oligosaccharide structures linked to asparagines of the scrapie prion protein. *Biochemistry* **28** 8380–8388.

Goldgaber, D., Goldfarb, L.G., Brown, P., Asher, D.M., Brown, W.T., Lin, S., Teener, J.W., Feinstone, S.M., Rubenstein, R., Kascsak, R.J., Boellaard, J.W. and Gajdusek, D.C. (1989) Mutations in familial Creutzfeldt–Jakob disease and Gerstmann–Sträussler-Scheinker's syndrome. *Exp. Neurol.* **106** 204–206.

Haraguchi, T., Fisher, S., Olofsson, S., Endo, T., Groth, D., Tarentino, A., Borchelt, D., Teplow, D., Hood, L., Burlingame, A., Lycke, E., Kobata, A. and Prusiner, S.B. (1989) Asparagine-linked glycosylation of the scrapie and cellular prion proteins. *Arch. Biochem. Biophys.* **274** 1–13.

Hatsuzawa, K., Hosaka, M., Nakagawa, T., Nagase, M., Shoda, A., Murakami, K. and Nakayama, K. (1990) Structure and expression of mouse furin, a yeast Kex2-related protease. Lack of processing of coexpressed prorenin in GH4C1 cells. *J. Biol. Chem.* **265** 22075–22078.

Hope, J., Morton, L.J.D., Farquhar, C.F., Multhaup, G. and Beyreuther, K. (1986) The major polypeptide of scrapie-associated fibrils (SAF) has the same size, charge distribution and *N*-terminal protein sequence as predicted for the normal brain protein (PrP). *EMBO J.* **5** 2591–2597.

Hope, J., Multhaup, G., Reekie, L.J.D., Kimberlin, R.H. and Beyreuther, K. (1988) Molecular pathology of scrapie-associated fibril protein (PrP) in mouse brain affected by the ME7 strain of scrapie. *Eur. J. Biochem.* **172** 271–277.

Hsiao, K., Baker, H.F., Crow, T.J., Poulter, M., Owen, F., Terwilliger, J.D., Westaway, D., Ott, J. and Prusiner, S.B. (1989) Linkage of a prion protein missense variant to Gerstmann–Sträussler syndrome. *Nature (London)* **338** 342–345.

Hsiao, K.K., Scott, M., Foster, D., Groth, D.F., DeArmond, S.J. and Prusiner, S.B. (1990) Spontaneous neurodegeneration in transgenic mice with mutant prion protein. *Science* **250** 1587–1590.

Hsiao, K.K., Cass, C., Schellenberg, G., Bird, T., Devine-Gage, E., Wisniewski, H. and Prusiner, S.B. (1991a) A prion protein variant in a family with the telencephalic form of Gerstmann–Sträussler–Scheinker syndrome. *Neurology* **41** 681–684.

Hsiao, K., Meiner, Z., Kahana, E., Cass, C., Kahana, I., Avrahami, D., Scarlato, G., Abramsky, O., Prusiner, S.B. and Gabizon, R. (1991b) Mutation of the prion protein in Libyan jews with Creutzfeldt–Jakob disease. *N. Engl. J. Med.* **324** 1091–1097.

Kaur, S., Medzihradszky, K.F., Yu, Z., Baldwin, M.A., Gillece-Castro, B.L., Walls, F.C., Gibson, B.W. and Burlingame, A.L. (1990) Strategies for protein sequencing and structural characterization by mass spectrometry. In: Burlingame, A.L. and McCloskey, J.A. (eds) *Biological Mass Spectrometry*. Elsevier, New York, pp. 285–313.

Laemmli, U.K. (1970) Cleavage of structural proteins during the assembly of the head of bacteriophage T-4 *Nature (London)* **227** 680–685.

McKinley, M.P., Meyer, R.K., Kenaga, L., Rahbar, F., Cotter, R., Serban, A. and

Prusiner, S.B. (1991) Scrapie prion rod formation *in vitro* requires both detergent extraction and limited proteolysis. *J. Virol.* **65** 1440–1449.

Meyer, N., Rosenbaum, V., Schmidt, B., Gilles, K., Mirenda, C., Groth, D., Prusiner, S.B. and Riesner, D. (1991) Search for a putative scrapie genome in purified prion fractions reveals a paucity of nucleic acids. *J. Gen. Virol.* **72** 37–49.

Milner, J. and Medcalf, E.A. (1991) Cotranslation of activated mutant p53 with wild type drives the wild-type p53 protein into the mutant conformation. *Cell* **65** 765–774.

Misumi, Y., Oda, K., Fujiwara, T., Takami, N., Tashiro, K. and Ikehara, Y. (1991) Functional expression of furin demonstrating its intracellular localization and endoprotease activity for processing of proalbumin and complement pro-C3. *J. Biol. Chem.* **266** 16954–16959.

Oesch, B., Westaway, D., Walchli, M., McKinley, M.P., Kent, S.B.H., Aebersold, R., Barry, R.A., Tempst, P., Teplow, D.B., Hood, L.E., Prusiner, S.B. and Weissmann, C. (1985) A cellular gene encodes scrapie PrP 27-30 protein. *Cell* **40** 735–746.

Oesch, B., Groth, D.F., Prusiner, S.B. and Weissmann, C. (1988) Search for a scrapie-specific nucleic acid: a progress report. In: Bock, G. and March, J. (eds) *Novel Infectious Agents and the Central Nervous System*, Ciba Foundation Symposium 135, Wiley, Chichester, pp. 209–223.

Owen, F., Poulter, M., Lofthouse, R., Collinge, J., Crow, T.J., Risby, D., Baker, H.F., Ridley, R.M., Hsiao, K. and Prusiner, S.B. (1989) Insertion in prion protein gene in familial Creutzfeldt–Jakob disease. *Lancet* **1** 51–52.

Palmer, M.S., Dryden, A.J., Hughes, J.T. and Collinge, J. (1991) Homozygous prion protein genotype predisposes to sporadic Creutzfeldt–Jakob disease. *Nature (London)* **352** 340–342.

Prusiner, S.B. (1991) Molecular biology of prion diseases. *Science* **252** 1515–1522.

Prusiner, S.B., Groth, D.F., Bolton, D.C., Kent, S.B. and Hood, L.E. (1984) Purification and structural studies of a major scrapie prion protein. *Cell* **38** 127–134.

Safar, J., Wang, W., Padgett, M.P., Ceroni, M., Piccardo, P., Zopf, D., Gajdusek, D.C. and Gibbs, C.J., Jr (1990) Molecular mass, biochemical composition, and physicochemical behavior of the infectious form of the scrapie precursor protein monomer. *Proc. Natl. Acad. Sci. USA* **87** 6373–6377.

Shively, J.E. (ed.) (1968) *Methods of Protein Microcharacterization: a Practical Handbook*. Humana Press, Clifton, NJ, pp. 41–87.

Stahl, N. and Prusiner, S.B. (1991) Prions and prion proteins. *FASEB J.* **5** 2799–2807.

Stahl, N. Borchelt, D.R., Hsiao, K. and Prusiner, S.B. (1987) Scrapie prion protein contains a phosphatidylinositol glycolipid. *Cell* **51** 229–240.

Stahl, N., Baldwin, M.A., Burlingame, A.L. and Prusiner, S.B. (1990a) Identification of glycoinositol phospholipid linked and truncated forms of the scrapie prion protein *Biochemistry* **29** 8879–8884.

Stahl, N., Borchelt, D.R. and Prusiner, S.B. (1990b) Differential release of cellular and scrapie prion proteins from cellular membrane by phosphatidylinositol-specific phospholipase C. *Biochemistry* **29** 5405–5412.

Stahl, N., Borchelt, D.R. and Prusiner, S.B. (1990c) Glycolipid anchors of the cellular and scrapie prion proteins. In: Turner, A.J. (ed.) *Molecular and Cell Biology of Membrane Proteins*. Ellis Horwood, Chichester pp. 189–216.

Stahl, N., Baldwin, M.A. and Prusiner, S.B. (1991) Electrospray mass spectrometry of the glycosylinositol phospholipid of the scrapie prion protein. *Cell Biol. Int. Rep.* **15** 853–862.

Stahl, N., Baldwin, M.A., Hecker, R., Pan, K.-M., Burlingame, A.L. and Prusiner, S.B. (1992) Glycoinositol phospholipid anchors of the scrapie and cellular prion proteins contain sialic acid. *Biochemistry* **31**: 5043–5053.

Stultz, J.T., Halualani, R. and Wetzel, R. (1989) Amino terminal derivatization of peptides yields improved CAD speactra. In: *Proc. 37th Annual Conference on Mass Spectrometry and Allied Topics, Miami Beach, Florida.* pp. 856–857.

Taraboulos, A., Rogers, M., Borchelt, D.R., McKinley, M.P., Scott, M., Serban, D. and Prusiner, S.B. (1990) Acquisition of protease resistance by prion proteins in scrapie-infected cells does not require asparagine-linked glycosylation. *Proc. Natl. Acad. Sci. USA* **87** 8262–8266.

Turk, E., Teplow, S.B., Hood, L.E. and Prusiner, S.B. (1988) Purification and properties of the cellular and scrapie hamster prion proteins. *Eur. J. Biochem.* **176** 21–30.

Weissman, C. (1991) A 'unified theory' of prion propagation. *Nature (London)* **352** 679–683.

Wold, F. (1981) *In vivo* chemical modification of proteins: post-translational modification. *Annu. Rev. Biochem.* **50** 783–814.

33

Glycosylinositol phospholipid anchors of prion proteins

Michael A. Baldwin, Neil Stahl, Rofl Hecker, Keh-Ming Pan, Alma L. Burlingame and Stanley B. Prusiner

ABSTRACT

The only identified component of the scrapie prion is PrP^{Sc}, a glycosylinositol phospholipid (GPI)-linked protein that is derived from the cellular isoform (PrP^C) by an as-yet unknown post-translational event. Analysis of the PrP^{Sc} GPI revealed six different glycoforms, three of which are unprecedented. Two of the glycoforms contain N-acetylneuraminic acid (Sia), which has not been previously reported as a component of any GPI. The largest form of the GPI is proposed to have a glycan core consisting of Manα–Manα–Man–(Sia–Gal–GaINAc–)Man–GlcN–Ino. Identical PrP^{Sc} GPI structures were found for two distinct isolates or 'strains' of prions which specify different incubation times, neuropathology, and PrP^{Sc} distribution in brains of Syrian hamsters. Limited analysis of the PrP^C GPI reveals that it also has sialylated glycoforms, arguing that the presence of this monosaccharide does not distinguish PrP^C from PrP^{Sc}.

INTRODUCTION

Research reports from the last decade have substantiated the central role of the prion protein (PrP^{Sc}) in the pathogens causing scrapie and other transmissible neurodegenerative disorders (Bolton et al., 1982; Gabizon et al., 1988; Hsiao et al., 1989, 1990; McKinley et al., 1983; Prusiner, 1982, 1991; Prusiner et al., 1990; Scott, et al., 1989). A crucial goal in acheiving an understanding of the causes of these diseases is to establish whether there are chemical differences between PrP^{Sc} and the normal cellular protein (PrP^{C}) (Stahl and Prusiner, 1991). The two isoforms have different physical properties: PrP^{C} is soluble in detergents and is sensitive to digestion with proteases (Meyer *et al.*, 1986; Oesch *et al.*, 1985), while PrP^{Sc} aggregates in the presence of detergent and gives rise to a resistant core called PrP 27-30 through loss of the *N*-terminus upon limited proteolysis (McKinley *et al.*, 1991). Experiments in scrapie-infected neuroblastoma cells reveal that PrP^{Sc} is formed with a half-time of 2 h from a protease-sensitive precursor by a post-translational event (Borchelt *et al.*, 1990). This post-translational event could be a chemical modification, a conformational change, or tight association with other cellular components. Although numerous post-translational chemical modifications of PrP^{Sc} have been found (Bolton *et al.*, 1985; Endo *et al.*, 1989; Hope *et al.*, 1988; Stahl *et al.*, 1990a, 1987; Stahl and Prusiner, 1991; Turk, *et al.*, 1988), they all appear to be present in PrP^{C} as well. While only detailed comparison of the chemical structure of PrP^{Sc} and PrP^{C} will reveal any differences, the structures known for the PrP^{Sc} modifications do not appear to be unique to PrP and there is no obvious candidate that would give altered physical properties and make PrP^{Sc} a prion component.

PrP^{Sc} was shown to be modified by the presence of a glycosylinositol phospholipid (GPI) anchor which replaces a hydrophobic peptide at the *C*-terminus (Stahl *et al.*, 1987), the point of attachment in Syrian hamster PrP^{Sc} being at Ser-231 (Stahl 1990a). That PrP^{C} is also GPI anchored is demonstrated by its release from cellular membranes upon incubation with the enzyme phosphatidylinositol-specific phospholipase C (PIPLC). By contrast, PrP^{Sc} is not released from cells by PIPLC, despite its sensitivity to this enzyme after denaturation (Caughey *et al.*, 1990; Stahl *et al.*, 1990b). It is unknown whether the exact structures of the GPIs differ and could thus play a role in the pathogenesis or transmission of scrapie, but the known differences in the cellular locations of these different forms of the protein makes the GPI an interesting candidate for study. We have therefore undertaken structural analysis of the PrP 27-30 GPI and established that it contains some compositional elements never previously reported for any other GPI, including the presence of sialic acid. Our studies on the PrP GPIs have also focused on two distinct Syrian hamster scrapie 'strains' (Bruce and Dickinson, 1987; Carp and Callahan, 1991; Fraser and Dickinson, 1968; Hecker *et al.*, 1992). No significant differences have been found, indicating that scrapie incubation times are not GPI related. Preliminary results on PrP^{C} also show the presence of sialic acid although purified PrP^{C} is not yet available in sufficient amounts to permit complete characterization.

RESULTS AND DISCUSSION

The experimental procedures used in this study have been published elsewhere in detail (Baldwin *et al.*, 1990a, b; Stahl *et al.*, 1990a, 1991, 1992). PrP 27-30 was derived from Syrian hamster PrP^{Sc} by limited proteolysis with proteinase K. The data presented here were derived from scrapie strain Sc237 (Scott *et al.*, 1989) having a characteristic incubation time of about 75 days after intracerebral inoculation, but

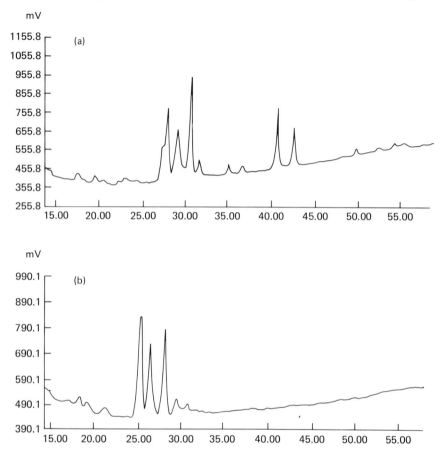

Fig. 1.   Dionex HPAE chromatography of PrP 27–30 GPI. (A) elution of the GPI; (B) chromatogram observed when the sample is predigested with neuraminidase. The horizontal axis is time in minutes.

parallel studies using a longer incubation time strain 139H (Hecker *et al.*, 1992) revealed no differences.

Neuraminidase sensitivity
Enzymatic digestion of PrP 27-30 with endoproteinase Lys-C and PIPLC followed by reverse-phase HPLC separation of the resulting peptides together with amino acid analysis and analysis by capillary electrophoresis (CE) were employed to isolate and purify the *C*-terminal peptide GPI (Stahl *et al.*, 1990a). After pronase treatment to remove all but the *C*-terminal amino acid, heterogeneous anchor structures were observed by Dionex high pH anion exchange (HPAE) chromatography (isocratic NaOH, linear gradient of sodium acetate), and sensitivity of the GPI to neuraminidase was also established by the same technique (Fig. 1(A) and 1(B). Figure 1(A) shows at least five distinct species whereas neuraminidase removed the two peaks eluting at

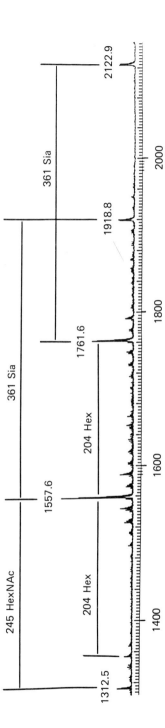

Fig. 2. LSIMS analysis of permethylated GPI glycans. The difference between the various peaks with indicated masses corresponds to the permethylated saccharides: sialic acid (Sia), hexose (Hex), or N-acetylhexosamine (HexNAc).

40.5 and 42.5 min, indicating the presence of sialic acid acid, an unprecedented finding for a GPI. This was confirmed by CE analysis of the intact C-terminal peptide GPI with and without neuraminidase treatment (see below). The neuraminidase-released monosaccharide was also analysed by HPAE (isocratic NaOH: sodium acetate 2:1) and was shown to co-elute with an authentic sample of N-acetylneuraminic acid (data not shown).

The structure of the glycan

The glycan core was released by treatment with 50% aqueous HF (Stahl *et al.*, 1990a). Permethylation, liquid secondary ion mass spectrometry (LSIMS) and tandem mass spectrometry revealed the composition and branching pattern (Baldwin *et al.*, 1990a, b; Stahl *et al.*, 1992). Fig. 2 shows the LSIMS spectrum of the permethylated glycan species. In addition to enhancing the surface activitity, permethylation causes formation of a quaternary ammonium cation which further enhances the MS sensitivity. The molecular masses reveal the sugar compositions e.g. the ion of nominal $m/z = 1312$ corresponds to an oligosaccharide containing four hexoses, hexosamine, and inositol, $m/z = 1557$ has an additional N-acetylhexosamine and $m/z = 1761$ has a further hexose, but they do not identify the way in which these species are connected. A comparison of the tandem spectra of these species (Fig. 3) shows that they correspond to Hex–Hex–Hex–Hex–HexNH$_2$–Ino, Hex–Hex–Hex–(HexNAc–)Hex–HexNH$_2$–Ino, and Hex–Hex–Hex–(Hex–HexNAc–)Hex–HexNH$_2$–Ino respectively, plus an isomeric species for $m/z = 1557$, Hex–Hex–(Hex–HexNAc–)HexNH$_2$–Ino. Other minor components can be identified similarly. The interpretation of these spectra is based on assignment of the peaks shown in Fig. 3 to cross-ring ^{1,5}X cleavages with charge retention at the hexosamine, and the lower mass peaks (not shown) to oxonium or B cleavage ions with charge retention at the non-reducing termini (Baldwin *et al.*, 1990b). As indicated in Fig. 2, the species of nominal masses 1918 and 2122 carry the sialic acid referred to above. These species were also studied by tandem mass spectrometry in order to ascertain the attachment point of the sialic acid. Fig. 4 shows the tandem spectrum including the B ions for $m/z = 2122$. The presence of ions for Sia$^+$ and Sia·Hex$^+$ and Sia·HexNAc$^+$ identifies the sequence Sia–Hex–HexNAc. Some of the species thus identified are similar to those reported for rat brain Thy-1 (Homans *et al.*, 1988). Additional forms carry either hexose or sialic acid–hexose attached to the N-acetylhexosamine.

Electrospray (ESP) mass spectrometry was carried out on the underivatized intact C-terminal peptide GPI (Stahl *et al.*, 1991). The heterogeneous species were separated by reverse-phase HPLC at pH 7.5 (Fig. 5(A)). Separate ESP spectra were obtained on the fraction eluting at 34 min (Fig. 5(B)) and the fraction eluting at 38 min (Fig. 5(C)). From earlier LSIMS studies on the peptide released by 50% aqueous HF, its sequence is known to be GluSerGlnAlaTyrTyrAspGlyArgArgSer-Ea (Stahl *et al.*, 1990a). In electrospray mass spectrometry molecular ions having n excess protons and thus n positive charges are observed at m/z values corresponding to $(M_r + n)/n$. By fitting different possible integer values of n to this equation, series of ions can be identified originating from a common molecular mass M_r. The spectra illustrated in Fig. 5 were transformed to show the distribution of molecular species. The observed

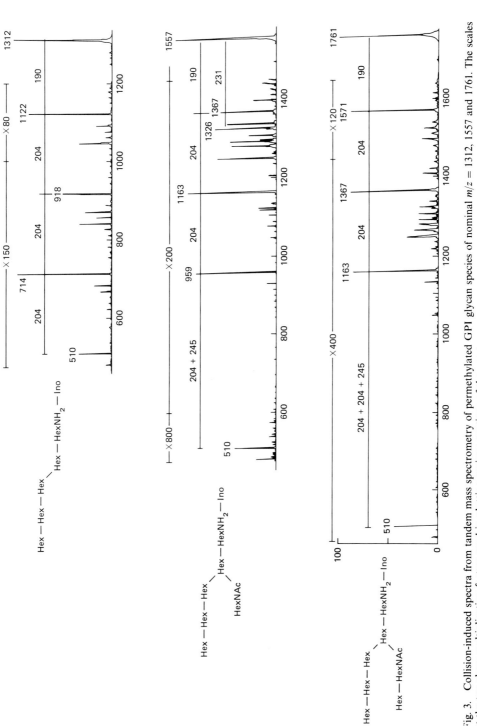

Fig. 3. Collision-induced spectra from tandem mass spectrometry of permethylated GPI glycan species of nominal m/z = 1312, 1557 and 1761. The scales at the top show multiplication factors used in plotting various portions of the spectrum.

molecular masses are consistent with the known peptide linked through phosphoethanolamine to the heterogeneous glycans already identified. In addition to the phosphate required to join the inositol to the diradylglycerol (removed by PIPLC), the spectra show that, like Thy-1 (Homans *et al.*, 1988) and acetylcholinesterase (Roberts *et al.*, 1988) but unlike trypanosome variable surface glycoproteins (Ferguson *et al.*, 1988), PrP^{Sc} also carries a second phosphoethanolamine as suggested previously by amino acid analysis (Stahl *et al.*, 1987). The earlier eluting components (Fig. 5(A)) are revealed by ESP to contain the sialic acid (Fig. 5(B)). Some of the minor components in the LSIMS spectrum of the glycan are shown to be artefacts resulting from the aqueous HF, e.g. ESP shows that all species possess the *N*-acetylhexosamine, thus the ion of $m/z = 1312$ in Fig. 2 is due to a hydrolysis product.

ESP analysis of the PrP 27-30 GPI clearly reveals masses corresponding to an inositol phosphate monoester, while the product of the reaction catalysed by the bacterial PIPLC is thought to be a mixture of inositol 1-phosphate and 1,2-cyclic inositol phosphate (Taguchi *et al.*, 1980). Conversion of the cyclic product to a phosphate monoester is accelerated under acidic conditions and has been described previously (Ferguson *et al.*, 1985). CE analysis of freshly isolated peptide-GPI reveals shoulders eluting after the main peaks. Briefly boiling the HPLC fraction (which contains $\sim 0.06\%$ TFA) before analysis shifts the elution time of the main peaks to match those of the shoulders, consistent with the opening of a cyclic phosphate ring (Stahl *et al.*, 1992). Furthermore, the process of drying and resuspension as carried out before analysis by ESP gave electropherograms identical to that seen after boiling. We also observed that this brief boiling, but not drying and resuspension of the peptide, resulted in complete loss of the sialic acid from the glycan (Stahl *et al.*, 1992).

Carbohydrate composition analysis
The identities of the sugar residues in the glycan were investigated by acid hydrolysis and HPAE analysis of the resulting monosaccharides, comparing the elution profiles with authentic samples of mannose, galactose, glucosamine and galactosamine. The molar yield of mannose proved to be significantly different for the intact peptide-GPI (1.1) compared with that for the glycan released by aqueous HF (5.4), indicating that several of the mannose units were modified. Within the accuracy of the experiment this result is consistent with the anticipated value of 4. Treatment with α-mannosidase and mass spectrometric analysis confirmed that the the third mannose from the glucosamine carries an aqueous HF-sensitive modification. This is consistent with the fact that in other GPIs this is the site of attachment of the protein. Galactose was found to be present and was relatively unaffected by incubation with 50% HF (0.6–0.9). No other GPI from a mammalian source has been shown to contain this monosaccharide. It is compatible, however, with the observation of an Sia–Hex–HexNAc branch of the GPI as indicated by the high energy collision-induced dissociation mass spectrum in Fig. 4. This trisaccharide is reminiscent of the common *O*-linked trisaccharide Siaα(2–3)Galβ(1–3)GalNAc which is attached to serine (Beyer et al., 1981), as well as structures found on some gangliosides (Feizi, 1985). Evidence from lectin binding indicated the location of the galactose on the *N*-acetyl hexosamine branch, and the variable presence of the hexose at this site as shown by mass

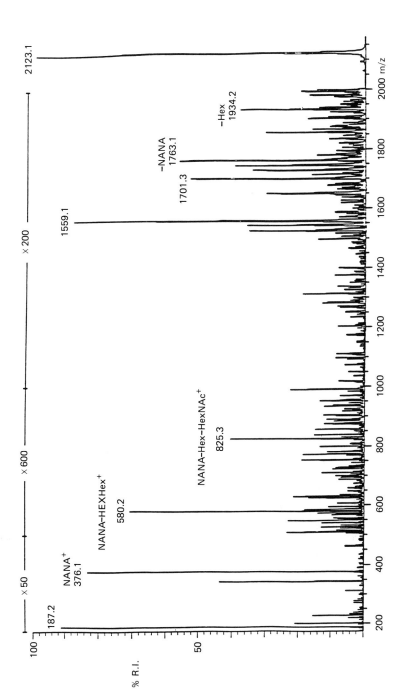

Fig. 4. Tandem mass spectrometry resulting from collision-induced dissociation of permethylated GPI glycan containing *N*-acetylneuraminic acid (NANA) or sialic acid of mass 2123.1. The scales at the top show multiplication factors used in plotting various portions of the spectrum.

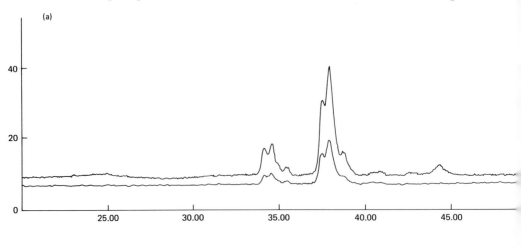

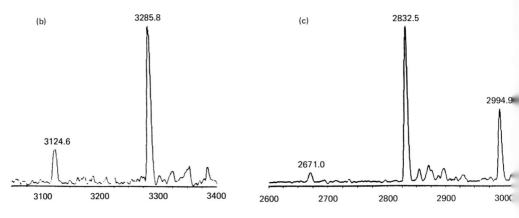

Fig. 5. Separation and electrospray mass spectrometry of various peptide-GPI glycoforms. (A) HPLC at pH 7: the solid line indicates the absorbance observed at 214 nm, while the dotted line shows the absorbance observed at 280 nm (5-fold more sensitive). The electrospray mass spectra of the fractions eluting at 34 and 38 min are shown in (B) and (C) respectively. The peak intensities in (B) and (C) were scaled independently and are not quantitative.

spectrometry is consistent with a molar ratio of < 1. Galactosamine (almost certainly formed by hydrolysis of N-acetylgalactosamine) was assayed at a level of 1.1. Glucosamine was found only at an approximate molar ratio of 0.3, despite mass spectrometry indicating that every glycan species contained a free hexosamine which by analogy with other GPIs is almost certainly glucosamine. The amount of sialic acid was determined by neuraminidase treatment and by mild acid hydrolysis, both methods giving 0.25 mol. Quantitation of the monosaccharides was by comparison with the response of standards and was relative to quantitation of the peptide based on amino acid analysis (Stahl et al., 1992). Myo-inositol was identified in previous experiments involving prolonged acid hydrolysis and GCMS analysis of the residue

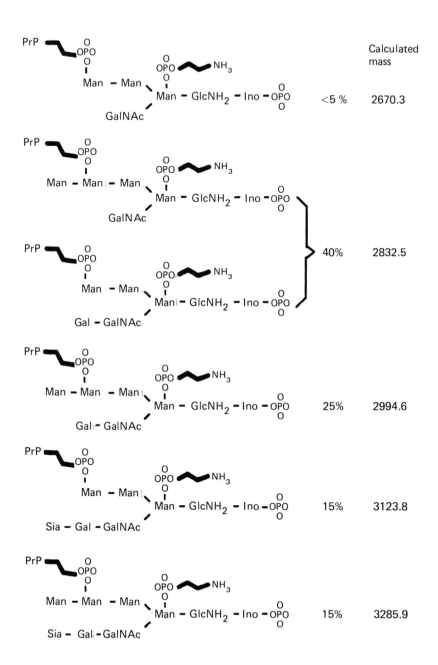

Fig. 6.   Proposed structures for six glycoforms of the PrP 27–30 GPI. The percentages indicate an estimate of the relative abundance of each glycoform.

following trimethylsilylation (Stahl *et al.*, 1987, 1990c).

On the basis of the results described above we have identified the six separate glycan species illustrated in Fig. 6. This degree of glycan heterogeneity is somewhat greater than that reported for other mammalian GPI anchors (Homans *et al.*, 1988; Roberts *et al.*, 1988). Although the assignment of glucosamine and not galactosamine as the unacetylated amino sugar was made by analogy to other GPI structures, substantive evidence shows the order and branching of the sugar units by mass, the position of the two α-mannose groups, and the lack of acetylation on the inositol-linked hexosamine. Furthermore, ricin binding data suggest that the galactose residue is β-linked. The positions of the protein attachment sites and the additional phosphoethanolamine are also shown by analogy with other GPI glycans.

Comparison of the GPIs PrP 27-30 and PrP^C

PrP^C was purified by a new procedure (Pan *et al.*, 1992). The *C*-terminal peptide–GPI from an endoproteinase Lys-C digest was isolated using the same method as that employed for PrP 27-30. Initial comparison of the two different forms was effected by performing two-dimensional gel electrophoresis followed by specific detection of the peptide–GPI on immunoblots with the α-P3 antibody. Analysis of the PrP 27-30 GPI indicated two sets of GPI species with different isoelectric points, the most acidic of which could be eliminated by pretreatment with neuraminidase. Similar analysis of a hamster brain fraction enriched for PrP^C revealed a similar staining pattern with a faint, but detectable, fraction of acidic peptide–GPI species. More compelling evidence was obtained by CE analysis using phosphate buffer (pH 2.5) in an open tubular column with UV detection at 200 nm Figs 7(A) and 7(B) show the electropherograms of the peptide GPIs of PrP 27-30 and PrP^C respectively. The former is highly purified whereas the latter is impure but, despite this, peaks arising from the peptide–GPI of PrP^C are clearly visible. Treatment of the peptide–GPI with 50% aqueous HF hydrolyses the phosphodiester bonds and releases the glycan from the peptide, which is seen as a new peak in the traces for PrP^C (Fig. 7(C)) and PrP 27-30 (Fig. 7(D)). As was stated above, the PrP 27-30 GPI is neuraminidase sensitive. Figs 7(E) and 7(F) show the effect of neuraminidase on the GPIs of PrP^C and PrP 27-30 to be identical. These data confirm that the GPI anchor of PrP^C is broadly similar to that of PrP 27-30 and that it too possesses sialic acid.

The addition of sialic acid to prion protein GPIs probably occurs in the Golgi as this is the purported localization of known sialyltransferases (Paulson and Colley, 1989), and sialylated proteins are apparently not found in earlier regions of the protein export pathway (Tartakoff and Vassalli, 1983). However, it should be noted that a cell surface *trans*-sialidase has recently been identified in *Trypanosoma cruzi* (Schenkman *et al.*, 1991; Zingales *et al.*, 1987). Although there is as yet no information regarding the purpose, if any, of the sialic acid attached to the prion protein GPI, there is a growing appreciation of the important regulatory roles served by sialic acid on glycoproteins and glycolipids (Rademacher *et al.*, 1988). Examples include lymphocyte migration and homing (Rosen *et al.*, 1989; Samlowski *et al.*, 1984) regulation of cell adhesion by polysialic acid on NCAM (Acheson *et al.*, 1991), and

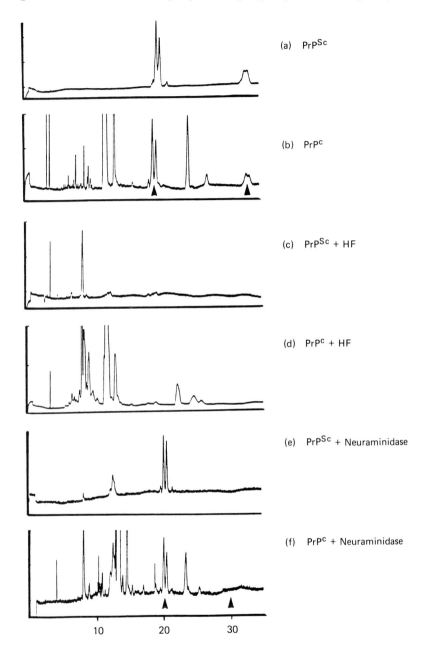

Fig. 7.   Capillary electrophoresis of HPLC fractions containing the *C*-terminal peptide–GPI from PrP 27–30 and PrP^C. Electropherograms of HPLC-purified fractions following endoproteinase Lys-C and PIPLC digests of (A), (C), (E) purified PrP 27–30 and (B), (D), (F) partially purified PrP^C are shown. The samples in (C) and (D) were incubated with 50% aqueous HF before analysis, while the samples in (E) and (F) were digested with neuraminidase before electrophoresis.

modulation by gangliosides of basic fibroblast growth factor efficacy (De Cristan *et al.*, 1990), CD4 function (Offner *et al.*, 1987), neurite outgrowth in neuroblastoma cells (Tsuji *et al.*, 1983), PDGF receptor phosphorylation (Bremer *et al.*, 1984), and insulin receptor tyrosine kinase activity (Nojiri *et al.*, 1991). The identification of sialic acid on a GPI anchor now presents the possibility for participation of GPI anchors in similar functions. There is also the opportunity for distinctive specificities if protein-derived GPI moieties function as growth factor second messengers (Low and Saltiel, 1988; Romero *et al.*, 1988).

CONCLUSIONS

The process by which PrP^{Sc} is formed post-translationally from a protease-sensitive precursor remains uncertain. The elucidation of the chemical structure of PrP^{Sc} has thus far revealed that N-linked glycosylation (Haraguchi *et al.*, 1989) and the GPI anchor (Stahl *et al.*, 1987) constitute the bulk of the covalent post-translational modifications to both the cellular and abnormal isoforms of PrP (Stahl and Prusiner, 1991). Partial structures of the N-linked carbohydrate are known for PrP^{Sc}, and while extremely heterogenous, there is no obvious way in which they could contribute to the altered physical properties exhibited by PrP^{Sc} (Endo *et al.*, 1989). Furthermore, ablation of the N-linked sites in PrP by site-directed mutagenesis does not eliminate the acquisition of protease resistance upon expression of the protein in scrapie-infected neuroblastoma cells (Taraboulos *et al.*, 1990). Although it is not yet known for certain whether the protease-resistant PrP without N-linked sites is still associated with scrapie infectivity, there are no reports of samples containing protease-resistant PrP^{Sc} in the absence of scrapie infectivity. This question should soon be answered through experiments with transgenic mice expressing the mutated PrP (Prusiner *et al.*, 1990). We therefore examined the structure of the GPI anchor to determine whether a scrapie-specific structural feature exists that might play a role in giving PrP^{Sc} its abnormal properties.

It is evident that the presence of the sialic acid in the PrP^{Sc} GPI is not responsible for the strain-specific characteristics of 139H prions. Analysis of the 139H-derived GPI revealed that it is identical in mass, retention in reverse-phase HPLC and capillary electrophoresis, and level of heterogeneity, although it has not been excluded that the GPI anchors from the two distinct prion isolates could differ by glycosidic linkage or the type of lipid attached. We are continuing the detailed comparison of PrP^{Sc} from Both the 139H and SC237 strains to identify chemical differences that could account for their distinct scrapie incubation times, neuropathology, and distribution of PrP^{Sc} in the brain (Hecker *et al.*, 1992). If no chemical differences are found, then either prions possess as-yet unidentified non-PrP components or PrP^{Sc} exists in different conformations which determine the properties of each distinct isolate (Prusiner, 1991).

The PrP 27-30 GPI structures identified here all contain a common core glycan also found in the six other GPI moieties that have been analysed in detail (Ferguson *et al.*, 1988; Homans *et al.*, 1988; Mayor *et al.*, 1990; Roberts *et al.*, 1988; Schmitz *et al.*, 1987; Schneider *et al.*, 1990). This core consists of three mannose residues, glucosamine, and inositol. This matches the structure of the putative GPI precursor characterized for trypanosomes (Mayor *et al.*, 1990); the structure of mammalian GPI precursors is not known (Singh *et al.*, 1991). All of the carbohydrate heterogeneity would be consistent with secondary modifications of a common core matching the trypanosome GPI precursor. Evidence consistent with secondary modification of the GPI after its attachment to the protein has been observed (Bangs *et al.*, 1988).

Our findings substantially narrow the search for chemical modifications that might feature in post-translational conversion of PrP^C to PrP^{Sc}. Prion proteins are the first molecules described with sialic acid residues attached to GPI glycans; in addition, earlier studies showed that sialic acids are attached to some of the *N*-linked complex type oligosaccharides of PrP^{Sc} (Endo *et al.*, 1989). Thus, some of the size and charge heterogeneity exhibited by PrP^{Sc} (Bolton *et al.*, 1985) is due to sialic acids. Whether the number and position of sialic acids attached to PrP glycans influences the fate of the cellular or scrapie PrP isoforms remains to be established.

ACKNOWLEDGEMENTS

This work was supported by grants from the National Institutes of Health (NS14069, AG02132, NS22786, AG08967) and the American Health Assistance Foundation, as well as by gifts from the Sherman Fairchild Foundation and National Medical Enterprises. We acknowledge the Bio-organic, Biomedical Mass Spectrometry Resource (A. L. Burlingame, Director), supported by the National Institutes of Health National Center for Research Resources Grant RR01614, and a National Science Foundation Biological Instrumentation Program Grant DIR8700766.

REFERENCES

Acheson, A. Sunshine, J.L. and Rutishauser, U. (1991) NCAM polysialic acid can regulate both cell–cell and cell–substrate interactions. *J. Cell Biol.* **114** 143–153.

Baldwin, M.A., Stahl, N., Burlingame, A.L. and Prusiner, S.B. (1990a) Structure determination of glycoinositol phospholipid anchors by permethylation and tandem mass spectrometry. *Methods: Companion Method Enzymol.* **1** 306–314.

Baldwin, M.A., Stahl, N., Reinders, L.G., Gibson, B.W., Prusiner, S.B. and Burlingame, A.L. (1990b) Permethylation and tandem mass spectrometry of oligosaccharides having free hexosamine: analysis of the glycoinositol phospholipid anchor glycan from the scrapie prion protein. *Anal. Biochem.* **191** 174–182.

Bangs, J.D., Doering, T.L., Englund, P.T. and Hart, G.W. (1988) Biosynthesis of a variant surface glycoprotein of *Trypanosoma brucei*: processing the glycolipid membrane anchor and *N*-linked oligosaccharides. *J. Biol. Chem.* **263** 17697–17705.

Beyer, T.A., Sadler, J.A., Rearick, J.I., Paulson, J.C. and Hill, R.L. (1981) Glycosyltransferases and their use in assessing oligosaccharide structure and structure–function relationships. *Adv. Enzymol. Relat. Areas Mol. Biol.* **52** 23–175.

Bolton, D.C., McKinley, M.P. and Prusiner, S.B. (1982) Identification of a protein that purifies with the scrapie prion. *Science* **218** 1309–1311.

Bolton, D.C., Meyer, R.K. and Prusiner, S.B. (1985) Scrapie PrP 27-30 is a sialoglycoprotein. *J. Virol.* **53** 596–606.

Borchelt, D.R., Scott, M., Taraboulos, A., Stahl, N. and Prusiner, S.B. (1990) Scrapie and cellular prion proteins differ in their kinetics of synthesis and topology in cultured cells. *J. Cell Biol.* **110** 743–752.

Bremer, E.G., Hakomori, S., Bowen-Pope, D.F., Raines, E. and Ross, R. (1984) Ganglioside-mediated modulation of cell growth, growth factor binding, and receptor phosphorylation. *J. Biol. Chem.* **259** 6818–6825.

Bruce, M.E. and Dickinson, A.G. (1987) Biological evidence that the scrapie agent has an independent genome. *J. Gen. Virol.* **68** 79–89.

Carp, R.I. and Callahan, S.M. (1991) Variation in the characteristics of 10 mouse-passaged scrapie lines derived form five scrapie-positive sheep. *J. Gen. Virol.* **72** 293–298.

Caughey, B., Neary, K., Butler, R., Ernst, D., Perry, L., Chesebro, B. and Race, R.E. (1990) Normal and scrapie-associated forms of prion protein differ in their sensitivities to phospholipase and proteases in intact neuroblastoma cells. *J. Virol.* **64** 1093–1101.

De Cristan, G., Morbidelli, L., Alessandri, G., Ziche, M., Cappa, A.P. and Gullino, P.M. (1990) Synergism between gangliosides and basic fibroblastic growth factor in favouring survival, growth, and motility of capillary endothelium. *J. Cell Physiol.* **144** 505–510.

Endo, T., Groth, D., Prusiner, S.B. and Kobata, A. (1989) Diversity of oligosaccharide structures linked to asparagines of the scrapie prion protein. *Biochemistry* **28** 8380–8388.

Feizi, T. (1985) Demonstration by monoclonal antibodies that carbohydrate structures of glycoproteins and glycolipids are onco-developmental antigens. *Nature (London)* **314** 53–57.

Ferguson, M.A.J., Low, M.G. and Cross, G.A.M. (1985) Glycosyl-sn-1,2-dimyristyl-phosphatidylinositol is covalently linked to *Trypanosoma brucei* variant surface glycoprotein. *J. Biol. Chem.* **260** 14547–14555.

Ferguson, M.A.J., Homans, S.W., Dwek, R.A. and Rademacher, T.W. (1988) Glycosyl-phosphatidylinositol moiety that anchors *Trypanosoma brucei* variant surface glycoprotein to the membrane. *Science* **239** 753–759.

Fraser, H. and Dickinson, A.G. (1968) The sequential development of the brain lesions of scrapie in three strains of mice. *J. Comp. Pathol.* **78** 301–311.

Gabizon, R., McKinley, M.P., Groth, D.F. and Prusiner, S.B. (1988) Immunoaffinity purification and neutralization of scrapie prion infectivity. *Proc. Natl. Acad. Sci. USA* **85** 6617–6621.

Haraguchi, T., Fisher, S., Olofsson, S., Endo, T., Groth, D., Tarantino, A., Borchelt, D.R., Teplow, D., Hood, L., Burlingame, A., Lycke, E., Kobata, A. and Prusiner, S.B. (1989) Asparagine-linked glycosylation of the scrapie and cellular prion proteins. *Arch. Biochem. Biophys.* **274** 1–13.

Hecker, R., Taraboulos, A., Scott, M., Pan,, K.-M., Jendroska, K., DeArmond, S.J. and Prusiner, S.B. (1992) Replication of distinct prion isolates is region specific in brains of transgenic mice and hamsters. *Genes and Development* **6** 1213–1228.

Homans, S.W., Ferguson, M.A., Dwek, R.A., Rademacher, T.W., Anand, R. and Williams, A.F. (1988) Complete structure of the glycosly-phosphatidylinositol membrane anchor of rat brain Thy-1 glycoprotein. *Nature (London)* **333** 269–272.

Hope, J., Multhaup, G., Reekie, L.J.D., Kimberlin, R.H. and Beyreuther, K. (1988) Molecular pathology of scrapie-associated fibril protein (PrP) in mouse brain affected by the ME7 strain of scrapie. *Eur. J. Biochem.* **172** 271–277.

Hsiao, K., Baker, H.F., Crow, T.J., Poulter, M., Owen, F., Terwilliger, J.D., Westaway, D., Ott, J. and Prusiner, S.B. (1989) Linkage of a prion protein missense variant to Gerstmann–Sträussler syndrome. *Nature (London)* **338** 342–345.

Hsiao, K.K., Scott, M., Foster, D., Groth, D.F., DeArmond, S.J. and Prusiner, S.B. (1990) Spontaneous neurodegeneration in transgenic mice with mutant prion protein of Gerstmann–Sträussler syndrome. *Science* **250** 1587–1590.

Low, M.G. and Saltiel, A.R. (1988) Structural and functional roles of glycosylphosphatidylinositol in membrane. *Science* **239** 268–275.

Mayor, S., Menon, A.K., Cross, G.A.M., Ferguson, M.A.J., Dwek, R.A. and Rademacher, T.W. (1990) Glycolipid precursors for the membrane anchor of *Trypanosoma brucei* variant surface glycoproteins. I. Can structure of the phosphatidylinositol-specific phospholipase C sensitive and resistant glycolipids. *J. Biol. Chem.* **265** 6164–6173.

McKinley, M.P., Bolton, D.C. and Prusiner, S.B. (1983) A protease-resistant protein is a structural component of the scrapie prion. *Cell* **35** 57–62.

McKinley, M.P., Meyer, R., Kenaga, L., Rahbar, F., Cotter, R., Serban, A. and Prusiner, S.B. (1991) Scrapie prion rod formation *in vitro* requires both detergent extraction and limited proteolysis. *J. Virol.* **65** 1440–1449.

Meyer, R.K., McKinley, M.P., Bowman, K.A., Braunfeld, M.B., Barry, R.A. and Prusiner, S.B. (1986) Separation and properties of cellular and scrapie prion proteins, *Proc. Natl. Acad. Sci. USA* **83** 2310–2314.

Nojiri, H., Stroud, M. and Hakomori, S. (1991) A specific type of ganglioside as a modulator of insulin-dependent cell growth and insulin receptor tyrosine kinase activity. Possible association of ganglioside-induced inhibition of insulin receptor function and monocytic differentiation induction in HL-60 cells. *J. Biol. Chem.* **266** 4531–4537.

Oesch, B., Westaway, D., Wälchli, M., McKinley, M.P., Kent, S.B.H., Aebersold, R., Barry, R.A., Tempst, P., Teplow, D.B., Hood, L.E., Prusiner, S.B. and Weissmann, C. (1985) A cellular gene encodes scrapie PrP 27-30 protein. *Cell* **40** 735–746.

Offner, H., Thieme, T. and Vandenbark, A.A. (1987) Gangliosides induce selective modulation of CD4 from helper T lymphocytes. *J. Immunol.* **139** 3295–3305.

Pan, K.-M., Stahl, N. and Prusiner, S.B. (1992) Purification and properties of the cellular prion protein from Syrian hamster brain. *Protein Sci.*, in press.

Paulson, J.C. and Colley, K.J. (1989) Glycosyltransferases: structure, localization and control of cell type-specific glycosylation. *J. Biol. Chem.* **264** 17615–17618.

Prusiner, S.B. (1982) Novel proteinaceous infectious particles cause scrapie. *Science* **216** 136–144.

Prusiner, S.B. (1991) Molecular biology of prion diseases. *Science* **252** 1515–1522.

Prusiner, S.B., Scott, M., Foster, D., Pan, K.-M., Groth, D., Mirenda, C., Torchia, M., Yang, S.-L., Serban, D., Carlson, G.A., Hoppe, P.C., Westaway, D. and DeArmond, S.J. (1990) Transgenetic studies implicate interactions between homologous PrP isoforms in scrapie prion replication. *Cell* **63** 673–686.

Rademacher, T.W., Parekh, R.B. and Dwek, R.A. (1988) Glycobiology. *Annu. Rev. Biochem.* **57** 785–838.

Roberts, G.W., Lofthouse, R., Allsop, D., Landon, M., Kidd, M., Prusiner, S.B. and Crow, T.J. (1988) CNS amyloid proteins in neurodegenerative diseases. *Neurology* **38** 1534–1540.

Romero, G., Luttrell, L., Rogol, A., Zeller, K., Hewlett, E. and Larner, J. (1988) Phosphatidylinositol–glycan anchors of membrane proteins: potential precursors of insulin mediators. *Science* **204** 509–511.

Rosen, S.D., Chi, S.I., True, D.D., Singer, M.S. and Yednock, T.A. (1989) Intravenously injected sialidase inactivates attachment sites for lymphocytes on high endothelial venules. *J. Immunol.* **142** 1895–1902.

Samlowski, W.W., Spangrude, G.J. and Daynes, R.A. (1984) Studies on the liver sequestration of lymphocytes bearing membrane-associated galactose-terminal glycoconjugates: reversal with agents that effectively compete for the asialoglycoprotein receptor. *Cell. Immunol.* **88** 309–322.

Schenkman, S., Jiang, M.-S., Hart, G.W. and Nussenzweig, V. (1991) A novel cell surface *trans*-sialidase of *Trypanosoma cruzi* generates a stage-specific eptiope required for invasion of mammalian cells, *Cell* **65** 1117–1125.

Schmitz, B., Klein, R.A., Duncan, I.A., Egge, H., Gunawan, J., Peter-Katalinic, J., Dabrowski, U. and Dabrowski, J. (1987) MS and NMR analysis of the cross-reacting determinant glycan from *Trypanosoma burcei* MITat 1.6 variant specific glycoprotein. *Biochem. Biophys. Res. Commun.* **146** 1055–1063.

Schneider, P., Ferguson, M.A.J., McConville, M.J., Mehlert, A., Homans, S.W. and Bordier, C. (1990) Structure of the glycosyl-phosphatidylinositol membrane anchor of the Leshmania major promastigote surface protease. *J. Biol. Chem.* **265** 16955–16964.

Scott, M., Foster, D., Mirenda, C., Serban. D., Coufal. F., Wälchli, M., Torchia, M., Groth, D., Carlson, G., DeArmond, S.J., Westaway, D. and Prusiner, S.B. (1989) Transgenic mice expressing hamster prion protein produce species-specific scrapie infectivity and amyloid plaques. *Cell* **59** 847–857.

Singh, N., Singleton, D. and Tartakoff, A.M. (1991) Anchoring and degradation of glycolipid-anchored membrane proteins by L929 versus by LM − TK − mouse fibroblasts: implications for anchor biosynthesis. *Mol. Cell. Biol.* **11** 2362–2374.

Stahl, N. and Prusiner, S.B. (1991) Prions and prion proteins. *FASEB J.* **5** 2799–2807.

Stahl, N., Borchelt, D.R., Hsiao, K. and Prusiner, S.B. (1987) Scrapie prion protein contains a phosphatidylinositol glycolipid. *Cell* **51** 229–240.

Stahl, N., Baldwin, M.A., Burlingame, A.L. and Prusiner, S.B. (1990a) Identification of glycoinositol phospholipid-linked and truncated forms of the scrapie prion protein. *Biochemistry* **29** 8879–8884.

Stahl, N., Borchelt, D.R. and Prusiner, S.B. (1990b) Differential release of cellular and scrapie prion proteins from cellular membranes by phosphatidylinositol-specific phospholipase C. *Biochemistry* **29** 5405–5412.

Stahl, N., Borchelt, D.R. and Prusiner, S.B. (1990c) Glycolipid anchors of the cellular and scrapie prion proteins. In: Turner, A.J. (ed.) *Molecular and Cell Biology of membrane Proteins*. Ellis Horwood Chichester, pp. 189–216.

Stahl, N., Baldwin, M.A. and Prusiner, S.B. (1991) Electrospray mass spectrometry of glycoinositol phospholipid of the scrapie prion protein, *Cell Bio. Int. Rep.* **15** 853–862.

Stahl, N., Baldwin, M.A., Hecker, R., Pan, K.-M., Burlingame, A.L. and Prusiner, S.B. (1992) Glycoinositol phospholipid anchors of the scrapie and cellular prion proteins contain sialic acid. *Biochemistry* **31** 5043–5053.

Taraboulos, A., Rogers, M., Borchelt, D.R., McKinley, M.P., Scott, M., Serban, D. and Prusiner, S.B. (1990) Acquisition of protease resistance by prion proteins in scrapie-infected cells does not require asparagine-linked glycosylation. *Proc. Natl. Acad. Sci. USA* **87** 8262–8266.

Tartakoff, A.M. and Vassalli, P. (1983) Lectin-binding sites as markers of golgi subcompartments: proximal-to-distant maturation of oligosaccharides. *J. Cell Biol.* **97** 1243–1248.

Tsuji, S., Arita, M. and Nagai, Y. (1983) GQ1b, a bioactive ganglioside that exhibits novel nerve growth factor (NGF)-like activities in the two neuroblastoma cell lines. *J. Biochem.* **94** 303–306.

Turk, E., Teplow, D.B., Hood, L.E. and Prusiner, S.B. (1988) Purification and properties of the cellular and scrapie hamster prion proteins. *Eur. J. Biochem.* **176** 21–30.

Zingales, B., Carniol, C., de Lederkremer, R.M. and Colli, W. (1987) Direct sialic acid transfer from a protein donor to glycolipids of trypomastigote forms of *Trypanosoma cruzi. Mol. Biochem. Parasitol,* **26** 135–144.

34

Interaction of the prion protein with cellular proteins

Bruno Oesch and Stanley B. Prusiner

ABSTRACT

The scrapie and normal prion proteins interact with other cellular proteins on ligand blots. A 45 kDa PrP ligand (Pli 45) which was more abundant in scrapie-infected brain has been identified as glial fibrillary acidic protein (GFAP). A 110 kDa Pli was present equally in scrapie and normal brain, lung, liver, pancreas and spleen but not in heart or skeletal muscle. The binding site of PrP has been studied using peptides corresponding to different regions of the PrP amino acid sequence. A peptide corresponding to amino acids 140 to 174 (denominated P5) was able to reproduce the binding pattern on ligand blots previously observed with intact PrP as a probe. The region of PrP corresponding to peptide P5 may therefore be involved in the interaction of the PrP isoforms with their ligands in the normal and scrapie-infected animal.

INTRODUCTION

Interactions of the prion protein (PrP) with itself or with other proteins appear to be important for the role of PrP in prion diseases (Prusiner *et al.*, 1990; Prusiner, 1991). Models of direct interaction of PrP with itself have been proposed to explain phenomena such as the species barrier for prion transmission or conversion of the cellular form of PrP (PrP^C) to the scrapie isoform (PrP^{Sc}) (Scott *et al.*, 1989; Prusiner *et al.*, 1990; Weissmann, 1991; Oesch *et al.*, 1988; Hope *et al.*, 1986). Other models suggest interaction of PrP with other cellular proteins as mediators for the above phenomena (Oesch *et al.*, 1988; Bolton and Bendhein, 1988). There are a number of structural features within the PrP polypeptide that might mediate protein–protein interactions. An amphipathic helix in the central portion of PrP may interact with other such helices resulting in homologous or heterologous complexes (Bazan *et al.*, 1987). Another binding site may be the glycolipid anchor; a similar structure has been implicated in binding and internalization of heparan sulphate proteoglycan (Ishihara *et al.*, 1987). Processing of the C-terminus reminiscent of neuroactive peptide activation suggests a role for PrP as a neurotrophic factor (Stahl *et al.*, 1990a).

A fast and sensitive assay for protein–protein interaction denominated ligand blot has been developed to detect receptors for a variety of proteins such as low density lipoprotein, interferons, lactotransferrin, interleukin I, lutropin as well as alpha-bungarotoxin (Wilson *et al.*, 1984; Critchley *et al.*, 1985; Mazurier *et al.*, 1985; Soutar *et al.*, 1986; Bird *et al.*, 1988; Keinanen, 1988; Schwabe *et al.*, 1988). Applying this method to probe the interaction of purified PrP with other proteins two PrP ligands of 45 and 110 kDa have been identified (Oesch *et al.*, 1990). In this paper evidence is presented that a region of the PrP polypeptide which may form an amphipathic helix is responsible for the interaction with other cellular proteins.

RESULTS AND DISCUSSION

Detection of PrP ligands

PrP is expressed in a variety of tissues; the highest levels of mRNA are found in brain (Oesch *et al.*, 1985; Robakis *et al.*, 1986). Various tissues were therefore analysed for PrP binding proteins by the ligand blot technique (Fig. 1; Oesch *et al.*, 1990). Briefly, total proteins were separated by SDS polyacrylamide gel electrophoresis followed by transfer to nitrocellulose (for details of the procedure see Oesch *et al.*, 1990). Blots were then incubated with [^{125}I]PrP 27-30 dispersed into detergent–lipid–protein complexes (DLPCs). After washing, bound radioactivity was visualized by autoradiography. Two proteins of 45 and 110 kDa were detected in brain (denominated Pli 45 and Pli 110) using PrP 27-30 as a probe (Fig.1(A), lanes 1). Pli 110 was also found in other tissues, namely lung, liver, pancreas and spleen (Fig. 1(B), lanes 2, 3, 5, and 6) but not in heart or skeletal muscle. Plis of different molecular weights were observed in lung (56 kDa and 170 kDa), heart (52 kDa), spleen (56 kDa), and skeletal muscle (52 kDa).

Pli 45 was particularly interesting in that it was more abundant in scrapie-infected than in normal brain (Fig. 1(A), lanes 1). Subsequently, Pli 45 was purified and sequenced. Comparison with known sequences revealed its identity with glial fibrillary acidic protein (GFAP) (Oesch *et al.*, 1990). The specificity of the interaction of PrP and Plis has been shown by competition of binding by unlabelled PrP 27-30 (Oesch *et al.*, 1990).

This raises the question whether the interaction of PrP and GFAP occurs also *in vivo*. Cellular PrP and GFAP copurified in the cytoskeletal fraction (B. Oesch, unpublished results) as has been observed for GFAP and prions which contain PrP^{Sc} (VanAlstyne *et al.*, 1987). PrP^{Sc} and GFAP have been colocalized in astrocytes prior

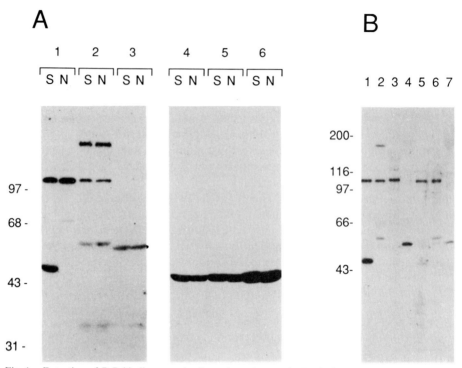

Fig. 1.　Detection of PrP binding proteins in various tissues. (A) Equivalent amounts of brain (lanes 1 and 4), lung (lanes 2 and 5), or heart (lanes 3 and 6) extracts from scrapie-infected (S) or normal (N) hamsters were ligand blotted (lanes 1–3). The same nitrocellulose filter was probed with anti-actin antibodies to reveal loading of individual lanes (lanes 4–6). (B) Ligand blot scrapie-infected hamster organs: brain (lane 1), lung (lane 2), liver (lane 3), heart (lane 4), pancreas (lane 5), spleen (lane 6), and skeletal muscle (lane 7). For details of experimental procedures see Oesch et al. (1990).

to any visible pathological changes (Diedrich et al., 1991), suggesting that the scrapie-specific isoform of PrP is either produced by astrocytes or released from producing cells followed by uptake by astrocytes (for a detailed model see below). These findings indicate that astrocytes may play an important role in prion diseases.

Identification of the PrP binding site
Both isoforms of PrP contain post-translational modifications. There are two attachment sites for N-linked glycosylation at positions 181 and 197 of the hamster PrP amino acid sequence. Analysis of the attached oligosaccharides revealed great diversity (Endo et al., 1989; Haraguchi et al., 1989). The C-terminal 23 amino acids are cleaved off and a glycolipid anchor is attached (Stahl et al., 1987, 1990a). It has also been reported that the arginine residues at positions 25 and 37 are modified (Hope et al., 1986; Turk et al., 1988) but the nature of the modification is unknown. It was therefore important to determine whether the interaction of PrP with Plis was mediated through the polypeptide or its modifications.

　　Preliminary results showed that enzymatically deglycosylated PrP 27-30 could still bind to ligand blots (B. Oesch, unpublished results), suggesting that the PrP

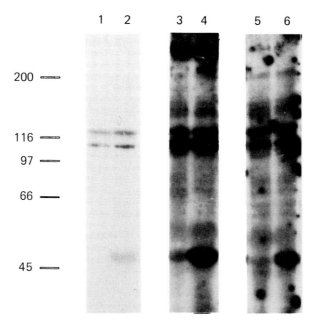

Fig. 2. Analysis of the PrP binding site. Normal (lanes 1, 3, 5) and scrapie-infected (lanes 2, 4, 6) hamster brain homogenates were ligand blotted with radiolabelled probes (PrP 27–30, lanes 1 and 2; peptide P5, lanes 3–6). For competition unlabelled P5 was added to 1 μM (lanes 5 and 6). PrP 27–30 was reconstituted into DLPC as described (Oesch *et al.*, 1990). Peptide P5 was radioiodinated using iodobeads. Labelled peptide was purified by gel filtration.

polypeptide might mediate the interaction. To test this hypothesis two peptides denominated P5 (corresponding to amino acid positions 140 to 174 of the hamster sequence) and P3 (amino acid positions 220 to 233) were radiolabelled at tyrosine residues and used as probes for ligand blotting. Peptide P5 showed the same binding patern as did radiolabelled PrP 27-30 (Fig. 2, lanes 1–4). Comparison with Fig. 1 revealed an additional band at approximately 125 kDa. A PrP ligand at very high molecular weight (> 200 kDa) can only be detected with peptide P5 (Fig. 2, lanes 3 and 4). Competition of labelled P5 with unlabelled P5 mostly reduced binding to Pli 45, Pli 110 and the high molecular weight Pli (Fig. 2, lanes 5 and 6) while binding to the 125 kDa protein was not reduced. In contrast to peptide P5 no binding was observed with peptide P3 (amino acid residues 220–233; results not shown). The similarity of the binding patterns of PrP 27-30 and peptide P5 suggests that both probes bind to the same proteins, i.e. Pli 45 (GFAP) and Pli 110. The high molecular weight Pli appears to bind P5 strongly while only a faint signal can be detected with PrP 27-30 as a probe (Fig. 2, lanes 1–4). The difference in binding may be attributed to differing affinities of the individual probes for the corresponding ligand. Alternatively, accessibility may be restricted for PrP 27-30 as a DLPC while peptide P5 is comparatively small.

Peptide P5 corresponds to a region of the PrP polypeptide which has been postulated to form an amphipathic helix (residues 138–174; Bazan *et al.*, 1987). A

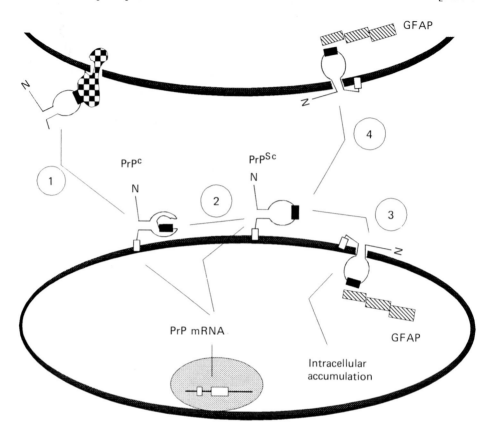

Fig. 3.   Model for the interaction of PrP isoforms with GFAP. The black box in PrP^C represents the binding site corresponding to peptide P5, N delineates the *N*-terminus of the PrP polypeptide, and the white box represents the glycolipid anchor attached at the *C*-terminus. For detailed explanations see text.

second amphipathic structure is positioned at residues 216–239. Peptide P3 is contained within this second region. Both peptides have therefore similar physical characteristics but only P5 binds to Plis. In other proteins amphipathic helices have been postulated to mediate protein–protein interactions. Peptide hormones have been found to adopt amphipathic structure effective for recognition of a cell surface receptor (Kaiser and Kezdy, 1983, 1984; Musso *et al.*, 1984; Lau *et al.*, 1983). These predictions agree with the finding that peptide P5 apparently can interact with other cellular proteins. Preliminary results also show that P5 can bind to intact cells (B. Oesch, unpublished results). The characterization of such cell surface Plis will reveal whether they correspond to the proteins binding to PrP on ligand blots.

MODEL FOR THE INTERACTION OF PrP WITH GFAP

The model outlined in Fig. 3 is based on the following observations.

(i)   PrP^C is found as a cell surface protein anchored to the membrane by a glycolipid (Stahl *et al.*, 1987, 1990b).

(ii)   The scrapie-specific isoform PrP^{Sc} differs from PrP^C in that it cannot be released from the cell surface of primary brain cells even though it is accessible (Stahl *et al.*, 1990b). In cultured cells PrP^{Sc} accumulates intracellularly (Taraboulos *et al.*, 1990).

(iii)   Two forms of PrP have been characterized by *in vitro* translation: a secreted form corresponding to PrP^C on the cell surface and a transmembrane form (denominated tm-PrP) spanning the membrane twice (Hay *et al.*, 1987; Yost *et al.*, 1990). The cytoplasmic portion of tm-PrP is bordered by transmembrane region I (aa 112–135) and a non-hydrophobic transmembrane domain II (aa 157–180) as proposed by Hay *et al.* (1987). Domain II is restricted at the carboxy terminal side by the first glycosylation attachment site (position 181–183) while the *N*-terminal boundary is uncertain.

(iv)   The binding site of PrP^{Sc} to Plis is contained within the sequence of peptide P5.

In a normal animal PrP^C is a cell surface protein which may be released to serve as a neuroactive peptide (Fig. 3, step 1). Spontaneous release of PrP^C from cultured cells as well as *C*-terminal processing at the glycolipid anchor similar to neuropeptides have been described (Borchelt *et al.*, 1990; Stahl *et al.*, 1990a). Release of PrP^C may be accompanied by a conformational change which would expose the binding site. In a scrapie-infected cell PrP^{Sc} is produced from PrP^C or a precursor (Fig. 3, step 2). It has been speculated that a conformational change may be responsible for the conversion of PrP^C to PrP^{Sc} (Basler *et al.*, 1986; Weissmann, 1991). Similar to the normal situation, the binding site may become exposed even though it is still attached to the cell surface by the glycolipid anchor. Binding to other proteins would then trigger the post-translational insertion into the membrane (Fig. 3, step 3) as has been observed *in vitro* with dog pancreas microsomes (Lopez *et al.*, 1990). Translocated PrP^{Sc} may then interact with cytoplasmic GFAP which may lead to intracellular accumulation. If PrP^{Sc} were not produced by astrocytes but by neurons, translocation of PrP^{Sc} into the membrane of neighbouring astrocytes may be envisaged (Fig. 3, step 4). Binding to intermediate filament proteins would also result in intracellular accumulation of PrP^{Sc} in chronically infected cells (Taraboulos *et al.*, 1990). Transmission of scrapie by cell-to-cell contact may also explain the observation that prions spread along anatomical pathways (Fraser and Dickinson, 1985; Foster *et al.*, 1990).

CONCLUSIONS

We have identified cellular proteins which interact with the prion protein isoforms. A peptide located within a putative amphipathic helix of PrP contains the binding site of PrP to its ligands. The position of this binding site in a putative cytoplasmic loop prompted the speculation that the scrapie-specific and the cellular isoform of PrP may differ in their membrane topology. The significance of the interaction of PrP^{Sc} with GFAP and the proposed involvement of astrocytes in prion diseases remains to be examined further.

REFERENCES

Basler, K., Oesch, B., Scott, M., Westaway, D., Walchli, M., Groth, D.F., McKinley, M.P., Prusiner, S.B. and Weissmann, C. (1986) Scrapie and cellular PrP isoforms are encoded by the same chromosomal gene. *Cell* **46** 417–428.

Bazan, J.F., Fletterick, R.J., McKinley, M.P. and Prusiner, S.B. (1987) Predicted secondary structure and membrane topology of the scrapie prion protein. *Protein Eng.* **1** 125–135.

Bird, T.A., Gearing, A.J. and Saklatvala, J. (1988) Murine interleukin 1 receptor. Direct identification by ligand blotting and purification to homogeneity of an interleukin 1-binding glycoprotein. *J. Biol. Chem.* **263** 12063–12069.

Bolton, D.C. and Bendheim, P.E. (1988) A modified host protein model of scrapie. *Ciba. Found. Symp.* **135** 164–181.

Borchelt, D.R., Scott, M., Taraboulos, A., Stahl, N. and Prusiner, S.B. (1990) Scrapie and cellular prion proteins differ in their kinetics of synthesis and topology in cultured cells. *J. Cell Biol.* **110** 743–752.

Critchley, D.R., Nelson, P.G., Habig, W.H. and Fishman, P.H. (1985) Fate of tetanus toxin bound to the surface of primary neurons in culture: evidence for rapid internalization. *J. Cell Biol.* **100** 1499–1507.

Diedrich, J.F., Bendheim, P.E., Kim, Y.S., Carp, R.I. and Haase, A.T. (1991) Scrapie-associated prion protein accumulates in astrocytes during scrapie infection. *Proc. Natl. Acad. Sci. USA* **88** 375–379.

Endo, T., Groth, D., Prusiner, S.B. and Kobata, A. (1989) Diversity of oligosaccharide structures linked to asparagines of the scrapie prion protein. *Biochemistry* **28** 8380–8388.

Foster, J.D., Scott, J.R. and Fraser, H. (1990) The use of monosodium glutamate in identifying neuronal populations in mice infected with scrapie. *Neuropathol. Appl. Neurobiol.* **16** 423–430.

Fraser, H. and Dickinson, A.G. (1985) Targeting of scrapie lesions and spread of agent via the retino-tectal projection. *Brain Res.* **346** 32–41.

Haraguchi, T., Fisher, S., Olofsson, S., Endo, T., Groth, D., Tarentino, A., Borchelt, D.R., Teplow, D., Hood, L. and Burlingame, A. (1989) Asparagine-linked glycosylation of the scrapie and cellular prion proteins. *Arch. Biochem. Biophys.* **274** 1–13.

Hay, B., Barry, R.A., Lieberburg, I., Prusiner, S.B. and Lingappa, V.R. (1987) Biogenesis and transmembrane orientation of the cellular isoform of the scrapie prion protein. *Mol. Cell Biol.* **7** 914–920.

Hope, J., Morton, L.J., Farquhar, C.F., Multhaup, G., Beyreuther, K. and Kimberlin, R.H. (1986) The major polypeptide of scrapie-associated fibrils (SAF) has the same size, charge distribution and *N*-terminal protein sequence as predicted for the normal brain protein (PrP). *EMBO J.* **5** 2591–2597.

Ishihara, M., Fedarko, N.S. and Conrad, H.E. (1987) Involvement of phosphatidylinositol and insulin in the coordinate regulation of proteoheparan sulfate metabolism and hepatocyte growth. *J. Biol. Chem.* **262** 4708–4716.

Kaiser, E.T. and Kezdy, F.J. (1983) Secondary structures of proteins and peptides in amphiphilic environments. (a review). *Proc. Natl. Acad. Sci. USA* **80** 1137–1143.

Kaiser, E.T. and Kezdy, F.J. (1984) Amphiphilic secondary structure: design of peptide hormones. *Science* **223** 249–255.

Keinanen, K.P. (1988) Effect of deglycosylation on the structure and hormone-binding activity of the lutropin receptor. *Biochem. J.* **256** 719–724.

Lau, S.H., Rivier, J., Vale, W., Kaiser, E.T. and Kezdy, F.J. (1983) Surface properties of an amphiphilic peptide hormone and of its analog: corticotropin-releasing factor and sauvagine. *Proc. Natl. Acad. Sci. USA* **80** 7070–7074.

Lopez, C.D., Yost, C.S., Prusiner, S.B., Myers, R.M. and Lingappa, V.R. (1990) Unusual topogenic sequence directs prion protein biogenesis. *Science* **248** 226–229.

Mazurier, J., Montreuil, J. and Spik, G. (1985) Visualization of lactotransferrin brush-border receptors by ligand-blotting. *Biochim. Biophys. Acta* **821** 453–460.

Musso, G.F., Assoian, R.K., Kaiser, E.T., Kezdy, F.J. and Tager, H.S. (1984) Heterogeneity of glucagon receptors of rat hepatocytes: a synthetic peptide probe for the high affinity site. *Biochem. Biophys. Res. Commun.* **119** 713–719.

Oesch, B., Westaway, D., Walchli, M., McKinley, M.P., Kent, S.B., Aebersold, R., Barry, R.A., Tempst, P., Teplow, D.B. and Hood, L.E. (1985) A cellular gene encodes scrapie PrP 27-30 protein. *Cell* **40** 735–746.

Oesch, B., Groth, D.F., Prusiner, S.B. and Weissmann, C. (1988) Search for a scrapie-specific nucleic acid: a progress report. *Ciba. Found. Symp.* **135** 209–223.

Oesch, B., Teplow, D.B., Stahl, N., Serban, D., Hood, L.E. and Prusiner, S.B. (1990) Identification of cellular proteins binding to the scrapie prion protein. *Biochemistry* **29** 5848–5855.

Prusiner, S.B. (1991) Molecular biology of prion diseases. *Science* **252** 1515–1522.

Prusiner, S.B., Scott, M., Foster, D., Pan, K.M., Groth, D., Mirenda, C., Torchia, M., Yang, S.L., Serban, D. and Carlson, G.A. (1990) Transgenetic studies implicate interactions between homologous PrP isoforms in scrapie prion replication. *Cell* **63** 673–686.

Robakis, N.K., Sawh, P.R., Wolfe, G.C., Rubenstein, R., Carp, R.I. and Innis, M.A. (1986) Isolation of a cDNA clone encoding the leader peptide of prion protein and expression of the homologous gene in various tissues. *Proc. Natl. Acad. Sci. USA* **83** 6377–6381.

Schwabe, M., Princler, G.L. and Faltynek, C.R. (1988) Characterization of the human type I interferon receptor by ligand blotting. *Eur. J. Immunol.* **18** 2009–2014.

Scott, M., Foster, D., Mirenda, C., Serban, D., Coufal, F., Walchli, M., Torchia, M., Groth, D., Carlson, G. and DeArmond, S.J. (1989) Transgenic mice expressing hamster prion protein produce species-specific scrapie infectivity and amyloid plaques. *Cell* **59** 847–857.

Soutar, A.K., Harders, S.K., Wade, D.P. and Knight, B.L. (1986) Detection and quantitation of low density lipoprotein (LDL) receptors in human liver by ligand blotting, immunoblotting, and radioimmunoassay. LDL receptor protein content is correlated with plasma LDL cholesterol concentration. *J. Biol. Chem.* **261** 17127–17133.

Stahl, N., Borchelt, D.R., Hsiao, K. and Prusiner, S.B. (1987) Scrapie prion protein contains a phosphatidylinositol glycolipid. *Cell* **51** 229–240.

Stahl, N., Baldwin, M.A., Burlingame, A.L. and Prusiner, S.B. (1990a) Identification of glycoinositol phospholipid linked and truncated forms of the scrapie prion protein. *Biochemistry* **29** 8879–8884.

Stahl, N., Borchelt, D.R. and Prusiner, S.B. (1990b) Differential release of cellular and scrapie prion proteins from cellular membranes by phosphatidylinositol-specific phospholipase. *Biochemistry* **29** 5405–5412.

Taraboulos, A., Serban, D. and Prusiner, S.B. (1990) Scrapie prion proteins accumulate in the cytoplasm of persistently infected cultured cells. *J. Cell Biol.* **110** 2117–2132.

Turk, E., Teplow, D.B., Hood, L.E. and Prusiner, S.B. (1988) Purification and properties of the cellular and scrapie hamster prion proteins. *Eur. J. Biochem.* **176** 21–30.

VanAlstyne, D., DeCamillis, M., Sunga, P., and Marsh, R.F. (1987) *Mol. Cell. Biol.* **54** 519–536.

Weissmann, C. (1991) Spongiform encephalopathies. The prion's progress (news). *Nature (London)* **349** 569–571.

Wilson, P.T., Gershoni, J.M., Hawrot, E. and Lentz, T.L. (1984) Binding of alpha-bungarotoxin to proteolytic fragments of the alpha subunit of *Torpedo* acetylcholine receptor analyzed by protein transfer on positively charged membrane filters. *Proc. Natl. Acad. Sci. USA* **81** 2553–2557.

Yost, C.S., Lopez, C.D., Prusiner, S.B., Myers, R.M. and Lingappa, V.R. (1990) Non-hydrophobic extracytoplasmic determinant of stop transfer in the prion protein. *Nature (London)* **343** 669–672.

35

Molecular cloning and structural analysis of a candidate chicken prion protein

Jean-Marc Gabriel, Bruno Oesch, Hans Kretzschmar, Michael Scott and Stanley B. Prusiner

ABSTRACT

The cellular isoform of the prion (PrP) is a sialoglycoprotein bound to the external surface of cells by a glycosylinositol phospholipid (GPI). An abnormal isoform of PrP designated PrP^{Sc} accumulates in transmissible neurodegenerative diseases of the central nervous system which include scrapie of sheep and goats, bovine spongiform encephalopathy (BSE) as well as kuru, Creutzfeldt–Jakob disease (CJD) and Ger-stmann–Sträussler–Scheinker (GSS) syndrome of humans. Because the function of the cellular PrP isoform (PrP^C) is unknown, studies of the acetylcholine receptor inducing activity (ARIA) may be of considerable interest with respect of PrP^C. Enrichment of fractions for ARIA from chicken brain led to the isolation of a protein (Harris et al., 1991). The N-terminal sequence of this protein was determined and used to prepare isocoding mixtures of oligonucleotides which were employed in the identification of cognate cDNA clones. The translated cDNA sequence was found to be $\sim 30\%$ homologous with mammalian prion proteins. All of the structural features of mammalian PrP were found in the chicken protein. Current knowledge is insufficient to assess whether this candidate chicken PrP possesses ARIA. To extend these observations, we recovered genomic clones encoding chicken PrP. Like mammalian PrP molecules, chicken PrP is encoded by a single copy gene and the entire open reading frame (ORF) is found within a single exon.

INTRODUCTION

Prions are a novel family of slow transmissible pathogens causing degenerative diseases of the central nervous system in both humans and animals (Prusiner, 1991). Kuru, Creutzfeldt–Jakob diseases (CJD) and Gerstmann–Sträussler–Scheinker (GSS) syndrome of humans, scrapie of sheep and goats as well as bovine spongiform encephalopathy (BSE) of cattle are all prion disease. Prions differ from conventional pathogens such as viruses, plasmids and viroids in that they appear to be devoid of a nucleic acid genome. Numerous attempts to identify a nucleic acid carried with the prion particle have been unsuccessful and procedures known to modify or hydrolyse specifically nucleic acids failed to inactivate the scrapie agent. In contrast, the scrapie isoform of the prion protein (PrP^{Sc}) is inseparable from prion infectivity. PrP is encoded by a single copy chromosomal gene which encodes both PrP^{Sc} and the cellular isoform of PrP designated PrP^C. The PrP gene is expressed at the same level in normal and scrapie-infected animals (Chesebro et al., 1985; Kretzschmar et al., 1986; Oesch et al., 1985).

Cloning of the PrP gene has been reported for three species of hamster (Lowenstein et al., 1990; Oesch et al., 1985), mouse (Chesebro et al., 1985; Locht et al., 1986; Westaway et al., 1987), rat (Liao et al., 1987), sheep (Goldmann et al., 1990b), bovine (Goldmann et al., 1991), and human (Kretzschmar et al., 1986; Liao et al., 1986; Puckett et al., 1991). The organization and the structure of the PrP gene are similar in all of the mammals examined, to date. The entire open reading frame (ORF) of the PrP gene is contained within a single exon which is separated from one or two exons encoding the 5′ untranslated region of PrP mRNA by a large intron. Sequence homology among mammalian PrP molecules ranges from 85% to 97%. All the PrP gene ORFs sequenced encode prion proteins ranging in length from 253 to 264 amino acids as well as N-terminal signal peptides and C-terminal hydrophobic peptides both of which are removed as the proteins mature. All mammalian PrP molecules contain two consensus sites for N-linked glycosylation and a C-terminal GPI anchor and all possess five or six glycine and proline-rich octapeptide repeats near the N-terminus of the mature protein.

The cell surface localization of PrP^C together with its highly regulated expression in the brains of neonatal hamsters suggested that PrP^C might function in cell recognition (Mobley et al., 1988; Stahl et al., 1987). Of particular interest with respect to the possible function of mammalian PrP^C are studies of a putative chicken brain factor called acetylcholine receptor inducing activity or ARIA (Usdin and Fischbach, 1986) which shares many structural features with mammalian PrP (Falls et al., 1990; Harris et al., 1991). ARIA increases the rate of nicotinic acetylcholine receptor (nAchR) incorporation on the surface of cultured chicken myotubes by 3- to 5-fold when compared to untreated cultures (Usdin and Fischbach, 1986). This effect is mediated by an increase in the transcription rate of the mRNA coding for the nAchR alpha subunit in chicken myotubes (Harris et al., 1988) or for the nAchR epsilon subunit in mouse myotube (Falls et al., 1990; Martinou et al., 1991). ARIA has been recovered form SDS–PAGE in a region where proteins of 35–45 kDa migrate. Based on the N-terminal sequence of a putative ARIA protein purified by several steps of column chromatography and/or SDS–PAGE, cDNA clones were recovered and sequenced. The translated cDNA sequence revealed an ORF encoding a protein of 267 amino acids. On the amino acid level, this protein has ~33% sequence identity with mouse PrP (Falls *et al.*, 1990; Harris *et al.*, 1991). Moreover, there is a hydrophobic region where 24 consecutive amino acids are identical between the putative ARIA protein and the PrP of the I/Ln mouse (Westaway *et al.*, 1987). Based on this intriguing similarity, it was suggested that the putative ARIA protein may represent the chicken homologue of PrP (Falls *et al.*, 1990; Harris *et al.*, 1991). Whether ARIA and a candidate chicken PrP molecule copurify coincidentally or they are one and the same remains uncertain (Harris *et al.*, 1991). No ARIA has been demonstrated in extracts prepared from cells expressing the recombinant cDNA clone and attempts to precipitate or block ARIA bioactivity with antibodies raised to synthetic peptides, corresponding to the candidate chicken PrP, have been unsuccessful to date (Fall *et al.*, 1990; Harris *et al.*, 199). Purified mouse PrP^C has no ARIA bioactivity on chicken myotube (S. B. Prusiner, unpublished results). These findings prompted us to search for clones encoding a putative prion protein in chicken. We report here studies on a genomic clone encoding a candidate chicken PrP and describe its relationship to the mammalian PrP molecules.

ISOLATION OF GENOMIC CLONES ENCODING PrP

A chicken genomic DNA library was screened with the cP1 oligonucleotide sense probe depicted in Fig. 1 (Sambrook *et al.*, 1989). This probe encodes the first five codons of the signal peptide sequence predicted from the chicken PrP cDNA (Harris *et al.*, 1989). The initial screening was performed at 13°C below the calculated melting temperature (T_m) in order to detect imperfect hybridizing DNA sequences. During plaque purification of the clones, the filters of two clones consistently displayed a much weaker hybridizing signal than the filters of the other six clones. Each spot on the autoradiogram matched with a plaque on the master plate. The weak signal did not improve with confluent plaques when more DNA was bound to the filter. In contrast, the autoradiogram of all the clones isolated in a second experiment, where a more stringent temperature was chosen, showed an intense signal.

IDENTIFICATION OF TWO DISTINCT POPULATIONS OF CLONES

The purified λDNAs of six clones from the first experiment were restricted with SalI and the digestion product electrophoresed on a 0.35% agarose gel overnight. The insert varied in size from 11 to 20 kb.

Restriction enzyme analysis with EcoRI or NcoI followed by Southern blotting with either the 5′ end cP1 oligonucleotide sense probe or the 3′ end cP2 oligonucleotide antisense probe (Fig. 1) showed that five of the clones contained sequences homologous to both probes and revealed a similar restriction pattern. However, another clone (λ8 AA.1) did not contain a DNA sequence homologous to the 3′ end cP2

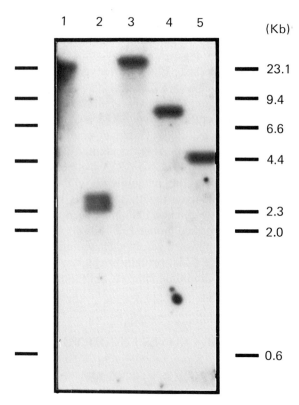

Fig. 1.   Southern blot of genomic DNA from white leghorn chicken probed with a 2.7 kb SalI: HindIII fragment containing the putative chicken PrP ORF. The fragment was radiolabelled by random nucleotide incorporation with the Klenow fragment of DNA polymerase and [^{32}P]dATP to a specific activity of 1.5×10^9 dpm/μg. The chicken DNA was restricted with BamHI (lane 1), BglI (lane 2), EcoRI (lane 3), HindIII (lane 4) and PstI (lane 5) as indicated.

oligonucleotide probe. It is unlikely that this clone was truncated at the 3′ end because Southern blotting with the 5′ end cP1 probe revealed a distinct restriction pattern. This was also confirmed with the HindIII restriction enzyme pattern as described below.

In order to screen all 18 clones for the presence of DNA sequences homologous with both oligonucleotide probes and to detect the presence of possible mismatches, plaque lifts were washed after hybridization at increasing stringency in the presence of 3 M TMACI (tetramethylammonium chloride).

The autoradiogram revealed that 16 clones contained DNA sequence which bind uniformly both probes at a temperature very close to their calculated melting temperatures (Jacobs et al., 1988; Wood et al., 1985). Two clones (λ18 AA.1 and λ18 BA.1) were shown to bind the 5′ end cP1 oligonucleotide at a much lower temperature than the other 16 positive clones. These findings suggest the presence of some mismatches in the formation of the heteroduplex between the probe and the target sequence. The two clones did not bind to the 3' end cP2 oligonucleotide probe and both gave similar patterns by restriction enzyme analysis with HindIII (results not shown).   To investigate further whether those clones with or without cP1 sequences were related to the putative chicken ARIA sequence found by Harris *et al.*, (1989), restriction analyses of clone *λ*7 AB.1 and clone *λ*8 AA.1 were followed by Southern blotting with the I-L/F antisense oligonucleotide probe of Westaway *et al.*, (1987) corresponding to the mouse I/Ln sequence encoding phenylalanine at codon 108. This probe corresponds to a part of the region where 24 residues are identical in the putative ARIA protein and I/Ln mouse PrP. Post-hybridization washes at 40, 48 and 53°C for 10 minutes showed binding for clone *λ*7 AB.1, but not for clone *λ*8 AA.1. Thus, clone *λ*7 AB.1 was thought to be related to both the putative chicken ARIA and I/Ln mouse PrP sequences. In contrast, clone *λ*8 AA.1 was found to be neither related to the putative ARIA nor similar to mouse PrP.

THE CHICKEN PrP GENE IS A SINGLE COPY GENE

Fig. 1 shows a Southern blot with chicken genomic DNA probed with the radiolabelled 2.7 kb insert fragment of 1 of the 16 candidate chicken PrP clones, *λ*7 AB.1. This probe detected a single band migrating at 23, >23, 8.1 or 4.6 kb of DNA digested with the restriction enzymes BamHI, EcoRI, HindIII or PstI respectively. In contrast, digestion with BglII gave two bands migrating at 2.7 or 2.3 kb. Identical Southern blots were obtained by washing at either low or high stringency. A Southern blot of one of the 16 candidate chicken PrP clones (*λ*10 BA.1) with the 5' end cP1 probe or the 3' end cP2 probe revealed an 8 kb HindIII fragment. These results are consistent with the results of the genomic Southern blot.

SUBCLONING AND SEQUENCING OF THE CHICKEN PrP CLONE pB7HS

Escherichia coli cells were transformed with the HindIII–SalI digested 2.7 kb fragment of clone *λ*7 AB.1 which was ligated into the HindIII–SalI digested pBluescript KS vector. Selection of colonies with the insert was not made by the blue–white selection method because formation of recombinant dimeric vectors lacking the insert is possible and would give a false positive signal. Instead, subclone selection was performed by restriction mapping miniprep DNAs from all the resulting colonies with ApaI, BamHI or XhoI restriction endonucleases. Both the supercoiled and the relaxed forms of recombinant plasmids with three subclones migrated as single bands of 5.6 kb after digestion. Southern blots with cP1 and cP2 probes confirmed the presence of the expected insert. Sequencing of the subclone pB7HS.1 was performed by primer walking and ambiguities were solved by sequencing the corresponding opposite strand. A single open reading frame (ORF) of 819 nucleotides (Fig. 2) encodes a protein of 273 amino acids, the deduced molecular weight of which is 29 912 daltons. Two translational initiator methionines are present at nucleotides 25 and 37; however, only the second seems to be functional. By comparing the 5' end cDNA sequence

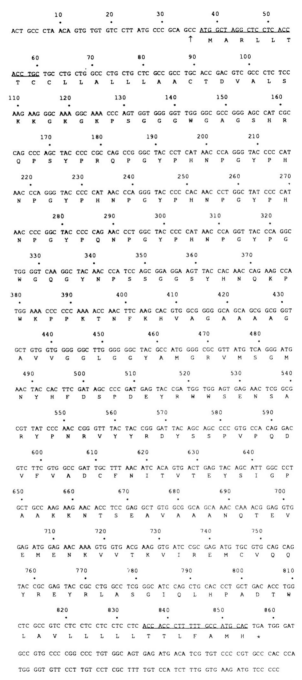

```
                10          20          30          40          50
                 •           •           •           •           •
        ACT GCC CTA ACA GTG TGT GTC CTT ATG CCC GCA GCC ATG GCT AGG CTC CTC ACC
                                                    ↑   M   A   R   L   L   T

          60          70          80          90          100
           •           •           •           •           •
        ACC TGC TGC CTG CTG GCC CTG CTG CTC GCC GCC TGC ACC GAC GTC GCC CTC TCC
         T   C   C   L   L   A   L   L   L   A   A   C   T   D   V   A   L   S

    110          120          130          140          150          160
     •            •            •            •            •            •
        AAG AAG GGC AAA GGC AAA CCC AGT GGT GGG GGT TGG GGC GCC GGG AGC CAT CGC
         K   K   G   K   G   K   P   S   G   G   G   W   G   A   G   S   H   R

              170          180          190          200          210
               •            •            •            •            •
        CAG CCC AGC TAC CCC CGC CAG CCG GGC TAC CCT CAT AAC CCA GGG TAC CCC CAT
         Q   P   S   Y   P   R   Q   P   G   Y   P   H   N   P   G   Y   P   H

          220          230          240          250          260          270
           •            •            •            •            •            •
        AAC CCA GGG TAC CCC CAT AAC CCA GGG TAC CCC CAC AAC CCT GGC TAT CCC CAT
         N   P   G   Y   P   H   N   P   G   Y   P   H   N   P   G   Y   P   H

              280          290          300          310          320
               •            •            •            •            •
        AAC CCC GGC TAC CCC CAG AAC CCT GGC TAC CCC CAT AAC CCA GGT TAC CCA GGC
         N   P   G   Y   P   Q   N   P   G   Y   P   H   N   P   G   Y   P   G

          330          340          350          360          370
           •            •            •            •            •
        TGG GGT CAA GGC TAC AAC CCA TCC AGC GGA GGA AGT TAC CAC AAC CAG AAG CCA
         W   G   Q   G   Y   N   P   S   S   G   G   S   Y   H   N   Q   K   P

    380          390          400          410          420          430
     •            •            •            •            •            •
        TGG AAA CCC CCC AAA ACC AAC TTC AAG CAC GTG GCG GGG GCA GCA GCG GCG GGT
         W   K   P   P   K   T   N   F   K   H   V   A   G   A   A   A   A   G

              440          450          460          470          480
               •            •            •            •            •
        GCT GTG GTG GGG GGC TTG GGG GGC TAC GCC ATG GGG CGC GTT ATG TCA GGG ATG
         A   V   V   G   G   L   G   G   Y   A   M   G   R   V   M   S   G   M

    490          500          510          520          530          540
     •            •            •            •            •            •
        AAC TAC CAC TTC GAT AGC CCC GAT GAG TAC CGA TGG TGG AGT GAG AAC TCG GCG
         N   Y   H   F   D   S   P   D   E   Y   R   W   W   S   E   N   S   A

              550          560          570          580          590
               •            •            •            •            •
        CGT TAT CCC AAC CGG GTT TAC TAC CGG GAT TAC AGC AGC CCC GTG CCA CAG GAC
         R   Y   P   N   R   V   Y   Y   R   D   Y   S   S   P   V   P   Q   D

          600          610          620          630          640
           •            •            •            •            •
        GTC TTC GTG GCC GAT TGC TTT AAC ATC ACA GTG ACT GAG TAC AGC ATT GGC CCT
         V   F   V   A   D   C   F   N   I   T   V   T   E   Y   S   I   G   P

    650          660          670          680          690          700
     •            •            •            •            •            •
        GCT GCC AAG AAG AAC ACC TCC GAG GCT GTG GCG GCA GCA AAC CAA ACG GAG GTG
         A   A   K   K   N   T   S   E   A   V   A   A   A   N   Q   T   E   V

              710          720          730          740          750
               •            •            •            •            •
        GAG ATG GAG AAC AAA GTG GTG ACG AAG GTG ATC CGC GAG ATG TGC GTG CAG CAG
         E   M   E   N   K   V   V   T   K   V   I   R   E   M   C   V   Q   Q

          760          770          780          790          800          810
           •            •            •            •            •            •
        TAC CGC GAG TAC CGC CTG GCC TCG GGC ATC CAG CTG CAC CCT GCT GAC ACC TGG
         Y   R   E   Y   R   L   A   S   G   I   Q   L   H   P   A   D   T   W

              820          830          840          850          860
               •            •            •            •            •
        CTC GCC GTC CTC CTC CTC CTC CTC ACC ACC CTT TTT GCC ATG CAC TGA TGG GAT
         L   A   V   L   L   L   L   L   L   T   T   L   F   A   M   H   *

        GCC GTG CCC CGG CCC TGT GGC AGT GAG ATG ACA TCG TGT CCC CGT GCC CAC CCA

        TGG GGT GTT CCT TGT CCT CGC TTT TGT CCA TCT TTG GTG AAG ATG TCC CCC
```

Fig. 2. Nucleotide sequence of chicken genomic clone pB7HS. The translated sequence is shown with the one letter amino acid code. Arrow denotes the 3′ splice junction. Star denotes translation termination codon. The underlined 5′ and 3′ sequences represent the oligonucleotide probes cP1 and cP2 derived from the cDNA sequence (Harris *et al.*, 1991) that were used for the initial screening.

(Harris *et al.*, 1991) with our genomic sequence, we identified a 3′ splice acceptor consensus site between nucleotide 28 and 34, where the sequences diverge. In chicken, the coding exon begins two nucleotides before the first functional methionine initiator codon; whereas, the hamster, mouse and sheep coding exons begin ten nucleotides before the methionine initiation codon (Basler *et al.*, 1986; Goldmann *et al.*, 1990a; Westaway *et al.*, 1987).

The protein sequence deducted from the chicken genomic clone is identical to that deduced from the cDNA clone (Harris *et al.*, 1991) except at two sites. First, the protein deduced from genomic DNA contains a serine at position 156 (AGC) instead of the arginine (AGA) encoded by the cDNA. Second, the genomic DNA pB7HS clone encodes nine imperfect hexapeptide repeats instead of eight encoded by the cDNA clone. At the nucleotide level, only three of the hexapeptide repeats deduced from the genomic DNA clone are identical while two repeats of the cDNA clone are identical. This finding suggests that the extra repeated element denoted HR3 (see Fig. 4) is located somewhere between residues 54 and 71 of the pB7HS clone. At the amino acid level, all of the hexapeptide repeats are identical, suggesting that these repeated elements have arisen by a duplication phenomenon which may have occurred quite recently in the White Leghorn chicken.

The second population of chicken genomic DNA clones was found to be unrelated to the clones described above. *E.coli* cells were transformed with the HindIII digested 3 kb fragment of clone λ8 AA.1 which has been previously ligated into the HindIII digested and dephosphorylated vector. DNAs were prepared form 12 colonies and aliquots digested with the SalI endonuclease. Southern blots with the cP1 probe identified one subclone (pB8H.8) displaying a single band migrating at 6 kb. Sequencing primed with the cP1 probe gave a deduced amino acid sequence that was less than 30 residues in length.

SEQUENCE HOMOLOGY

Significant sequence homology was found only between the chicken sequence and mammalian prion proteins that have been sequenced, to date when searches of the GenBank (release 68) or the PIR (release 28) databases were performed using the Fasta (Pearson and Lipman, 1988) or the Blast (Altschul *et al.*, 1990) programs. The chicken amino acid and DNA sequences were then aligned with those of 11 mammalian prion proteins (Fig. 3). Amino acid sequences were aligned by the Feng and Doolittle algorithm (Feng and Doolittle, 1987). On the nucleotide level, the chicken clone and mammalian PrP sequences were identical at 39–43% of the positions, whereas on the amino acid level, the chicken and mammalian PrP sequences were identical at 31–34% of the residues. When conservative substitutions are taken into account, the similarity increased to 48–50%. Most of the homology (88%) is derived from the C-terminal half of the chicken protein, which corresponds largely to the protease-resistant core of PrP[Sc], i.e. PrP 27-30.

```
               1          15           30            45            60            75            90

CkPrP    MARLLTTCCLLALLLAACTDVALSKKGKGKPSGGGWGAGSHRQPSYPRQPGYPHNPGYPHNPGYPHNPGYPHNPGYPHNPGYPQNPGYPH          90
Bov6PrP          VKSHIGSWI V FV MWS   G C  RPKPGG WNT GSRYPGQGS GGNR  PQG GGWGQPHGGGW Q  GG WGQPH GGWGQPHGG        90
Bov5PrP          VKSHIGSWI V FV MWS   G C  RPKPGG WNT GSRYPGQGS GGNR  PQG GGWGQPHGGGW Q--------PH GGWGQPHGG        82
ShePrP           VKSHIGSWI V FV MWS   G C  RPKPGG WNT GSRYPGQGS GGNR  PQG GGWGQPHGGGW Q--------PH GGWGQPHGG        82
Rt*PrP           NL--GYW   FVTMW     G C  RPKP-G WNT GSRYPGQGS GGNR  PQS GTWGQPHGGGW Q--------PH GGWGQPHGG       (59)*
MoPrP-A          NL--GYW   FVTMW     G C  RPKP-G WNT GSRYPGQGS GGNR  PQG -TWGQPHGGGW Q--------PH GSWGQPHGG        86

MoPrP-B          NL--GYW   FVTMW     G C  RPKP-G WNT GSRYPGQGS GGNR  PQG -TWGQPHGGGW Q--------PH GSWGQPHGG        86
AHaPrP           NL--SYW   FV TW     G C  RPKP-G WNT GSRYPGQGS GGNR  PQG GTWGQPHGGGW Q--------PH GGWGQPHGG        87
CHaPrP           NL--SYW   FV TW     G C  RPKP-G WNT GSRYPGQGS GGNR  PQG GTWGQPHGGGW Q--------PH GGWGQPHGG        87
SHaPrP           NL--SYW   FV MW     G C  RPKP-G WNT GSRYPGQGS GGNR  PQG GTWGQPHGGGW Q--------PH GGWGQPHGG        87
MinkPrP          VKSHIGSW V FV TWS IGFC  RPKPGG WNT GSRYPGQGS GGNR  PQG GGWGQPHGG-W Q--------PH GGWGQPHGG        81
HuPrP            NL--G WM V FV TWS LG C  RPKP-G WNT GSRYPGQGS GGNR  PQG GGWGQPHGGGW Q--------PH GGWGQPHGG        87

               91        105          120           135           150           165           180

CkPrP    NPGYPGWGQGYNPSSGGSYHNQKPWKPPKTNFKHVAGAAAAGAVVGGLGGYAMGRVMSGMNYHFDSPDEYRWWSENSARYPNRVYYRDYS         180
Bov6PrP          GW Q HG G WGQGG-THGQWN   S   -    M              ML SA  RPLI  G DY D YYR  MH      Q    PVD        178
Bov5PrP          GW Q HG G WGQGG-THGQWN   S   -    M              ML SA  RPLI  G DY D YYR  MH      Q    PVD        170
ShePrP           GW Q HG G WGQGG-SHSQWN   S   -    M              ML SA  RPLI  GNDY D YYR  MY      Q    PVD        170
Rt*PrP           GW Q HG - WGQGG THNQWN   S   -    L              ML SA  RPML  GNDW D YYR  MY      Q    PVD       (139)*
MoPrP-A          SW Q HG - WGQGG THNQWN   S   -    L              ML SA  RPMI  GNDW D YYR  MY      Q    PVD        166

MoPrP-B          SW Q HG - WGQGG THNQWN   S   -                   ML SA  RPMI  GNDW D YYR  MY      Q    PVD        166
AHaPrP           GW Q HG - WGQGG THNQWN   N   - SM M              ML SA  RPML  GNDW D YYR  MN      Q    PVD        167
CHaPrP           GW Q HG - WGQGG THNQWN   S   -    M              ML SA  RPML  GNDW D YYR  MN      Q    PVD        167
SHaPrP           GW Q HG - WGQGG THNQWN   S   -    M M            ML SA  RPMM  GNDW D YYR  MN      Q    PVD        167
MinkPrP          GW Q HG G WGQGG SHGQWG   S   -    M              ML SA  RPLI  GNDY D YYR  MY      Q    KPVD       170
HuPrP            GW Q HG - WGQGG THSQWN   S   -    M M            ML SA  RPII  G DY D YYR  MH      Q    PMD        167

               181       195          210           225           240           255           270

CkPrP    SPVPQDVFVADCFNITVTEYSIGPAAKKNTSEAVAAANQTEVEMENKVVTKVIREMCVQQY-REY-------RLASGIQLHPADTWLAVL         262
Bov6PrP          QYSN NN  H  V    K HTVTTTT ----------GENFT TDI MMER VEQ  IT Q  SQAYYQ-- G  V LFSSPP---VI         256
Bov5PrP          QYSN NN  H  V    K HTVTTTT ----------GENFT TDI MMER VEQ  IT Q  SQAYYQ-- G  V LFSSPP---VI         248
ShePrP           RYSN NN  H  V    KQHTVTTTT ----------GENFT TDI IMER VEQ  IT Q  SQAYYQ-- G  V LFSSPP---VI         248
Rt*PrP           QYSN NN  H  V    IKQHTVTTTT ---------GENFT TDV MMER VEQ     T QK SQAYYDGR -S AVLFSSPP---VI       (218)*
MoPrP-A          QYSN NN  H  V    IKQHTVTTTT ---------GENFT TDV MMER VEQ     T QK SQAYYDGR SS TVLFSSPP---VI        246

MoPrP-B          QYSN NN  H  V    IKQHTVVTTT ---------GENFT TDV MMER VEQ     T QK SQAYYDGR SS TVLFSSPP---VI        246
AHaPrP           QYNN NN  H  V    IKQHTVTTTT ---------GENFT TDV MMER VEQ     T QK SQAYYDGR -S AVLFSSPP---VI        246
CHaPrP           QYNN NN  H  V    IKQHTVTTTT ---------GENFT TDV MMER VEQ     T QK SQAYYDGR -S AVLFSSPP---VI        246
SHaPrP           QYNN NN  H  V    IKQHTVTTTT ---------GENFT TDI IMER VEQ  TT QK SQAYYDGR -S AVLFSSPP---VI         246
MinkPrP          QYSN NN  H  V    KQHTVTTTT ----------GENFT TDM IMER VEQ     T Q  SQAYYQ-- G  A LFS PP---VI        248
HuPrP            EYSN NN  H  V    IKQHTVTTTT ---------GENFT TDV MMER VEQ  IT E  SQAYYQ-- GS MVLFSSPP---VI         245

               271       285

CkPrP    LLLLTTLFAMH     273
Bov6PrP          ISFLIFLIVG     264
Bov5PrP          ISFLIFLIVG     256·
ShePrP           ISFLIFLIVG     256
Rt*PrP           ISFLIFLIVG    (226)*
MoPrP-A          ISFLIFLIVG     254

MoPrP-B          ISFLIFLIVG     254
AHaPrP           ISFLIFLIVG     254
CHaPrP           ISFLIFLIVG     254
SHaPrP           ISFLIFLMVG     254
MinkPrP          IS LI LIVG     256
HuPrP            ISFLIFLIVG     253
```

Fig. 3.   Alignment of the putative chicken PrP amino acid sequence with 11 mammalian PrP sequences.
*, missing rat sequence replaced by mouse.

GLYCINE–PROLINE TANDEM REPEATS

Bovine PrP is encoded by two distinct alleles which contain either five (OR1, OR2, OR4, OR5, OR6) or six (OR1, OR2, OR3, OR4, OR5, OR6, Figure 4) glycine–proline-rich nona- or octapeptide repeats (M. Scott *et al.*, unpublished results; Goldmann *et al.*, 1991). In contrast, the mink, sheep, human, hamster, rat and mouse

```
repeats#     OR'                  HR1    HR2    HR3    HR3    HR3    HR4    HR5    HR6    HR7

CkPrP        K PSGGGWGA GSH RQPSYP RQPGYP HNPGYP HNPGYP HNPGYP HNPGYP HNPGYP QNPGYP HNPGYP GWGQGYNPSSGGSYHNQ-KP

Bov6PrP      K PGGGWNT GGSRYP GQGSP GGNRYP PQGGGGWGQ PHGGGWGQ PHGGGWGQ PHGGGWGQ PHGGGWGQ PHGGGGWGQ GG-THGQWNKP
Bov5PrP      K PGGGWNT GGSRYP GQGSP GGNRYP PQGGGGWGQ PHGGGWGQ -------- PHGGGWGQ PHGGGWGQ PHGGGGWGQ GG-THGQWNKP
ShePrP       K PGGGWNT GGSRYP GQGSP GGNRYP PQGGGGWGQ PHGGGWGQ -------- PHGGGWGQ PHGGGWGQ PHGGGGWGQ GG-SHSQWNKP
Rt*PrP       K P-GGWNT GGSRYP GQGSP GGNRYP PQSGGTWGQ PHGGGWGQ -------- PHGGGWGQ PHGGGWGQ PHGG-GWSQ GGGTHNQWNKP
MoPrP-A      K P-GGWNT GGSRYP GQGSP GGNRYP PQGG-TWGQ PHGGGWGQ -------- PHGGGWGQ PHGGGWGQ PHGG-GWGQ GGGTHNQWNKP
MoPrP-B      K P-GGWNT GGSRYP GQGSP GGNRYP PQGG-TWGQ PHGGGWGQ -------- PHGGGWGQ PHGGGWGQ PHGG-GWGQ GGGTHNQWNKP
AHaPrP       K P-GGWNT GGSRYP GQGSP GGNRYP PQGGGTWGQ PHGGGWGQ -------- PHGGGWGQ PHGGGWGQ PHGG-GWGQ GGGTHNQWNKP
CHaPrP       K P-GGWNT GGSRYP GQGSP GGNRYP PQGGGTWGQ PHGGGWGQ -------- PHGGGWGQ PHGGGWGQ PHGG-GWGQ GGGTHNQWNKP
SHaPrP       K P-GGWNT GGSRYP GQGSP GGNRYP PQGGGTWGQ PHGGGWGQ -------- PHGGGWGQ PHGGGWGQ PHGG-GWGQ GGGTHNQWNKP
MinkPrP      K PGGGWNT GGSRYP GQGSP GGNRYP PQGGGGWGQ PHGG-WGQ -------- PHGGGWGQ PHGGGWGQ PHGGGGWGQ GGGSHGQWGKP
HuPrP        K P-GGWNT GGSRYP GQGSP GGNRYP PQGGGTWGQ PHGGGWGQ -------- PHGGGWGQ PHGGGWGQ PHGG-GWGQ GGGTHSQWNKP

repeats#     OR"     HR'          HR"     OR1       OR2      OR3      OR4      OR5      OR6
```

Fig. 4. Organization of glycine–proline rich hexa- and octarepeats in vertebrate prion proteins. Boxed residues delimit the extent of the repeats. HR, hexapeptide repeats; OR, octa- or nonapeptide repeats ;–, gap; boldfaced GWGQ, residues encoded by identical nucleotide sequences across species.

prion proteins have five repeats (Fig. 4). All of the repeats contain a trypotophan residue. To assess where the extra repeat of the bovine PrP might have arisen, we manipulated the alignment and found it to be optimal when a gap of 24 nucleotides is introduced between the second and the third sheep repeats. The prototype structure of the bovine repeats is P(Q/H)GGG(–/G)WGQ, as shown in Fig. 4. Interestingly, the first OR1 and last OR6 repeats of the bovine or sheep sequences contain four glycines, whereas the other three or four middle repeats contain three glycines. The fact that these three or four middle tandem repeats are identical suggests that these duplications might have arisen more recently. In contrast, the first OR1 and last OR6 repeats are not identical. These findings suggest that an ancestral sequence containing four glycines may have been duplicated first, and subsequent duplication or unequal crossing-over events occurred between the two 'anchoring' repeats producing tandem repeats containing three glycines. Thus the extra OR3 repeat present in the Bov6PrP allele must have occurred more recently, because all other mammal prion proteins contain only five repeats (OR1, OR2, OR4, OR5, OR6).

It seems likely that the situation is similar in the chicken. The chicken PrP genomic sequence contains nine hexapeptide repeats and all of them contain a characteristic tyrosine, which appears to be a conservative substitution for the tryptophan residue of the mammalian repeats. Since the chicken repeats are nearly identical with each other, this suggests that the same genetic events responsible for the mammalian repeats occurred independently in the chicken. However, when inspecting the sequence preceding the nine chicken glycine–proline hexapeptide repeats, there is a PSGGGWGA sequence (residues 31 to 38, designated OR′) which is reminiscent of the bovine OR2–5 octarepeats. There is a P–(–/G)GGWNT sequence designated OR″ at a similar location in the mammalian PrP. Furthermore, the chicken GWGQ sequence (boldface residues 96 to 99, Fig. 4) following the ninth HR7 hexapeptide repeat is identical on the nucleotide level with the end of the sixth bovine octarepeat OR6), or with the end of the fifth octarepeat (OR6) of mink, sheep, human, Armenian, Chinese and Syrian hamster octarepeat (Fig. 4). Because the greatest differences between the avian and the mammalian prion proteins occurred in the hexapeptide repeat region (Fig. 3), the conserved nucleotides encoding the PrP sequence, i.e. a breaking–joining site which has been conserved during vertebrate evolution. In support of this hypothesis is the deletion of a glycine downstream of this site in both bovine and sheep PrP, whereas a glycine has been deleted upstream of this site in the other mammalian prion proteins except for the mink which has no deletions around this site. These findings suggest that the entire nine avian-type hexapeptide repeat domain of the chicken may have been independently created between the bovine-like 'ancestor' octapeptide repeat (OR′) and GWGQ sequence during the speciation of birds.

Although distinct in amino acid sequence and thus independently created, both the avian hexarepeat and the mammalian octarepeat have similar physicochemical consensus sequences: for birds, polar–small–small–aromatic –small–polar (typified by NPGYPH) and for mammals, small–polar–small–small–small–aromatic–small–polar (typified by PHGGGWGQ) respectively. Thus these repeats may have the propensity to form a closely related structure rich in coil and turn, as shown by the roughly

```
                          10              14
                        spacer       hydrophobic
                   ************^^^^^^^^^^^^^^^
  CkPrP            Y---RLASGIQLHPADTWLAVLLLLLTTLFAMH    273

  Bov6PrP          YQ--RGASVILFSSPP---VILLISFLIFLIVG    264
  Bov5PrP          YQ--RGASVILFSSPP---VILLISFLIFLIVG    256
  ShePrP           YQ--RGASVILFSSPP---VILLISFLIFLIVG    256
  RtPrP            YDGRR-SSAVLFSSPP---VILLISFLIFLIVG    226
  MoPrP-A          YDGRRSSSTVLFSSPP---VILLISFLIFLIVG    254
  MoPrP-B          YDGRRSSSTVLFSSPP---VILLISFLIFLIVG    254
  AHaPrP           YDGRR-SSAVLFSSPP---VILLISFLIFLIVG    254
  CHaPrP           YDGRR-SSAVLFSSPP---VILLISFLIFLIVG    254
  SHaPrP           YDGRR-SSAVLFSSPP---VILLISFLIFLMVG    254
  MinkPrP          YQ--RGASAILFSPPP---VILLISLLILLIVG    256
  HuPrP            YQ--RGSSMVLFSSPP---VILLISFLIFLIVG    253
                    **********    ^^^^^^^^^^^^^^^
                          10              14
                        spacer       hydrophobic
```

Fig. 5. Conserved signal sequence for GPI anchor addition in vertebrate prion proteins. The boldfaced serine is the GPI anchor addition site in the SHaPrP. The italicized glycine or the preceding serine in chicken PrP is predicted to be the GPI anchor addition site.

homologous Robson and Garnier structure prediction patterns (see Fig. 7). Although the overall homology between the chicken and mammalian prion proteins is low, the remarkable similarity raises the possiblity that this may be an example of structural convergence during evolution.

While chicken genomic DNA clones described here encode nine hexarepeats, the cDNA clone was found to encode eight hexarepeats (Harris *et al.*, 1991). Whether this difference reflects genetic variation amongst chickens similar to that seen with cattle or it is due to a cloning artefact remains to be established.

C-TERMINAL SIGNAL SEQUENCE AND GPI ANCHORS

The signal sequence at the *C*-terminus of chicken protein differs from the mammalian sequences which are highly conserved (Fig. 5). An optimum alignment of the sequences surrounding the GPI anchor addition site which has been demonstrated to occur on the first serine (boldfaced residue) following two arginine residues in the Syrian hamster is shown in Fig. 5 (Stahl *et al.*, 1990). If one assumes that the rules of adding a GPI anchor are the same between mammalian and avian prion proteins (i.e. 14 hydrophobic residues at the *C*-terminus and a linker of 10 amino acids), we predict that the GPI anchor of the chicken protein will be added to either the glycine (italicized) or the preceding serine. Interestingly, the chicken RLAS sequence preceding the potential GPI anchor addition site is more homologous corresponding domain of the bovine or sheep RGAS than to the corresponding domains of other mammals. In general,

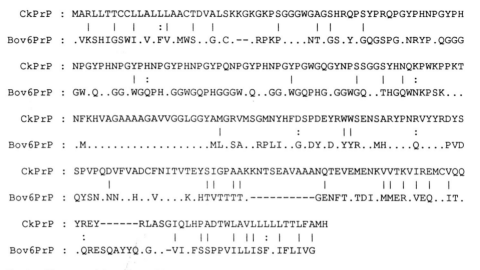

```
CkPrP   : MARLLTTCCLLALLLAACTDVALSKKGKGKPSGGGWGAGSHRQPSYPRQPGYPHNPGYPH
            |   |   |   :|   |   |       |    |  |   |
Bov6PrP : .VKSHIGSWI.V.FV.MWS..G.C.--.RPKP....NT.GS.Y.GQGSPG.NRYP.QGGG


CkPrP   : NPGYPHNPGYPHNPGYPHNPGYPQNPGYPHNPGYPGWGQGYNPSSGGSYHNQKPWKPPKT
                    |  :                      |          |    |   | :
Bov6PrP : GW.Q..GG.WGQPH.GGWGQPHGGGW.Q..GG.WGQPHG.GGWGQ..THGQWNKPSK...


CkPrP   : NFKHVAGAAAAGAVVGGLGGYAMGRVMSGMNYHFDSPDEYRWWSENSARYPNRVYYRDYS
                            |              :        ||            :
Bov6PrP : .M.................ML.SA..RPLI..G.DY.D.YYR..MH....Q....PVD


CkPrP   : SPVPQDVFVADCFNITVTEYSIGPAAKKNTSEAVAAANQTEVEMENKVVTKVIREMCVQQ
                 |              ||   ||           |        || | | |  |
Bov6PrP : QYSN.NN..H..V....K.HTVTTTT.---------GENFT.TDI.MMER.VEQ..IT.


CkPrP   : YREY------RLASGIQLHPADTWLAVLLLLLTTLFAMH
            :              |    ||    | || : |   | |
Bov6PrP : .QRESQAYYQ.G..-VI.FSSPPVILLISF.IFLIVG
```

Fig. 6.   Alignment of the amino acid sequence of putative chicken PrP with the bovine PrP allele encoding 6 octarepeats. Dots (.) denote identical residues; dashes (–) signify gaps; vertical lines (|) designate strongly conservative changes; colons (:) denote weakly conservative substitutions. Scores are 94 identical + 39 strong + 6 weak conservative over 279 aligned residues, i.e. 33.7% identical and 49.8% conservative homologous, with 1 gap in chicken and 3 in bovine PrP.

GPI anchors seem to be covalently bound to amino acids with small side chains such as G, A, or S, or bulky side chains such as N, D, or C (see Ferguson and Williams, 1988; Moran *et al.*, 1991).

COMPARISON OF CHICKEN AND BOVINE PROTEINS

The high identity score was obtained when the deduced amino acid sequence of chicken PrP clone was aligned by the pairwise alignment program of Altschul and Erickson (1986) with the amino acid sequence of bovine PrP containing six octarepeats. With one gap introduced in the chicken and three gaps in the bovine PrP sequences, the two species are identical at 33.7% over their ORFs. The degree of similarity increased to 49.8% when conservative substitutions are taken into account (Fig. 6). The protease-resistant core of PrP^{Sc} designated PrP 27-30 extends form codon 90 to 231 in the Syrian hamster. If we consider only this region and, thus, exclude the signal peptide, the proline–glycone-rich repeats and the hydrophobic *C*-terminus signal sequence of both the chick and bovine sequences, then the homology increases to 42.8% and the degree of similarity considering conservative substitutions rises to 55.4% between the avian and the bovine molecules over the 159 aligned residues.

To characterize further chicken PrP, hydropathy analyses were performed and compared with the bovine PrP, as shown in Fig. 7. Both the chicken and bovine proteins display three major hydrophobic peaks which are almost superimposable. These three hydrophobic domains represent the *N*-terminal signal peptide, the middle

(a)

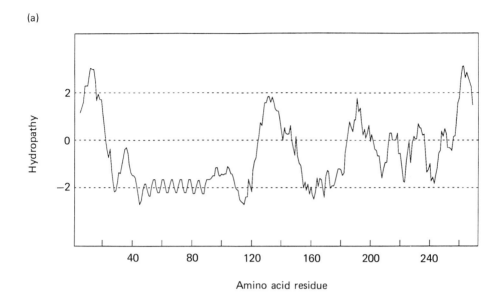

(b)

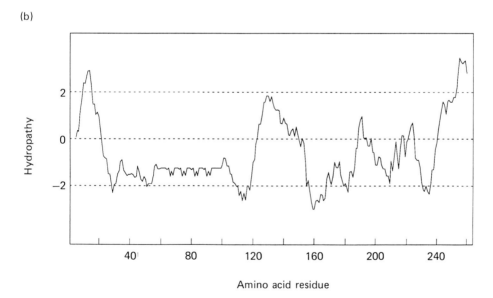

Fig. 7. Secondary structure predictions for (A) the putative chicken PrP and (B) bovine PrP using the Kyte and Doolittle algorithm.

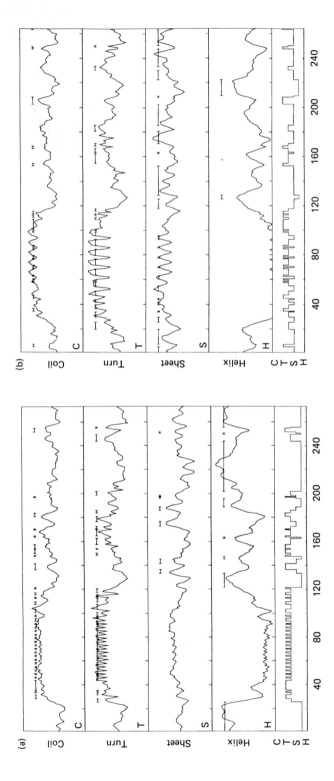

Fig. 8. Secondary structure predictions for (A) the putative chicken PrP and (B) bovine PrP using the Robson and Garnier algorithm. C, coil; H, helix; S, sheet; T, turn.

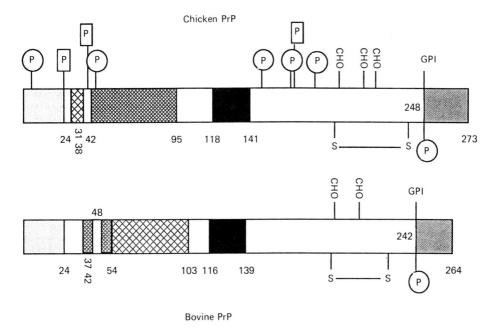

Fig. 9.   Structural features of the putative chicken PrP and bovine PrP. Ⓟ, cam kinase II phosphorylation site; ☐P, protein kinase C phosphorylation site; GPI, GPI anchor addition site; ▢, N-terminal hydrophobic signal peptide; ▨, C-terminal hydrophobic signal peptide; ■, domain with 23 identical residues; ▨, Pro–Gly-rich hexapeptide repeats; ⊗, Pro–Gly-rich octa- or nonapeptide repeats.

hydrophobic domain which is long enough to span the membrane bilayer and the C-terminal signal sequence. Moreover, both the repeat domain and the two small hydrophilic peaks containing the cysteines show a similar pattern.

In order to refine the analysis, the secondary structure of the chicken protein was predicted by the algorithm of Garnier *et al.*, (1978) and was compared with bovine PrP, as shown in Fig. 8. Although not perfectly superimposable, the patterns of avian and bovine proteins share many common structural motifs. Although the sequences of the glycine and proline-rich repeats of the chicken are different from those of bovine PrP, both domains are predicted to be rich in beta turns and random coils. The domains bounded by the disulphide loops have roughly the same predicted structure for both the chicken and bovine molecules.

Both the chicken and bovine deduced proteins have N-terminal hydrophobic signal peptides of 24 amino acids (Fig. 9). The chicken molecule has nine glycine and proline-rich hexapeptide repeats (residues 42 to 95) while the bovine has six nona- or octapeptide repeats (residues 54 to 103 respectively). The chicken contains one bovine-like octapeptide repeat (between residues 31 and 38) whereas the bovine contains two avian-like hexarepeats (between residues 37 and 42 or 48 and 53). A domain of 23 hydrophobic residues is identical in the chicken (residues 118–141) and bovine (residues 116–139) PrP molecules. Both sequences encode a single, conserved disulphide bond (chicken 192–237 cysteines and bovine 190–225 cysteines)

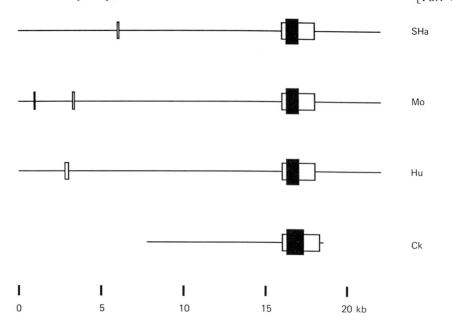

Fig. 10.   Organization of PrP genes. Exons (□) and introns shown with the PrP open reading frame (■).

and C-terminal hydrophobic signal sequences which are presumably cleaved upon GPI anchor addition. There are three avian consensus sites for N-linked glycosylation at Asn residues 194, 209 and 218 and two bovine sites at Asn residues 192 and 208. There is one potential calcium dependent calmodulin (Cam) kinase II phosphorylation site (consensus: RxxS/T (Payne *et al.*, 1983; Pearson *et al.*, 1985)) at serine 248 of the avian sequence and serine 242 of the bovine sequence. It seems unlikely that these residues are phosphorylated since these amino acids or the succeeding residues are covalently attached to GPI anchors. Chicken PrP may contain five potential Cam kinase II phosphorylation sites at serines 6, 45, 148, 164 and 180 and three potential protein kinase C phosphorylation sites at serines 24, 40 and 167 (consensus: SxR/K (Woodgett *et al.*, 1986)). Protein chemistry studies have failed to reveal any phosphorylation of serine, threonine or tyrosine residues in SHa PrP (N. Stahl *et al.*, in preparation). Whether the chicken protein undergoes phosphorylation remains to be determined.

GENE STRUCTURE AND ORGANIZATION

The entire chicken ORF was found to be contained within a single intronless exon like several mammalian prion protein genes (Fig. 10). The PrP gene structure and organization have been elucidated for the Syrian hamster (Basler *et al.*, 1986), the mouse (Westaway *et al.*, 1991) and the human (Puckett *et al.*, 1991). In all three cases, a splice junction lies near the 5′ end of the large exon containing the ORF. One or

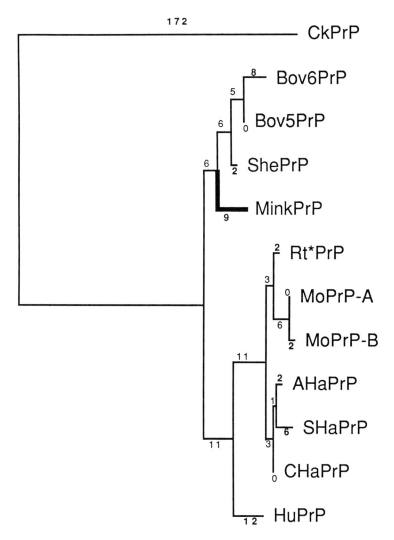

Fig. 11. Phylogram of 12 vertebrate prion proteins. Numbers calculated by the phylogenetic analysis using parsimony (PAUP) method for the amino acid sequence are represented by the horizontal distance between internal nodes or between a node and a terminal taxon. Thick line denotes a phylogenetic incongruency.

two small exons encode most of the 5′ untranslated region of the PrP mRNAs. While the exons encoding the 5′ untranslated segment of the chicken protein are unknown, the splice junction appears to occur 2 bp upstream from the first codon as deduced from comparing the cDNA sequence (Harris et al., 1991) with the genomic sequence (Fig. 2).

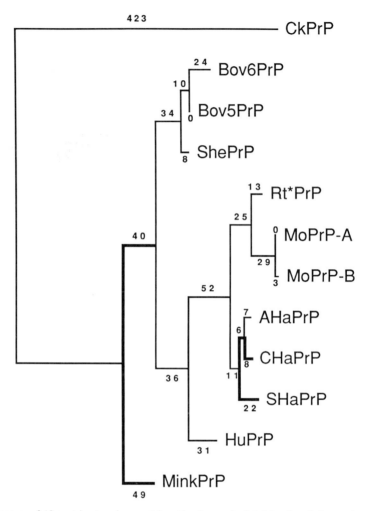

Fig. 12. Phylogram of 12 vertebrate prion proteins. Numbers calculated by the phylogenetic analysis using parsimony (PAUP) method for the nucleotide sequence are represented by the horizontal distance between internal nodes or between a node and a terminal taxon. Thick line denotes major differences with Fig. 11.

MOLECULAR PHYLOGENY OF AVIAN AND MAMMALIAN PRION PROTEINS

The vertebrate amino acid and DNA sequence alignments (Fig. 3) obtained by the conventional distance matrix method of Feng and Doolittle (1990) were then used to construct phylogenetic trees. The distance matrix-based trees were then compared with the trees obtained by either the conventional nearest-neighbour character-based matrix PAPA method of Doolittle and Feng (1990) or by the PAUP maximum parsimony method of Swofford (1991). All three methods gave identical topologies, differing primarily in the lengths of various branches. The general consequence of the phylogenetic trees formed by these three different algorithms indicates that the proposed relationships are likely to be valid. Fig. 11 shows a phylogram of the chicken protein and the 11 mammalian prion proteins obtained by PAUP analysis of the amino acid sequences. The horizontal distances with their values indicated between internal nodes or between a node and a terminal taxon provide a measure of relatedness. Vertical distances are not meaningful and were created only for clarity. With the exception of the mink sequence clustering with the artiodactyla, the molecular evolution of the vertebrate prion proteins recapitulates classical phylogeny. However, the relationship between all the mammalian taxa cannot be definitively resolved owing to the high degree of sequence similarity (85–97%). This is illustrated by the fact that phylograms obtained by PAUP method performed at the DNA level were similar except for the branching order of mink, Chinese hamster and Syrian hamster, as illustrated in Fig. 12. At the protein level, the Chinese hamster seems to be the ancestor of the clustered Armenian and Syrian hamster; whereas, on the DNA level the Syrian hamster branches off earlier than the Armenian or Chinese hamsters.

MOLECULAR EVOLUTION OF AVIAN AND MAMMALIAN PRION PROTEINS

To trace the evolutionary history of the vertebrate PrP genes, the sequence divergence of the coding region was evaluated by measuring two distinct types of substitutions: those leading to amino acid change or replacement (d_r) and those leading to synonymous or silent change (d_s). Obviously, the former is under the influence of selective constraints at both protein and RNA levels, whereas the latter is under the influence of selective constraint at the RNA level alone. The divergence is defined as the number of mismatches per nucleotide site in a pair of aligned sequence. Thus a gap is counted as a substitution, reflecting either an insertion or a deletion, and was excluded from calculation. Table 1 shows the divergence values at silent or replacement sites (Perler et al., 1980). Divergence values at silent sites are always greater than those at replacement sites. The chicken protein shows considerably higher values for d_s and d_r when compared with the mammalian prion proteins than among the mammals. Greater silent divergence was observed between chicken and artiodactyla (1,76–1.81) than between chicken and rodents (1.10–1.41), but replacement divergence is unchanged. This is surprising because artiodactyla are thought to be closer to avians than the rodents are to avains. In this respect, it has been observed that silent nucleotide substitutions in different mammalian genes do not have the same molecular clock (Bulmer et al., 1991). Moreover, it has been proposed that the rate of substitution is much slower in human than in other mammals, and that the rodent rate is higher than that in human or in artiodactyla (Li et al., 1987; Li and Tanimura 1987). Another study points out the slowest rates in higher primates and some bird lineages, while faster rates are seen in rodents (Britten, 1986). Functional constraints imposed upon the translation product of a gene can be estimated by the ratio of silent to replacement substitutions, The ratio d_s/d_r for the chicken protein compared with mammalian prion proteins is between 1.5 (mouse) and 2.5 (bovine; mean $\pm$ 0.39). In contrast, the mean

Table 1. Corrected divergence of vertebrate PrP genes[a]

	CkPrP		ShePrP		Bov5PrP		Bov6PrP		RtPrP[b]		MoPrPA		MoPrPB		ShaPrP		AHaPrP		CHaPrP		MinkPrP	
	d_s	d_r	d_s	d_r	d_s	d_r	d_s	d_r	d_s	d_r	d_s	d_r	d_s	d_r	d_s	d_r	d_s	d_r	d_s	d_r	d_s	d_r
CkPrP	—	—	—	—	—	—	—	—	—	—	—	—	—	—	—	—	—	—	—	—	—	—
ShePrP	1.79	0.71	—	—	—	—	—	—	—	—	—	—	—	—	—	—	—	—	—	—	—	—
Bov5PrP	1.81	0.71	0.12	0.02	—	—	—	—	—	—	—	—	—	—	—	—	—	—	—	—	—	—
Bov6PrP	1.76	0.71	0.12	0.02	0.00	0.00	—	—	—	—	—	—	—	—	—	—	—	—	—	—	—	—
RtPrP[b]	1.36	0.72	0.83	0.09	0.86	0.08	0.86	0.08	—	—	—	—	—	—	—	—	—	—	—	—	—	—
MoPrpA	1.10	0.72	0.98	0.09	1.02	0.08	1.02	0.08	0.21	0.01	—	—	—	—	—	—	—	—	—	—	—	—
MoPrPB	1.11	0.71	0.97	0.09	1.01	0.08	1.01	0.08	0.21	0.01	0.00	0.01	—	—	—	—	—	—	—	—	—	—
SHaPrP	1.24	0.73	0.81	0.09	0.80	0.11	0.80	0.11	0.47	0.07	0.61	0.11	0.60	0.10	—	—	—	—	—	—	—	—
AHaPrP	1.41	0.72	0.95	0.08	0.94	0.07	0.94	0.07	0.32	0.02	0.45	0.02	0.45	0.03	0.27	0.05	—	—	—	—	—	—
CHaPrP	1.27	0.71	0.85	0.08	0.84	0.07	0.84	0.07	0.35	0.02	0.43	0.02	0.43	0.02	0.21	0.05	0.11	0.01	—	—	—	—
MinlPrP	1.19	0.72	0.75	0.04	0.76	0.05	0.76	0.05	0.70	0.08	0.70	0.08	0.70	0.09	0.61	0.10	0.67	0.08	0.60	0.08	—	—
HuPrP	1.29	0.72	0.59	0.05	0.60	0.04	0.60	0.04	0.57	0.06	0.64	0.05	0.64	0.06	0.50	0.09	0.53	0.05	0.49	0.05	0.60	0.07

[a] Corrected (for multiple hits) divergence values for silent sites (d_s) or replacement site (d_r) were calculated for pairs of genes according to Perler *et al.*, (1980).
[b] The missing N-terminal sequence of the rat PrP has been substituted with that of the mouse PrP sequence to provide a realistic comparison index.

d_s/d_r ratios for human–mink–artiodactyla, artiodactyla–rodents, among rodent, and human–mink–rodents are 15.17 ± 2.2, 11.1 ± 1.9, 13.9 ± 7.2 and 8.9 ± 2, respectively.

It has been previously observed that the rate of silent substitutions is constant for different genes and may therefore be used as an evolutionary clock (Miyata and Hayashida, 1982). On one hand, calculation of divergence time between the hamsters and human gave a value of 77–90 million years (MYr). On the other hand, calculation of divergence time between the mouse and human gave a value of 141 MYr. This last value is twice the estimated time of divergence of 75 MYr obtained by the classical paleontological data. With the value of 75 MYr, we calculated a rate of silent substitution of about 1×10^9 substitution per site per year. Similar results were obtained by comparing the chicken with the human, with an estimated time of divergence of about 250–300 Myr. This is 5 times lower than the seemingly constant value of Miyata and Hayashida (1982). However, as mentioned above, the molecular clock (rate of silent substitution) is not constant between different species. The examples described above suggest that the rate of substitution present in the hamster PrP gene is approximately equal to the human rate. In contrast to the evolution of the cricetidae (hamsters), the rate of substitution in mouse and rat PrP genes seem to increase greatly during the recent evolution of the muridae. Calculation of divergence time between the mouse and the Armenian, Chinese and Syrian hamster gave a value of 67, 62 and 118 MYr, respectively. This is 2–3 times the estimated time of rodent divergence of 35–45 MYr obtained by the paleontological data. This is also 5 times higher than the value obtained by Locht et al., (1986) for the mouse and Syrian hamster pair. This discrepency is explained by the following. According to their figure, they first miscalculated the frequency of synonymous changes as 57/254 = 0.22 (it should be 53/254 = 0.21). Second, they only considered substitution at the third base of a codon as silent substitution. This underestimates degenerate codons such as serine, arginine or leucine. Third, they calculated the absolute rate of silent substitutions at the third nucleotide level, with no corrections for multiple hits such as insertion, deletion or transversion. This is fundamentally different to the procedure of Perler et al., (1980) which calculates synonymous changes at the codon level with several correction factors applied for the 'path' to the codon change (degeneracy of the genetic code).

DOES THE CHICKEN GENE ENCODE A PRION PROTEIN?

The data presented here and in another study (Falls et al., 1990; Harris et al., 1991) contend that the chicken protein which purifies in fractions enriched for ARIA is chicken PrP. Against this hypothesis is the relatively low degree of homology between the putative chicken PrP and mammalian PrP molecules. In large part, this appears to result from the sequence difference between the glycine–proline hexarepeats of chicken and the octarepeats of mammals. When these repeats along with the N- and C-terminal signal sequences are excluded from the analysis, the degree of similarity considering conservative substitutions is as high as ∼ 55% for avian and mammalian sequences.

Other similarities between the chicken and mammalian PrP genes include similar genomic structure and organization with the entire putative chicken PrP ORF in a single exon. Both the chicken and mammalian PrP genes are single copy genes as judged by Southern blotting. All of the structural features of mammalian PrP are found in the putative chicken molecule. The avian and mammalian prion proteins have N-terminal signal peptides of similar length adjacent to a glycine protein rich repeat region. Near the mid-point, both the chicken and mammalian molecules have hydrophobic sequences of sufficient length to span the membrane (Hay *et al.*, 1987a, b). Both mature proteins have only two cysteines which presumably form a disulphide bond that encompass the three consensus sites in the chicken and two sites in mammals for N-linked glycosylation. Both proteins have similar signal sequences at the C-termini and both proteins have been shown to possess GPI anchors (Stahl *et al.*, 1987; Harris *et al.*, 1991).

All of these considerations taken together argue persuasively that the chicken molecule is an avian prion protein. Whether avian PrP has the same function as mammalian PrP^C is unknown. Whether avian PrP has a disease causing isoform analogous to mammalian PrP^{Sc} is also unknown. All of the prion diseases in mammals are degenerative disorders of the CNS. Whether an avian prion disease would be manifest in the CNS or another organ similarly is uncertain.

When there is uncertainty with respect to whether chicken PrP possesses ARIA (Harris *et al.*, 1991), the function of mammalian PrP remains unknown. Suggestions concerning the function of mammalian PrP range from trophic factors, to receptors, to signal transduciton molecules, to cell adhesion molecules (Cashman *et al.*, 1990; Stahl *et al.*, 1987). Several valuable contributions may arise from studies of the chicken prion protein. These include (1) molecular cloning of PrP related sequences from lower organisms, (2) defining functional domains of PrP and (3) identifying a function for mammalian PrP^C.

REFERENCES

Altschul, S.F. and Erickson, B.W. (1986) Optimal sequence alignment using affine gap costs. *Bull. Math. Biol.* **48** 603–616.

Altschul, S.F., Gish, W., Miller, W., Myers, E.W. and Lipman. D.J. (1990) Basic local alignment search tool. *J. Mol. Biol.* **215** 403–410.

Basler, K., Oesch, B., Scott, M., Westaway, D., Wälchli, M., Groth, D.F., McKinley, M.P., Prusiner, S.B. and Weissmann, C. (1986) Scrapie and cellular PrP isoforms are encoded by the same chromosomal gene. *Cell* **46** 417–428.

Britten, R.J. (1986) Rates of DNA sequence evolution differ between taxonomic groups. *Science* **231** 1393–1398.

Bulmer, M., Wolfe, K.H., and Sharp, P.M. (1991) Synonymous nucleotide substitution rates in mammalian genes: implications for the molecular clock and the relationship of mammalian orders. *Proc. Natl. Acad. Sci. USA* **88** 5974–5978.

Cashman, N.R., Loertscher, R., Nalbantoglu, J., Shaw, I., Kascsak. R.J., Bolton, D.C. and Bendheim, P.E. (1990) Cellular isoform of the scrapie agent protein participates in lymphocyte activiation. *Cell* **61** 185–192.

Chesebro, B., Race, R., Wehrly, K., Nishio, J., Bloom, M., Lechner, D., Bergstrom, S., Robbiins, K., Mayer, L., Keith, J.M., Garon, C. and Hasse, A. (1985) Identification of scrapie prion protein-specific mRNA in scrapie-infected and uninfected brain. *Nature (London)* **315** 331–333.

Doolittle, R.F. and Feng, D.F. (1990) Nearest neighbor procedure for relating progressively aligned amino acid sequences. *Mehtod Enzymol.* **183** 659–669.

Falls, D.L., Harris, D.A., Johnson, F.A., Morgan, M.M., Corfas, G. and Fischbach, G.D. (1990) 42 kD ARIA: a protein that may regulate the accumulation of acetylcholine receptors at developing chick neuromuscular junctions. *Cold Spring Harbor Symp. Quant. Biol.* **55** 397–406.

Feng, D.F. and Doolittle, R.F. (1987) Progressive sequence alignment as a prerequisite to correct phylogenetic trees. *J. Mol. Evol.* **25** 351–360.

Feng, D.F. and Doolittle, R.F. (1990) Progressive alignment and phylogenetic tree construction of protein sequences. *Methods Enzymol.* **183** 375–387.

Ferguson, M.A.J. and Williams, A.F. (1988) Cell-surface anchoring of protein via glycosyl-phosphatidylinositol structures. *Annu. Rev. Biochem.* **57** 285–320.

Garnier, J., Osquthorpe, D.J. and Robson, B. (1978) Analysis of the accuracy and implications of simple methods of predicting the secondary structure of globular proteins. *J. Mol. Biol.* **102** 97–120.

Goldmann, W., Hunter, N., Foster, J.D., Salbaum, J.M., Beyreuther, K. and Hope, J. (1990a) Two alleles of a neural protein gene linked to scrapie in sheep. *Proc. Natl. Acad. Sci. USA* **87** 2476–2480.

Goldmann, W., Hunter, N., Manson, J. and Hope, J. (1990b) The PrP gene of the sheep, a natural host of scrapie. In: *Proc. VIIIth International Congress of Virology, Berlin, August 26–31.* Abstract, p. 284.

Goldmann, W., Hunter, N., Martin, T., Dawson, M. and Hope, J. (1991) Different forms of the bovine PrP gene have five or six copies of a short, G–C–rich element with the protein-coding exon. *J. Gen. Virol.* **72** 201–204.

Harris, D.A., Falls, D.L., Dill-Devor, R.M. and Fischbach, G.D. (1988) Acetylcholine receptor-inducing factor from chicken brain increases the level of mRNA encoding the receptor alpha subunit. *Proc. Natl. Acad. Sci. USA* **85** 1983–1987.

Harris, D.A., Falls, D.L., Walsh, W. and Fischbach, G.D. (1989) Molecular cloning of an acetylcholine receptor-inducing protein. *Soc. Neurosci.* **15** 70.7.

Harris, D.A., Falls, D.L., Johnson, F.A. and Fischbach, G.D. (1991) A prion-like protein from chicken brain copurifies with an acetylcholine receptor-inducing activity. *Proc. Natl. Acad. Sci. USA* **88** 7664–7668.

Hay, B., Barry, R.A., Lieberburg, I., Prusiner, S.B. and Lingappa, V.R. (1987a) Biogenesis and transmembrane orientation of the cellular isoform of the scrapie prion protein. *Mol. Cell. Biol.* **7** 914–920.

Hay, B., Prusiner, S.B. and Lingappa, V.R. (1987b) Evidence for a secretory form of the cellular prion protein. *Biochemistry* **26** 8110–8115.

Jacobs, K.A., Rudersdorf, R., Neill, S.D., Dougherty, J.P., Brown, E.L. and Fritsch, E.F. (1988) The thermal stability of oligonucleotide duplexes is sequence independent in tetraalkylammonium salt solutions: application to identifying recombinant DNA clones. *Nucleic Acids Res.* **16** 4637–4650.

Kretzschmar, H.A., Prusiner, S.B., Stowring, L.E. and DeArmond, S.J. (1986) Scrapie prion proteins are synthesized in neurons. *Am. J. Pathol.* **122** 1–5.

Li, W.H. and Tanimora, M. (1987) The molecular clock runs more slowly in man than in apes and monkeys. *Nature (London)* **326** 93–96.

Li, W.H., Tanimora, M. and Sharp, P.M. (1987) An evaluation of the molecular clock hypothesis using mammalian DNA sequences. *J. Mol. Evol.* **25** 330–342.

Liao, Y.-C., Lebo, R.V., Clawson, G.A. and Smuckler, E.A. (1986) Human prion protein cDNA: molecular cloning, chromosomal mapping, and biological implication. *Science* **233** 364–367.

Liao, Y.-C., Tokes, Z., Lim, E., Lackey, A., Woo, C.H., Button, J.D. and Clawson, G.A. (1987) Cloning of rat 'prion-related protein' cDNA. *Lab. Invest.* **57** 370–374.

Locht, C., Chesebro, B., Race, R. and Keith, J.M. (1986) Molecular cloning and complete sequence of prion cDNA from mouse brain infected with the scrapie agent. *Proc. Natl. Acad. Sci. USA* **83** 6372–6376.

Lowenstein, D.H., Butler, D.A., Westaway, D., McKinley. M.P., DeArmond, S.J. and Prusiner, S.B. (1990) Three hamster species with different scrapie incubation times and neuropathological features encode distinct prion proteins. *Mol. Cell Biol.* **10** 1153–1163.

Martinou, J.C., Falls, D.L., Fischbach, G.D. and Merlie, J.P. (1991) Acetylcholine receptor-inducing activity stimulates expression of the epsilon-subunit gene of the muscle acetylcholine receptor. *Proc. Natl. Acad. Sci. USA* **88** 7669–7673.

Miyata, T. and Hayashida, H. (1982) Recent divergence from a common ancestor of human IFN-α genes. *Nature (London)* **295** 165–168.

Mobley, W.C., Neve, R.L., Prusiner, S.B. and McKinley, M.P. (1988) Nerve growth factor increases mRNA levels for the prion protein and the beta-amyloid protein precursor in developing hamster brain. *Proc. Natl. Acad. Sci. USA* **85** 9811–9815.

Moran, P., Raab, H., Kohr, W.J. and Caras, I.W. (1991) Glycophospholipid membrane anchor attachment. Molecular analysis of the cleavage/attachment site. *J. Biol. Chem.* **266** 1250–1257.

Oesch, B., Westaway, D., Wälchli, M., McKinley, M.P., Kent, S.B.H., Aebersold, R., Barry, R.A., Tempest, P., Teplow, D.B., Hood, L.E., Prusiner, S.B. and Weissmann, C. (1985) A cellular gene encodes scrapie PrP 27-30 protein. *Cell.* **40** 735–746.

Payne, M.E., Schwover, C.M. and Soderling, T.R. (1983) Purification and characterization of rabbit liver calmodulin-dependent glycogen synthase kinase. *J. Biol. Chem.* **258** 2376–2382.

Pearson, R.B., Woodgett, J. R., Cohen, P. and Kemp, B.E. (1985) Substrate specificity of a multifunctional calmodulin-dependent protein kinase. *J. Biol. Chem.* **260** 14471–14476.

Pearson, W.R. and Lipman, D.J. (1988) Improved tools for biological sequence comparison. *Proc. Natl. Acad. Sci. USA* **85** 2444–2448.

Perler, F., Efstratiadis, A., Lomedico, P., Gilbert, W., Kolodner, R. and Dodgson, J. (1980) The evolution of genes: the chicken preproinsulin gene. *Cell* **20** 555–566.

Prusiner, S.B. (1991) Molecular biology of prion diseases. *Science* **252** 1515–1522.

Puckett, C., Concannon, P., Casey, C. and Hood, L. (1991) Genomic structure of the human prion protein gene. *Am. J. Hum. Genet.* **49** 320–329.

Sambrook, J., Fritsch, E.F. and Maniatis, T. (1989) *Molecular Cloning — a Laboratory Manual*, 2nd edn. Cold Spring Harbour Laboratory Press, Cold Spring Harbor, NY.

Stahl, N., Borchelt, D.R., Hsiao, K. and Prusiner, S.B. (1987) Scrapie prion protein contains a phosphatidylinositol glycolipid. *Cell* **51** 229–240.

Stahl, N., Borchelt, D.R. and Prusiner, S.B. (1990) Differential release of cellular and scrapie prion from cellular membranes by phosphatidylinositol-specific phospholipase C. *Biochemistry* **29** 5405–5412.

Swofford, D.L. (1991) *Phylogenetic Analysis Using Parsimony*. Computer program distributed by the Illinois Natural History Survey, Champaign, Illinois.

Usdin, T.B. and Fischbach, G.D. (1986) Purification and characterization of a polypeptide from chick brain that promotes the accumulation of acetylcholine receptors in chick myotubes. *J. Cell Biol.* **103** 493–507.

Westaway, D., Goodman, P.A., Mirenda, C.A., McKinley, M.P., Carlson, G.A. and Prusiner, S.B. (1987) Distinct prion proteins in short and long scrapie incubation period mice. *Cell* **51** 651–662.

Westaway, D., Mirenda, C.A., Foster, D., Zebarjadian, Y., Scott, M., Torchia, M., Yang, S.-L., Serban, H., DeArmond, S.J., Ebeling, C., Prusiner, S.B. and Carlson, G.A. (1991) Paradoxical shortening of scrapie incubation times by expression of prion protein transgenes derived from long incubation period mice. *Neuron* **7** 59–68.

Wood, W.I., Gitschier, J., Lasky, L.A. and Lawn, R.M. (1985) Base composition-independent hybridization in tetramethylammonium chloride: a method for oligonucleotide screening of highly complex gene libraries. *Proc. Natl. Acad. Sci. USA* **82** 1585–1588.

Woodgett, J.R., Gould, K.L. and Hunter, T. (1986) Substrate specificity of protein kinase C. *Eur. J. Biochem.* **161**. 177–184.

Part VI   Cell biology of prion proteins

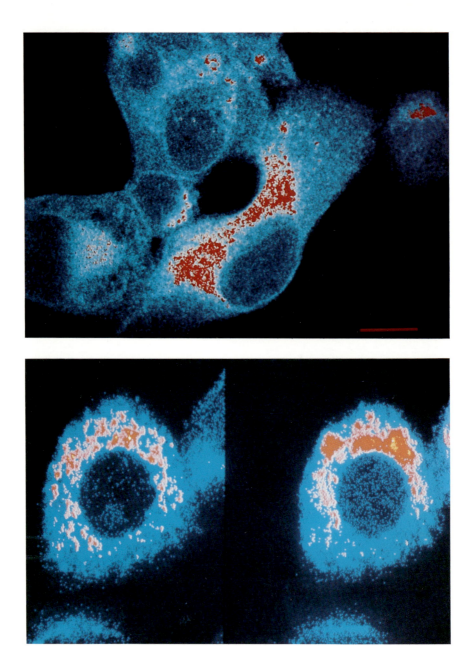

Fig. 1. PrP^{Sc} accumulates in the cytoplasm of ScN$_2$a cells and partially colocalizes with WGA binding sites. Upper panel. ScN$_2$a cells were fixed, denatured with GdnSCN, and then processed for the immunodetection of PrP by indirect immunofluorescence (Taraboulos *et al.*, 1990b) using the PrP antiserum RO73 (Serban *et al.*, 1990). The speckled staining suggests the association of PrP^C with cytoplasmic vesicles. Lower panel. The cells were doubly stained with RO73 (FITC channel, right-hand photograph) and WGA (Texas red channel, left-hand photograph). The cells were observed by computerized laser confocal microscopy. Modified, with permission, from *J. Cell Biol.* **110** 2117–2132, 1990.

36

Dissecting the pathway of scrapie prion synthesis in cultured cells

Albert Taraboulos, David R. Borchelt, Michael P. McKinley, Alex Raeber, Dan Serban, Stephen J. DeArmond and Stanley B. Prusiner

ABSTRACT

Defining the subcellular site(s) at which PrP^C or a precursor is converted to PrP^{Sc} may give insight into the molecular processes involved in the formation of PrP^{Sc}. We identified the subcellular sites of PrP^{Sc} accumulation in ScN_2a and ScHaB cells by immunofluorescence and by immunoelectron microscopy. In contrast to PrP^C, PrP^{Sc} accumulated in the cytoplasm, in part within secondary lysosomes. The subcellular pathways involved in the synthesis of protease-resistant PrP^{Sc} were defined. Tunicamycin did not prevent formation of PrP^{Sc}, indicating that the acquisition of protease resistance is independent of Asn-linked carbohydrates. Brefeldin A added throughout the incubation prevented the formation of PrP^{Sc}, suggesting that the endoplasmic reticulum is not competent for the synthesis of PrP^{Sc}. PrP^{Sc} underwent a NH_4Cl-sensitive N-terminal trimming 1–2 h after acquiring protease resistance. These and other results suggest that PrP^{Sc} becomes protease resistant during transit between the mid-Golgi and lysosomes.

INTRODUCTION

Scrapie prions are composed largely, if not entirely, of PrP^{Sc} molecules (Hsiao *et al.*, 1990; Prusiner, 1982, 1991). Delineating the molecular events involved in PrP^{Sc} synthesis as well as the subcellular sites at which this process occurs is important. The advent of tissue culture systems infectable with scrapie prions (Borchelt *et al.*,

1990; Butler *et al.*, 1988; Caughey *et al.*, 1990; Race *et al.*, 1987, 1988; Taraboulos *et al.*, 1990b) has made possible the detailed study of PrP^{Sc} biosynthesis. Using immunocytochemical and biochemical approaches we have started to delineate the subcellular pathways utilized in the formation of PrP^{Sc}.

We utilized two cell lines in our study. ScN$_2$a is a subclone of the mouse neuroblastoma cells persistently infected with scrapie (Butler *et al.*, 1988). HaB is a novel cell line isolated from the brain of a Syrian hamster. ScHaB is a subclone of these cells persistently infected with Sc237 scrapie prions (Taraboulos *et al.*, 1990b).

RESULTS AND DISCUSSION

PrP^{Sc} accumulates intracellularly in scrapie-infected cells

The ability to localize the PrP isoforms *in situ* in cultured cells is essential to the study of these proteins. Traditional immunofluorescence methods readily detected PrP^{C} on the surface of cultured cells (Stahl *et al.*, 1987). However, when applied to scrapie-infected cells these methods failed to yield an additional, scrapie-specific signal. The discovery that denaturation of PrP^{Sc} greatly enhanced its immunoreactivity (Serban *et al.*, 1990) led us to develop a protocol for the detection of PrP^{Sc} in ScN$_2$a and ScHaB cells (Taraboulos *et al.*, 1990b). Cells were fixed with formaldehyde, permeabilized with Triton X-100, incubated with 3 M guanidine thiocyanate (GdnSCN) and then processed for the immunofluorescent detection of PrP using antibodies directed against the protease-resistant core of PrP^{Sc} (PrP 27-30). A speckled cytoplasmic signal was detected in scrapie-infected cells but not in control, uninfected cells. The PrP^{Sc} signal was resistant to proteolysis conditions that digested PrP^{C} to completion in these cells and depended on pretreating the cells with GdnSCN (Taraboulos *et al.*, 1990b). The intracellular distribution of PrP^{Sc} was best observed using laser confocal microscopy (White *et al.*, 1987) (Fig. 1, upper panel) demonstrating its cytoplasmic location. In double staining experiments, PrP^{Sc} colocalized with ligands of wheat germ agglutinin, a marker for the *trans*-Golgi network (TGN) and lysosomes (Virtanen *et al.*, 1980) (Fig. 1, lower panels). The intracellular accumulation of PrP^{Sc} contrasts sharply with the subcellular localization of PrP^{C}, which is bound to the plasma membrane by a glycoinositol phospholipid (GPI) anchor (Stahl *et al.*, 1987).

At least some PrP^{Sc} accumulates within secondary lysosomes

To obtain a more precise definition of the compartments in which PrP^{Sc} accumulates, we adapted the GdnSCN denaturation procedures described above to pre-embedding immunoelectron microscopy (McKinley *et al.*, 1991b). Cells were treated as for the immunofluorescence except that detergent incubation was omitted and PrP was detected by immunoperoxidase (Fig. 2, upper panels) or immunogold (Fig. 2, lower panels). Following incubation with the antibodies, the cells were postfixed with glutaraldehyde, osmicated and processed for electron microscopy (Brown and Farquhar, 1989). Numerous immunoperoxidase-positive vacuoles and round bodies, ranging in diameter from 1 to 2 μm, were scattered throughout the cytoplasm (Fig. 2, upper panel). Diaminobenzidine deposition was often observed in small vesicles that

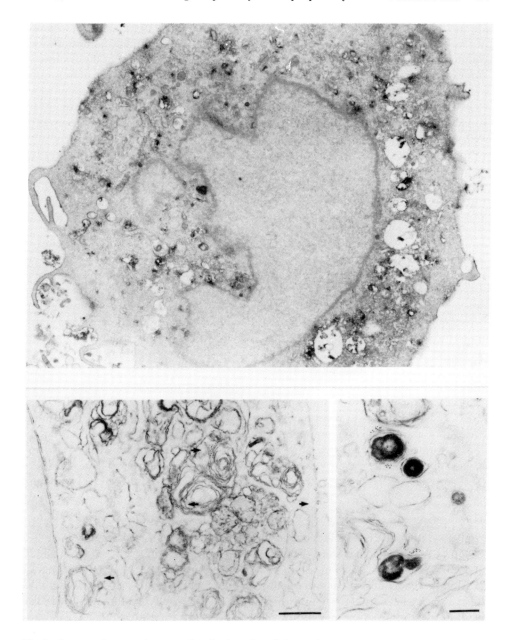

Fig. 2.   Immunoelectron microscopy localization of PrP^Sc in ScN2a and ScHaB cells. Cells were stained for PrP using a pre-embedding protocol that included protein denaturation with GdnSCN (McKinley *et al.*, 1991a). The cells were fixed, treated with GdnSCN, immunolabelled for PrP using immunoperoxidase (upper panel) or 5 μm immunogold (lower panels), and then processed for electron microscopy. Colocalization of immunogold with histochemical stain for acid phosphatase (lower panel, right-hand photograph) confirms the accumulation of at least some PfP^Sc in secondary lysosomes. ScN₂a cells in upper panel. ScHaB cells in lower panel. Bars in lower panel: left hand photograph, 0.5 μm and right hand photograph, 0.2 μm. Modified, with permission, from *Lab. Invest.* **65** 622–630 1991.

appeared to merge into larger vesicular structures, reminiscent of secondary lysosomes. Moreover, the lysosomal marker acid phosphatase could be found in some vesicles containing PrP^{Sc} (Fig. 2, lower panels). No PrP^{Sc} could be detected in the nucleus or the Golgi stacks (Fig. 2, upper panel). The modest ultrastructural preservation obtained with this pre-embedding procedure prevented the identification of other PrP^{Sc} accumulation sites. In particular, whether some PrP^{Sc} was present in the TGN is unclear. The PrP staining present on the plasma membrane probably represented PrP^{C} (Stahl *et al.*, 1987).

The detection of PrP^{Sc} in secondary lysosomes is consistent with the finding that metabolically radiolabelled PrP^{Sc} becomes exposed to lysosomal hydrolases shortly after becoming protease resistant (Caughey *et al.*, 1991; Taraboulos *et al.*, 1992 see Fig. 5). Such an accumulation could conceivably contribute to the extensive neuronal degeneration observed during prion diseases.

Biosynthesis of the PrP isoforms

In uninfected N$_2$a cells, PrP^{C} is processed and transported to the plasma membrane within 30 min of its synthesis (Caughey *et al.*, 1989). PrP^{C} synthesis and processing appears to follow the secretory pathway. In contrast to PrP^{C}, pulse-chase experiments determined that PrP^{Sc} acquires protease resistance as a result of a slow post-translational process in both ScN$_2$a ($t_{1/2} \sim 3$ h) Borchelt *et al.*, 1990; Caughey and Raymond, 1991) and in ScHaB cells ($t_{1/2} \sim 1$ h) (Taraboulos *et al.*, 1990a; Borchelt *et al.*, 1992). These results raise several questions. First, what are the pathways through which PrP^{Sc} transits during its biosynthesis? Second, does PrP^{Sc} diverge from the secretory pathway during its synthesis? Third, what is the compartment in which PrP^{Sc} acquires protease resistance? We have utilized inhibitors of glycoprotein processing and trafficking to address these questions. To discriminate between the PrP isoforms, we used the resistance of PrP^{Sc} to proteolysis (Oesch *et al.*, 1985) and its insolubility in detergents (Meyer *et al.*, 1986). Protein denaturation using GdnSCN (Serban *et al.*, 1990) was used to enhance PrP^{Sc} recognition by antibodies (Taraboulos *et al.*, 1990a).

N-glycosylation of PrP is not required for acquisition of protease resistance

Complex Asn-linked glycosylation accounts for up to 30% of the mass of both PrP^{C} and PrP^{Sc} (Bolton *et al.*, 1985; Endo *et al.*, 1989; Haraguchi *et al.*, 1989), raising the possibility that Asn-linked carbohydrates may harbour the critical information differentiating between the PrP isoforms (Basler *et al.*, 1986). To explore this possibility, we studied the formation of PrP^{Sc} in cells treated with tunicamycin (Takatsuki and Tamura, 1971; Taraboulos *et al.*, 1990a). In both ScN$_2$a and ScHaB cells, addition of tunicamycin to the pulse radiolabelling period resulted in the formation of an unglycosylated PrP^{Sc} with an M_r of 19 kDa (Fig. 3). This 19 kDa protein was insoluble in detergent, resistant to digestion by proteinase K, and was absent from uninf cted cells similarly treated with the inhibitor (not shown), thus displaying the hallmarks of PrP^{Sc}. These results argue that Asn-linked glycosylation is not necessary for the synthesis of PrP^{Sc} and that structural differences unrelated to these carbohydrates

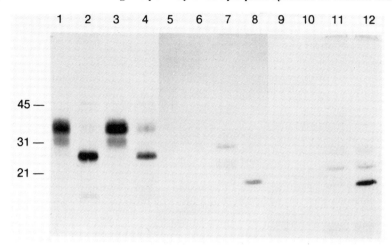

Fig. 3. Asn-linked glycosylation is not necessary for the formation of protease-resistant PrP^{Sc}. HaB (lanes 1, 2, 5 and 6), ScHaB (lanes 3, 4, 7 and 8), N$_2$a (lanes 9 and 10) and ScN$_2$a (lanes 11 and 12) cells were radiolabelled for 2 h in the presence (even lanes) or absence of tunicamycin. In lanes 5–12, the cell lysates were treated with proteinase K (20 μg/ml 37°C, 1 h) prior to immunoprecipitation (Taraboulos *et al.*, 1990a). Total PrP extracted at the end of the pulse displayed the expected reduction in M_r for the unglycosylated PrP (lanes 2 and 4). Following a 16 h chase unglycosylated, protease-resistant PrP^{Sc} ($M_r = 19$ kDa) was synthesized in scrapie-infected cells (lanes 8 and 12) but not in uninfected cells (lanes 6 and 10). Modified from *Proc. Natl. Acad. Sci. USA* **87** 8262–8266, 1990.

must exist between two PrP isoforms. Interestingly, the synthesis of PrP^{Sc} was more rapid in the presence of tunicamycin than in its absence (Taraboulos *et al.*, 1990a).

Brefeldin A reversibly inhibits the formation of PrP^{Sc} in ScN$_2$a and ScHaB cells

The fungal metabolite brefeldin A causes the redistribution of the Golgi stacks and resident Golgi proteins into the endoplasmic reticulum (ER), blocks the export of proteins into the secretory pathway and inhibits transport of proteins through endosomes (Lippincott-Schwartz *et al.*, 1989, 1991). In contrast, other cellular processes are not directly inhibited (Misumi *et al.*, 1986). We studied the effects of brefeldin A on the synthesis of PrP^{Sc} and verified that the drug has the expected effect on the cells (Taraboulos *et al.*, 1991; Taraboulos *et al.*, 1992). Treatment of ScHaB cells with 10 μg/ml brefeldin A resulted in the redistribution of the Golgi marker MG160 into a reticulated pattern characteristic of the ER (not shown). In contrast, the pattern of intracellular PrP^{Sc} accumulation was not modified by the drug, consistent with the localization of PrP^{Sc} outside of the Golgi stacks (not shown). Presence of BFA throughout the pulse and the chase periods completely inhibited the formation of protease-resistant PrP in both cell types while absence of the drug from either incubation period permitted PrP^{Sc} synthesis (Fig. 4). This did not result from a rapid degradation of the PrP^{Sc} precursor since upon removal of the drug the cells resumed the synthesis of protease-resistant PrP^{Sc} (not shown).

These results suggest that the ER is not competent for the formation of PrP^{Sc}. It is possible that this organelle lacks the conditions necessary for this synthesis.

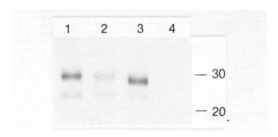

Fig. 4.   Brefeldin A prevents synthesis of protease-resistant PrP^{Sc}. ScHaB cells were pulse radiolabelled for 1 h in the presence (lanes 3 and 4) or absence of brefeldin A (5 μg/ml) (Lippincott-Schwartz et al., 1989; Misumi et al., 1986) and then chased in unlabelled medium for 6 h in the presence (lanes 2 and 4) or absence of the inhibitor. The cell lysates were treated with proteinase K (20 μg/ml, 37°C, 1 h) prior to immunoprecipitation. Presence of brefeldin A throughout the pulse and the chase completely inhibited the formation of protease-resistant PrP^{Sc} (lane 4) (Taraboulos et al., 1991).

Alternatively, ER components such as chaperonins may actively prevent the formation of PrP^{Sc}. Another possibility is that an informational molecule or template, presumably PrP^{Sc} itself (Prusiner et al., 1990), is missing from the ER. Further studies will be necessary to discriminate between these possibilities.

Our findings argue that PrP^{Sc} synthesis occurs after PrPO has excited the ER. Results from other studies with monensin suggest that PrP^{Sc} synthesis occurs after passage of PrP through the *mid*-Golgi apparatus (Taraboulos et al., 1991; Taraboulos et al., 1992).

Does PrP^{Sc} synthesis occur in endosomes?

Recent studies argue that the release of PrP^{C} from the cell surface by digestion with phosphatidylinositol-specific phospholipase C (PIPLC) or its hydrolysis catalysed by dispase prevents PrP^{Sc} synthesis (Borchelt et al., 1992; Caughey and Raymond, 1991). Whether the conversion of PrP^{C} to PrP^{Sc} occurs on the surface of the cells or within their interior is unclear. The synthesis of PrP^{Sc} was found to be reversibly inhibited by incubation of the cells at 18°C (Borchelt et al., 1992). Many studies have shown that exposure to 18°C selectively inhibits the endosomal pathway.

Alternatively, PrP^{Sc} might occur in a non-endosomal pathway where PrP^{C} recycles into the cell through caveolae. Recent studies have shown that many GPI-anchored proteins on the surface of cells reenter through the caveolae (Anderson et al., 1991). Whether exposure of the cells to 18°C also inhibits caveolae-dependent transport to PrP^{C} is unknown.

PrP^{Sc} becomes exposed to lysosomal enzymes several hours after its synthesis

Since the immunoelectron microscopy studies indicated that PrP^{Sc} accumulates at least in part in secondary lysosomes (McKinley et al., 1991b), we explored the possible consequences of exposure of this protein to lysosomal hydrolases (Taraboulos et al., 1991; Taraboulos et al., 1992)). A variety of proteases are able to hydrolyse the *N*-terminal portion of PrP^{Sc}, leaving the protease-resistant core PrP 27-30 (Oesch et al., 1985); for unglycosylated PrP^{Sc}, this core has an M_r of 19 kDa (Fig. 3). ScN$_2$a

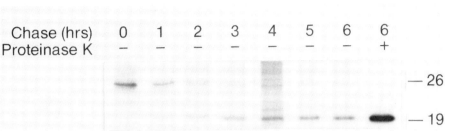

Fig. 5. PrP^{Sc} becomes exposed to lysosomal hydrolases after 1–2 hof its synthesis. ScN$_2$a cells were radiolabelled for 1 h in the presence of tunicamycin (30 μg/ml) (Taraboulos *et al.*, 1990a) and then chased for the indicated period without tunicamycin. In lanes 1–6, total PrP was analysed at the end of the pulse. In lane 7, the cell lysates were first subjected to proteolysis prior to analysis. PrP^{Sc} underwent a spontaneous degradation (lanes 1–6) that was inhibited by the lysosomal amine NH$_4$Cl (not shown).

cells were radiolabelled for 1 h in the presence of tunicamycin and then chased for the indicated period (Fig. 5). Since unglycosylated PrP^C is unstable ($t_{1/2}$ for degradation $\sim$ 1 h) (Taraboulos *et al.*, 1990a) and removed from the cell, we analysed total PrP without separating PrP^{Sc} from PrP^C.

PrP was first synthesized as a 26 kDa species, an M_r consistent with the full-length, unglycosylated PrP molecule (Taraboulos *et al.*, 1990a). However, within a 1 h chase a shift to an M_r of 19 kDa could already be observed, and within a 3 h chase all the PrP^{Sc} molecules seemed similarly processed (Fig. 5). No similar 19 kDa band was observed in uninfected cells (not shown). To ascertain whether this degradation was lysosomal, we used the lysosomotropic amines NH$_4$Cl and chloroquine, as well as the ionophore monensin (not shown) that inhibit the action of lysosomal hydrolases by raising the pH of lysosomes (Caughey *et al.*, 1991; Taraboulos *et al.*, 1992). These drugs all prevented the trimming of the 26 kDa species, suggesting that this trimming is indeed lysosomal. This result is consistent with electron microscopic findings described above but it indicates that the large majority of, if not all, PrP^{Sc} molecules transit or accumulate in lysosomes. We can also conclude for these experiments that acidic vacuolar pH is not essential for the synthesis of PrP^{Sc}.

CONCLUSIONS

In conclusion, the investigations summarized here have begun to delineate the intracellular pathways involved in PrP^{Sc} synthesis (Fig. 6). A precursor of PrP^{Sc}, presumably PrP^C, appears to exit the ER, pass through the *mid*-Golgi and exit to the cell surface within a brief period of time. The precise compartment in which PrP^C or a precursor is converted into PrP^{Sc} is unknown but, 1–2 h after becoming resistant to proteases, PrP^{Sc} is found in an acidic degradative compartment, probably secondary lysosomes. Since PrP^{Sc} of $M_r = 33$–35 kDa is found in scrapie-infected brains (Meyer *et al.*, 1986), the N-terminal trimming observed in cultured cells is not an essential feature of PrP^{Sc} synthesis. Identifying the precise subcellular compartments in which

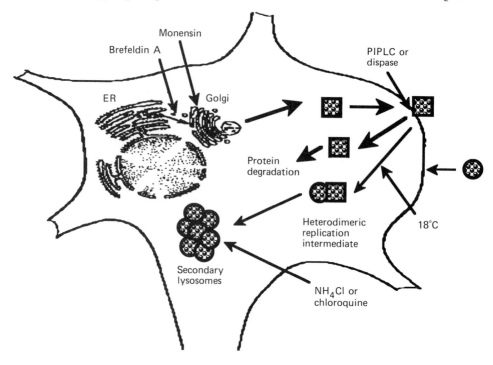

Fig. 6.   Cell biology of PrP^Sc synthesis in a cultured cell. Squares denote PrP^C, circles PrP^Sc. Inhibitors and modifiers of PrP^Sc synthesis are shown in italics.

the formation of PrP^Sc occurs may help in deciphering the molecular mechanism which features in this process.

REFERENCES

Anderson, R.G.W., Kamen, B.A., Rothberg, K.G. and Lacey, S.W. (1991) Potocytosis: sequestration and transport of small molecules by caveolae. *Science* **255** 410–411.

Basler, K., Oesch, B., Scott, M., Westaway, D., Wälchli, M., Groth, D.F., McKinley, M.P., Prusiner, S.B. and Weissmann, C. (1986) Scrapie and cellular PrP isoforms are encoded by the same chromosomal gene. *Cell* **46** 417–428.

Bolton, D.C., Meyer, R.K. and Prusiner, S.B. (1985) Scrapie PrP 27-30 is a sialoglycoprotein. *J. Virol.* **53** 596–606.

Borchelt, D.R., Scott, M., Taraboulos, A., Stahl, N. and Prusiner, S.B. (1990) Scrapie and cellular prion proteins differ in their kinetics of synthesis in cultured cells. *J. Cell Biol.* **110** 743–752.

Borchelt, D.R., Taraboulos, A. and Prusiner, S.B. (1992) Evidence for synthesis of scrapie prion proteins in the endocytic pathway. *J. Biol. Chem.*, in press.

Brown, W.J. and Farquhar, M.G. (1989) Immunoperoxidase methods for the localization of antigens in cultured cells and tissue sections by electron microscopy. *Methods Cell Biol.* **31** 553–569.

Butler, D.A., Scott, M.R.D., Bockman, J.M., Borchelt, D.R., Taraboulos, A., Hsiao, K.K., Kingsbury, D.T. and Prusiner, S.B. (1988) Scrapie-infected murine neuroblastoma cells produce protease-resistant prion proteins. *J. Virol.* **62** 1558–1564.

Caughey, B., Race, R.E., Ernst, D., Buchmeier, M.J. and Chesebro, B. (1989) Prion protein biosynthesis in scrapie-infected and uninfected neuroblastoma cells. *J. Virol.* **63** 175–181.

Caughey, B., Neary, K., Butler, R., Ernst, D., Perry, L., Chesebro, B. and Race, R.E. (1990) Normal and scrapie-associated forms of prion protein differ in their sensitivities to phospholipase and proteases in intact neuroblastoma cells. *J. Virol.* **64** 1093–1101.

Caughey, B. and Raymond, G.J. (1991) The scrapie-associated form of PrP is made from a cell surface precursor that is both protease- and phospholipase-sensitive. *J. Biol. Chem.* **266** 18217–18233.

Caughey, B., Raymond, G.J., Ernst, D. and Race, R.E. (1991) N-terminal truncation of the scrapie-associated form of PrP by lysosomal protease(s): implications regarding the site of conversion of PrP to the protease-resistant state. *J. Virol.* **65** 6597–6603.

Endo, T., Groth, D., Prusiner, S.B. and Kobata, A. (1989) Diversity of oligosaccharide structures linked to asparagines of the scrapie prion protein. *Biochemistry* **28** 8380–8388.

Haraguchi, T., Fisher, S., Olofsson, S., Endo, T., Groth, D., Tarantino, A., Borchelt, D.R., Teplow, D., Hood, L., Burlingame, A., Lycke, E., Kobata, A. and Prusiner, S.B. (1989) Asparagine-linked glycosylation of the scrapie and cellular prion proteins. *Arch. Biochem. Biophys.* **274** 1–13.

Hsiao, K.K., Scott, M., Foster, D., Groth, D.F., DeArmond, S.J. and Prusiner, S.B. (1990) Spontaneous neurodegeneration in transgenic mice with mutant prion protein of Gerstmann–Sträussler syndrome. *Science* **250** 1587–1590.

Lippincott-Schwartz, J., Yuan, L.C., Bonifacino, J.S. and Klausner, R.D. (1989) Rapid redistribution of Golgi proteins into the ER in cells treated with Brefeldin A: evidence for membrane cycling from the Golgi to ER. *Cell* **56** 801–813.

McKinley, M.P., Meyer, R., Kenaga, L., Rahbar, F., Cotter, R., Serban, A. and Prusiner, S.B. (1991a) Scrapie prion rod formation *in vitro* requires both detergent extraction and limited proteolysis. *J. Virol.* **65** 1440–1449.

McKinley, M.P., Taraboulos, A., Kenaga, L., Serban, D., Stieber, A., DeArmond, S.J., Prusiner, S.B. and Gonatas, N. (1991b) Ultrastructural localization of scrapie prion proteins in cytoplasmic vesicles of infected cultured cells. *Lab. Invest.* (in press).

Meyer, R.K., McKinley, M.P., Bowman, K.A., Braunfeld, M.B., Barry, R.A. and Prusiner, S.B. (1986) Separation and properties of cellular and scrapie prion proteins. *Proc. Natl. Acad. Sci. USA* **83** 2310–2314.

Misumi, Y., Misumi, Y., Miki, K., Takatsuki, A., Tamura, G. and Ikehara, Y. (1986) Novel blockade by brefeldin A of intracellular transport of secretory proteins in cultured rat hepatocytes. *J. Biol. Chem.* **261** 11398–11403.

Oesch, B., Westaway, D., Wälchli, M., McKinley, M.P., Kent, S.B.H., Aebersold, R., Barry, R.A., Tempst, P., Teplow, D.B., Hood, L.E., Prusiner, S.B. and Weissmann, C. (1985) A cellular gene encodes scrapie PrP 27-30 protein. *Cell* **40** 735–746.

Prusiner, S.B. (1982) Novel proteinaceous infectious particles cause scrapie. *Science* **216** 136–144.

Prusiner, S.B. (1991) Molecular biology of prion diseases. *Science* **252** 1515–1522.

Prusiner, S.B., Scott, M., Foster, D., Pan, K.-M., Groth, D., Mirenda, C., Torchia, M., Yang, S.-L., Serban, D., Carlson, G.A., Hoppe, P.C., Westaway, D. and DeArmond, S.J. (1990) Transgenetic studies implicate interactions between homologous PrP isoforms in scrapie prion replication. *Cell* **63** 673–686.

Race, R.E., Fadness, L.H. and Chesebro, B. (1987) Characterization of scrapie infection in mouse neuroblastoma cells. *J. Gen. Virol.* **68** 1391–1399.

Race, R.E., Caughey, B., Graham, K., Ernst, D. and Chesebro, B. (1988) Analyses of frequency of infection, specific infectivity, and prion protein biosynthesis in scrapie-infected neuroblastoma cell clones. *J. Virol.* **62** 2845–2849.

Serban, D., Taraboulos, A., DeArmond, S.J. and Prusiner, S.B. (1990) Rapid detection of Creutzfeldt–Jakob disease and scrapie prion proteins. *Neurology* **40** 110–117.

Stahl, N., Borchelt, D.R., Hsiao, K. and Prusiner, S.B. (1987) Scrapie prion protein contains a phosphatidylinositol glycolipid. *Cell* **51** 229–240.

Takatsuki, A. and Tamura, G. (1971) Effect of tunicamycin on the synthesis of macromolecules in cultures of chick embryo fibroblasts infected with Newcastle disease virus. *J. Antibiot. Ser. A* **24** 785–794.

Taraboulos, A., Rogers, M., Borchelt, D.R., McKinley, M.P., Scott, M., Serban, D. and Prusiner, S.B. (1990a) Acquisition of protease resistance by prion proteins in scrapie-infected cells does not require asparagine-linked glycosylation. *Proc. Natl. Acad. Sci. USA* **87** 8262–8266.

Taraboulos, A., Serban, D. and Prusiner, S.B. (1990b) Scrapie prion proteins accumulate in the cytoplasm of persistently-infected cultured cells. *J. Cell Biol.* **110** 2117–2132.

Taraboulos, A., Raeber, A., Borchelt, D., McKinley, M.P. and Prusiner, S.B. (1991) Brefeldin A inhibits protease resistant prion protein synthesis in scrapie-infected cultured cells. *FASEB J.* **5** A1177.

Taraboulos, A., Raeber, A., Borchelt, D.R., Serban, D. and Prusiner, S.B. (1992) Synthesis and trafficking of prion proteins in cultured cells. *Mol. Biol. Cell*, in press.

Virtanen, I., Ekblom, P. and Laurila, P. (1980) Subcellular compartmentalization of saccharide moieties in cultured normal and malignant cells. *J. Cell Biol.* **85** 429–434.

White, J.G., Amos, W.B. and Fordham, M. (1987) An evaluation of confocal versus conventional imaging of biological structures by fluorescence light microscopy. *J. Cell Biol.* **105** 41–48.

37

Effects of scrapie infection on cellular PrP metabolism

B. Caughey, R. Race and B. Chesebro

ABSTRACT

Given the apparent role of the PrP protein in the transmissible spongiform encephalopathies, it is critical to understand the cellular events and structural changes which cause the formation of the protease-resistant PrP from its normal, protease-sensitive isoform. Studies in scrapie-infected murine neuroblastoma cells have compared the metabolism and membrane topologies of the normal and protease-resistant PrP forms. We have recently determined that protease-resistant PrP is made from a cell surface precursor that has been indistinguishable from the normal isoform. The protease-resistant PrP appears to be translocated to the lysosomes, where it is N-terminally truncated but resistant to complete degradation. These observations imply that the conversion of PrP to the protease-resistant state is a post-translational event that occurs at the plasma membrane or along an endocytic pathway before exposure to proteases. Although it is not clear what structural features account for the metabolic stability of the protease-resistant PrP, our infrared spectroscopic studies have indicated that the protease-resistant core of the molecule has a high β-sheet content, like other stable amyloids, and lacks random coil structures which could serve as favourable sites for proteolysis.

INTRODUCTION

A prime event in the study of scrapie and related transmissible spongiform encephalopathies (TSEs) was the discovery of disease-specific fibrils in brain extracts of scrapie-infected mice (Merz *et al.*, 1981). Soon it was determined that preparations of these scrapie-associated fibrils were highly infectious and composed primarily of a protease-resistant protein which was first named PrP (Bolton *et al.*, 1982; McKinley *et al.*, 1983; Diringer *et al.*, 1983). This discovery led to proposals that this protein or fibril is the scrapie agent (Bolton *et al.*, 1982; Diringer *et al.*, 1983; McKinley *et al.*, 1983). Subsequent studies revealed that protease-resistant PrP is common to other TSEs (Manuelidis *et al.*, 1985; Bendheim *et al.*, 1985; Bockman *et al.*, 1985; Brown *et al.*, 1986; Hope *et al.*, 1988b) and is an abnormal isoform of a normally protease-sensitive endogenous host protein (Chesebro *et al.*, 1985; Oesch *et al.*, 1985; Hope *et al.*, 1986; Rubenstein *et al.*, 1986; Meyer *et al.*, 1986; Cho, 1986). While the protease-resistant PrP isoform is found only in TSE-infected tissues, the normal isoform is found in a wide variety of both infected and uninfected tissues and cell types (Chesebro *et al.*, 1985; Hope *et al.*, 1986; Meyer *et al.*, 1986; Rubenstein *et al.*, 1986; Cho, 1986; Hope *et al.*, 1988a; Caughey *et al.*, 1988a; Race *et al.*, 1988; Caughey *et al.*, 1988b; Cashman *et al.*, 1990).

Several names have been given to these two PrP isoforms (McKinley *et al.*, 1983; Diringer *et al.*, 1983; Bolton *et al.*, 1984; Sklaviadis *et al.*, 1986; Meyer *et al.*, 1986; Bolton *et al.*, 1987). Since they are usually discriminated experimentally on the basis of their sensitivity to proteases, we have referred to them simply as protease-resistant PrP (PrP-res) and protease-sensitive PrP (PrP-sen) (Caughey *et al.*, 1990). Although the precise relationship of PrP-res to the TSE agents is not yet clear, the importance of PrP in the TSEs has been underscored by molecular genetic data indicating that variations in the endogenous PrP gene, which encodes both PrP-sen and PrP-res, appear to influence incubation time of scrapie (Carlson *et al.*, 1986; Westaway *et al.*, 1987; Hunter *et al.*, 1987; Carlson *et al.*, 1988; Race *et al.*, 1990) and host susceptibility to TSEs (Hsiao *et al.*, 1989; Scott *et al.*, 1989; Doh-ura *et al.*, 1989; Goldgaber *et al.*, 1989).

The chemical basis for the difference between PrP-sen and PrP-res remains a mystery. A number of studies have suggested that the differences arise at the post-translational level (Chesebro *et al.*, 1985; Oesch *et al.*, 1985; Basler *et al.*, 1986; Caughey *et al.*, 1988b; Borchelt *et al.*, 1990). Strong evidence for this has been obtained recently (Caughey and Raymond, 1991). Several covalent post-translational modifications of PrP are known to occur; however, none has been identified as TSE specific (Bolton *et al.*, 1985; Manuelidis *et al.*, 1985; Stahl *et al.*, 1987; Caughey *et al.*, 1988b, 1989; Stahl *et al.*, 1990a). Thus, it is possible that a conformational abnormality accounts for the aberrant properties of PrP-res (Hope *et al.*, 1986).

Since PrP-res appears to play a central role in the TSEs, it is important to understand the cellular and structural changes which cause the formation of PrP-res. Although the scrapie-dependant mechanism by which PrP is converted to the protease-resistant state is not known, considerable progress has been made in understanding PrP biosynthesis and turnover and how it is influenced by scrapie infection. The development of scrapie-infected mouse neuroblastoma cells (Race *et al.*, 1987) made much of this progress possible because it provided scrapie-competent cells which could be metabolically labelled and conveniently manipulated *in vitro*. In this paper, we review studies of PrP metabolism in normal and scrapie-infected tissue culture cells.

PrP precursors

Since the nascent PrP polypeptide has a hydrophobic amino-terminal signal sequence (Locht *et al.*, 1986) and becomes glycosylated (Manuelidis *et al.*, 1985; Bolton *et al.*, 1985; Caughey *et al.*, 1989), its biosynthetic processing presumably begins within the endoplasmic reticulum. When either scrapie-infected or uninfected mouse neuroblastoma cells are labelled with [^{35}S]methionine for short periods (2–10 min), PrP precursors of 25, 28 and 33 kDa are identified (Caughey *et al.*, 1989). Similar precursors have been observed in mouse C127 cells (Caughey *et al.*, 1988b), rat PC12 pheochromocytoma cells (B. Caughey and R. Rubenstein, manuscript in preparation) and in a cell-free PrP translation system in the presence of microsomes (Hay *et al.*, 1987). Treatment of the neuroblastoma cell precursors with endoglycosidase H converts the larger two precursors to species that comigrate with the 25 kDa precursor, indicating that the precursors differ in the amount of *N*-linked high mannose glycan they contain (Caughey *et al.*, 1989). This is confirmed by the fact that the drug tunicamycin, which prevents *N*-linked glycosylation, allows the synthesis of only the 25 kDa PrP. Since there are two potential *N*-linked glycosylation sites (Asn–X–Thr) on the mouse PrP polypeptide (Locht *et al.*, 1986), it is likely that the 28 and 33 kDa precursors represent the addition of high mannose glycan chains to one and both the glycosylation sites, respectively, of the 25 kDa unglycosylated precursor.

Treatment of these early PrP precursors with phosphatidylinositol-specific phospholipase C (PIPLC) reduces their SDS–PAGE mobilities, indicating that they contain a phosphatidylinositol (PI) moiety (Caughey *et al.*, 1989). The rapid and simultaneous labelling of the three PrP precursors is consistent with studies of other proteins showing that the addition of glycosyl-PI (Low and Saltiel, 1988) and high mannose glycans (Kornfeld and Kornfeld, 1985) occurs concurrent with or soon after translation and translocation of polypeptides into the lumen of the endoplasmic reticulum. The addition of a glycosyl-PI moiety to proteins is typically associated with the removal of a hydrophobic *C*-terminal domain of the polypeptide (Ferguson and Williams, 1988). Stahl and colleagues have recently shown that a similar *C*-terminal truncation of PrP-res occurs in scrapie-infected brain tissue (Stahl *et al.*, 1990a). Amino acid sequencing of PrP derived from brain tissue has also indicated that the *N*-terminal signal sequence is removed (Hope *et al.*, 1986; Bolton *et al.*, 1987; Hope *et al.*, 1988a; Turk *et al.*, 1988).

PrP maturation

The 28 and 33 kDa PrP precursors labelled in the neuroblastoma and PC12 cells are post-translationally processed to 30 and 35–41 kDa species within 10–30 min (Race *et al.*, 1988; Caughey *et al.*, 1989; B. Caughey and R. Rubenstein, unpublished data). These mature PrP species are no longer susceptible to endoglycosidase H, indicating that the high mannose glycans present on the precursors are converted to hybrid or complex glycans, presumably within the Golgi apparatus. This is consistent with studies indicating that PrP of hamster brain, at least, contains complex carbohydrate moieties (Bolton *et al.*, 1985; Manuelidis *et al.*, 1985; Haraguchi *et al.*, 1989; Endo *et al.*, 1989).

Cell surface localization of PrP-sen

PrP-sen can be labelled on the surface of a variety of cells by membrane immuno-fluorescence (Stahl *et al.*, 1987; Caughey *et al.*, 1988b, 1990; Borchelt *et al.*, 1990: Cashman *et al.*, 1990), biotinylation (Borchelt *et al.*, 1990) and radioiodination (Caughey and Raymond, 1991). Cell surface PrP-sen is also susceptible to treatments of intact cells by proteases (Stahl *et al.*, 1987; Caughey *et al.*, 1989, 1990). Pulse [^{35}S]methionine-labelling experiments with mouse neuroblastoma cells have shown that PrP-sen starts to become exposed to extracellular proteases within 60 min of the initiation of biosynthesis (Caughey *et al.*, 1989). Within 90 min, approximately 90% of pulse-labelled 30 and 35–41 kDa PrP species can be digested by extracellular trypsin, demonstrating that they have reached the cell surface (Caughey *et al.*, 1990; Caughey and Raymond, 1991). Once at the cell surface, the labelled PrP-sen has a half-life of 3–6 h in the absence of exogenous proteases (Caughey *et al.*, 1989; Borchelt *et al.*, 1990; Caughey and Raymond, 1991). Since only a small proportion of the PrP-sen is recovered from the medium after extended chase periods, it appears that catabolism, rather than secretion or release from the cell surface, is the major fate of PrP in these cells (Caughey *et al.*, 1989; Borchelt *et al.*, 1990).

Treatments of intact cells with PIPLC have shown that most of the PrP-sen on the surface of several cell types is anchored by PI (Stahl *et al.*, 1987; Caughey *et al.*, 1989, 1990; Cashman *et al.*, 1990). Such experiments have indicated that all three mature PrP species (25, 30 and 35–41 kDa) of mouse neuroblastoma cells can be released from the cell surface (Caughey *et al.*, 1989).

EFFECTS OF SCRAPIE INFECTION

Detection of PrP-res in scrapie-infected cells

None of the parameters of PrP-sen biosynthesis that have been analysed to date in neuroblastoma cells appears to be generally affected by scrapie infection (Race *et al.*, 1988; Caughey *et al.*, 1989, 1990). Nonetheless, PrP-res has been identified in scrapie-infected but not uninfected clones by a variety of methods (Butler *et al.*, 1988; Caughey *et al.*, 1990; Borchelt *et al.*, 1990; Taraboulos *et al.*, 1990b; Caughey and Raymond, 1991). The PrP-res species observed are similar in apparent molecular mass (19, 23 and ~28 kDa) to the proteinase K-treated, *N*-terminally truncated PrP-res species from mouse brain (Hope *et al.*, 1988a) but differ in the relative intensities of the bands (Caughey, 1991). The multiple PrP-res species result from varying amounts of *N*-linked glycosylation (Taraboulos *et al.*, 1990a), as is the case with the PrP-sen species (Caughey *et al.*, 1989). Although the PrP-res isolated from brain tissue is primarily full length unless treated with proteases *in vitro* (Hope *et al.*, 1986; Bolton *et al.*, 1987; Hope *et al.*, 1988a), the neuroblastoma cell PrP-res is quantitatively converted to the *N*-terminally truncated forms within the cell (Caughey *et al.*, 1991a). Thus, the neuroblastoma PrP-res species are, as a group, approximately 6–7 kDa lower in apparent molecular mass than the corresponding PrP-sen bands of the same cells. Like the PrP-res of scrapie brain tissue, the neuroblastoma cell PrP-res forms large aggregates in detergent lysates, providing evidence that the tendency of PrP-res to aggregate is an intrinsic property and not merely a consequence of degenerative brain pathology (Taraboulos et al., 1990a; Caughey et al., 1991a). The scrapie infectivity in neuroblastoma cell extracts also aggregates and is relatively proteinase K-resistant which is consistent with, but does not prove, an association between PrP-res and the scrapie agent (Neary et al., 1991).

Kinetics of PrP-res biosynthesis

Once PrP-res was identified in scrapie-infected neuroblastoma cells, the problem became one of determining how its biosynthesis differs from that of the normal PrP-sen. The insolubility of PrP-res in most detergent lysates of scrapie-infected neuroblastoma cells prevented the detection of metabolically labelled PrP-res by immunoprecipitation until it was shown that PrP-res could be immunoprecipitated after being sonicated into phospholipid–detergent micelles (Borchelt et al., 1990). Pulse-chase metabolic labelling studies of PrP-res then indicated that PrP-res is labelled much more slowly than PrP-sen and only after a lag period of ∼ 1 h (Borchelt et al., 1990; Caughey and Raymond, 1991). These relative labelling kinetics of PrP-sen and PrP-res suggested that PrP-res is synthesized from a protease-sensitive precursor such as PrP-sen. However, one could not exclude the possibility that PrP-res was made from a protease-resistant precursor that was not immunoprecipitable, especially given the difficulties originally encountered in solubilizing PrP-res aggregates. Thus, for more definitive evidence of whether or not PrP-res is made from a PrP–sen-like precursor, topological studies were required.

Intracellular site(s) of PrP-res formation and accumulation

Membrane topology studies showed that, unlike PrP-sen, PrP-res is resistant to removal from both intact and permeabilized cells and membranes with PIPLC or proteases (Caughey et al., 1990; Borchelt et al., 1990; Stahl et al., 1990b; Safar et al., 1991). However, treatment of intact, pulse-labelled neuroblastoma cells with PIPLC or trypsin prevents the subsequent incorporation of label into PrP-res (Caughey and Raymond, 1991). This demonstrated that PrP-res is made from a PIPLC- and trypsin-sensitive precursor. Additionally, the accessibility of the PrP-res precursor to PIPLC, trypsin and surface radioiodination indicated that it resides, at least transiently, on the cell surface. These properties make the PrP-res precursor similar, if not identical, to the normal PrP-sen of uninfected cells. Thus, although some investigators have argued that transit of PrP to the cell surface is not likely to be necessary for PrP-res formation (Taraboulos et al., 1990a), these topological studies indicate clearly that the conversion of PrP to the protease-resistant state is a post-translational event that occurs after the precursor (presumably normal PrP-sen) reaches the cell surface.

The PIPLC sensitivity of the PrP-res precursor also eliminates the possibility that formation of an integral transmembrane topology by nascent PrP polypeptides in the endoplasmic reticulum (Hay et al., 1987) accounts for the aberrant properties of PrP-res in the neuroblastoma cell model. If such a topological change is associated with PrP-res formation, it must occur after the precursor reaches the plasma membrane (Caughey and Raymond, 1991).

A number of studies have provided evidence that PrP-res accumulates in an intracellular compartment (Caughey et al., 1990; Borchelt et al., 1990; Stahl et al., 1990b; Wiley et al., 1987; Diedrich et al., 1991) and appears to be partially colocalized with markers of the Golgi apparatus (Taraboulos et al., 1990b). Since the PrP-res precursor is located, at least transiently, on the cell surface, the intracellular deposition of PrP-res must result from the internalization of PrP-res or its precursor from the plasma membrane rather than the retention of PrP-res in the Golgi apparatus during biosynthesis (Caughey and Raymond, 1991). Most components of the plasma membrane are continuously internalized into endosomes (Thilo, 1985) which are then commonly fused with Golgi-derived vesicles containing hydrolytic enzymes and acidified to form lysosomes (Kelly, 1990). Proteases within the endolysosomes and lysosomes then digest susceptible proteins. This process results in the normal turnover of plasma membrane-derived proteins (Hare, 1990) and thus is likely to account for PrP-sen catabolism. PrP-res is also translocated to the lysosomes in neuroblastoma cells as indicated by the fact that the N-terminal truncation of PrP-res is performed in part by ammonia- and leupeptin-sensitive proteases (Caughey et al., 1991a). The accumulation of PrP-res in the lysosomes could explain the perinuclear staining of PrP-res reported by Taraboulos and colleagues (Taraboulos et al., 1990b) since endosome-derived transport vesicles are taken to a perinuclear location before the prelysosomal and lysosomal compartments are formed (Kelly, 1990; Caughey and Raymond, 1991). Since PrP-res is not completely degraded in the lysosomes, its conversion to the protease-resistant state probably occurs prior to its exposure to proteases within endolysosomes and lysosomes (Caughey and Raymond, 1991; Caughey et al., 1991a). Thus PrP-res is likely to be formed at the plasma membrane, where the protease- and PIPLC-sensitive precursor is found, or along the endocytic pathway before exposure to proteases (Fig. 1).

The stability of PrP-res

Although PrP-res comes into contact with cellular proteases in neuroblastoma cells, we and other investigators have obtained evidence that it is not completely digested and has a half-life of $\gg 48$ h (Borchelt et al., 1990; Caughey and Raymond, 1991). The inability of cells to degrade PrP-res might explain its steady build-up in the non-dividing cells of the central nervous system of an infected host. This accumulation of PrP-res may be harmful and ultimately lead to neurodegeneration.

The structural features that account for the extraordinary stability of PrP-res have not been ascertained. Conformational studies by conventional methods have been hampered by the insoluble, amyloid-like properties of PrP-res. However, a recently developed infrared spectroscopy technique for analyzing protein secondary structure in aqueous media has allowed us to determine the secondary structure composition of the protease-resistant core of PrP-res (PrP-res 27-30) (Caughey et al., 1991b). These studies have revealed that PrP-res 27-30 has a high β-sheet content as do other highly stable amyloids. In addition, PrP-res 27-30 has very little random coil to provide favourable sites for proteolytic attack. Since PrP is a glycoprotein, one might expect that the N-linked glycans could help to make the molecule resistant to proteases. However, analyses of PrP-res synthesized in scrapie-infected neuroblastoma cells

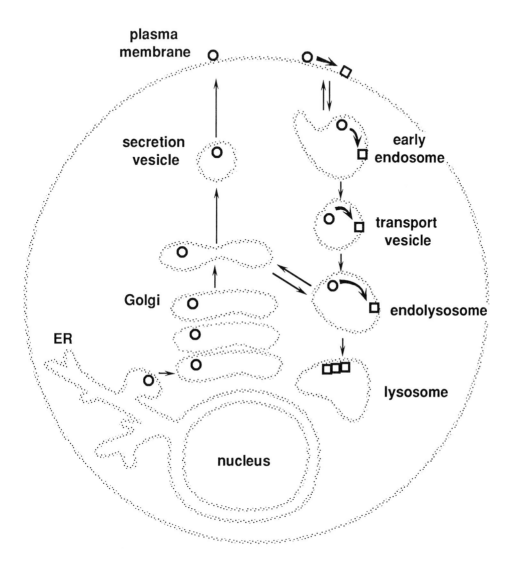

Fig. 1.   Model of probable subcellular sites of PrP-res formation. The major routes of cellular membrane trafficking in the biosynthetic pathway (endoplasmic reticulum to plasma membrane) and degradative pathway (plasma membrane to lysosome) of plasma membrane proteins are indicated by the thin arrows after a previous description (Kelly, 1990). The circles show the likely presence of both the normal PrP-sen and, if different, the PIPLC-sensitive PrP-res precursor which are anchored to the membranes by phosphatidylinositol. Based on these fundamental pathways of membrane trafficking and the results of studies in scrapie-infected neuroblastoma cells (Caughey and Raymond, 1991; Caughey et al., 1991a), the likely sites of PrP-res formation and deposition are designated by the heavy arrows and squares, respectively.

indicate that glycosylation is not required for PrP-res formation (Taraboulos *et al.*, 1990a) and may actually disfavour either PrP-res formation or stability (Caughey and Raymond, 1991; Caughey *et al.*, 1991a). Interestingly, one of the potential *N*-linked glycosylation sites (at Asn-181) is located in an especially hydrophobic portion of what is predicted to be a long β-sheet strand and, thus, might be expected to have a profound influence on the folding and stability of the molecule (Caughey *et al.*, 1991b).

CONCLUSIONS

Our studies of PrP-res metabolism in scrapie-infected neuroblastoma cells implicate the plasma membrane and the endocytic pathway as likely sites for the formation of PrP-res. In the future, experiments which investigate the effects of treatments that influence intracellular membrane trafficking on the formation, proteolysis and accumulation of PrP-res should be revealing. Furthermore, it now seems possible to investigate the effects of recombinant PrP gene expression on these events in scrapie-infected neuroblastoma cells. Such studies may ultimately provide information about whether perturbations of PrP-res biosynthesis affect the replication of scrapie infectivity.

The nature of the difference between the normal PrP-sen and the scrapie-associated PrP-res which accounts for their disparate physical and metabolic properties remains a central question in the transmissible spongiform encephalopathy field. Our infrared spectroscopic studies of PrP-res 27-30 have provided insight into the folding of the PrP molecule. The infrared technique we have demonstrated should allow the direct comparison of the conformations of the normal and scrapie-associated forms of PrP and their variants once sufficient quantities of the purified PrP species become available. Such information should shed light on whether it is likely that a conformational change in PrP can account for PrP-res formation and, perhaps, the initiation of TSE pathogenesis.

REFERENCES

Basler, K., Oesch, B., Scott, M., Westaway, D., Walchli, M., Groth, D.F., McKinley, M.P., Prusiner, S.B. and Weissman, C. (1986) Scrapie and cellular PrP isoforms are encoded by the same chromosomal gene. *Cell* **46** 417–428.

Bendheim, P.E., Bockman, J.M., McKinley, M.P., Kingsbury, D.T. and Prusiner, S.B. (1985) Scrapie and Creutzfeldt–Jakob disease prion proteins share physical properties and antigenic determinants. *Proc. Natl. Acad. Sci. USA* **82** 997–1001.

Bockman, J.M., Kingsbury, D.T., McKinley, M.P., Bendheim, P.E. and Prusiner, S.B. (1985) Creutzfeldt–Jakob disease prion proteins in human brains. *N. Engl. J. Med.* **312** 73–78.

Bolton, D.C., McKinley, M.P. and Prusiner, S.B. (1982) Identification of a protein that purifies with the scrapie prion. *Science* **218** 1309–1311.

Bolton, D.C., McKinley, M.P. and Prusiner, S.B. (1984) Molecular characteristics of the major scrapie prion protein. *Biochemistry* **23** 5898–5906.

Bolton, D.C., Meyer, R.K. and Prusiner, S.B. (1985) Scrapie PrP 27-30 is a sialoglycoprotein. *J. Virol.* **53** 596–606.

Bolton, D.C., Bendheim, P.E., Marmostein, A.D. and Potempska, A. (1987) Isolation and structural studies of the intact scrapie agent protein. *Arch. Biochem. Biophys.* **258** 579–590.

Borchelt, D.R., Scott, M., Taraboulos, A., Stahl, N. and Prusiner, S.B. (1990) Scrapie and cellular prion proteins differ in the kinetics of synthesis and topology in cultured cells. *J. Cell Biol.* **110** 743–752.

Brown, P., Coker-Vann, M., Pomeroy, K., Franko, M., Asher, D.M., Gibbs, C.J., Jr, and Gajdusek, D.C. (1986) Diagnosis of Creutzfeldt–Jakob disease by Western blot identification of marker protein in human brain tissue. *N. Engl. J. Med.* **314** 547–551.

Butler, D.A., Scott, M.R.D., Bockman. J.M., Borchelt, D.R., Taraboulos, A., Hsiao, K.K., Kingsbury, D.T. and Prusiner, S.B. (1988) Scrapie-infected murine neuroblastoma cells produce protease-resistant prion proteins. *J. Virol.* **62** 1558–1564.

Carlson, G.A., Kingsbury, D.T., Goodman, P.A., Coleman, S., Marshall, S.T., DeArmond, S., Westaway, D. and Prusiner, S.B. (1986) Linkage of prion protein and scrapie incubation time gene. *Cell* **46** 503–511.

Carlson, G.A., Goodman, P.A., Lovett, M., Taylor, B.A., Marshall, S.T., Peterson-Torchia, M., Westaway, D. and Prusiner, S.B. (1988) Genetics and polymorphism of the mouse prion gene complex: control of scrapie incubation time. *Mol. Cell. Biol.* **8** 5528–5540.

Cashman, N.R., Loertscher, R., Nalbantoglu, J., Shaw, I., Kascsak, R.J., Bolton, D.C. and Bendheim, P.E. (1990) Cellular isoform of the scrapie agent protein participates in lymphocyte activation. *Cell* **61** 185–192.

Caughey, B. (1991) *In vitro* expression and biosynthesis of prion protein. *Curr. Top. Microbiol. Immunol.* **172** 93–107.

Caughey, B. and Raymond, G.J. (1991) The scrapie-associated form of PrP is made from a cell surface precursor that is both protease- and phospholipase-sensitive. *J. Biol. Chem.* **266** 18217–18213.

Caughey, B., Race, R.E. and Chesebro, B. (1988a) Detection of prion protein mRNA in normal and scrapie-infected tissues and cell lines. *J. Gen. Virol.* **69** 711–716.

Caughey, B., Race, R.E., Vogel, M., Buchmeier, M.J. and Chesebro, B. (1988b) *In vitro* expression in eukaryotic cells of the prion protein gene cloned from scrapie-infected mouse brain. *Proc. Natl. Acad. Sci. USA* **85** 4657–4661.

Caughey, B., Race, R.E., Ernst, D., Buchmeier, M.J. and Chesebro, B. (1989) Prion protein (PrP) biosynthesis in scrapie-infected and uninfected neuroblastoma cells. *J. Virol.* **63** 175–181.

Caughey, B., Neary, K., Buller, R., Ernst, D., Perry, L., Chesebro, B. and Race, R. (1990) Normal and scrapie-associated forms of prion protein differ in their sensitivities to phospholipase and proteases in intact neuroblastoma cells. *J. Virol.* **64** 1093–1101.

Caughey, B., Raymond, G.J., Ernst, D. and Race, R.E. (1991a) *N*-terminal truncation of the scrapie-associated form of PrP by lysosomal protease(s): implications regarding the site of conversion of PrP to the protease-resistant state. *J. Virol.* **65** 6597–6603.

Caughey, B.W., Dong, A., Bhat, K.S., Ernst, D., Hayes, S.F. and Caughey, W.S. (1991b)

Secondary structure analysis of the scrapie-associated protein PrP 27-30 in water by infrared spectroscopy. *Biochemistry* **30** 7672–7680.

Chesebro, B., Race, R., Wehrly, K., Nishio, J., Bloom, M., Lechner, D., Bergstrom, S., Robbins, K., Mayer, L., Keith, J.M., Garon, C. and Haase, A. (1985) Identification of scrapie prion protein-specific mRNA in scrapie-infected and uninfected brain. *Nature (London)* **315** 331–333.

Cho, H.J. (1986) Antibody to scrapie-associated fibril protein identifies a cellular antigen. *J. Gen. Virol.* **67** 243–253.

Diedrich, J.F., Bendheim, P.E., Kim, Y.S., Carp, R.I. and Haase, A.T. (1991) Scrapie-associated prion protein accumulates in astrocytes during scrapie infection. *Proc. Natl. Acad. Sci. USA* **88** 375–379.

Diringer, H., Gelderblom, H., Hilmert, H., Ozel, M., Edelbluth, C. and Kimberlin, R.H. (1983) Scrapie infectivity, fibrils and low molecular weight protein. *Nature (London)* **306** 476–478.

Doh-ura, K., Tateishi, J., Sasaki, H., Kitamoto, T. and Sakaki, Y. (1989) Pro→leu change at position 102 of prion protein is the most common but not the sole mutation related to Gerstmann–Sträussler syndrome. *Biochem. Biophys. Res. Common.* **163** 974–979.

Endo, T., Groth, D., Prusiner, S.B. and Kobata, A. (1989) Diversity of oligosaccharide structures linked to asparagines of the scrapie prion protein. *Biochemistry* **28** 8380–8388.

Fergusin, M.A.J. and Williams, A.F. (1988) Cell-surface anchoring of proteins via glycosylphosphatidylinositol structure. *Annu. Rev. Biochem.* **57** 285–320.

Goldgaber, D., Goldfarb, L.G., Brown, P. *et al.*, (1989) Mutations in familial Creutzfeldt–Jakob disease and Gerstmann–Sträussler–Scheinker's syndrome. *Exp. Neurol.* **106** 204–206.

Haraguchi, T., Fisher, S., Olofsson, S., Endo, T., Groth, D., Tarentino, A., Borchelt, D.R., Teplow, D., Hood, L., Burlingame, A., Lycke, E., Kobata, A. and Prusiner, S.B. (1989) Asparagine-linked glycosylation of the scrapie and cellular prion proteins. *Arch. Biochem. Biophys.* **274** 1–13.

Hare, J.F. (1990) Mechanisms of membrance protein turnover. *Biochim. Biophys. Acta* **1031** 71–90.

Hay, B., Barry, R.A., Lieberburg, I., Prusiner, S.B. and Lingappa, V.R. (1987) Biogenesis and transmembrane orientation of the cellular isoform of the scrapie prion protein. *Mol. Cell. Biol.* **7** 914–920.

Hope, J., Morton, L.J.D., Farquhar, C.F., Multhaup, G., Beyreuther, K. and Kimberlin, R.H. (1986) The major polypeptide of scrapie-associated fibrils (SAF) has the same size, charge distribution and *N*-terminal protein sequence as predicted for the normal brain protein (PrP). *EMBO J.* **5** 2591–2597.

Hope, J., Multhaup, G., Reekie, L.J.D., Kimberlin, R.H. and Beyreuther, K. (1988a) Molecular pathology of scrapie-associated fibril protein (PrP) in mouse brain affected by the ME7 strain of scrapie. *Eur. J. Biochem.* **172** 271–277.

Hope, J., Reekie, L.J.D., Hunter, N., Multhaup, G., Beyreuther, K., White, H., Scott, A.C., Stack, M.J., Dawson, M. and Wells, G.A.H. (1988b) Fibrils from brains of cows with new cattle disease contain scrapie-associated protein. *Nature (London)*

336 390–392.

Hsiao, K., Baker, H.F., Crow, T.J., Poulter, M., Owen, F., Terwilliger, J.D., Westaway, D., Ott, J. and Prusiner, S.B. (1989) Linkage of a prion protein missense variant to Gerstmann–Sträussler syndrome. *Nature (London)* **338** 342–345.

Hunter, N., Hope, J., McConnell, I., and Dickenson, A.G. (1987). Linkage of the scrapie-associated fibril protein (PrP) gene and sinc using congenic mice and restriction fragment length polymorphism analysis. *J. gen. Virol.* **68** 2711–2716.

Kelly, R.B. (1990) Microtubules, membrane traffic, and cell organization. *Cell* **61** 5–7.

Kornfeld, R. and Kornfeld, S. (1985) Assembly of asparagine-linked oligosaccharides. *Annu. Rev. Biochem.* **54** 631–664.

Locht, C., Chesebro, B., Race, R. and Keith, J.M. (1986) Molecular cloning and complete sequence of prion protein cDNA from mouse brain infected with the scrapie agent. *Proc. Natl. Acad. Sci. USA* **83** 6372–6376.

Low, M.G. and Satiel, A.R. (1988) Structural and functional roles of glycosyl-phosphatidylinositol in membranes. *Science* **239** 268–275.

Manuelidis, L., Valley, S. and Manuelidis, E.E. (1985) Specific proteins associated with Creutzfeldt–Jakob disease and scrapie share antigenic and carbohydrate determinants. *Proc. Natl. Acad. Sci. USA* **82** 4263–4267.

McKinley, M.P., Bolton, D.C. and Prusiner, S.B. (1983) A protease-resistant protein is a structural component of the scrapie prion. *Cell* **35** 57–62.

Merz, P.A., Somerville, R.A., Wisniewski, H.M. and Iqbal, K. (1981) Abnormal fibrils from scrapie-infected brain. *Acta Neuropathol.* **54** 63–74.

Meyer, R.K., McKinley, M.P., Bowman, K.A., Braunfeld, M.B., Barry, R.A. and Prusiner, S.B. (1986) Separation and properties of cellular and scrapie prion protein. *Proc. Natl. Acad. Sci. USA* **83** 2310–2314.

Neary, K., Caughey, B., Ernst, D., Race, R.E. and Chesebro, B. (1991) Protease sensitivity and nuclease resistance of the scrapie agent propagated *in vitro* in neuroblastoma cells. *J. Virol.* **65** 1031–1034.

Oesch, B., Westaway, D., Walchli, M., McKinley, M.P., Kent, S.B.H., Aebersold, R., Barry, R.A., Tempst, P., Teplow, D.B., Hood, L.E., Prusiner, S.B. and Weissmann, C. (1985) A cellular gene encodes scrapie PrP 27-30 protein. *Cell* **40** 735–746.

Race, R.E., Fadness, L.H. and Chesebro, B. (1987) Characterization of scrapie infection in mouse neuroblastoma cells. *J. Gen. Virol.* **68** 1391–1399.

Race, R.E., Caughey, B., Graham, K., Ernst, D. and Chesebro, B. (1988) Analysis of frequency of infection, specific infectivity, and prion protein biosynthesis in scrapie-infected neuroblastoma cell clones. *J. Virol.* **62** 2845–2849.

Race, R.E., Graham, K., Ernst, D., Caughey, B. and Chesebro, B. (1990) Analysis of linkage between scrapie incubation period and the prion protein gene in mice. *J. Gen. Virol.* **71** 493–497.

Rubenstein, R., Kascsak, R.J., Merz, P.A., Papini, M.C., Carp, R.I., Robakis, N.K. and Wisniewski, H.M. (1986) Detection of scrapie-associated fibril (SAF) proteins using anti-SAF antibody in non-purified tissue preparations. *J. Gen. Virol.* **67** 671–681.

Safar, J., Ceroni, M., Gajdusek, D.C. and Gibbs, C.J., Jr (1991) Differences in the membrane interaction of scrapie amyloid precursor proteins in normal and scrapie-

or Creutzfeldt–Jakob disease-infected brains. *J. Infect. Dis.* **163** 488–494.

Scott, M., Foster, D., Mirenda, C., Serban, D., Coufal, F., Walchli, M., Torchia, M., Groth, D., Carlson, G., DeArmond, S.J., Westaway, D. and Prusiner, S.B. (1989) Transgenic mice expressing hamster prion protein produce species-specific scrapie infectivity and amyloid plaques. *Cell* **59** 847–857.

Sklaviadis, T., Manuelidis, L. and Manuelidis, E.E. (1986) Characterization of major peptides in Creutzfeldt–Jakob disease and scrapie. *Proc. Natl. Acad. Sci. USA* **83** 6146–6150.

Stahl, N., Borchelt, D.R., Hsiao, K. and Prusiner, S.B. (1987) Scrapie prion protein contains a phosphatidylinositol glycolipid. *Cell* **51** 229–240.

Stahl, N., Baldwin, M.A., Burlingame, A.L. and Prusiner, S.B. (1990a) Identification of glycoinositol phospholipid linked and truncated forms of the scrapie prion protein. *Biochemistry* **29** 8879–8884.

Stahl, N., Borchelt, D.R. and Prusiner, S.B. (1990b) Differential release of cellular and scrapie prion proteins from cellular membranes by phosphatidylinositol-specific phospholipase C. *Biochemistry* **29** 5405–5412.

Taraboulos, A., Rogers, M., Borchelt, D.R., McKinley, M.P., Scott, M., Serban, D. and Prusiner, S.B. (1990a) Acquisition of protease resistance by prion proteins in scrapie-infected cells does not require asparagine-linked glycosylation. *Proc. Natl. Acad. Sci. USA* **87** 8262–8266.

Taraboulos, A., Serban, D. and Prusiner, S.B. (1990b) Scrapie prion proteins accumulate in the cytoplasm of persistently infected cultured cells. *J. Cell Biol.* **110** 2117–2132.

Thilo, L. (1985) Qualification of endocytosis-derived membrane traffic. *Biochim. Biophys. Acta* **822** 243–266.

Turk, E., Teplow, D.B., Hood, L.E. and Prusiner, S.B. (1988) Purification and properties of the cellular and scrapie hamster prion proteins. *Eur. J. Biochem.* **176** 21–30.

Westaway, D., Goodman, P.A., Mirenda, C.A., McKinley, M.P., Carlson, G.A. and Prusiner, S.B. (1987) District prion proteins in short and long scrapie incubation period mice. *Cell* **51** 651–662.

Wiley, C.A., Burrola, P.G., Buchmeier, M.J., Wooddell, M.K., Barry, R.A., Prusiner, S.B. and Lampert, P.W. (1987) Immuno-gold localization of prion filaments in scrapie-infected hamster brains. *Lab Invest.* **57** 646–655.

38

Modification and expression of prion proteins in cultured cells

Mark Rogers, Albert Taraboulos, Michael Scott, David Borchelt, Dan Serban, Tibor Gyuris and Stanley B. Prusiner

ABSTRACT

A single copy host gene encodes both the scrapie and cellular isoforms of the prion protein. These isoforms differ strikingly in their biochemical properties. Expression of recombinant PrP^C and PrP^{Sc} should help determine the basis of this difference. Using a variety of vector systems, the PrP gene from a number of species has been expressed and recombinant PrP^C produced. However, in all but one case this recombinant protein has been processed aberrantly and has failed to produce PrP^{Sc}. Recombinant vaccinia virus provides a method to express transiently PrP genes in a variety of cell types. Syrian hamster (SHa)–mouse (Mo) chimeric PrP constructs were expressed using recombinant vaccinia virus. These chimeric SHa–MoPrP species were used to map the binding sites for a number of SHaPrP specific monoclonal antibodies. Mutant SHaPrP genes altered at the potential sites for the addition of Asn-linked oligosaccharides have also been expressed using this approach. Both potential Asn glycosylation sites are modified by oligosaccharides as determined by a reduction in the apparent molecular weight and size heterogeneity on SDS–PAGE. The absence of Asn-linked carbohydrate has been confirmed using PNGaseF and anhydrous HF. Furthermore, the mutation of either site prevented the transport of the mutant PrP species to the cell surface. Stable expression of recombinant PrP genes has been achieved in scrapie-infected mouse neuroblastoma cells (ScN$_2$a). This recombinant protein is fully competent for the formation of PrP^{Sc}. Mutant PrP genes were altered in the sequences required to specify the addition of Asn-linked oligosaccharides and used to study the role of Asn-linked oligosaccharides in the formation of PrP^{Sc}. In ScN$_2$a cells, the recombinant protein was competent to form PrP^{Sc} demonstrating that the presence of Asn-linked oligosaccharides in *cis* is not an essential requirement in the synthesis of PrP^{Sc}. The other known post-translational modifications present in PrP^{Sc} may be studied in a similar manner using this approach. The construction of transgenic mice expressing these modified PrP genes may allow assessment of the role of these changes on the formation of infectious prions.

INTRODUCTION

The scrapie prion protein, PrP^{Sc}, is the only known component of the infectious agent causing transmissible neurodegenerative diseases (Gabizon and Prusiner, 1990; Scott *et al.*, 1989; Prusiner *et al.*, 1990). PrP^{Sc} accumulates in the brain of humans and animals with prion diseases as well as persistently infected cell cultures.

A single copy chromosomal gene encodes both the normal cellular prion protein, PrP^{C}, and PrP^{Sc}. The observation that the PrP open reading frame in this gene is contiguous and is not interrupted by any introns argues that the differences between these isoforms do not occur at the level of transcription (Oesch *et al.*, 1985; Basler *et al.*, 1986). Furthermore, the data obtained on the kinetics of synthesis of both PrP^{C} and PrP^{Sc} show that the formation of PrP^{Sc} occurs post-translationally (Caughey *et al.*, 1989; Borchelt *et al.*, 1990).

While antigenically related, PrP^{C} and PrP^{Sc} may be distinguished by a number of criteria.

(1) PrP^{Sc} possesses a protease-resistant core termed PrP 27-30 while PrP^{C} is completely sensitive to proteolysis (Oesch *et al.*, 1985).
(2) PrP^{C} is soluble in the presence of detergent while PrP^{Sc} forms insoluble aggregates (Meyer *et al.*, 1986).
(3) PrP^{Sc} immunoreactivity is enhanced after treatment with denaturants (Serban *et al.*, 1990); PrP^{C} is not.
(4) PrP^{C} is located on the cell surface where it is attached by a glycosyl phosphatidylinositol anchor (GPI); PrP^{Sc} is found primarily within the interior of cells (Taraboulos *et al.*, 1990b; Stahl *et al.*, 1987, 1990).
(5) The GPI anchor of PrP^{C} is sensitive to cleavage by the enzyme phosphatidylinositol-specific phospholipase C (PIPLC); the GPI anchor attached to PrP^{Sc} is only sensitive to PIPLC after the protein has been denatured (Caughey *et al.*, 1990; Stahl *et al.*, 1990).
(6) PrP^{C} synthesis and degradation are rapid while the synthesis of PrP^{Sc} is slow and degradation has yet to be observed (Borchelt *et al.*, 1990).

Resistance to proteolysis and a propensity to aggregate in the presence of detergents represent the most commonly used criteria for distinguishing PrP^{Sc} from PrP^{C}.

To date no structural or conformational difference between PrP^{C} and PrP^{Sc} has been detected. Whether a covalent chemical modification is responsible for the conversion of PrP^{C} to PrP^{Sc} is unknown. This putative modification need not be in the form of an additional chemical group but could be due to the removal of a modification normally found on PrP^C. Alternatively, a stable conformational change either independent of other components or dependent upon the tight association of other, as-yet undefined, components could also account for PrP^{Sc}. It has been shown that both PrP^C and PrP^{Sc} are modified extensively post-translationally (Oesch et al., 1985). Both possess an N-terminal signal sequence removed during translocation into the endoplasmic reticulum (Hope et al., 1986), an intramolecular disulphide bond, a GPI anchor (Stahl et al., 1987; Turk et al., 1988) and extensive Asn-linked oligosaccharides (Endo et al., 1989; Bolton et al., 1985; Haraguchi et al., 1989).

We review here the use of expression systems in the study of PrP^C and PrP^{Sc}. The efficient and high level expression of authentic PrP^C and/or PrP^{Sc} would facilitate the identification of any structural difference between these molecules if one exists. To date, such expression has proved elusive. Instead these studies have provided novel approaches towards understanding the processes involved in the synthesis of PrP^C and PrP^{Sc}.

RESULTS AND DISCUSSION

Attempts to express recombinant PrP genes are summarized in Table 1. All the

Table 1.   Expression of recombinant PrP

System–vector	Molecular weight (kDa)	Cell surface	Reference
Yeast (S. cerevisiae)–pTT1	28–45	NA	T. Torchia et al. (unpublished observation)
Baculovirus–pAC611	26–28	NA	Scott et al. (1988)
BPV–p341-1-PrP-BPV	35–41	+	Caughey et al. (1988)
Vaccinia–vTF7-3 and Cop3-6	26–42	+	Rogers et al. (1991)
SV40–SPOXIIneo	33–35[a]	+	M. Scott et al. (unpublished observation)
CHO–pMT2	32–42	+	M. Rogers et al. (unpublished observation)
Transient–pSVL, pHSP70, pMX1132	28–45	+	Scott et al. (1988)

[a] The pattern observed includes bands at 26 kDa and 28 kDa as well as 33–35 kDa. This is identical to the pattern observed for the endogenous protein.

methods used have resulted in some expression of recombinant prion protein. In most cases this protein is expressed at modest levels and does not migrate with the same molecular weight (M_r) profile as endogenous PrP. This is most likely due to aberrant post-translational processing mainly associated with the Asn-linked oligosaccharides (Scott *et al.*, 1988; Caughey *et al.*, 1988; Rogers *et al.*, 1990, 1991) and has limited the use of many of these expression systems. Where 'authentic' processing of the recombinant protein has occurred, the levels of expression have not been sufficient to provide PrP^C or PrP^{Sc} in the quantities required for direct biochemical comparison.

Extracts containing recombinant PrP derived from the hamster (Scott *et al.*, 1988) or mouse (Caughey *et al.*, 1988) PrP gene did not contain any infectious prions. This result is expected as none of the recombinant prion proteins was exposed to infectious prions.

Two expression systems have proved especially useful in the characterization of mutated PrP genes. The first is a transient expression system that uses recombinant Vaccinia virus to express mutant PrP genes directly from plasmids (Fuerst *et al.*, 1986). This system cannot be used to study directly the synthesis of PrP^{Sc}. Infection of cells by Vaccinia virus rapidly results in an inhibition of host cell protein synthesis and eventually cell death; however, this system is useful as a rapid method of screening and characterizing mutant PrP species to assess the role of the mutation upon PrP^C biosynthesis. The second system comprises scrapie-infected cell cultures that express recombinant PrP genes using SV40 based expression vectors. This system has provided the first opportunity to study the effect of *in vitro* generated mutations in the PrP gene on the biosynthesis of PrP^{Sc} (Scott *et al.*, 1992; Taraboulos *et al.*, 1990a).

Expression using recombinant Vaccinia virus

The broad host range of Vaccinia virus makes it a versatile vector for expression of recombinant genes. We have utilized the plasmid–virus expression protocol described by Feurst *et al.* (1986) since it does not require the construction of recombinant vaccinia virus for the expression of each mutant PrP gene. Fig. 1 outlines the essential features of this system. Plasmids containing the PrP gene of interest located 3;pr to a T7 promoter element are transfected into the host cells that have previously been infected with the recombinant virus vTF7-3. vTF7-3 supplies T7 RNA polymerase which is required to transcribe the PrP gene from the T7 promoter.

Mapping epitopes in SHaPrP

The limited number of amino acid differences between the Mo and SHaPrP protein sequences makes possible the mapping of epitopes within the SHaPrP sequence recognized by monoclonal antibodies (mAbs) raised against SHaPrP 27-30 (Barry *et al.*, 1985; Barry and Prusiner, 1986; Kascsak *et al.*, 1987; Rogers *et al.*, 1991). Chimeric Mo–SHaPrP genes were constructed in which segments of the MoPrP open reading frame were replaced with homologous regions of the SHaPrP gene. This results in a chimeric PrP gene that is essentially Mo in origin but which has a few (two or three) SHaPrP specific amino acid substitutions (Fig. 2, part I). These recombinant PrP genes were expressed in CV1 cells , extracts run on SDS–PAGE and immunoblotted against a series of mAbs. Fig. 2 shows the results obtained for the polyclonal antibody R073 and three representative mAbs. The results of a similar screening of a total of 18 mAbs are summarized in Table 2. All the antibodies tested could be placed in one

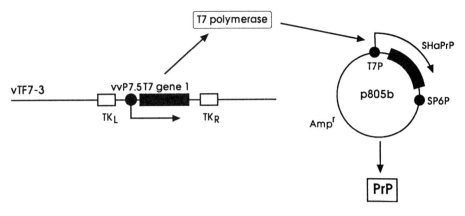

Fig. 1.   Scheme for Vaccinia virus and plasmid constructs. Components for the expression of PrP are shown. The plasmid plus virus expression systems require the expression of the T7 gene 1 protein which is supplied from the recombinant Vaccinia virus vTF7-3 (Fuerst *et al.*, 1986). This T7 RNA polymerase can direct the expression *in trans* of any gene cloned 3′ to the appropriate T7 promoter (T7P). p805b is the prototypic plasmid encoding SHaPrP under the control of a T7 promoter (Scott *et al.*, 1988). It is based upon pSP72. The expression of PrP in the plasmid system requires the infection of CV1 cells with vTF7-3 and the subsequent transfection of these cells with the appropriate plasmid.

of three distinct classes of antibodies that recognize a specific epitope within PrP that is defined by specific amino acid differences between MoPrP and SHaPrP.

The PrP mAb 27-2 is the prototype of the class of mAbs that bind an epitope present in MHM2 PrP. The mAb 3F4 falls into this class (Kascsak *et al.*, 1987). MHM2 PrP differs from MoPrP by two amino acids at positions 108 and 111 (see Fig. 2, part I). One or both of these differences define this epitope. Antibodies typified by 13A5 and 7D4 recognize M2HM which has three SHaPrP specific amino acid substitutions (Fig. 2, part I). However, these changes defined two distinct epitopes recognized by these 13A5 and 7D4-like antibodies. This was shown by constructing and expressing a mutant MoPrP gene which has a single amino acid substitution at codon 138 that converts an Ile to a Met. PrP derived from this construct (PrP MoMet138) is specifically recognized by 13A5-like antibodies but not by 7D4-like antibodies (Fig. 2, part II). This result defines the essential feature of the 13A5-specific epitope as the Met at position 139 in the SHaPrP sequence. Finally, the 7D4-specific epitope is defined by one or both of the Asn residues at codons 155 or 170 in the SHaPrP sequence.

Characterization of mutant PrP genes by transient expression in CV1 cells
As a prelude to studies on the role of post-translational modifications in the synthesis of PrP^{Sc} and infectious prions, modified SHaPrP genes were expressed transiently and characterized using the recombinant Vaccinia virus system (Rogers *et al.*, 1990). Initially we chose to study the major post-translational modification present on PrP, i.e. the Asn-linked glycosylation. PrP has two potential sites for the addition of Asn-linked oligosaccharides defined by the consensus sequence motif Asn–X–Ser/Thr. Mutations which altered these sites were introduced into the SHaPrP gene by site

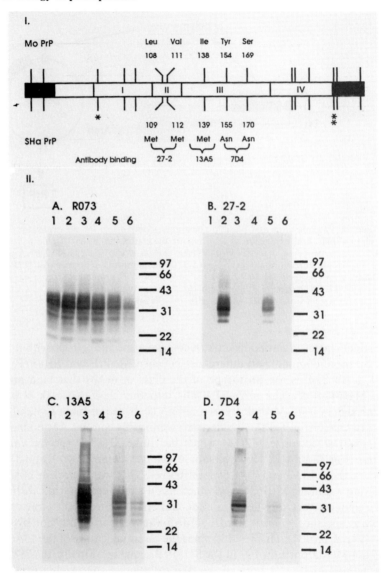

Fig. 2.   Mapping epitopes within SHaPrP that are recognized by mouse monoclonal antibodies raised against SHaPrP 27–30. Part I. Linear map of PrP gene showing the regions used in constructing chimeric genes (Rogers *et al.*, 1991). Amino acid differences between MoPrP and SHaPrP sequences are shown by lines. 7D4-like antibodies may bind an epitope defined by either or both the asparagines at position 155 and 170 in SHaPrP. The amino acid differences associated with specific epitopes are shown. The regions labelled I, II, III, and IV define the segments of the MoPrP gene replaced by homologous segments of the SHaPrP gene in HM3, MHM2, M2HM, and M3H PrP molecules respectively. The MoPrP sequence is shown above and the SHaPrP sequence is given below. * indicates an amino acid insertion in the SHaPrP sequence, ** an amino acid insertion in the MoPrP sequence. 27–2, 13A5 and 7D4 represent prototypic antibodies binding to the epitopes shown. Antibody 354 (Kascsak *et al.*, 1987) shows an identical binding pattern to the antibody 27–2. Part II. Immunoblots of chimeric and mutant PrP proteins stained with (A) R073, (B) 27–2, (C) 13A5 and (D) 7D4. Each antibody was incubated with filters with the following lanes: lane 1, M4 (MoPrP^C); lane 2, MHM2; lane 3, M2HM; lane 4, M3H; lane 5, H4 (SHaPrP^C); lane 6, PrP MoMet138. The Met at position 138 in MoPrP is equivalent to the Met at position 139 in SHaPrP. Adapted, with permission, from *J. Immunol.* **147** 3568–3574, 1991.

Table 2.   Distribution of antibodies binding particular epitopes[a]

Antibody Class	27–2	13A5	7D4
IgG1	3	2	4
IgG2a	0	6	0
Total	3	11	4

[a] 27–2, 13A5 and 7D4 are prototypic antibodies for these epitopes which map to amino acids 108 and 111, 138, and 154 and 174, respectively. The mAb 3F4 (Kascsak *et al.*, 1988) falls into the 27–2 class.

directed mutagenesis and the resulting mutant PrP genes expressed (Fig. 3(A)). The presence of the correct mutation was confirmed by restriction analysis and by DNA sequencing. Three mutants were constructed: PrP_{Ala183}, PrP_{Ala199} and $PrP_{Ala183/199}$, which alter a Thr to an Ala at codons 183 or 199. These changes disrupt the first, second, or both potential sites for the addition of Asn-linked carbohydrates respectively.

Fig. 3(B) is an immunoblot of extracts generated after transient expression of these mutant PrP genes in CV1 cells stained with the mAb 13A5. Broad size heterogeneity was obtained for wild-type (*wt*) PrP expressed by this method which probably reflects different degrees of oligosaccharide processing in these cells. PrP_{Ala183}, PrP_{Ala199} and $PrP_{Ala183/199}$ did not exhibit this heterogeneity but showed an apparent reduction in M_r consistent with the loss of one (PrP_{Ala183} and PrP_{Ala199}) or both $PrP_{Ala183/199}$) Asn-linked oligosaccharides. This interpretation was confirmed by chemically or enzymatically removing the Asn-linked oligosaccharides using anhydrous HF or PNGase F, respectively (Fig. 3(C)). *wt* PrP, PrP_{Ala183} and PrP_{Ala199} showed a decrease in M_r to a single band of 26 kDa equivalent to the M_r of $PrP_{Ala183/199}$ or *wt* PrP synthesized in the presence of tunicamycin which specifically inhibits the addition of Asn-linked carbohydrates. Neither anhydrous HF nor PNGase F affected the M_r of $PrP_{Ala183/199}$ or *wt* PrP synthesized in the presence of tunicamycin, indicating that $PrP_{Ala183/199}$ possessed no Asn-linked oligosaccharide. It should also be noted that none of the PrP species expressed using this system was resistant to proteolysis (Fig. 3(B)). This was expected as none of these recombinant prion proteins was exposed to scrapie prions.

It has been shown for a number of proteins that cellular targeting is affected by Asn-linked glycosylation. In view of this finding, indirect immunofluoresence was used to determine if the mutant PrP species were localized to the cell surface. Fig. 4 demonstrates that while *wt* PrP^C was localized to the cell membrane where it is attached by a GPI anchor, mutant PrP species remain intracellular.

These and other studies (Haraguchi *et al.*, 1989; Endo *et al.*, 1989) indicate that PrP^C contain Asn-linked oligosaccharides at both potential addition sites. Mutant PrP proteins that have lost one or both of these sites do not reach the cell membrane but remain intracellular. This observation places some limits on the essential cellular steps involved in the synthesis of PrP^{Sc} (see below).

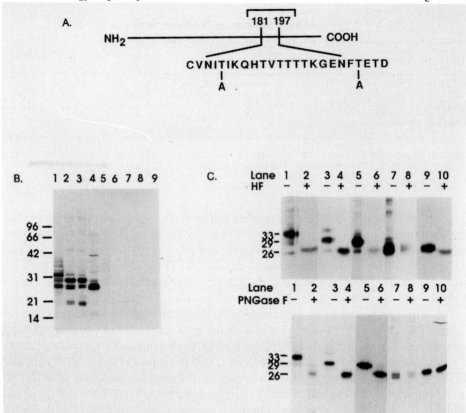

Fig. 3. Expression of SHaPrP genes mutated at potential sites for the addition of asparagine-linked carbohydrates. (A) Representation of the SHaPrP gene showing the amino acid changes made to ablate the sequences used as sites for the addition of Asn-linked oligosaccharides. Theonines at amino acids 181 and 197 were changed to alanines disrupting the consensus sequence Asn–X–Ser/Thr. (B) Immunoblot of total cell extracts of CV1 (lanes 1, 2, 3, 4, and 5 without proteinase K treatment and lanes 6, 7, 8, and 9 after digestion by proteinase K) cells expressing the following proteins; *wt* PrP (lanes 1 and 6), PrP_{Ala183} (lane 2 and 7), PrP_{Ala199} (lane 3 and 8), $PrP_{Ala183/199}$ (lane 4 and 9),mock-transfected cells (lane 5). M_r are shown in kilodaltons. (C) Fluorograph of *wt* and mutant PrP^C species after treatment with anhydrous HF (upper), or PNGase F (lower). Samples were labelled for 30 min and immunoprecipitated using the polyclonal antibody RO73. *wt* PrP (lanes 1 and 2), PrP_{Ala183} (lanes 3 and 4); PrP_{Ala199} (lanes 5 and 6); $PrP_{Ala183/199}$ (lanes 7 and 8); *wt* PrP synthesized in the presence of tunicamycin (lanes 9 and 10). Lanes 1, 3, 5, 7, and 9 are untreated and lanes 2, 4, 6, 8, and 10 are treated with anhydrous HF (upper panel) or PNGase F (lower panel).

Expression using SV40 based vectors in mammalian cells

To study the effects of mutations within the PrP gene upon its competence to form PrP^{Sc}, scrapie-infected cells in culture as well as a method of expressing and distinguishing the recombinant PrP from the endogenous background are essential. Cell culture models that faithfully replicate infectious prions and PrP^{Sc} have been established in a number of cell types (Clarke, 1979; Clarke and Millson, 1976; Rubenstein *et al.*, 1984; Butler *et al.*, 1988; Race *et al.*, 1988). However, the failure to

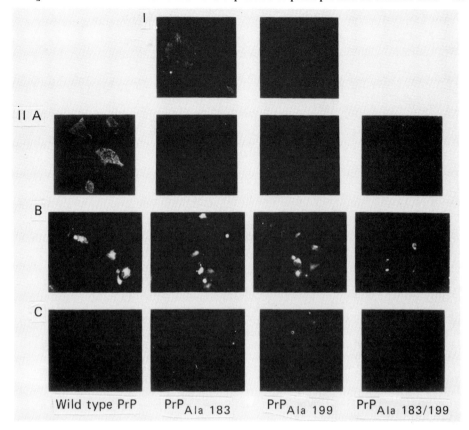

Fig. 4. Mutant PrP species accumulate intracellularly. Indirect immunofluorescence of CV1 cells expressing PrP species stained with the polyclonal antisera RO73 (parts I and II(A)) and monospecific polyclonal antisera raised against a synthetic PrP peptide (P1) spanning amino acids 90–102 (parts II(B) and II(C)). Part I shows unifixed cells stained with RO73. The left-hand panel shows untreated cells expressing wild-type PrP^C. The right-hand panel shows cells expressing *wt* PrP^C treated with PIPLC. Part II(A) shows unfixed cells expressing *wt* or mutant PrP proteins stained with RO73. Parts II(B) and II(C) show CV1 cells expressing *wt* or mutant proteins fixed with formaldehyde, permeabilized with Triton-X 100 and stained with anti-P1 in the absence (part II(B)) or presence (part II(C) of excess P1 peptide. Reproduced, with permission, from *Glycobiology* **1** 101–109, 1990.

clone infected cells from infected pools and the low levels of infectivity observed have limited the usefulness of these systems. Recently, cloned infected N2a cells have been described which faithfully replicate both infectious prions and PrP^{Sc} (Butler *et al.*, 1988). The levels of PrP^{Sc} and infectivity are sufficient (albeit low) for both kinetic studies and the molecular genetic studies described here.

The expression of mutated PrP genes was accomplished using a novel expression vector (SPOXII.NEO; Scott *et al.*, 1992) that employs the SV40 promoter and enhancer to drive expression of the PrP gene cloned downstream. Since the cell culture model is derived from Mo cells, a MoPrP gene forms the template for the mutagenesis. This

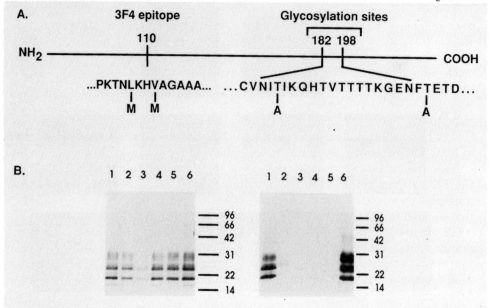

Fig. 5.   Stable expression of mutant MHM2 PrP genes in ScN$_2$A cells. (A) Representation of the MHM2 PrP gene showing the alterations made at the potential sites for the addition of Asn-linked oligosaccharides at amino acids 182 and 98. The SHaPrP amino acids that define the 27–2 epitope (also recognized by the more avidly binding antibody 3F4 (Kascsak *et al.*, 1987)). (B) Immunoblots showing the pattern of staining of protease resistant PrP in ScN$_2$a cells expressing *wt* and mutant MHM2 PrP genes. Left-hand panel, stained with polyclonal antibody RO73. Right-hand panel, stained with monoclonal antibody 3F4. Lane 1, MHM2; lane 2, MHM2$_{Ala182}$; lane 3, MHM2$_{Ala198}$; lane 4, MHM2$_{Ala182/198}$; lane 5, ScN$_2$A (untransfected); lane 6, ScHAB cells (scrapie-infected Syrian hamster brain cells grown in culture).

gene has been modified such that two amino acid changes are included that allow the recombinant protein to be specifically recognized by the SHaPrP specific mAb 3F4 (Kascsak *et al.*, 1987; Rogers *et al.*, 1991). These amino acid changes are outlined in Fig. 5(A). While they confer upon recombinant PrP, termed MHM2 PrP, the ability to be recognized by 3F4 and similar antibodies, they do not affect the ability of this protein to form PrP^{Sc}. This modification allows the recombinant PrP mutants to be distinguished from the endogenous MoPrP by differential screening with 3F4 and a polyclonal anti-PrP antiserum that recognizes both Mo and MHM2 PrP species (see Fig. 2(B)).

We have introduced mutations that disrupt the two sites for the addition of Asn-linked oligosaccharides into MHM2 PrP by *in vitro* mutagenesis (Fig. 5(A)). These mutant PrP species (MHM2 PrP, MHM2 PrP$_{Ala182}$, PrP$_{Ala198}$, and PrP$_{Ala182/198}$, mutated at glycosylation site 1, site 2 and both sites respectively) were transfected into ScN2a cells and clones stably expressing the recombinant prion proteins isolated. Immunoblot analysis was carried out using the polyclonal antibody R073 and the mAb 3F4 (Fig. 5(B)). Extracts derived from ScN$_2$a cells expressing the mutant MHM2 PrP genes were treated with proteinase K (20 μg/ml for 1 hr at 37°C) prior to SDS–PAGE and the proteinase K-resistant PrP species concentrated by ultracentrifugation. R073 reveals PrP^{Sc} in all extracts indicating that each cell line was synthesizing PrP^{Sc}. Staining with 3F4 also revealed staining for MHM2 PrP and much reduced staining of bands in lanes derived from extracts expressing the mutant PrP genes. The M_r of the bands present in these lanes are consistent with the M_r of PrP 27-30 that is deglycosylated or glycosylated at only one site (19 kDa and 23 kDa respectively). No 3F4 reactive PrP was observed in the lane derived from extracts of ScN$_2$A cells, ruling out the possibility that the low level staining observed was due to cross-reactivity with endogenous MoPrP^{Sc}.

These results indicate that the Asn-linked oligosaccharides present in both PrP^{Sc} and PrP^C are not required for the formation of PrP^{Sc}. This interpretation has been confirmed by an independent and complementary approach to the analysis of PrP^{Sc} biosynthesis. Taraboulos et al. (1990a) have shown that the de novo synthesis of PrP^{Sc} is not affected by tunicamycin, which inhibits the addition of Asn-linked oligosaccharides. Since Asn-linked carbohydrates account for about 30% of the mass of PrP^C and PrP^{Sc} and probably the vast majority of the structural heterogeneity found in these proteins, these results significantly reduce the possible sources for the potential difference between these isoforms. Furthermore, the observation that the SHaPrP glycosylation mutants are not on the cell surface but remain intracellular coupled with the competence of unglycosylated PrP to form PrP^{Sc} suggests that the normal targeting of PrP^C to the cell surface may not be essential for the formation of PrP^{Sc}.

CONCLUDING REMARKS

The study of recombinant PrP in cell culture provides a powerful approach toward mapping the features within the PrP gene that are required for the formation of PrP^{Sc}. The inability to study the infectivity of the modified PrP species directly in these cultures is the major limitation of this approach. However, the construction of transgenic mice expressing modified PrP species allows a direct assessment of the role of PrP mutations on prion infectivity. Such studies are currently underway. This approach is not limited to the study of Asn-linked oligosaccharides. Other known modifications to both PrP^C and PrP^{Sc} may also be examined. For example, the GPI anchor, disulphide bond and the N-terminal region removed during proteolysis of PrP^{Sc} to form PrP 27-30 are all potential targets for this approach.

REFERENCES

Barry, R.A. and Prusiner, S.B. (1986) Monoclonal antibodies to the cellular and scrapie prion proteins. J. Infect. Dis. **154** 518–521.

Barry, R.A., Kent, S.B.H., McKinley, M.P., Meyer, R.K., DeArmond, S.J., Hood, L.E. and Prusiner, S.B. (1986) Scrapie and cellular prion proteins share polypeptide epitopes. J. Infect. Dis. **153** 848–854.

Basler, K., Oesch, B., Scott, M., Westaway, D., Walchli, M., Groth, D.F., McKinley, M.P., Prusiner, S.B. and Weissmann, C. (1986) Scrapie and cellular PrP isoforms are encoded by the same chromosomal gene. Cell **46** 417–428.

Bolton, D.C., McKinley, M.P. and Prusiner, S.B. (1984) Molecular characteristics of the major scrapie prion protein. *Biochemistry* **23** 5893–5906.

Bolton, D.C., Meyer, R.K. and Prusiner, S.B. (1985) Scrapie PrP 27-30 is a sialoglycoprotein. *J. Virol.* **53** 596–606.

Borchelt, D.R., Scott, M., Taraboulos, A., Stahl, N. and Prusiner, S.B. (1990) Scrapie and cellular prion proteins differ in their kinetics of synthesis and topology in cultured cells. *J. Cell. Biol.* **110** 743–752.

Butler, D.A., Scott, M.R., Bockman, J.M., Borchelt, D.R., Taraboulos, A., Hsaio, K., Kingsbury, D.T. and Prusiner, S. B. (1988) Scrapie-infected murine neuroblastoma cells produce protease-resistant prion proteins. *J. Virol.* **62** 1558–1564.

Caughey, B., Race, R.E., Vogel, M., Buchmeier, M.J. and Chesebro, B. (1988) *In vitro* expression in eukaryotic cells of a prion protein gene cloned from scrapie-infected mouse brain. *Proc. Natl. Acad. Sci. USA* **85** 4657–4661.

Caughey, B., Race, R.E., Ernst, D., Buchmeier, M.J. and Chesebro, B. (1989) Prion protein biosynthesis in scrapie-infected and uninfected neuroblastoma cells. *J. Virol.* **63** 175–181.

Caughey, B., Neary, K., Buller, R., Ernst, D., Perry, L.L., Chesebro, B. and Race, R. E. (1990) Normal and scrapie-associated forms of prion proteins differ in their sensitivities to phospholipase and proteases in intact neuroblastoma cells. *J. Virol.* **64** 1093–1101.

Clarke, M.C. (1979) Infection of cell cultures with scrapie agent. In: Prusiner, S.B. and Hadlow, W.J. (eds) *Slow Transmissible Diseases of the Nervous System*, Vol. 2. Academic Press, New York, pp. 225–234.

Clarke, M.C. and Millson (1976) Infection of a cell line of mouse L fibroblasts with scrapie agent. *Nature (London)* **261** 144–145.

Endo, T., Groth, D., Prusiner, S.B. and Kobata, A. (1989) Diversity of oligosaccharides structure linked to asparagines of the scrapie prion protein. *Biochemistry* **28** 8380–8388.

Fuerst, T.R., Niles, E.G., Studier, F.W. and Moss, B. (1986) Eukaryotic transient-expression system based on recombinant vaccinia virus that synthesizes bacteriophase T7 RNA polymerase. *Proc. Natl. Acad. Sci. USA* **83** 8122–8126.

Gabizon, R. and Prusiner, S.B. (1990) Prion liposomes. *Biochem. J.* **166** 1–14.

Haraguchi, T., Fisher, S., Olofsson, S., Endo, T., Groth, D., Tarentino, A., Teplow, D., Hood, L., Burlingame, A., Lycke, E., Kobata, A. and Prusiner, S.B. (1989) Asparagine-linked glycosylation of the scrapie and cellular prion proteins. *Arch. Biochem. Biophys.* **274** 1–13.

Hope, J., Morton, L.J., Farquhar, C.F., Multhaup, G., Beyreuther, K. and Kimberlin, R.H. (1986) The major polypeptide of scrapie-associated fibrils (SAF) has the same size, charge distribution and *N*-terminal protein sequence as predicted for the normal brain protein (PrP). *EMBO J.* **5** 2591–2597.

Kascsak, R.J., Rubenstein, R., Merz, P.A., Tonna-DeMasi, M., Fersko, R., Carp, R.I., Wisniewski, H.M. and Diringer, H. (1987) Mouse polyclonal and monoclonal antibody to scrapie-associated fibril proteins. *J. Virol.* **61** 3688–3693.

Laemmli, U.K. (1970) Cleavage of structural proteins during the assembly of the head of bacteriophage T4. *Nature (London)* **227** 680–685.

Meyer, R.K., McKinley, M.P., Bowman, K.A., Braufeld, M.B., Barry, R.A. and Prusiner, S.B. (1986) Separation and properties of the cellular and scrapie prion proteins. *Proc. Natl. Acad. Sci. USA* **83** 2310–2314.

Oesch, B., Westaway, D., Wälchli, M., McKinley, M.P., Kent, S.B.H., Aebersold, R., Barry, R.A., Tempst, P., Teplow, D.B., Hood, L.E., Prusiner, S.B. and Weismann, C. (1985) A cellular gene encodes PrP 27-30 protein. *Cell* **40** 735–746.

Race, R.E., Caughey, B., Graham, K., Ernst, D. and Chesebro, B. (1988) Analysis of frequency of infection, specific infectivity, and prion protein biosynthesis in scrapie-infected neuroblastoma cell clones. *J. Virol.* **62** 2845–2849.

Rogers, M., Taraboulos, A., Scott, M., Groth, D. and Prusiner, S.B. (1990) Intracellular accumulation of the cellular prion protein after mutagenesis of its Asn-linked glycosylation sites. *Glycobiology* **1** 101–109.

Rogers, M., Serban, D., Gyuris, T., Scott, M., Torchia, T. and Prusiner, S.B. (1991) Epitope mapping of the Syrian hamster prion protein utilizing chimeric and mutant genes in a vaccinia virus expression vector. *J. Immunol.* **147** 3568–3574.

Rubenstein, R., Carp, R.I. and Callahan, S.M. (1984) *In vitro* replication of scrapie agent in a neurol model: infection of PC12 cells. *J. Gen. Virol.* **65** 2191–2198.

Sambrook, J., Fritsch, E.F. and Maniatis, T. (1989) *Molecular Cloning—a Laboratory Manual*, 2nd edn. Cold Spring Harbor Laboratory Press, Cold Spring Harbor, NY.

Scott, M., Butler, D., Bredesen, D., Walchli, M., Hsiao, K. and Prusiner, S.B. (1988) Prion protein gene expression in cultured cells. *Protein Eng.* **2** 69–76.

Scott, M., Foster, D., Mirenda, C., Serban, D., Coufal, F., Walchli, M., Torchia, M., Groth, D., Carlson, G., DeArmond, S.J., Westaway, D. and Prusiner, S.B. (1989) Transgenic mice expressing hamster prion protein produce species specific scrapie infectivity and amyloid plaques. *Cell* **59** 847–857.

Scott, M.R., Kohler, R., Foster, D. and Prusiner, S.B. (1992) Chimeric prion protein expression in cultured cells and transgenic mice. Protein Sci., in press.

Stahl, N., Borchelt, D.R., Hsaio, K. and Prusiner, S.B. (1987) Scrapie prion protein contains a phosphatidylinositol glycolipid. *Cell* **51** 229–240.

Stahl, N., Borchelt, D.R. and Prusiner, S.B. (1990) Differential release of cellular and scrapie prion protein from cellular membranes by phosphatidylinositol-specific phospholipase C. *Biochemistry* **29** 5405–5412.

Taraboulos, A., Rogers, M., Borchelt, D.R., McKinley, M.P., Scott, M., Serban, D. and Prusiner, S.B. (1990a) Acquisition of protease resistance by prion proteins in scrapie-infected cells does not require asparagine-linked glycosylation. *Proc. Natl. Acad. Sci. USA* **87** 8262–8266.

Taraboulos, A., Serban, D. and Prusiner, S.B. (1990b) Scrapie prion proteins accumulate in the cytoplasm of persistently infected cultured cells. *J. Cell. Biol.* **110** 2117–2132.

Turk, E., Teplow, D.B., Hood, L.E. and Prusiner, S.B. (1988) Purification and properties of the cellular and scrapie hamster prion proteins. *Eur. J. Biochem.* **176** 21–30.

Part VII Transgenetics and animal models

39

Transgenic approaches to experimental and natural prion diseases

David Westaway, Sara Neuman, Vincent Zuliani, Carol Mirenda, Dallas Foster, Linda Detwiler, George Carlson and Stanley B. Prusiner

ABSTRACT

Prolonged incubation times for experimental scrapie in I/LnJ mice are dictated by a dominant gene designated *Prn-i* linked to the prion protein gene (*Prn-p*). *Prn-i* may correspond to the scrapie incubation time gene *Sinc* defined by analysis of the VM strain of mice (Dickinson and Meikle, 1971). Transgenic (Tg) mice were analysed to discriminate between an effect of the I/LnJ *Prn-p^b* allele and a distinct incubation time locus. Unexpectedly, five independent Tg(*Prn-p^b*) mouse (Mo) lines had scrapie incubation times shorter than non-Tg controls, instead of the anticipated prolonged incubation periods (Westaway *et al.*, 1991). Aberrant expression of the *Prn-p^b* transgenes may dictate abbreviated incubation times, masking genuine *Prn-p/Prn-i* congruence; alternatively, a discrete *Prn-i* gene lies adjacent to *Prn-p*. In another series of experiments, expression of a Syrian hamster (SHa) prion protein renders Tg mice susceptible to prions passaged in Syrian hamsters (Prusiner *et al.*, 1990). Incubation times in these mice were inversely correlated with the levels of transgene expression, indicating that both the quantity and the primary structure of PrP^C molecules modulate disease susceptibility. These studies may be germane to understanding natural scrapie. Natural scrapie is widely held to be a maternally and horizontally infectious disease (Dickinson *et al.*, 1974) with a host genetic component corresponding to segregation of alleles of autosomal susceptibility genes (perhaps corresponding to the *Sip* gene, a potential analogue of *Prn-i/Sinc* (Dickinson and Fraser, 1979; Hunter *et al.*, 1989)). However, the mechanism of contagious spread is not clearly defined. To attain a better understanding of natural scrapie, we have created Tg mice harbouring sheep (She) PrP genes. By analogy to the behaviour of Tg(SHaPrP) mice, Tg(ShePrP) mice may be highly susceptible to sheep scrapie prions. A sensitive bioassay exploiting these mice may in turn lead to an understanding of how natural scrapie is propagated.

INTRODUCTION

Transgenetic studies have converged with biochemistry and cell biology in assigning the scrapie prion protein (PrP^{Sc}) as a component of the infectious particle or prion of experimental scrapie (Carlson *et al.*, 1991; Prusiner, 1991). Expression of a heterologous Syrian hamster (SHa) PrP gene renders transgenic (Tg) mice susceptible to prions passaged in Syrian hamsters (Prusiner *et al.*, 1990; Scott *et al.*, 1989). Incubation times in these mice were inversely correlated with the levels of transgene expression. These experiments indicate that both the quantity and the primary structure of PrP^C molecules modulate disease susceptibility. Prolonged scrapie incubation times in I/LnJ mice are dictated by a dominant gene linked to the prion protein gene (*Prn-p*). Tg mice were analysed to discriminate between an effect of the I/LnJ *Prn-p^b* allele (which encodes a variant prion protein differing at two codons from the wild-type Mo *Prn-p^a* allele), and a distinct incubation time locus designated *Prn-i*. Paradoxically, five independent Tg(*Prn-p^b*) mouse (Mo) lines had incubation times shorter than non-Tg controls, instead of the anticipated ca 200 day incubation periods (Westaway *et al.*, 1991; G. A. Carlson and D. Westaway, unpublished data). Aberrance or overexpression of the *Prn-p^b* transgenes may dictate abbreviated incubation times, masking *Prn-p/Prn-i* congruence; alternatively, a discrete *Prn-i* gene lies adjacent to *Prn-p*. Both of these studies may be relevant to understanding the pathogenesis of natural scrapie.

FOUR LINES OF Tg MICE EXPRESSING SHaPrP mRNA

Four lines of Tg mice expressing SHaPrP^C were constructed and propagated. Southern blot analysis of the four lines suggested that the transgenes are integrated at one chromosomal site in a tandem array as has been reported for many Tg mice harbouring other foreign genes (Scott *et al.*, 1989) Northern blots showed that Tg69 mice with two to four copies of the transgene expressed the lowest levels of SHaPrP mRNA while Tg71 with a similar number of transgenes expressed slightly higher levels of SHaPrP mRNA. Tg81 mice with 30 to 50 copies of the transgene expressed substantially higher levels of SHaPrP mRNA.

The Tg20 line was produced in the same series of microinjections into (C57BL/6 × LT/Sv)F2 fertilized eggs that yielded Tg7. Neither SHaPrP^C nor SHaPrP mRNA was detectable in Tg20 mice. Digestion with Xbal yielded an aberrant 3 kb hybridizing fragment rather than the 3.8 kb fragment seen in Tg7 and in hamsters. Whether a rearrangement or deletion within the SHaPrP insert in Tg20 is responsible for the lack of expression is unknown.

A range of scrapie incubation times varying from 277 ± 6.7 to 48 ± 1.0 days were

Table 1. Scrapie incubation times in Tg mice expressing SHaPrP after inoculation with either mouse or hamster prions

Species	Line	SHa prions[a]			Mo prions[b]		
		n^c	Illness (days $\pm$ SE)	Death	n^c	Illness (days $\pm$ SE)	Death
Mo	Non 69[d]	0/20	>370		17/17	143 $\pm$ 2.9	154 $\pm$ 3.6
Mo	Non 71	0/18	>270		14/14	148 $\pm$ 4.8	162 $\pm$ 5.7
Mo	Non 81	0/24	>460		21/21	128 $\pm$ 1.7	142 $\pm$ 2.6
TgMo	Tg20	0/6	>305		21/21	134 $\pm$ 3.1	154 $\pm$ 2.7
TgMo	Tg69	18/18	277 $\pm$ 6.7	293 $\pm$ 8.1	17/17	166 $\pm$ 4.7	173 $\pm$ 5.5
TgMo	Tg71	15/17[e]	176 $\pm$ 2.7	182 $\pm$ 2.7	19/19	165 $\pm$ 3.4	179 $\pm$ 2.3
TgMo	Tg81	24/24	75 $\pm$ 1.1	75 $\pm$ 1.1	20/20	194 $\pm$ 3.5	200 $\pm$ 3.2
TgMo	Tg7	26/26	48 $\pm$ 1.0	51 $\pm$ 0.8	25/25	173 $\pm$ 4.8	180 $\pm$ 4.8
SHa	LVG:Lak	32/32	89 $\pm$ 0.9	101 $\pm$ 1.4	0/32	>360	

[a] Animals were inoculated intracerebrally with 30 μl containing $\sim 10^7$ ID$_{50}$ units of SHa prions in crude extracts prepared from scrapie-infected SHa brains.
[b] Animals were inoculated intracerebrally with 30 μl containing $\sim 10^6$ ID$_{50}$ units of Mo prions in crude extracts prepared from scrapie-infected Mo brains.
[c] n is the number of animals developing clinical signs of scrapie divided by the total number of animals inoculated. Mice dying atypically were virtually always <5% of the total number of animals inoculated and thus they were excluded (Prusiner, 1987).
[d] Non-Tg mice are littermates of the Tg animals.
[e] Two mice are alive and well at 250 days after inoculation with SHa prions. Whether these mice were mistyped and they are not transgenic remains to be determined.

recorded for the four Tg(HaPrP)Mo lines expressing SHaPrP mRNA inoculated with SHa prions (Table 1). Tg69 mice with lowest steady-state levels of SHaPrP mRNA had the longest incubation times while Tg7 mice with the highest levels of SHaPrP mRNA had the shortest incubation times. The Tg71 and 81 mice had intermediate levels of SHaPrP mRNA and displayed incubation times of intermediate length. The Tg20 mice which failed to express SHaPrP mRNA have scrapie incubation times after inoculation with SHs prions exceeding 300 days. These observations demonstrate an inverse relationship between the level of transgene SHaPrP mRNA and the length of the incubation time after inoculation with SHa prions.

While non-Tg mice developed scrapie between 128 and 148 days after inoculation with Mo prions, Tg69 and Tg71 mice exhibited incubation times of 166 $\pm$ 4.7 and 165 $\pm$ 3.4 days, respectively. Even greater prolongation of the incubation times after Mo prion inoculation was seen with Tg81 and Tg7 mice where periods of 194 $\pm$ 3.5 and 173 $\pm$ 4.8 days were observed, respectively. Tg20 mice developed scrapie 134 $\pm$ 3.1 days after inoculation with Mo prions consistent with their failure to express the SHaPrP transgene. These observations argue that expression of the SHaPrP transgene impedes Mo prion synthesis.

SPECIES-SPECIFIC INOCULA DIRECT SYNTHESIS OF SHa OR Mo PRIONS

Inoculation of Tg mice expressing SHaPrP genes with SHa prions resulted in the formation of SHa prions as determined by bioassay in hamsters. As shown in Table 2, inoculation of either Tg71 or 81 mice with SHa prions produced high levels of SHa prions, but virtually no Mo prions were detected by bioassay. Conversely, inoculation with Mo prions generated substantial levels of Mo prions, but virtually no SHa prions were found. These findings argue that the origin of the prion inoculum determines whether SHa or Mo prions are produced by Tg(SHaPrP) mice which are capable of supporting the replication of either prion.

Table 2. Production of SHa or Mo prions in the brains of Tg mice

Transgenic mice				Bioassays of prions		
Inoculum[a]	Tg line	n^b	Incubation time[c] (days ± SE)	Species	n^d	log(titre)[e] (ID_{50}/ml ± SE)
SHa	71	3	163 ± 1.5	Mo	0/30	<1
SHa	71	4	161 ± 1.3	Ha	39/39	8.0 ± 0.20
SHa	81	6	75 ± 0.9	Mo	0/40	<1
SHa	81	6	75 ± 0.9	Ha	47/47	7.5 ± 0.20
Mo	71	3	158 ± 1.3	Mo	14/14	5.5 ± 0.41
Mo	71	3	155 ± 1.6	Ha	0/16	<1
Mo	81	3	173 ± 3.2	Mo	30/30	5.2 ± 0.34
Mo	81	3	173 ± 3.2	Ha	0/24	<1

[a] SHa inoculum was a 10% (w/v) homogenate of scrapie-infected SHa brain designated Sc237C diluted 10-fold. Mo inoculum was a 10% (w/v) homogenate of scrapie-infected Swiss CD-1 mouse brain designated RML. See Experimental Procedures for passage history.
[b] n is the number of Tg mice on which individual bioassays of their brains were performed.
[c] Mean incubation times for the Tg mice on which bioassays were performed.
[d] n is the number of mice or hamsters developing scrapie divided by the total number of all animals used for the bioassays. Typically between 5 and 10 animals were used for each bioassay.
[e] Titres were calculated from published curves relating the doseof prions inoculated to theO resulting incubation time, as described in Experimental Procedures. Titres are given in ID_{50} units/ml of 10% (w/v) brain homogenate.

GENETIC CONTROL OF SCRAPIE INCUBATION TIME IN MICE

The impact of a *Prn-p* linked gene upon scrapie incubation period is a striking and widely confirmed observation (Carlson *et al.*, 1986, 1988; Hunter *et al.*, 1987; Race *et al.*, 1990). An attempt was made to resolve the issue of whether prolongation of scrapie incubation time is due to the amino acid substitutions in the *Prn-p^b* open reading frame or to a genetically distinct *Prn-i* locus. Our results revealed a paradoxical effect of expression of the cos6.1/LnJ-4 $Prn-p^b$ cosmid transgene in shortening scrapie incubation time, superficially paralleling previous results from segregating crosses in which 1.5% to 5.9% of the offspring were discordant for $Prn-p$ genotype and incubation time phenotype (Carlson et al., 1986, 1988; Race et al., 1990). However, these putative $Prn-i/Prn-p$ recombinants were not progeny tested, and typing for $Prn-p$ flanking markers in these crosses indicated that a minimum of two cross-over events would be required to account for the deviant mice (G. Carlson et al., unpublished data). Similarly, recombinant chromosomes defined by an interval extending 1.4 ± 9.5 cM proximal and 4.4 ± 0.9 cM distal from $Prn-p$ were identified and propagated. As these map distances are comparable with the putative $Prn-p/Prn-i$ interval, at least some single cross-over events between the two loci should have been captured. In fact, none of the offspring of these recombinant mice showed separation of incubation time phenotype and $Prn-p$ genotype. It is likely that deviant mice in earlier crosses were not meiotic recombinants but were due to convergence of non-$Prn-p$ genes that influence incubation time (Carlson et al., 1988; Race et al., 1990; G. Carlson et al., unpublished data). It is important to stress, however, that these results against meiotic recombination between $Prn-i$ and $Prn-p$ cannot be taken as proof that the two genes are identical.

The incubation time of all the Tg($Prn-p^b$) lines derived from microinjection of the cos6.1/LnJ-4 cosmid clone is far shorter than that of $Prn-p^a/Prn-p^b$ F_1 mice. Even the Tg93L line with only four copies of the $Prn-p^b$ transgene per diploid genome has a 97 day incubation time, considerably shorter than the 223 days for $Prn-p^{a/b}$ heterozygous mice (Table 3) (Carlson et al., 1988). The observed incubation periods must reflect an increase in the rate of prion replication, but do not rule out the possibility that $Prn-p^b$ prolongs scrapie incubation time in non-Tg mice. For example, aberrant transgene expression or subtle alterations in transgene structure could mask a genuine dictation of long scrapie incubation periods by the PrP-B protein. However, the possibility that our unexpected results derive from 'position effects' on transgene expression can most likely be excluded because all four independent Tg lines exhibit abbreviated incubation times (Palmiter and Brinster, 1986).

With regard to the issue of transgene expression, all Tg($Prn-p^b$) lines harbour higher steady-state levels of PrP^C than non-Tg mice. An inverse correlation of SHaPrP mRNA and SHaPrP^C levels with SHa scrapie prion incubation times is manifest in Tg(SHaPrP) mice (Prusiner et al., 1990; Scott et al., 1989). PrP^C expression in Tg93L is approximately 2- to 4-fold greater than in non-Tg animals. The Tg93H, Tg94 and 117 lines have shorter incubation times than the Tg93L line, and express more $Prn-p^b$ mRNA and total PrP^C (Westaway et al., 1991; D. Westaway, unpublished data). An increased supply of PrP^C or a precursor for conversion to infectious PrP^{Sc} in Tg($Prn-p^b$) mice may mask effects of PrP-B primary structure in prolonging incubation time in non-Tg mice. Tg mice that express comparable amounts of PrP-A are not currently available for comparison of incubation times with Tg($Prn-p^b$) mice.

Ectopic transgene expression also may be a plausible explanation for our results. Our analyses of purified brain RNA would not reveal whether neurons in a few critical nuclei express PrP at high levels making Tg($Prn-p^b$) mice particularly vulnerable to scrapie. While expression of PrP mRNA in SHa brain is known to be

Table 3.   Incubation times of Tg(Prn-p^b) lines

Mice	Ilness[a]	Death[b]
	Incubation times (days $\pm$ SE)	
Tg93L[c]	96.7 $\pm$ 3.0 (12)	107.1 $\pm$ 5.0 (3)
non-Tg[d]	125.3 $\pm$ 2.8 (16)	135.8 $\pm$ 3.3 (10)
Tg93H	79.4 $\pm$ 2.9 (14)	86.3 $\pm$ 3.0 (6)
non-Tg	129.9 $\pm$ 3.7 (9)	145.4 $\pm$ 5.7 (7)
Tg94	75.5 $\pm$ 1.8 (15)	83.4 $\pm$ 1.7 (10)
non-Tg	137.0 $\pm$ 2.1 (21)	145.9 $\pm$ 2.4 (15)
Tg117	78.5 $\pm$ 1.9 (13)	85.8 $\pm$ 2.7 (8)
non-Tg[e]	129.5 $\pm$ 4.6 (15)	140.5 $\pm$ 4.2 (11)

[a] Mice were inoculated intracerebrally with $\sim 10^7$ ID$_{50}$ units of RML extract prepared from the brains of NZW mice. In parentheses are the number of animals developing clinical signs of scrapie.
[b] In parentheses are the number of mice dying of scrapie. Mice sacrificed for pathologic examination were excluded in these calculations.
[c] One mouse with sickness and death times of 124 and 138 days, respectively, was presumably a mistyped non-Tg animal; however, no tissues were available for retyping.
[d] One mouse with sickness and death times of 93 and 96 days, respectively, was presumably a mistyped Tg animal, but no tissues were available for retyping.
[e] One mouse with sickness and death times of 77 and 85 days, respectively, was presumably a mistyped Tg animal; however, no tissues were available for retyping.

under developmental control (McKinley *et al.*, 1987; Mobley *et al.*, 1988), the ontogeny of PrP expression in Tg mice is unexplored, as is the possibility of ectopic expression in inappropriate organs.

Although it is plausible that overexpression of PrP^C is sufficient to account for the novel incubation time phenotype of Tg(Prn-p^b) mice, it is also possible that an undetected gene or genes within the cos6.1/LnJ-4 clone programs enhanced scrapie susceptibility. For example, the cosmid transgene may contain an 'activated' incubation time gene. A *Prn-i* gene close to or interdigitated with *Prn-p* may have been bisected by the boundaries of the genomic clone, leading to aberrant expression or coding sequence truncation, and correspondingly unusual scrapie incubation times. Translocations encompassing the proto-oncogenes *c-myc* and *bcl* exemplify this type of gene activation event (Croce, 1987) and the murine RAG-1 and RAG-2 genes provide a recent and striking precedent for a mammalian gene complex (Oettinger *et al.*, 1990).

Alternatively, a discrete *Prn-i* gene may not be included within the cosmid transgene described herein. PrP-A and PrP-B may be equally efficient in prion replication, with abbreviated incubation times arising from the elevated net levels of PrP^C expression in the Tg(Prn-p^b) mice. However, the possibility that PrP-B itself programs long incubation times in normal (non-Tg) *Prn-p^b* mice still exists, as discussed earlier.

NATURAL SCRAPIE OF SHEEP

Unlike natural (non-iatrogenic) human prion diseases, natural scrapie is thought to be a contagious disease, with both horizontal and vertical modes of infectious spread (Dickinson et al., 1974; Harries-Jones et al., 1988). The view has shaped the agricultural policy to scrapie in both Britain and the USA. Although a purely genetic aetiology (specifically an autosomal recessive disorder) was championed by James Parry (Parry, 1960, 1962, 1983), this view of the disease is not widely held, and the perceived genetic component to natural scrapie has been equated with the segregation of susceptibility loci. A locus, Sip, was deduced by Alan Dickinson and coworkers in the context of experimental scrapie of sheep (Dickinson and Fraser, 1979; Hunter et al., 1989) and may also be involved in the natural disease (Foster and Dickinson, 1988). While contagious transmission of natural scrapie has been described in several symposia proceedings, there are a few articles on this topic in refereed journals. Crosses between sheep of deduced parentage (scrapied or scrapie free) were performed by Dickinson and coworkers. These experiments failed to recapitulate the predictions of Parry, and revealed a bias towards scrapie in the offspring of scrapied ewes, indicating a maternally mediated vertical transmission. An increasing incidence of scrapie in 'scrapie-free' stock raised in various degrees of contact with scrapie-affected stock was also compatible with this hypothesis (Dickinson et al., 1965). A milder skew towards scrapie in the offspring of scrapied ewes was observed in a second set of experiments (Dickinson et al., 1974). In another study, the age of onset of scrapie in an endemically affected flock of Suffock sheep steadily declined, implying that the 'load' of infectious agent present in these animals was increasing with each generation (Foster and Dickinson, 1989).

Infectious transmission is the most obvious interpretation of this type of data, but there are some puzzles and inconsistencies in this scenario.

(1) For example, Parry also conducted analyses which addressed the possibility of contagious spread in natural scrapie, with negative results (Parry, 1983).

(2) There are conflicting reports as to the presence of infectious titre in the reproductive organs of scrapie ewes, which is predicted by the hypothesis of maternal transmission. While Pattison reported titre in the placentae of Swaledale ewes (Pattison et al., 1972), Hadlow and coworkers failed to detect titre in the reproductive organs of affected Suffolks (Hadlow et al., 1982).

(3) Maternal transmission of experimental scrapie in sheep does not appear to take place at a significant frequency in embryo transfer experiments performed by Warren Foote and colleagues at Utah State University.

(4) Bovine spongiform encephalopathy perhaps most closely resembles natural scrapie (and has been postulated as arising from natural scrapie), yet it is becoming increasingly unlikely that this syndrome does not feature a maternal route of transmission (see Chapter 25).

(5) The mooted maternal transmission of natural scrapie would be unique amongst prion diseases (summarized in Westaway and Prusiner (1990)).

(6) Finally, a spontaneous scrapie-like neurodegenerative disease programmed by a MoPrP transgene encoding leucine at codon 101 defines a purely genetic aetiology for a natural prion disease, Gerstmann–Sträussler–Scheinker syndrome (Hsiao et al., 1990).

While some of the discrepancies in the data on natural scrapie may of course be due to the genetic heterogeneity of the sheep populations under study (now documented in detail for the PrP gene (Goldmann *et al.*, 1991)), it is plausible that natural scrapie is a heterogeneous disease with regard to pathology (Parry, 1983) and possibly aetiology. From a practical point of view, defining the host genes which modulate the disease is a high priority and arguments about whether natural scrapie is purely genetic or infectious with the segregation of a susceptibility locus are a secondary consideration. In this regard, it is regrettable that Parry's practices of distinguishing different scrapie pathologies and keeping detailed pedigrees of affected flocks has been ignored, as an opportunity to define these loci by contemporary methods (Lander, 1988) has been missed.

To attain a better understanding of the aetiology of natural scrapie, we have created Tg mice harbouring sheep (She) PrP genes. By analogy to our experience with SHaPrP transgenes, such mice may be highly susceptible to sheep scrapie prions. We are focusing our study on natural scrapie of Suffolk sheep, which constitute >84% of the cases in the USA. Accordingly, Tg(ShePrP) mice were generated with a cosmid clone isolated from a 7.5-year-old phenotypically normal Suffolk ewe. This cosmid encodes glutamine at codon 171 (which we believe to correspond to the wild-type allele in Suffolk sheep) and encompasses the entire ShePrP transcription unit, as high levels of ~5.0 kb ShePrP mRNA are exhibited by at least one of the Tg lines. This is consistent with the analysis of full-length cDNA clones, which position two exons encoding the 5′ untranslated region of ShePrP mRNA within the boundaries of the cosmid. These Tg(ShePrP) mice have been inoculated with isolates of sheep experimental scrapie. Should they prove to have a heightened susceptibility to sheep prions, we may be in a position to assess prion titres in the reproductive organs of scrapied ewes, as well as in the environment of endemically affected flocks.

PROSPECTS

The 'species barrier' to transmission of prions is of theoretical and practical importance. Work with Tg(SHaPrP) mice has equated this effect with the primary sequence of PrP^C molecules encoded by the host, and it will be important to ascertain whether the same mechanism underlies 'short' and 'long' scrapie incubation time gene alleles in mice. A sensitive and reproducible bioassay for sheep prions exploiting Tg(ShePrP) mice may clarify the mechanism of spread of natural scrapie.

ACKNOWLEDGMENT

The authors thank James Foster for discussions on the aetiology of natural scrapie.

REFERENCES

Carlson, G.A., Kingsbury, D.T., Goodman, P.A., Coleman, S., Marshall, S.T., DeArmond, S.J., Westaway, D. and Prusiner, S.B. (1986) Linkage of prion protein and scrapie incubation time genes. *Cell* **46** 503–511.

Carlson, G.A., Goodman, P.A., Lovett, M., Taylor, B.A., Marshall, S.T., Peterson-Torchia, M., Westaway, D. and Prusiner, S.B. (1988) Genetics and polymorphism of the mouse prion gene complex: the control of scrapie incubation time. *Mol. Cell. Biol.* **8** 5528–5540.

Carlson, G.A., Hsiao, K., Oesch, B., Westaway, D. and Prusiner, S.B. (1991) Genetics of prion infections. *Trends Genet.* **7** 61–65.

Croce, C.M. (1987) Role of chromosome translocations in human neoplasia. *Cell* **49** 155–156.

Dickinson, A.G. and Fraser, H. (1979) An assessment of the genetics of scrapie in sheep and mice. In: Prusiner, S.B. and Hadlow, W.J. (eds) *Slow Transmissible Diseases of the Nervous System*, Vol. 1. Academic Press, New York, pp. 367–386.

Dickinson, A.G. and Meikle, V.M.H. (1971) Host-genotype and agent effects in scrapie incubation: change in allelic interaction with different strains of agent. *Mol. Gen. Genet.* **112** 73–79.

Dickinson, A.G., Young, G.B., Stamp, J.T. and Renwick, C.C. (1965) An analysis of natural scrapie in Suffolk sheep. *Heredity* **20** 485–503.

Dickinson, A.G., Stamp, J.T. and Renwick, C.C. (1974) Maternal and lateral transmission of scrapie in sheep. *J. Comp. Pathol.* **84** 19–25.

Foster, J.D. and Dickinson, A.G. (1988) Genetic control of scrapie in Cheviot and Suffolk sheep. *Vet. Rec.* **123** 159.

Foster, J.D. and Dickinson, A.G. (1989) Age at death from natural scrapie in a flock of Suffolk sheep. *Vet. Rec.* **125** 415–417.

Goldmann, W., Hunter, N., Foster, J.D., Salbaum, J.M., Beyreuther, K. and Hope, J. (1990) Two alleles of a neural protein gene linked to scrapie in sheep. *Proc. Natl. Acad. Sci. USA* **87** 2476–2480.

Goldmann, W., Hunter, N., Benson, G., Foster, J.D. and Hope, J. (1991) Different scrapie-associated fibril proteins (PrP) are encoded by lines of sheep selected for different alleles of the *Sip* gene. *J. Gen. Virol* **72** 2411–2417.

Hadlow, W.J., Kennedy, R.C. and Race, R.E. (1982) Natural infections of Suffolk sheep with scrapie virus. *J. Infect. Dis.* **146** 657–664.

Harries-Jones, R., Knight, R., Will, R.G., Cousens, S., Smith, P.G. and Matthews, W.B. (1988) Creutzfeldt–Jakob disease in England and Wales, 1980–1984: a case-control study of potential risk factors. *J. Neurol. Neurosurg. Psychiatry* **51** 1113–1119.

Hsiao, K.K., Scott, M., Foster, D., Groth, D.F., DeArmond, S.J. and Prusiner, S.B. (1990) Spontaneous neurodegeneration in transgenic mice with mutant prion protein of Gerstmann–Sträussler syndrome. *Science* **250** 1587–1590.

Hunter, N., Hope, J., McConnell, I. and Dickinson, A.G. (1987) Linkage of the scrapie-associated fibril protein (PrP) gene and Sinc using congenic mice and restriction fragment length polymorphism analysis. *J. Gen. Virol.* **68** 2711–2716.

Hunter, N., Foster, J.D., Dickinson, A.G. and Hope, J. (1989) Linkage of the gene for the scrapie-associated fibril protein (PrP) to the Sip gene in Cheviot sheep. *Vet. Rec.* **124** 364–366.

Lander, E.S. (1988) In: Davies, K. (ed.) *Genome Analysis, a Practical Approach.* IRL Press, Oxford, pp. 171–189.

McKinley, M.P., Hay, B., Lingappa, V.R., Lieberburg, I. and Prusiner, S.B. (1987) Developmental expression of prion protein gene in brain. *Dev. Biol.* **121** 105–110.

Mobley, W.C., Neve, R.L., Prusiner, S.B. and McKinley, M.P. (1988) Nerve growth factor increases mRNA levels for the prion protein and the beta-amyloid protein precursor in developing hamster brain. *Proc. Natl. Acad. Sci. USA* **85** 9811–9815.

Oettinger, M.A., Schatz, D.G., Carolyn, G. and Baltimore, D. (1990) RAG-1 and RAG-2, adjacent genes that synergistically activate V(D)J recombination. *Science* **248** 1517–1523.

Palmiter, R.D. and Brinster, R.L. (1986) Germ-line transformation of mice. *Annu. Rev. Genet.* **20** 465–499.

Parry, H.B. (1960) Scrapie: a transmissible hereditary disease of sheep. *Nature (London)* **185** 441–443.

Parry, H.B. (1962) Scrapie: a transmissible and hereditary disease of sheep. *Heredity* **17** 75–105.

Parry, H.B. (1983) *Scrapie Disease in Sheep.* Academic Press, New York, 192 pp.

Pattison, I.H., Hoare, M.N., Jebbett, J.N. and Watson, W.A. (1972) Spread of scrapie to sheep and goats by oral dosing with foetal membranes from scrapie-affected sheep. *Vet. Rec.* **90** 465–468.

Prusiner, S.B. (1987) The biology of prion transmission and replication. In: Prusiner, S.B. and McKinley, M.P. (eds) *Prions—Novel Infectious Pathogens Causing Scrapie and Creutzfeldt–Jakob Disease.* Academic Press, Orlando, FL, pp. 83–112.

Prusiner, S.B. (1991) Molecular biology of prion diseases. *Science* **252** 1515–1522.

Prusiner, S.B., Scott, M., Foster, D., Pan, K.-M., Groth, D., Mirenda, C., Torchia, M., Yang, S.-L., Serban, D., Carlson, G.A., Hoppe, P.C., Westaway, D. and DeArmond, S.J. (1990) Transgenetic studies implicate interactions between homologous PrP isoforms in scrapie prion replication. *Cell* **63** 673–686.

Race, R.E., Graham, K., Ernst, D., Caughey, B. and Chesebro, B. (1990) Analysis of linkage between scrapie incubation period and the prion protein gene in mice. *J. Gen. Virol.* 371 493–497.

Scott, M., Foster, D., Mirenda, C., Serban, D., Coufal, F., Wälchli, M., Torchia, M., Groth, D., Carlson, G., DeArmond, S.J., Westaway, D. and Prusiner, S.B. (1989) Transgenic mice expressing hamster prion protein produce species-specific scrapie infectivity and amyloid plaques. *Cell* **59** 847–857.

Westaway, D. and Prusiner, S.B. (1990) Infectious and genetic manifestations of prion diseases: implications for the BSE epidemic. *Nature (London)* **346** 113.

Westaway, D., Mirenda, C.A., Foster, D., Zebarjadian, Y., Scott, M., Torchia, M., Yang, S.-L., Serban, H., DeArmond, S.J., Ebeling, C., Prusiner, S.B. and Carlson, G.A. (1991) Paradoxical shortening of scrapie incubation times by expression of prion protein transgenes derived from long incubation period mice. *Neuron* 7 59–68.

40

Scrapie prion protein accumulation correlates with neuropathology and incubation times in hamsters and transgenic mice

Stephen J. DeArmond, Klaus Jendroska, Shu-Lian Yang, Albert Taraboulos, Rolf Hecker, Karen Hsiao, Linda Stowring, Michael Scott and Stanley B. Prusiner

ABSTRACT

The dynamic relationship between regional PrP^{Sc} accumulation in the brain and the development of neuropathology during prion diseases in different animal species and in transgenic mice expressing a variety of PrP constructs supports a unifying hypothesis that PrP^{Sc} is not only a required component of the scrapie prion but also that its structure may determine the properties of distinct scrapie prion isolates. The primary premise on which this is based is that PrP^{Sc} accumulation causes the clinically relevant neuropathology. It follows then that scrapie incubation time must be related to the rate of accumulation and anatomic distribution of PrP^{Sc}. Scrapie prion isolates or 'strains' are defined by specific patterns of neuropil vacuolation, length of the incubation time and specific patterns of PrP^{Sc} accumulation. Evidence is reviewed which indicates that the patterns of PrP^{Sc} accumulation in brain are unique for each prion strain and correlate with incubation time.

INTRODUCTION

Investigations of the relationship between regional PrP^{Sc} accumulation in the brain during scrapie and the characteristic neuropathology suggest that PrP^{Sc} is not only necessary for transmission of the disease but also the cause of the clinically relevant

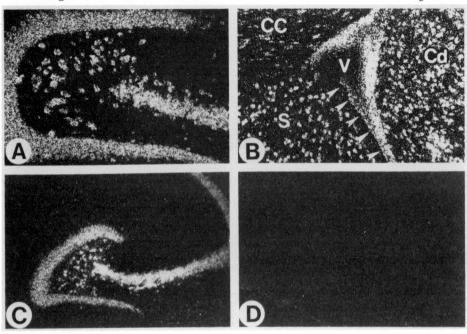

Fig. 1.   The largest number of PrP mRNA copies is consistently found in CNS neurons by both (A), (B) cDNA and (C), (D) RNA *in situ* hybridization techniques. (A) Chinese hamster hippocampus. Above-background numbers of silver grains overlie exclusively neurons of the dentate gyrus and Ammon's horn whereas none above background overlie astrocytes. A similar pattern was found in Syrian and Chinese hamsters. (B) Chinese hamster, region of caudate nucleus (Cd) and septal nuclei (S). Above-background numbers of silver grains overlie neurons in both the caudate nucleus and septum; however, they also overlie ependymal cells (arrows piont to the ependymal lining of the septum) and overlie astrocytes and/or oligodendrocytes in the corpus callosum (CC). A similar pattern was seen in Syrian and Armenian hamsters. V, lateral ventricle. (C) Tg(SHaPrP)81 transgenic mouse hippocampus incubated with the antisense PrP riboprobe. Like hamster, PrP mRNA in this transgenic mouse, which expresses both Syrian hamster and mouse PrP, is confined to neurons. (D) Tg(SHaPrP)81 hipocampus incubated with the sense PrP riboprobe. Darkfield photographs. PrP cDNA probes were prepared as previously published and consisted of 1950 base pairs encoding the open reading frame of Syrian hamster PrP (Jendroska *et al.*, 1991; Kretzschmar *et al.*, 1986). PrP riboprobes were constructed from a PrP-specific cDNA prepared from Swiss mouse (a gift from Dr Bruce Chesebro) (Locht *et al.*, 1986). It was subcloned into the plasmid vector pSP72. For the sense transcripts initiated at the T7 promotor, the pSP72 PrP template was double digested with BstE II (Westaway *et al.*, 1987) and Xbo I to yield a riboprobe encompassing about 80% of the PrP open reading frame beginning at the 5' end. For the antisense transcripts initiated at the SP6 promotor, the template was double digested with Sma I (Locht *et al.*, 1986) and BgI II toyield a riboprobe encompassing about 80% of the PrP open reading frame beginning at the 3' end.

neuropathology in both infectious and genetic forms of prion disease. Our results and those of others have expanded the foregoing premise in several respects. First. PrP^{Sc} accumulation during scrapie is the direct cause of the clinically relevant neuropathology. Second, the rate of accumulation and anatomic distribution of PrP^{Sc} and, therefore, the rate of formation of clinically relevant neuropathology, are the operational determinants of scrapie incubation time. Third, the rate and pattern of PrP^{Sc} accumulation are determined by both the strain of infecting prion and the host animal. The goal of this review is to outline the evidence which has led to these hypotheses.

ORIGIN OF PrP^{Sc}

PrP^{Sc} is only found in prion diseases. The highest concentration occurs in the brain where it can be 100 times higher than the cellular isoform, PrP^C (DeArmond et al., 1987; Jendroska et al., 1991; Meyer et al., 1986; Oesch et al., 1985). PrP^{Sc} is thought to be derived from pre-existing PrP^C. First, the entire open reading frame of the PrP gene is in a single exon, which excludes exon splicing as a factor in its synthesis (Basler et al., 1986). Second, PrP turnover studies in scrapie-infected neuroblastoma cells have revealed a precursor–product relationship between PrP^C and PrP^{Sc} (Borchelt et al., 1990; Caughey and Raymond, 1991).

The structural differences between PrP^C and PrP^{Sc} are unknown. How prions convert PrP^C to PrP^{Sc} is also not known; however, there are several reasons to believe that most of the clinically relevant conversion *in vivo* occurs in CNS neurons. First, neurons are consistently found to express significantly more PrP mRNA than any other cell types (Figs 1(A) and 1(C)) (Oesch et al., 1985; Kretzschmar et al., 1986). Second, the highest concentration of PrP^{Sc} occurs in the brain (Jendroska et al., 1991). Third, CNS neurons appear to be the only cell type targeted for injury in prion diseases. Fourth, PrP^{Sc} appears to be transported from one brain region to another primarily along axons (Jendroska et al., 1991). Astrocytes, oligodendrocytes and ependymal cells show variable expression of PrP mRNA which appears to be dependent on the host animal species and the brain region. The number of autoradiographic silver grains overlying glia vary from background levels (Figs 1(A), 1(C) and 1(D)) to levels equalling those in neurons in some regions, particularly those adjacent to the ependyma (Fig. 1(B)). Although neurons appear to be the major source of brain PrP^{Sc}, the possibility that glia contribute to the pathogenesis of prion diseases in some animal species cannot be excluded (Diedrich et al., 1991).

Evidence for axonal transport of PrP^{Sc} comes from investigations of the rate and pattern of PrP^{Sc} accumulation during scrapie. Two techniques were used to assess PrP^{Sc} kinetics: (1) by quantitative Western analysis of the relative concentration of PrP^{Sc} in homogenates of brain regions dissected and pooled from six to eight animals at varying times throughout the course of scrapie (DeArmond et al., 1987; Jendroska et al., 1991) and (2) by a newly developed and more sensitive histoblot technique (A. Taraboulos et al., in preparation). Histoblots reveal the location and relative concentration of PrP^{Sc} and are obtained by pressing frozen sections of unfixed brain to nitrocellulose paper followed by digestion with proteinase K to eliminate PrP^C and treatment with guanidine thiocyanate (GdnSCN) to enhance binding of PrP antibodies to PrP^{Sc}. The result is an exceptionally high signal-to-noise ratio with superb specificity. PrP^{Sc} was detected earlier during the course of scrapie with the histoblot technique than by Western analysis. The complementary data obtained by the two techniques have greatly aided the interpretation of PrP^{Sc} kinetics.

When Syrian hamsters were inoculated unilaterally in the thalamus with Sc237 prions, the first brain region in which PrP^{Sc} was detectable by Western blot analysis

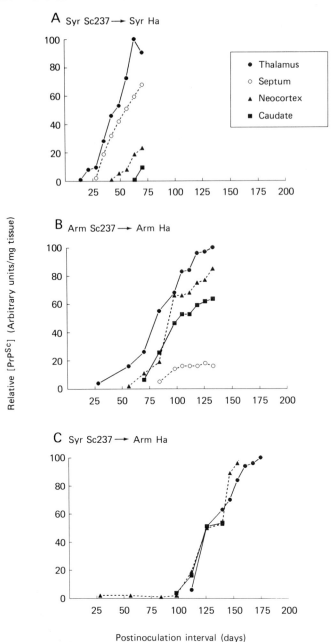

Fig. 2. Comparative regional kinetics of PrP^{Sc} in Syrian and Armenian hamsters. (A) Intrathalamic inoculation of Syrian hamsters (SHa) with Sc237 prions. (B) Intrathalamic inoculation of Armenian hamsters (AHa) with Sc237 prions passaged in AHa. (C) Intrathalamic inoculation of AHa with Sc237 prions passaged in SHa. PrP^{Sc} concentration was estimated by quantitative Western transfer analysis and is relative to the highest concentration in the thalamus which was given an arbitrary value of 100. The rate of PrP^{Sc} accumulation and its pattern of spread in the brain are unique for each prion strain and for each host animal species.

was the thalamus at 14 to 21 days post-inoculation (Fig. 2(A)) (Jendroska *et al.*, 1991). By histoblots, PrP^{Sc} was first detected unilaterally in the thalamus at 7 days (Fig. 3) (Hecker *et al.*, 1992). The size and staining intensity of the immunopositive region in the thalamus increased over the next two weeks and, by 28 days, PrP^{Sc} had spread to the contralateral half of the thalamus. The progressive spread of disease to other brain regions neuroanatomically interconnected with the thalamus argues that the PrP^{Sc} is transported by axons. The neocortex, which is richly interconnected with the thalamus, began to accumulate PrP^{Sc} after 42 days (Fig. 2(A)). The caudate nucleus, which is interconnected with the neocortex, began to accumulate PrP^{Sc} after 63 days. At 65 days, the most intense PrP^{Sc} immunostaining occurred in layers 3 and 6 of the neocortex (Fig. 3). This lends further support to an axonal transport mechanism since thalamocortical pathways terminate in these layers. Whether PrP^{Sc} accumulates in presynaptic boutons, in the dendritic tree, and/or in the CNS extracellular space cannot be determined by the histoblot method. Immunohistochemistry with PrP antibodies suggests that all these locations are possible (DeArmond *et al.*, 1987). It is well established that some PrP^{Sc} is released into the CNS extracellular space since extracellular aggregates of PrP^{Sc} in the form of amyloid filaments occur on subependymal, subpial, subcallosal and perivascular regions (DeArmond *et al.*, 1985).

Although axonal transport appears to account for most of the spread of Sc237 scrapie in the Syrian hamster, the early accumulation of PrP^{Sc} at 28 days in the septum–basal forebrain region, regions not richly interconnected with the thalamus, remained an enigma until the development of the histoblot technique (Fig. 2(A)). Histoblot analysis suggests that ependymal accumulation of PrP^{Sc} precedes and is an important step in early septal–basal forebrain involvement. PrP^{Sc} was detected in the ventrical walls of the septum and caudate nucleus 7 days after intrathalamic inoculation with prions (Fig. 3). While some of this PrP^{Sc} may have originated from the inoculum, the majority appears to be derived from newly formed PrP^{Sc} because the intensity and the thickness of the ventricular wall staining increased over the next two weeks. At 28 days, when PrP^{Sc} was first detected in the septum–basal forebrain homogenates, PrP^{Sc} was visible in the medial septal nucleus and the diagonal band of Broca which are functionally related (Fig. 3). There are two possible explanations for the early accumulation of PrP^{Sc} in the ventricular walls. First, it may have been released from the thalamus into the cerebrospinal fluid (CSF) followed by absorption by the ventricular wall. This is supported by the finding that the thalamus was the only grey matter region with any detectable PrP^{Sc} accumulation prior to involvement of the septum. Another possibility is that some of the original prion inoculum and/or newly formed prions in the thalamus infected the ependyma via the CSF followed by formation of PrP^{Sc} locally in the ependyma. The latter possibility is consistent with the relatively high PrP mRNA content of ependyma revealed by *in situ* hybridization (Fig. 1(B)). The early accumulation of PrP^{Sc} in the medial septal nucleus and diagonal band followed its deposition in the ventricular wall; this appears to be unique to these nuclei since the caudate nucleus, the ventricular surface of which was PrP^{Sc} immunopositive, did not accumulate PrP^{Sc} until much later, i.e. 63 days after inoculation when clinical signs of scrapie are present. The late appearance of pathology in the caudate nucleus, even though it is adjacent to the thalamus, argues that the

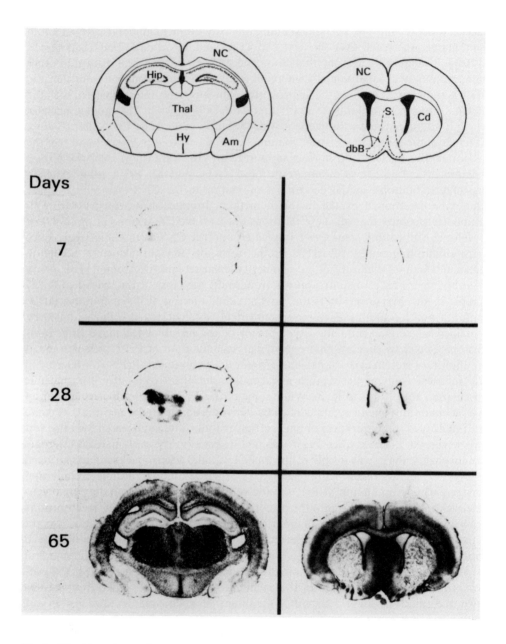

Fig. 3.   Histoblots of coronal sections through the thalamus–hippocampal and the caudate–septal regions of Syrian hamsters inoculated in the thalamus with Sc237 prions. Days post inoculation are indicated. Clinical signs of scrapie were present at 65 days. Am, amygdala; Cd, caudate nucleus: dbB, diagonal band of Broca; Hip, hippocampus; Hy, hypothalamus; NC, neocortex; S, septum; Thal, thalamus. Lateral ventricles are filled in black.

spread of scrapie prions did not occur through diffusion in the interstitial space.

A significant correlation between total brain PrP^{Sc} concentration and scrapie infectivity titre was found throughout the incubation period (Jendroska et al., 1991). This finding and many other lines of evidence argue that PrP^{Sc} is a component of the prion (Prusiner, 1991). Whether PrP^{Sc} alone constitutes the infectious prion or whether there is a second component remains to be established. Whether prion-induced conversion of PrP^{C} to PrP^{Sc} is initiated by PrP^{Sc} alone or a hypothetical second component is unknown.

EVIDENCE THAT PrP^{Sc} ACCUMULATION DURING SCRAPIE IS THE CAUSE OF NEUROPATHOLOGY

The primary neuropathological characteristics of prion diseases include spongiform degeneration of neurons, nerve cell loss, reactive astrocytic gliosis and amyloid plaque formation. There are several reasons to believe that these changes are caused by the local accumulation of PrP^{Sc}. First, PrP^{Sc} is specific for prion diseases including Creutzfeldt-Jakob disease (CJD) and Gerstmann–Sträussler–Scheinker (GSS) syndrome; it is not found in other neurodegenerative disorders (Bockman et al., 1985; Brown et al., 1986; Prusiner and DeArmond, 1987; Roberts et al., 1988; Serban et al., 1990). Second, the highest concentrations of PrP^{Sc} are found in the brain (Oesch et al., 1985) which is the only organ exhibiting recognizable pathological changes. Third, amyloid plaques characteristic of prion diseases contain PrP^{Sc} (DeArmond et al., 1985, 1987; Kitamoto et al., 1986; Prusiner and DeArmond, 1987) and do not contain the β-amyloid protein of Alzheimer's disease and aging (Roberts et al., 1988; Snow et al., 1989). Fourth, there is a temporal correlation between the accumulation of PrP^{Sc} and the development of pathology since spongiform degeneration of neurons and reactive astrocytic gliosis do not develop until one or two weeks after the detection of PrP^{Sc} within a brain region (Jendroska et al., 1991). Negative observations support the conclusions that pathology follows PrP^{Sc} accumulation and that it is specifically related to PrP^{Sc}. For example, both the caudate nucleus and cerebellum did not begin to accumulate detectable amounts of PrP^{Sc} until after 63 days, approximately one to two weeks before death from scrapie (Fig. 2(A)). No spongiform degeneration or reactive astrocytic gliosis were found in these structures at 63 days and only minimal changes were found at 75 days. Fifth, we found an excellent spatial correlation between spongiform degeneration of grey matter, reactive astrocytic gliosis and PrP^{Sc} immunoreactivity (DeArmond et al., 1987).

The correlation between the anatomic location of PrP^{Sc} and neuropathology was recently verified in a study comparing the effects of two prion isolates, Sc237 and 139H, in Syrian hamsters (Hecker et al., 1992). Syrian hamsters were given intrathalamic inoculations with equivalent doses of Sc237 and 139H prions. Clinical signs of scrapie presented at about 70 days after inoculation with Sc237 prions and 170 days with 139H (Table 1). Although a direct correlation between the mass of spongiform degeneration and clinical disease was not found, the anatomic distribution of spongiform degeneration and reactive astrocytic gliosis was virtually identical to the

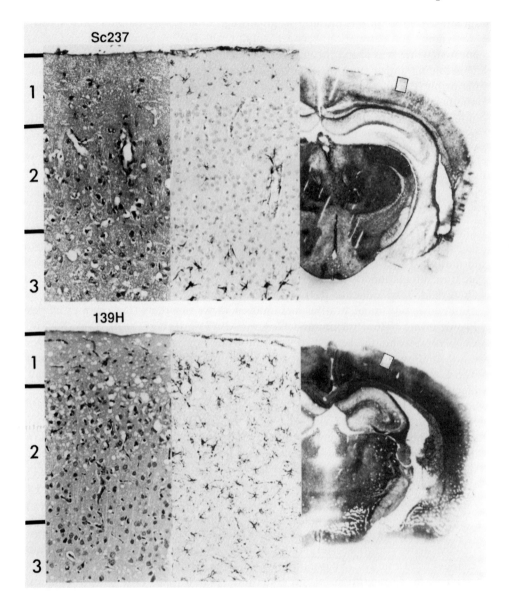

Fig. 4. Comparative distribution of spongiform degeneration, reactive astrocytic gliosis and PrP^Sc in Syrian hamsters receiving intrathalamic inoculations of Sc237 or 139H prions. Histoblots of coronal brain sections at the level of the thalamus–hippocampus are at the far right. The two panels to the left are photomicrographs of the outer half of the cerebral cortex. Their approximate location is indicated by the white rectangle on the histoblots. They are from formalin-fixed paraffin-embedded sections stained with haematoxylin and eosin (far left) to evaluate spongiform degeneration and immunostained for glial fibrillary acidic protein to reveal the degree of reactive astrocytic gliosis (middle). The hamsters were sacrificed at the time clinical signs of scrapie developed (70 days for Sc237 and 170 days for 139H).

Table 1. Scrapie incubation time among hamster species as a function of prion strain and passage history[a]

Passage[b] history	Hamster[c] inoculated	Incubation times (days $\pm$ SEM)[d]	
		Sc237 (n)	139H (n)
SHa→SHa	SHa	77 $\pm$ 1 (48)[e]	167 $\pm$ 1 (94)
SHa→SHa	AHa	174 $\pm$ 1 (4)	146 $\pm$ 1 (7)
SHa→SHa	CHa	344 $\pm$ 7 (4)	241 $\pm$ 1 (20)
SHa→AHa	AHa	128 $\pm$ 3 (8)	153 $\pm$ 0 (4)
AHa→AHa	AHa	125 $\pm$ 2 (16)	—
SHa→CHa	CHa	265 $\pm$ 2 (9)	226 $\pm$ 4 (8)
CHa→CHa	CHa	272 $\pm$ 3 (15)	—

[a] Data from Hecker *et al.* (1992) and Lowenstein *et al.* (1990).
[b] Passage history of the inoculum: SHa, Syrian hamster; AHa, Armenian hamster; CHa, Chinese hamster.
[c] Species of hamster inoculated intrathalamically.
[d] Time to clinical signs of scrapie $\pm$ standard error of the mean (SEM).
[e] *n* is the number of animals.

distribution of PrP^{Sc} with both prion isolates or strains (Fig. 4). For example, PrP^{Sc} was largely localized to the inner half of the neocortex in Sc237 scrapie whereas the entire thickness of the neocortex contained PrP^{Sc} in 139H scrapie at the time clinical signs developed. Both spongiform degeneration and reactive astrocytic gliosis colocalized to the inner half of the neocortex in Sc237 scrapie whereas the entire thickness of the neocortex exhibited these changes in 139H scrapie at the time clinical signs developed. Thus both spongiform degeneration and reactive astrocytic gliosis colocalized with PrP^{Sc}.

In summary, there are multiple lines of experimental evidence to support a cause and effect relationship between PrP^{Sc} and the neuropathology; however, the most convincing evidence comes from studies in transgenic (Tg) mice expressing foreign PrP molecules (Prusiner *et al.*, 1990).

EVIDENCE FROM MOLECULAR GENETIC AND TRANSGENIC MOUSE STUDIES THAT THE PRION PROTEIN DETERMINES THE CHARACTERISTICS OF THE NEUROPATHOLOGY AND SCRAPIE INCUBATION TIME

Recent studies indicate that the structure of the prion protein determines the specific characteristics of scrapie neuropathology and influences scrapie incubation times (Scott *et al.*, 1989; Prusiner *et al.*, 1990). Four transgenic mouse lines were created which express the Syrian hamster (SHa)PrP in addition to the endogenous mouse (Mo)PrP designated Tg(SHaPrP) mice nos. 69, 71, 81, and 7. When non-Tg littermates were inoculated intracerebrally with Sc237 SHa prions, scrapie incubation times were greater than 600 days. In contrast, when the Tg(SHaPrP)81 mouse line was inoculated with Sc237 prions, both the scrapie incubation time of ∼75 days and the neuropathology were similar to those found in the Syrian hamster. Spongiform degeneration was confined to the grey matter and numerous subependymal, subpial and subcallosal amyloid plaques were found. The plaques contained SHaPrP based on immunohistochemical staining with the 13A5 monoclonal antibody. When the Tg(SHaPrP) mice were inoculated with Mo prions, the scrapie incubation times were prolonged but neuropathology resembled that found in the non-Tg mouse. Spongiform degeneration was present in grey and white matter; no amyloid plaques were found. Thus, both scrapie incubation time and neuropathology are influenced by the primary structure of host PrP and PrP^{Sc} in the inoculum.

Additional evidence that PrP is the cause of the neuropathology comes from studies of spontaneous forms of prion disease. Approximately 15% of human CJD cases and 100% of GSS cases are dominantly inherited (for reviews, see Hsaio and Prusiner, 1990; Prusiner, 1991). Molecular genetic studies indicate that a single proline-to-leucine substitution at codon 102 in the human prion protein gene is genetically linked to GSS (Doh-ura *et al.*, 1989; Hsiao *et al.*, 1989). Additionally, at least two other point mutations appear to be linked to spontaneous prion disease in humans; a valine for alanine at codon 117 in a family with GSS (Doh-ura *et al.*, 1989; Hsiao *et al.*, 1991a) and a glutamate-to-lysine substitution at codon 200 in familial CJD (Goldgaber *et al.*, 1989; Hsiao *et al.*, 1991b).

Corroboration that the codon 102 proline-to-leucine substitution leads to spontaneous prion disease was found in Tg mice with a MoPrP transgene carrying the human codon 102 point mutation (Hsiao *et al.*, 1990). The neuropathological features in these Tg(GSSMoPrP) mice included both widespread spongiform degeneration of neurons and mild to moderate reactive astrocytic gliosis.

SUMMARY AND CONCLUSIONS

Multiple lines of experimental evidence presented here indicate that PrP^{Sc} accumulation in the brain is the cause of the pathological changes characteristic of prion disease including spongiform degeneration, reactive astrocytic gliosis and amyloid plaque formation. Studies of PrP^{Sc} accumulation described here give a new perspective to the early studies of Fraser and Dickinson (1968, 1973) who reported that the intensity and pattern of neuropil vacuolization is unique for each strain of scrapie agent and for each host animal species.

A correlation between PrP^{Sc} accumulation and spongiform degeneration has several implications. It supports a unifying hypothesis that PrP^{Sc} is central to both the aetiology and pathogenesis of scrapie. Moreover, it provides insight into the pathogenic mechanisms which ultimately determine scrapie incubation time. In the past, scrapie prion isolates or strains were identified by phenotypic differences in the patterns of neuropathology and scrapie incubation time. The studies of PrP^{Sc} kinetics presented here indicate that the pattern of PrP^{Sc} accumulation is also a phenotypic signature of distinct prion isolates. The diversity of kinetic patterns found indicates a complex relationship between the rates of synthesis and degradation or loss of PrP^{Sc}. The complex patterns of PrP^{Sc} accumulation suggest further that the development of clinically relevant neuropathology is also complex. The complexity was not predicted by molecular genetic studies in which the tacit assumption was that a single factor, amino acid sequence of PrP, alone determines clinical behaviour. For example, recent experiments demonstrated that long and short scrapie incubation times are tightly linked to the prion protein gene in NZW and I/LnJ mice (Carlson *et al.*, 1986; Hunter *et al.*, 1987; Race *et al.*, 1990); however, when the long incubation form of the PrP gene was constructed in Tg mice, paradoxical shortening of incubation times occurred (Westaway *et al.*, 1991). As more is learned with different prion isolates in different animal species, including Tg mice expressing foreign and mutant PrP molecules, we may be able to predict both the patterns of neuropathology and scrapie incubation times.

ACKNOWLEDGMENTS

The authors wish to thank Mr John McCulloch for photographic help and Ms Juliana Cayetano for neurohistological preparations. This work was supported by research grants from the National Institutes of Health (AG02132 and NS14069) as well as gifts from National Medical Enterprises and the Sherman Fairchild Foundation. Dr Jendroska was supported by a grant from Deutsche Forschungsgemeinschaft (JE145-1).

REFERENCES

Basler, K., Oesch, B., Scott, M., Westaway, D., Wälchli, M., Groth, D.F., McKinley, M.P., Prusiner, S.B. and Weissmann, C. (1986) Scrapie and cellular PrP isoforms are encoded by the same chromosomal gene. *Cell* **46** 417–428.

Bockman, J.M., Kingsbury, D.T., McKinley, M.P., Bendheim, P.E. and Prusiner, S.B. (1985) Creutzfeldt–Jakob disease prion proteins in human brains. *N. Engl. J. Med.* **312** 73–78.

Borchelt, D.R., Scott, M., Taraboulos, A., Stahl, N. and Prusiner, S.B. (1990) Scrapie and cellular prion proteins differ in their kinetics of synthesis and topology in cultured cells. *J. Cell Biol.* **110** 743–752.

Brown, P., Coker-Vann, M., Pomery, K., Franko, M., Asher, D.M., Gibbs, C.J., Jr, and Gajdusek, D.C. (1986) Diagnosis of Creutzfeldt–Jakob disease by Western blot identification of marker protein in human brain tissue. *N. Engl. J. Med.* **314** 547–551.

Carlson, G.A., Kingsbury, D.T., Goodman, P.A., Coleman, S., Marshall, S. T., DeArmond, S.J., Westaway, D. and Prusiner, S.B. (1986) Linkage of prion protein and scrapie incubation time genes. *Cell* **46** 503–511.

Caughey, B. and Raymond, G.J. (1991) The scrapie-associated form of PrP is made from a cell surface precursor that is both protease- and phospholipase-sensitive. *J. Biol. Chem.* **266** 18217–18223.

DeArmond, S.J., McKinley, M.P., Barry, R.A., Braunfeld, M.B., McCulloch, J.R. and Prusiner, S.B. (1985) Identification of prion amyloid filaments in scrapie-infected brain. *Cell* **41** 221–235.

DeArmond, S.J., Mobley, W.C., DeMott, D.L., Barry, R.A., Beckstead, J.H. and Prusiner, S.B. (1987) Changes in the localization of brain prion proteins during scrapie infection. *Neurology* **37** 1271–1280.

Diedrich, J.F., Bendheim, P.E., Kim, Y.S., Carp, R.I. and Haase, A.T. (1991) Scrapie-associated prion protein in astrocytes during scrapie infection. *Proc. Natl. Acad. Sci. USA* **88** 375–379.

Doh-ura, K., Tateishi, J., Sasaki, H., Kitamoto, T. and Sakaki, Y. (1989) Pro→Leu change at position 102 of prion protein is the most common but not the sole mutation related to Gerstmann–Sträussler syndrome. *Biochem. Biophys. Res. Commun.* **163** 974–979.

Fraser, H. and Dickinson, A.G. (1968) The sequential development of the brain lesions of scrapie on the three strains of mice. *J. Comp. Pathol.* **78** 301–311.

Fraser, H. and Dickinson, A.G. (1973) Scrapie in mice. Agent-strain differences in the distribution and intensity of grey matter vacuolation. *J. Comp. Pathol.* **83** 29–40.

Gabizon, R., McKinley, M.P., Groth, D. and Prusiner, S.B. (1988) Immunoaffinity purification and neutralization of scrapie prion infectivity. *Proc. Natl. Acad. Sci. USA* **85** 6617–6621.

Ghetti, B., Tagliavini, F., Masters, C.L., Beyreuther, K., Giaccone, G., Verga, L., Farlow, M.R., Conneally, P.M., Azzarelli, B. and Bugiani, O. (1989) Gerstmann–Sträussler–Scheinker disease. II. Neurofibrillary tangles and plaques with PrP-amyloid coexist in an affected family. *Neurology* **39** 1453–1461.

Goldgaber, D., Goldfarb, L.G., Brown, P., Asher, D.M., Brown, W.T., Lin, S., Teener, J.W., Feinstone, S.M., Rubenstein, R., Kascsak, R.J., Boellaard, J.W. and Gajdusek, D.C. (1989) Mutations in familial Creutzfeldt–Jakob disease and Gerstmann–Sträussler–Scheinker's syndrome. *Exp. Neurol.* **106** 204–206.

Hecker, R., Taraboulos, A., Scott, M., Pan, K.-M., Torchia, M., Jendroska, K., DeArmond, S.J. and Prusiner, S.B. (1992) Replication of distinct prion isolates is region specific in brains of transgenic mice and hamsters. *Genes and Development* **6** 1213–1228.

Hsiao, K., Baker, H.F., Crow, T.J., Poulter, M., Owen, F., Terwilliger, J.D., Westaway, D., Ott, J. and Prusiner, S.B. (1989) Linkage of a prion protein missense variant to Gerstmann–Sträussler syndrome. *Nature (London)* **338** 342–345.

Hsiao, K., Scott, M., Foster, D., Groth, D., DeArmond, S.J. and Prusiner, S.B. (1990) Spontaneous neurodegeneration in transgenic mice with mutant prion protein. *Science* **250** 1587–1590.

Hsiao, K., Cass, B.A., Schellenberg, G.D., Bird, T., Devine-Gage, E., Wisniewski, H. and Prusiner, S.B. (1991a) A prion protein variant in a family with the telencephalic form of Gerstmann–Sträussler–Scheinker syndrome. *Neurology* **41** 681–684.

Hsiao, K., Meiner, Z., Kahana, I., Avrahami, D., Scarlato, G., Abramsky, O., Prusiner, S.B. and Gabizon, R. (1991b) Mutation of the prion protein in Lybian Jews with Creutzfeldt–Jakob disease. *N. Engl. J. Med.* **324** 1091–1097.

Jendroska, K., Heinzel, F.P., Torchia, M., Stowring, L., Kretzschmar, H.A., Kon, A.,

Stern, A. and Prusiner, S.B. (1991) Proteinase-resistant prion protein accumulation in Syrian hamster brain correlates with regional pathology and scrapie infectivity. *Neurology* **41** 1482–1490.

Kitamoto, T., Tateishi, J., Tashima, T., Takeshita, I., Barry, R.A., DeArmond, S. J. and Prusiner, S.B. (1986) Amyloid plaques in Creutzfeldt–Jakob disease stain with prion protein antibodies. *Ann. Neurol.* **20** 204–208.

Kretzschmar, H.A., Prusiner, S.B., Stowring, L.E. and DeArmond, S.J. (1986) Scrapie prions are synthesized in neurons. *Am. J. Pathol.* **122** 1–5.

Locht, C., Chesebro, B., Race, R. and Keith, J.M. (1986) Molecular cloning and complete sequence of prion protein cDNA from mouse brain infected with the scrapie agent. *Proc. Natl. Acad. Sci. USA* **83** 6372–6376.

Lowenstein, D.H., Butler, D.A., Westaway, D., McKinley, M.P., DeArmond, S.J. and Prusiner, S.B. (1990) Three hamster species with different scrapie incubation times and neuropathological features encode distinct prion proteins. *Mol. Cell. Biol.* **10** 1153–1163.

Masters, C.L. and Rochardson, E.P. (1978) Subacute spongiform encephalopathy (Creutzfeldt–Jakob disease): the nature and progression of spongiform change. *Brain* **101** 333–344.

Meyer, R.K., McKinley, M.P., Bowman, K.A., Barry, R.A. and Prusiner, S.B. (1986) Separation and properties of cellular and scrapie prion proteins. *Proc. Natl. Acad. Sci. USA* **83** 2310–2314.

Oesch, B., Westaway, D., Walchli, M., McKinley, M.P., Kent, S.B.H., Aebersold, R., Barry, R.A., Tempst, P., Teplow, D.B., Hood, L.E., Prusiner, S.B. and Weissman, C. (1985) A cellular gene encodes scrapie PrP 27-30 protein. *Cell* **40** 735–746.

Prusiner, S.B. (1991) Molecular biology of prion diseases. *Science* **252** 1515–1522.

Prusiner, S.B. and DeArmond, S.J. (1987) Prions causing nervous system degeneration. *Lab. Invest.* **56** 349–363.

Prusiner, S.B., Scott, M., Foster, D., Pan, K.-M., Groth, D., Mirenda, C., Torchia, M., Yang, S.-L., Serban, D., Carlson, G.A., Hoppe, P.C., Westaway, D. and DeArmond, S.J. (1990) Transgenic studies implicate interactions between homologous PrP isoforms in scrapie prion replication. *Cell* **63** 673–686.

Roberts, G.W., Lofthouse, R., Allsop, D., Landon, M., Kidd, M., Prusiner, S.B. and Crow, T.J. (1988) CNS amyloid proteins in neurodegenerative diseases. *Neurology* **38** 1534–1540.

Scott, M., Foster, D., Mirenda, C., Serban, D., Coufal, F., Walchli, M., Torchia, M., Groth, D., Carlson, G., DeArmond, S.J., Westaway, D. and Prusiner, S.B. (1989) Transgenic mice expressing hamster prion protein produce species-specific scrapie infectivity and amyloid plaques. *Cell* **59** 847–857.

Serban, D., Taraboulos, A., DeArmond, S.J. and Prusiner, S.B. (1990) Rapid detection of Creutzfeldt–Jakob disease and scrapie prion protein. *Neurology* **40** 110–117.

Snow, A.D., Kisilevsky, R., Wilmer, J., Prusiner, S.B. and DeArmond, S.J. (1989) Sulfated glycosaminoglycans in amyloid plaques of prion diseases. *Acta Neuropathol. (Berlin)* **77** 337–342.

Taraboulos, A., Jendroska, K., Serban, D., Yang, S.-L., DeArmond, S.J. and Prusiner, S.B. (1992) Regional mapping of prion proteins in brain. *Proc. Natl. Acad. Sci.*

USA, in press.

Westaway, D., Goodman, P.A., Mirenda, C.A., McKinley, M.P., Carlson, G.A. and Prusiner, S.B. (1987) Distinct prions proteins in short and long scrapie incubation period mice. *Cell* **51** 651–662.

Westaway, D., Mirenda, C.A., Foster, D., Zebarjadian, Y., Scott, M., Torchia, M., Yang, S.-L., Serban, H., DeArmond, S.J., Ebeling, C., Prusiner, S.B. and Carlson, G.A. (1991) Paradoxical shortening of scrapie incubation times by expression of prion protein transgenes derived from long incubation period mice. *Neuron* **7** 59–68.

41

The basis of strain variation in scrapie

M.E. Bruce, H. Fraser, P.A. McBride, J.R. Scott and A.G. Dickinson

ABSTRACT

There are many strains of mouse-passaged scrapie, differing in their incubation periods and neuropathology in mice of defined *Sinc* genotypes. Numerous strains with stable properties have been isolated in the same mouse genotype, indicating that scrapie agents carry information which is independent of the host. For some strains a change in the *Sinc* genotype used for passage results in a change in properties. Such changes are consistent with the permissive selection of variants with shorter incubation periods in the new host, rather than active modifications of the scrapie informational molecule by the host. Biological cloning by serial passage at limiting dilution for infectivity also changes the properties of certain isolates, in a manner consistent with the removal of minor strains from a mixture.

INTRODUCTION

The molecular nature of the scrapie agent is still a matter for speculation. Various models have been proposed, the most controversial being that the agent consists solely of an altered host protein, PrP, and is devoid of nucleic acid (the 'prion' hypothesis) (Prusiner, 1982). This idea has sprung from the failure so far to detect scrapie-specific nucleic acids in infected tissues, the resistance of infectivity to treatments which would usually inactivate conventional viruses and the copurification of abnormal PrP and infectivity in tissue extracts. Other suggested models include a scrapie-specific nucleic acid, not yet detected because of the limitations of the techniques currently available. In the 'virino' hypothesis this nucleic acid is closely associated with and protected by host tissue components, possibly abnormal forms of PrP (Dickinson and Outram, 1988). One group has suggested that the scrapie agent contains aberrant mitochondrial DNA (Aiken *et al.*, 1990). Yet another view is that the properties of the scrapie agent, however unusual, do not exclude the possibility that it is a virus (Rohwer, 1984; Sklaviadis *et al.*, 1989).

The basis of scrapie strain variation is a crucial issue when considering the merits of the various molecular models. There are many strains of scrapie, each with its own distinct, reproducible disease characteristics which are stable on repeated experimental passage (Bruce *et al.*, 1991). Any valid model for the scrapie agent must include a molecule which carries information and a mechanism for the accurate replication of this information over many passages. It must be able to account, on the one hand, for the persistence of strain-specific properties on passage in a single host species or genotype and, on the other hand, for changes in the properties of some isolates when the host in which they are passaged is changed. In this chapter we describe strain variation, with particular emphasis on the behavior of isolates on passage in different hosts. We go on to discuss the constraints that these observations impose on molecular models of the scrapie agent.

RESULTS AND DISCUSSION

Characteristics of scrapie strains

Scrapie strain variation has been observed in a number of species, but by far the most detailed studies on strain discrimination have been performed in mice. In this species many strains of scrapie have been identified on the basis of their disease characteristics, in particular the incubation period between initial infection and clinical disease and the pathology they produce in the brain (Bruce and Fraser, 1991; Bruce *et al.*, 1991). The mouse *Sinc* gene, which has a major effect on incubation period, has been an invaluable tool in these studies. The action of the *Sinc* gene was first described over 20 years ago (Dickinson *et al.*, 1968); since then it has been shown that *Sinc* exerts a precise control on the incubation periods of all scrapie strains tested, but that the details of this control differ between scrapie strains (Dickinson and Meikle, 1971; Bruce *et al.*, 1991). Two alleles of *Sinc*, s7 and p7, have been identified on the basis of their biological effects; most scrapie strain characterization has been performed in the C57BL (*Sinc*[s7]) and VM (*Sinc*[p7]) inbred mouse strains and in the F_1 cross between these two strains.

Each scrapie strain, under standard conditions of dose and route of infection, has a characteristic, highly reproducible pattern of incubation periods in mice of the three possible *Sinc* genotypes (Table 1). Scrapie strains differ (1) in their incubation periods in any single genotype, (2) in which of the two homozygotes has the shorter incubation period and (3) in the dominance characteristics of the two alleles in the heterozygote. Recently some of these strains have been tested in *Sinc* congenic mouse lines, giving similar results to those in non-congenic mice (Bruce *et al.*, 1991).

Scrapie strains also differ dramatically in the type, severity and distribution of the neuropathological lesions they produce (Fraser, 1976). The most obvious change seen

Table 1.   Mean incubation period (days) $\pm$ SEM for 14 strains of scrapie or BSE,
injected intracerebrally at high dose, in the three *Sinc* genotypes of mouse

Mouse strain used for isolation and passage	Scrapie strain	Mouse strain or cross[a]		
		C57BL (*Sinc*[s7])	VM (*Sinc*[p7])	C57BL × VM (*Sinc*[s7p7])
C57BL	ME7[b]	171 $\pm$ 2	328 $\pm$ 4	251 $\pm$ 2
(*Sinc*[s7])	22C[b]	182 $\pm$ 1	458 $\pm$ 3	269 $\pm$ 4
	22F	293 $\pm$ 3	275 $\pm$ 2	348 $\pm$ 3
	22L	148 $\pm$ 1	208 $\pm$ 1	189 $\pm$ 1
	79A	158 $\pm$ 2	301 $\pm$ 6	280 $\pm$ 4
	87A	355 $\pm$ 5	602 $\pm$ 5	513 $\pm$ 11
	139A[b]	155 $\pm$ 1	201 $\pm$ 3	249 $\pm$ 3
	301C	207 $\pm$ 3	361 $\pm$ 8	547 $\pm$ 4
VM	22A	466 $\pm$ 4	203 $\pm$ 3	587 $\pm$ 7
(*Sinc*[p7])	22H	214 $\pm$ 6	350 $\pm$ 3	320 $\pm$ 7
	79V	242 $\pm$ 3	276 $\pm$ 6	317 $\pm$ 4
	87V	>700	290 $\pm$ 3	>700
	111A	>700	>700	>700
	301V	263 $\pm$ 4	115 $\pm$ 3	244 $\pm$ 2

[a] Number of mice/group = 10–43.
[b] Strains also maintained by passage in VM mice.

in routine histological sections is a vacuolation of neurons and neuropil, which is
targeted precisely and reproducibly to different areas of brain depending on the
scrapie strain. These differences have been exploited in a quantitative method of strain
discrimination in which vacuolar changes are scored from coded sections in nine grey
matter and three white matter areas of brain to produce a 'lesion profile' (Fraser and
Dickinson, 1968). Each scrapie strain–mouse strain combination has a characteristic
lesion profile which, unlike the incubation period, is independent of the initial dose
of infection.

Recently it has been shown by immunostaining that the distribution of PrP-related
pathology also depends on the scrapie strain (Fig. 1) (Bruce *et al.*, 1989). The most
usual change seen is a diffuse, rather granular accumulation of PrP in the neuropil,
occurring in the same areas as vacuolar degeneration, but pre-dating vacuolation by
several weeks. With certain selective scrapie strains, such as 87V, this pathology is
targeted with exquisite precision to particular anatomically defined neuronal groups.
Less selective strains, such as ME7, produce widespread diffuse changes involving
most parts of the brain, although even in these models there are clear and reproducible
regional differences in the intensity of PrP immunostaining. Although *Sinc* and other
mouse genes do influence the pathology to some extent, the major determinant of

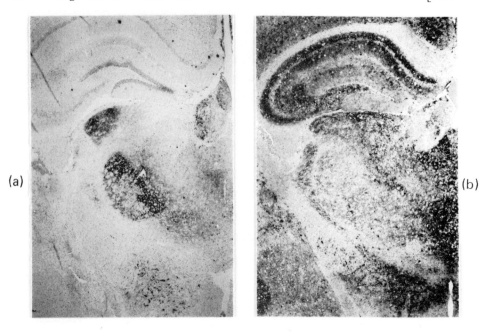

Fig. 1.   Patterns of immunostaining with antiserum to PrP in the brains of VM (*Sinc^(p7)*) mice infected with (a) 87V and (b) ME7 scrapie strains (PAP). Both scrapie strains were injected. i.c. as 1% brain homogenates and the mice were killed in the clinical phase of the disease.

lesion distribution is the strain of scrapie. These results suggest that a fundamental difference between scrapie strains is their ability to recognize and replicate in different neuronal populations.

This selectivity has been most elegantly demonstrated in a comparison of scrapie strains, injected by the intraocular route (J. R. Scott, unpublished results). Following intraocular infection, infectivity is transported via the optic nerve to the primary projection areas of the retinal ganglion cells, which are the superior colliculus and the dorsal lateral geniculate nucleus on the opposite side of the brain to the injected eye (Fraser and Dickinson, 1985). The first pathology is seen in one or the other of these areas. Scrapie strains show consistent preferences between the two areas, even though they receive collaterals from the same ganglion cell axons (Table 2).

Using the dual criteria of incubation periods and pathology, fourteen unequivocally different strains of mouse-passaged scrapie or BSE have been identified (Table 1). A further five isolates have unique properties, indicative of new strains, but have not yet been fully characterized (data not shown). The phenotypic properties of several well-characterized scrapie strains have survived exposure to high doses of ionizing (38 kGy) or UV radiation, treatment with formalin or ethanol and prolonged boiling or dry heating (see Dickinson *et al.*, 1986). These results show that strain characteristics are specified by the scrapie agent itself and not by a separate conventional virus or other microorganism, copassaged with it. Recently it has also been demonstrated

Table 2.   Sites of first vacuolar lesions following right intraocular infection with scrapie

Sinc genotype of mouse	Strain of scrapie	Site of first lesions
s7	ME7	Left superior colliculus
s7	22L	Left superior colliculus
s7	79A	Left dorsal LGN[a]
s7	22C	Left dorsal LGN
p7	ME7	Left superior colliculus
p7	87V	Left dorsal LGN
p7	22A	Left dorsal LGN

[a] Lateral geniculate nucleus

that strain-specific properties are maintained in scrapie-associated fibril preparations (J.R. Scott *et al.*, 1991).

Behaviour of scrapie strains on passage

Several different strains of scrapie have been isolated and maintained by passage in mice of each *Sinc* genotype (at least six in VM mice and eight in C57BL mice: see Table 1). Furthermore, three of the strains isolated in C57BL mice, ME7, 22C and 139A, are unchanged when subsequently passaged in VM mice. This indicates that scrapie agents carry their own information which is independent of the host. This agent-specified information interacts with genetic factors in the host, in particular the *Sinc* gene, to determine the phenotypic characteristics of the disease. For all but one scrapie strain this phenotype is stable indefinitely on serial passage in the *Sinc* genotype in which the strain was isolated. The one exception, 87A, often changes its properties when passaged at high dose in *Sinc*^{s7} mice, in a predictable way that is consistent with the selection of a new mutant strain with a shorter incubation period than the parent strain (Bruce and Dickinson, 1987).

When separate serial passage lines from a primary scrapie source have been set up in *Sinc*^{s7} and *Sinc*^{p7} mice, distinct strains have usually, but not always, been isolated in the two genotypes. There are two possible explanations. Either the host has imposed a change on the information carried by the agent, a change which has been different in the two *Sinc* genotypes, or there has been a host-permitted selection of variants with shorter incubation periods in the *Sinc* genotype used for passage.

This resolution of isolates into pairs of strains obeys a simple rule: the strain isolated in a particular mouse genotype has a shorter incubation period in that genotype than the strain isolated in the other genotype. For example, 79A and 79V were isolated from the 'drowsy' goat source by passage in *Sinc*^{s7} and *Sinc*^{p7} mice respectively; 79A is quicker than 79V in *Sinc*^{s7} mice, and 79V is quicker than 79A in *Sinc*^{p7} mice (see Table 1). For individual scrapie strains the incubation period is not necessarily shorter

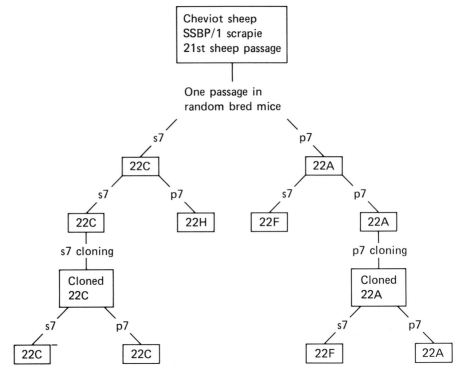

Fig. 2.   Passage history of four strains of scrapie derived from the SSBP 1 sheep source. Each time they appear in the diagram, the symbols s7 and p7 represent several serial passages in $Sinc^{s7}$ and $Sinc^{p7}$ mice respectively.

in the genotype used for passage; both 22H and 79V have shorter incubation periods in $Sinc^{s7}$ mice, despite having been isolated in $Sinc^{p7}$ mice.

The above relationships suggest a selection of variants based on incubation periods rather than a host-induced modification. This conclusion is reinforced by a large body of experimental data concerning the behaviour of scrapie strains when the passaging genotype is changed. To illustrate this we describe below the mouse passage lines derived from one particular primary source, a Cheviot sheep experimentally infected with the SSBP/1 sheep-passaged isolate (Fig. 2) (Dickinson *et al.*, 1986).

After one passage in random-bred mice of uncertain *Sinc* genotype, two separate passage lines were established in $Sinc^{s7}$ and $Sinc^{p7}$ mice. After another 3–4 high dose passages the incubation periods and pathological characteristics became stable and two distinct strains were identified, 22C in the s7 line and 22A in the p7 line. As in the examples already cited, the isolation of these two strains was consistent with the selection of variants with shorter incubation periods in the passaging genotypes (see Table 1). After the 22C passage line had stabilized, a further $Sinc^{p7}$ passage line taken from it resulted in a gradual change in properties (Fig. 3). The incubation period characteristics of this $Sinc^{p7}$-passaged isolate may not yet be fully stable, but the new predominant strain, 22H, is clearly different from 22A. This demonstrates that the

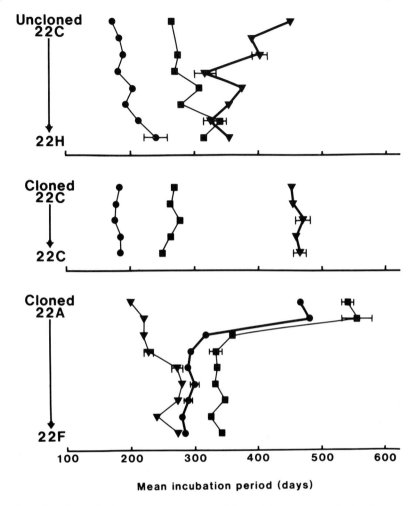

Fig. 3. Effect of a change in the mouse genotype used for serial passage on the incubation period characteristics of uncloned and cloned 22C and cloned 22A. Incubation periods $\pm$ SEM following intracerebral infection with 1% brain homogenates are shown in $Sinc^{s7}$ ($\bullet$), $Sinc^{p7}$ ($\blacktriangledown$) and $Sinc^{s7p7}$ ($\blacksquare$) mice. The new passaging genotype in each case is indicated by a thicker line.

properties of an isolated do not depend solely on the *Sinc* genotype in which it is passaged.

The stable 22C line was also biologically cloned in $Sinc^{s7}$ mice by performing three consecutive passages at the limiting dilution for infectivity (10^{-5} or 10^{-6}) (Bruce and Fraser, 1991). This procedure had no perceptible effect on the subsequent incubation periods and pathology of the isolate when passaged in $Sinc^{s7}$ mice. However, after cloning, 22C was completely unchanged by serial passage through $Sinc^{p7}$ mice (Fig. 3). This shows unequivocally that the change from 22C to 22H is not an obligatory host-induced modification. By far the simplest explanation is that 22C and 22H coexisted in the early mouse passages of the isolate and that 22H was removed from the isolate by cloning.

A similar passage strategy was employed for 22A. In this case a change in the passaging genotype from $Sinc^{p7}$ to $Sinc^{s7}$ resulted in a change to a new stable strain, 22F, which was unlike 22C (Fig. 3) (Bruce and Dickinson, 1979). The same shift in properties was seen in sixteen separate $Sinc^{s7}$ passage lines, starting with a variety of uncloned and cloned 22A sources. The fact that this occurred even after cloning suggests that 22F was generated from 22A, either by a host-induced modification or by mutation. In this context the word 'mutation' simply means a change in the scrapie-specific information of the agent, making no assumptions about the type of molecule involved. The change in properties has always been gradual. In the example shown there was no change whatsoever at the first passage from $Sinc^{s7}$ mice; stable incubation periods in all three genotypes were achieved only after the isolate had been passaged four times through $Sinc^{s7}$ mice. These results indicate that the phenomenon is not simply a result of a host-induced modification, which would be expected to show itself at the first passage through the new genotype. Rather, it is entirely consistent with the gradual selection of a mutant strain, 22F, which has a shorter incubation period than 22A in the passaging genotype. Previously we have shown that another strain of scrapie, 87A, generates a shorter incubation mutant strain, even when the $Sinc$ genotype used for passage remains unchanged (Bruce and Dickinson, 1987).

Implications for molecular models of the scrapie agent

It is easy to explain strain variation and the behaviour of isolates on passage in different hosts in terms of a scrapie-specific nucleic acid, with the generation of variants by mutation and their selection by passage in hosts which favour their replication. It is much more difficult to envisage how a protein alone could specify strain diversity. According to protein-only models the 'pathogen' is PrP which is modified in some specific but unknown way. It has been suggested that this abnormal protein induces the same modification in host protein molecules, either by direct interaction or by inducing mistranslation of the gene (Bolton and Bendheim, 1988; Prusiner, 1989; Wills, 1989). The modification would then be passed on to more PrP molecules, in a process of amplification. It may be possible to account for a limited number of scrapie strains in this model by postulating a number of different self-perpetuating modifications.

The recognition of polymorphic forms of the PrP protein has provided more scope for speculation. In mice two polymorphisms have been identified within the PrP coding region (Westaway et al., 1987). One of the most interesting developments in recent years has been the recognition that the PrP gene is closely linked to the $Sinc$ gene (Carlson et al., 1986; Westaway et al., 1987; Hunter et al., 1987). It has been shown that $Sinc^{s7}$ and $Sinc^{p7}$ mice consistently differ in their PrP amino acid sequence, suggesting that PrP is in fact the $Sinc$ gene product. More recent results using transgenic mice reinforce this view (Scott et al., 1989).

An extension to the 'prion' theory of strains is that the incubation period depends on the compatibility of PrPs between donor and recipient animals. This was suggested by Carlson *et al.*, (1989) to account for the change in properties of a particular 'Chandler'-derived scrapie isolate when the mouse genotype used for passage was changed from $Sinc^{s7}$ to $Sinc^{p7}$. In effect this would be a host-induced modification, an explanation which is unlikely for the reasons already given. However, there is a more direct reason to suspect that their interpretation is incorrect. The properties of the $Sinc^{s7}$- and $Sinc^{p7}$-passaged isolates described by Carlson are closely similar to those of two strains we have derived from the 'Chandler' isolate, 79A and 139A. As both of these strains were isolated by passage in $Sinc^{s7}$ mice the differences between them clearly do not depend on which type of PrP is present in the passaging genotype. On the other hand, Carlson's results are consistent with the selection from a mixture of a strain which replicates faster in the new genotype.

It has also be asserted that there is the equivalent of a species barrier between mice of different *Sinc* genotypes, based on the incompatibility of their PrPs (Carlson *et al.*, 1989). The 'species barrier' refers to the relatively long incubation periods often seen on interspecies transmission of scrapie, compared with later passages in the new species. In a series of experiments in which cloned strains were passaged between mice, hamsters and rats Kimberlin has shown the species barrier to have a number of components (Kimberlin *et al.*, 1987, 1989). The first is the selection of strains with shorter incubation periods in the new species, leading to a permanent change in properties when the isolate is repassaged in the original species. In this respect, changing the species is the equivalent of changing the mouse *Sinc* genotype. However, there are other factors which operate only at the first passage in the new species and do not lead to any permanent change in properties. The simplest is a reduced efficiency of infection, probably due to the association of infectivity with foreign host tissue in the inoculum. It has also been observed in some interspecies transmissions that the intracerebral and intraperitoneal incubation periods are not greatly different, suggesting that, under these circumstances, an intracerebral injection may not be capable of establishing infection directly in the brain (R. H. Kimberlin, personal communication; H. Fraser and M. E. Bruce, unpublished results). It is possible that these effects on efficiency of infection and pathogenesis depend on the incompatibility of PrPs from different species. However, when scrapie is transmitted between mouse *Sinc* genotypes there is little, if any, diminution in the efficiency of infection (Dickinson and Outram, 1988) and there is the usual large difference between intracerebral and intraperitoneal incubation periods (Bruce *et al.*, 1991). It is therefore misleading to regard this as a species barrier.

CONCLUSIONS

We have shown that scrapie agents carry their own information which specifies strain characteristics and that passage in different mouse *Sinc* genotypes selects different strains, rather than inducing strain differences. If the scrapie informational molecule is a nucleic acid, these results are easily explained in terms of classical genetics. On the other hand, most proposed protein-only models depend on host-induced modification to account for some aspects of strain diversity, a mechanism which is incompatible with our observations. Therefore, if scrapie agents do consist solely of protein, there must be as many distinct infection-specific modifications as there are strains. Each must be able to 'replicate' itself accurately over many passages, apart from the predictable generation of other specific modifications. Multiple forms of modified protein must be capable of being copassaged as mixtures and still retain their separate identities and these mixtures must be resolvable by biological cloning. We feel that it is unlikely that a protein alone can meet all these criteria.

Very recently, in an attempt to reconcile the 'prion' and 'virino' hypotheses, it has been suggested that the agent contains a modified protein which is necessary for transmission and a nucleic acid which specifies strain characteristics but is not essential for transmission (Weissmann, 1991). However, the failure of ionizing and UV radiation, heating and various chemical treatments to modify strain properties suggests that the informational molecule is just as tough as the infectious particle. We therefore prefer the simpler explanation for strain variation, that scrapie agents contain an essential small nucleic acid, protected by host tissue components, as proposed in the 'virino' hypothesis. However, of course, the issue will only be resolved finally with the direct identification of the scrapie informational molecule and the variations in it which lead to phenotypic diversity.

REFERENCES

Aiken, J.M., Williamson, J.L., Borchardt, L.M. and Marsh, R.F. (1990) Presence of mitochondrial D-loop DNA in scrapie-infected brain preparations enriched for the prion protein. *J. Virol.* **64** 3265–3268.

Bolton, D.C. and Bendheim, P.E. (1988) A modified host protein model of scrapie. In: Bock, G. and Marsh, J. (eds) *Novel Infectious Agents and the Central Nervous System.* Ciba Foundation Symposium 135, Wiley, Chichester, pp. 164–181.

Bruce, M.E. and Dickinson, A.G. (1979) Biological stability of different classes of scrapie agent. In: Prusiner, S.J. and Hadlow, W.J. (eds) *Slow Transmissible Diseases of the Nervous System*, Vol. 2. Academic Press, New York, pp. 71–86.

Bruce, M.E. and Dickinson, A.G. (1987) Biological evidence that scrapie agent has an independent genome. *J. Gen. Virol.* **68** 79–89.

Bruce, M.E. and Fraser, H. (1991) Scrapie strain variation and its implications. In: Chesebro, B.W. (ed) *Transmissible Spongiform Encephalopathies: Scrapie, BSE and Related Disorders.* Current Topics in Microbiology and Immunology, Vol. 172, Springer, Berlin, pp. 125–138.

Bruce, M.E., McBride, P.A. and Farquhar, C.F. (1989) Precise targeting of the pathology of the sialoglycoprotein, PrP, and vacuolar degeneration in mouse scrapie. *Neurosci. Lett.* **102** 1–6.

Bruce, M.E., McConnell, I., Fraser, H. and Dickinson, A.G. (1991) The disease characteristics of different strains of scrapie in *Sinc* congenic mouse lines: implications for the nature of the agent and host control of pathogenesis. *J. Gen. Virol.* **72** 595–603.

Carlson, G.A., Kingsbury, D.T., Goodman, P.A., Coleman, S., Marshall, S.T., DeArmond, S., Westaway, D. and Prusiner, S.B. (1986) Linkage of prion protein and scrapie incubation time genes. *Cell* **46** 503–511.

Carlson, G.A., Westaway, D., DeArmond, S.J., Peterson-Torchia, M. and Prusiner, S.B. (1989) Primary structure of prion protein may modify scrapie isolate properties. *Proc. Natl. Acad. Sci. USA* **86** 7475–7479.

Dickinson, A.G. and Meikle, V.M.H. (1971) Host-genotype and agent effects in scrapie incubation: change in allelic interaction with different strains of agent. *Mol. Gen. Genet.* **112** 73–79.

Dickinson, A.G. and Outram, G.W. (1988) Genetic aspects of unconventional virus infections: the basis of the virino hypothesis. In: Bock, G. and Marsh, J. (eds) *Novel Infectious Agents and the Central Nervous System*. Ciba Foundation Symposium 135, Wiley, Chichester, pp. 63–83.

Dickinson, A.G., Meikle, V.M.H. and Fraser, H. (1968) Identification of a gene which controls the incubation period of some strains of scrapie in mice. *J. Comp. Pathol.* **78** 293–299.

Dickinson, A.G., Outram, G.W., Taylor, D.M. and Foster, J.D. (1986) Further evidence that scrapie agent has an independent genome. In: Court, L.A., Dormont, D., Brown, P. and Kingsbury, D.T. (eds) *Unconventional Virus Diseases of the Central Nervous System*. Commissariat à l'Énergie Atomique, Fontenay-aux-Roses, pp. 446–460.

Fraser, H. (1976) The pathology of natural and experimental scrapie. In: Kimberlin, R.H. (ed.) *Slow Virus Diseases of Animals and Man*. North-Holland, Amsterdam, pp. 267–305.

Fraser, H. and Dickinson, A.G. (1968) The sequential development of the brain lesions of scrapie in three strains of mice. *J. Comp. Pathol.* **78** 301–311.

Fraser, H. and Dickinson, A.G. (1985) Targeting of scrapie lesions and spread of agent within the retino-tectal projection. *Brain Res.* **346** 32–41.

Hunter, H., Hope, J., McConnell, I. and Dickinson, A.G. (1987) Linkage of the scrapie-associated fibril protein (PrP) gene and *Sinc* using congenic mice and restriction fragment length polymorphism analysis. *J. Gen. Virol.* **68** 2711–2716.

Kimberlin, R.H., Cole, S. and Walker, C.A. (1987) Temporary and permanent modifications to a single strain of mouse scrapie on transmission to rats and hamsters. *J. Gen. Virol.* **68** 1875–1881.

Kimberlin, R.H., Walker, C.A. and Fraser, H. (1989) The genomic identity of different strains of mouse scrapie is expressed in hamsters and preserved on reisolation in mice. *J. Gen. Virol.* **70** 2017–2025.

Prusiner, S.B. (1982) Novel proteinaceous infectious particles cause scrapie. *Science* **216** 136–144.

Prusiner, S.B. (1989) Scrapie prions. *Annu. Rev. Microbiol.* **43** 345–374.

Rohwer, R.G. (1984) Scrapie infectious agent is virus-like in size and susceptibility to inactivation. *Nature (London)* **308** 658–662.

Scott, J.R., Reekie, L.J.D. and Hope, J. (1991) Evidence for intrinsic control of scrapie pathogenesis in the murine visual system. *Neurosci. Lett.* **133** 141–144.

Scott, M., Foster, D., Mirenda, C., Serban, D., Coufal, F., Walchii, M., Torchia, M., Groth, D., Carlson, G., DeArmond, S.J., Westaway, D. and Prusiner, S.B. (1989) Transgenic mice expressing hamster prion protein produce species-specific scrapie infectivity and amyloid plaques. *Cell* **59** 847–857.

Sklaviadis, T.K., Manuelidis, L. and Manuelidis, E.E. (1989) Physical properties of the Creutzfeldt–Jakob disease agent. *J. Virol.* **63** 1212–1222.

Weissmann, C. (1991). A 'unified theory' of prion propagation. *Nature (London)* **352** 679–683.

Westaway, D., Goodman, P.A., Mirenda, C.A., McKinley, M.P., Carlson, G.A. and Prusiner, S.B. (1987) Distinct prion proteins in short and long scrapie incubation period mice. *Cell* **51** 651–662.

Wills, P.R. (1989) Induced frameshifting mechanism of replication for an information-carrying scrapie prion. *Microb. Pathogen.* **6** 235–249.

42

The genetics of prion susceptibility in the mouse

George A. Carlson, David Westaway and Stanley B. Prusiner

ABSTRACT

Genetic analysis of experimental scrapie in mice is consistent with a pivotal role for prion protein. However, formal proof that the primary structure of PrP, rather than a linked locus, controls scrapie incubation time is lacking. We present the evidence for and against the postulate that control of scrapie incubation time is a consequence of PrP amino acid sequence. The prion protein gene (*Prn-p*) may exert two distinct effects on prion incubation time phenotype: (1) control of initiation of disease through interaction of the normal cellular (PrP^C) and malignant (PrP^{Sc}) isoforms, and (2) determination of the rate of prion replication.

INTRODUCTION

Natural and experimental scrapie in sheep indicated a strong genetic influence on disease susceptibility and course. However, the transmissibility of scrapie to mice (Chandler, 1961) provided a better-defined, more readily manipulable model. Dickinson and coworkers (Dickinson and MacKay, 1964) identified a stock of mice that had a greatly prolonged incubation period following inoculation with the ME7 scrapie isolate, and later demonstrated that this prolongation of incubation time was the effect of a single gene designated *Sinc* (Dickinson *et al.*, 1968). VM/Dk and the related IM/Dk mouse strains have the $Sinc^{p7}$ allele and ME7 scrapie incubation times of approximately 300 days; other mouse strains have the $Sinc^{s7}$ allele and short incubation periods of 150 days or less. The *Sinc* gene was found to control incubation period for all scrapie isolates tested, but scrapie strains could be categorized by their distinct incubation period profiles in mice of the three *Sinc* genotypes (Dickinson and Meikle, 1971).

The remarkable finding that the only known functional component of infectious scrapie prions, PrP^{Sc}, was encoded by a chromosomal gene rather than by agent-specific nucleic acid (Oesch *et al.*, 1985; Basler *et al.*, 1986) paved the way for molecular genetic analysis of scrapie. The first evidence that the PrP gene (*Prn-p* in mice) might control susceptibility to prion diseases came from genetic linkage studies in mice (Carlson *et al.*, 1986). In a survey of prion incubation times in inbred strains of mice, the I/LnJ strain was identified as having an exceptionally long incubation time (Kingsbury *et al.*, 1983). Classical backcross analysis involving I/LnJ and a short incubation time strain, NZW/LacJ, indicated that prolongation of Chandler scrapie isolate incubation time was due to the effects of a single gene designated *Prn-i* (VM/Dk mice were not available to test for allelism with *Sinc*). A *Prn-p* restriction fragment length variant that distinguished I/LnJ (*Prn-p^b*) from most other mouse strains (*Prn-p^a*) was used to demonstrate linkage of the prion incubation time gene and *Prn-p*, with 65 of 66 backcross offspring concordant for incubation time phenotype and *Prn-p* genotype (Carlson *et al.*, 1986). Subsequently, typing of *Sinc* congeneic mice for *Prn-p* demonstrated that *Prn-i* and *Sinc* are most likely the same locus (Hunter *et al.*, 1987).

Although the sequence of the open reading frames of the *Prn-p^a* and *Prn-p^b* alleles revealed that PrP-A and PrP-B differ at positions 108 (Leu/Phe) and 189 (Thr/Val) (Westaway *et al.*, 1987), it has not been proven that the prolongation of incubation time in *Prn-p^b* mice is an effect of the primary structure of PrP. The designation *Prn* refers to the prion gene complex, comprising *Prn-p* and its linked prion incubation time gene (*Sinc/Prn-i*). In this report we present the evidence for and against the hypothesis that control of scrapie incubation time is a consequence of PrP amino acid sequence. Results using distinct scrapie isolates raise the possibility that *Prn-p* may exert two distinct effects on prion incubation time: (1) control of initiation of disease through interaction of the normal cellular (PrP^C) and malignant (PrP^{Sc}) isoforms of PrP, and (2) determination of the rate of prion replication.

RESULTS AND DISCUSSION

Distribution of the long incubation time *Prn-p^b* haplotype among inbred strains is due to a 'founder effect'

All inbred strains of mice known to carry the *Prn-p^b* allele have long (over 200 days) incubation periods for the Chandler murine scrapie isolate, as shown in Table 1. The *Prn-p^b* haplotype is defined by flanking restriction enzyme site variations, absence of a BstE II site indicative of a threonine codon at position 189 that is present in mice of other haplotypes, and presence of a phenylalanine codon at position 108 detected with allele-specific oligonucleotides (Westaway *et al.*, 1987; Carlson *et al.*, 1988a). These multiple shared nucleotide sequence differences among mice with long scrapie incubation periods strongly suggest a common origin for the *Prn-p^b* allele. Indeed, the ancestry of mice carrying the *b* allele was compatible with all seven strains,

Table 1.   Inbred strains of mice carrying the *b* allele of *Prn-p* have long scrapie incubation times[a]

Mice	n	Onset of illness (days)	Death (days)
I/LnJ	21	255 + 14	265 + 11
BDP/J	3	270 + 5	291 + 12
P/J	8	295 + 9	306 + 19
JE/Le	4	322 + 28	363 + 29
IS/Cam	3	282 + 57	292 + 57
VM/Dk[b]	20	201 + 3	Not reported

[a] Mice were inoculated intracerebrally with 30 μl of 1% brain homogenate the Chandler scrapie isolate passaged in random-bred Swiss mice.
[b] Data for VM/Dk mice inoculated i.c. with scrapie isolate 139A (similar to the Chandler isolate) reported in Bruce *et al.* (1991).

including VM/Dk and IM/Dk mice in Edinburgh, having descended from the progenitor of the I/LnJ mouse strain (Carlson *et al.*, 1988a). A common origin and subsequent inbreeding following separation would increase the probability that genes closely linked to *Prn-p* are also shared by I/Ln, BDP/J, P/J, JE/Le, IS/Cam, IM/Dk and VM/Dk mice, dictating caution in concluding that the primary structure of PrP determines scrapie incubation time. This stands in contrast with linkage of genetic prion disease in humans that has been linked to missense mutations in PRNP. For example, the codon 102 mutation that has been linked to ataxic Gerstmann–Sträussler–Scheinker (GSS) syndrome apparently arose independently in several linkages (Hsiao *et al.*, 1989; Hsiao and Prusiner, 1989); spontaneous neurological disease in mice expressing GSS codon 102 mutant transgenes strongly argues that the mutation causes disease (Hsiao *et al.*, 1990).

Expression of *Prn-p^b* transgenes from long incubation time mice shortens, rather than prolongs, the scrapie incubation period

Five independent lines of transgenic (Tg) mice were produced by microinjection of *Prn-p^a* homozygous fertilized eggs with the insert from a *Prn-p^b*-containing cosmid clone (cos6.I/LnJ-4) (Westaway *et al.*, 1991). This cosmid insert contains the two 5′ untranslated exons, the open reading frame, promoter sequences and approximately 6 kb of 5′ and more than 15.5 kb of 3′ flanking sequences. All five lines of Tg (*Prn-p^b*) mice express transgene mRNA and have more PrP^C in their brains than non-Tg mice. If the primary structure of PrP-B were sufficient to program long scrapie incubation times, these Tg lines would be expected to have prolonged incubation times, similar to those of *Prn-p^a/Prn-p^b* heterozygous mice (> 200 days). Paradoxically, all lines of Tg mice expressing PrP-B had significantly shorter scrapie incubation times than their non-Tg controls. Chandler scrapie isolate incubation times for Tg (*Prn-p^b*) 15 and non-Tg mice are illustrated in Fig. 1; only a slight shortening of incubation time was observed in this line which contains three copies of the transgene.

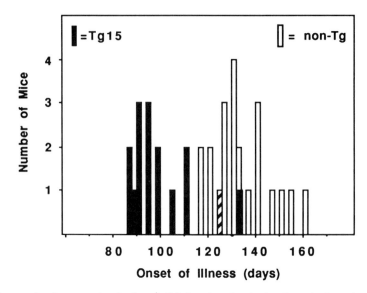

Fig. 1. Transgenic mice expressing the *Prn-p^b* allele from long incubation time mice have shorter Chandler scrapie isolate incubation times than non-transgenic mice. Tg (*Prn-p^b*) mice and non-Tg mice were inoculated intracerebrally with 30 μl of a 10-fold dilution of 10% brain homogenate from clinically ill Swiss mice that had been inoculated with the Chandler murine scrapie isolate (kindly provided by Dr W. Hadlow). Results from both hemizygous (three transgene copies) and homozygous (six copies) Tg 15 mice are shown. The solid bars represent Tg 15 mice and the open bars non-Tg mice; the hatched bar represents 1 Tg 15 and 1 non-Tg mouse that became ill 123 days after inoculation.

Incubation times in the four lines, which carry higher transgene copy numbers, were shorter (Westaway *et al.*, 1991); for example, Tg 117 (> 30 copies) had incubation times of 78.5 ± 2 days (*n* = 13). However, no clear differences in Chandler isolate incubation times between hemizygous (three transgene copies) and homozygous (six transgene copies) Tg 15 mice were observed; Fig. 1 includes both hemizygous and homozygous animals.

In contrast, a quantitative and possibly allele-specific effect of *Prn-p^b* transgene expression was observed in Tg 15 mice inoculated with the 22A scrapie isolate. In contrast to most common scrapie isolates, mice with the *Prn-p^a*-linked *Sincs^{s7}* allele have much longer scrapie incubation times than *Sinc^{p7}* mice (Dickinson and Meikle, 1969). Heterozygous mice have a longer 22A incubation time than either parent; the long incubation times of F1 hybrids between VM and its *Sinc^{s7}* congeneic partner indicate true overdominance (Bruce *et al.*, 1991), rather than simple hybrid vigour as suggested earlier (Carlson *et al.*, 1988b). A clear difference in 22A incubation times was observed between transgene hemizygous and homozygous Tg 15 mice (Table 2). Transgene homozygous mice, which harboured six transgene copies in addition to the two endogenous copies of *Prn-p^a*, had incubation times of 286 ± 6 days, significantly shorter than those of hemizygous animals (395 ± 12 days). Both hemizygous and homozygous Tg mice had incubation times shorter than those of *Prn-p^a* or *Prn-p^a*/*Prn-p^b* mice, but longer than those of *Prn-p^b* homozygous mice. We do not yet know whether the transgene-mediated mitigation of the effects of $Sinc^{s7}$ in prolonging 22A incubation time is simply a manifestation of transgene copy number or reflects alteration of the ratio of PrP-A to PrP-B proteins. In their 'replication site' hypothesis, Dickinson and Outram (1979) proposed interaction between allotypic

Table 2.   Effects of $Prn-p^b$ transgene expression on 22A incubation time

Mice	$Prn-p^a$:$Prn-p^b$	Illness	Death
VM/Dk[a]	0:2	199 ± 3 $(39)^b$	n.a.[c]
VM/Sinc^{s7}	2:0	441 ± 2 (84)	n.a.
(VM × VM/Sinc^{s7})F1	1:1	504 ± 5 (18)	n.a.
B6.I-$Prn-p^b$	0:2	194 ± 10 (7)	214 ± 8 (7)
B6	2:0	402 ± 2 (6/8)	408 (2/8)
(B6 × B6.1-$Prn-p^b$)F1	1:1	>412	n.a.
Tg ($Prn-p^b$) 15 hemizygous	2:3	395 ± 12 (6)	443 ± 12 (6)
Tg ($Prn-p^b$) 15 homozygous	2:6	286 ± 6 (5)	302 ± 5 (6)

[a] Data from VM and *Sinc* congeneic mice as reported by Bruce *et al.* (1991).
[b] Number of mice is indicated in parentheses. If not all mice have been affected to date the number ill or dead over the number inoculated is indicated.
[c] Not available.

forms of proteins necessary for agent replication to account for *Sinc* overdominance.

A quantitative effect of the amount of PrP^C on incubation time has been previously observed in mice carrying Syrian hamster (Ha) PrP genes, with incubation time for Ha prions inversely correlated with steady-state levels of HaPrP mRNA and HaPrP^C (Prusiner *et al.*, 1990). An increased supply of PrP^C or a precursor for conversion to infectious PrP^{Sc} in Tg ($Prn-p^b$) mice may mask effects of PrP-B primary structure in prolonging incubation time in non-Tg mice. Unfortunately, transgenic mice that express comparable levels of PrP-A are not currently available for comparison of incubation times with Tg ($Prn-p^b$) mice.

Although it is plausible that overexpression of PrP^C may be the cause of the abbreviated incubation times in Tg ($Prn-p^b$) mice, it is also conceivable that a functional incubation time locus was not included or was bisected in the cosmid clone. One candidate gene is the large overlapping open reading fame in the DNA strand opposite to the PrP transcriptional unit found in the PrP genes of all species that have been sequenced (Goldgaber, 1991). However, there is no convincing evidence for transcripts from this 'anti-PrP' gene (Manson and Hope, 1991; Hewinson *et al.*, 1991). It is also possible that the normal chromosomal position of *Prn-p* may affect aspects of its expression essential for control of scrapie incubation time. Expression of the $Prn-p^b$ transgene is position independent, presumably because of elements included within the cosmid insert that insulate it from positive or negative effects of regulators flanking the insertion site (Eissenberg and Elgin, 1991). For example, high level ectopic expression of transgene-encoded Prp in neurons in a 'clinical target area' could lead to abbreviated incubation times.

Nonetheless, it is clear that PrP-B expression in and of itself is not sufficient to prolong incubation time—a fact previously revealed by occurrence in segregating crosses of occasional *Prn-p^b* offspring that had short scrapie incubation times.

Discordance of scrapie incubation time phenotype and *Prn-p* genotype among offspring of segregating crosses

In the (NZW × I/LnJ)F1 × NZW backcross that was first used to establish linkage between *Prn-p* and the prion incubation time gene, 65 of 66 mice were concordant for incubation time phenotype and *Prn-p* genotype; if the single *Prn-p* heterozygous mouse with a short incubation time was a true meiotic recombinant, an interval of 1.5 ± 1.5 (SE) cM would separate *Sinc/Prn-i* and *Prn-p* (Carlson *et al.*, 1986). We also observed four discordant mice among offspring of an (NZW × I/Ln)F2 cross consistent with a map distance of 4.8 ± 1.3 cM separating the two genes (Carlson *et al.*, 1988a). Similar results were obtained by Race and coworkers (Race *et al.*, 1990) who had observed putative recombinants in NZW × (NZW × I/Ln)F1 and NZW × (P/J × NZW)F1 backcrosses, yielding intervals between *Sinc/Prn-i* and *Prn-p* of 2.3 ± 2.2 and 5.9 ± 4 cM. However, it is important to stress that the lethal nature of the scrapie bioassay precluded progeny testing to determine whether any of the deviant mice actually resulted from meiotic recombination between two distinct loci in the prion gene complex.

Discordant mice in *Prn-p^a* × *Prn-p^b* crosses are not meiotic recombinants

Results from restriction fragment length polymorphism (RFLP) typing of 102 (NZW × I/LnJ)F2 offspring (Carlson *et al.*, 1988a), including four putative *Sinc/Prn-i–Prn-p* recombinant mice, for loci flanking *Prn-p* indicated that a minimum of two crossover events would be required to account for the deviant mice. For example, the lack of recombinants between *Il-1a* and *Prn-p* among the four discordant mice, along with the fact that incubation times of six mice that were *Il-1a–Prn-p* recombinants were concordant with *Prn-p* genotype, suggests that the prion incubation time gene does not lie proximal to *Prn-p*. Similarly, only two of the four discordant mice were recombinants between *Pax-1* (see below) and agouti (*A*); in this F2 cross *A* mapped 10.9 ± 2.3 cM distal from *Prn-p*.

In order to capture recombinants between *Sinc/Prn-i* and *Prn-p*, should they exist, we used the wellhärig (*we*) and undulated (*un*) mutations which are easily scored by their phenotypes. Homozygous *we/we* mice have a wavy first coat and *un/un* homozygotes have a kinky tail, likely due to a point mutation in the paired-rule homeobox containing gene, *Pax-1* (Balling *et al.*, 1988). An overview of the 'recombinant capture' experiment is given in Fig. 2. (B6.I-*Il-1a^d Prn-p^b* × B10.UW/Sn)F1 mice were backcrossed to B10.UW/Sn and the offspring scored for separation of the wavy coat and kinky tail phenotypes. The genotype of B10.UW/Sn mice is *B2m^a we Prn-p^a un* and the genotype of the B6.I congenic mice is

Recombinant Capture Protocol

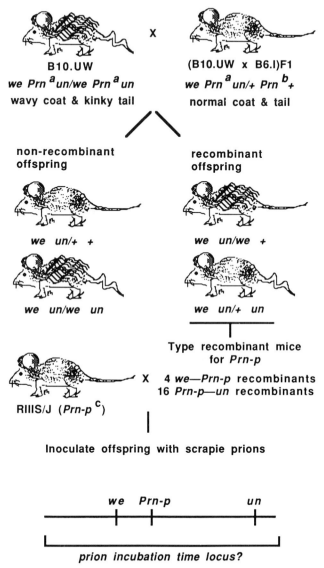

Fig. 2. Summary of the experimental protocol designed to identify and perpetuate chromosomes on which recombination between *Prn-p* and its linked incubation time locus (*Sinc/Prn-i*) had occurred. *Prn-p* is located on chromosome 2 between the *we* and *un* loci. Mice homozygous for the *we* (wellhärig) mutation have a wavy first coat and *un* (undulated) homozygous mice have a kinky tail. The wild type alleles at these loci are indicated as '+'. Recombination between *we* and *un* could be scored visually, obviating the need for analysis of DNA from all backcross offspring to identify recombination near *Prn-p*. To preserve recombinant chromosomes, recombinant mice were crossed to RIIIS/J, which has a short scrapie incubation time (the *Prn-p^c* haplotype is identified by restriction site variants flanking the open reading frame, but the encoded protein does not differ from that of *Prn-p^a*).

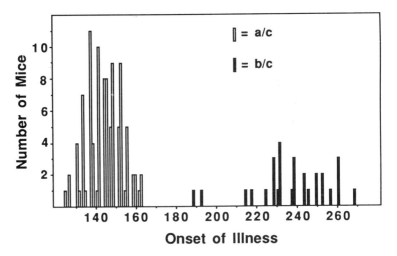

Fig. 3. Scrapie incubation time phenotype and *Prn-p* genotype were concordant among offspring of *we–un* recombinant mice. One hundred and twenty six offsrpingof 4 *we–Prn-p* and 16 *Prn-p–un* recombinant mice crossed with RIIS/J (*Prn-p^c*). All *Prn-p^a/Prn-p^c* mice had short scrapie incubation times while all *Prn-p^b/Prn-p^c* mice had long incubation times.

B2m^b + Prn-p^b +. RFLP typing of DNA from 155 of these backcross mice for *B2m* and *Prn-p*, in addition to the visible phenotypes, indicated that the most probable gene order is *B2m–we–Prn-p–un*. A total of 482 backcross mice were observed for recombination in the *we–un* interval, and all recombinants were also typed for *Prn-p*. Seven mice were recombinants between *we* and *Prn-p* (1.45 ± 0.5 cM) and 21 were *Prn-p–un* recombinants (4.4 ± 0.9 cM). Sixteen of the *Prn-p–un* recombinants and four of the *we–Prn-p* recombinants were crossed to the short incubation time strain RIIIS/J (*Prn-p^c*) to preserve the recombinant chromosomes. The progeny were inoculated with scrapie prions to determine whether separation of the incubation time phenotype and *Prn-p* genotype occurred in offspring carrying recombinant chromosomes. As the map distance separating *we* and *un* is comparable with the putative *Prn-p–un* interval deduced in earlier crosses (Carlson *et al.*, 1986, 1988a; Race *et al.*, 1990), as least some single crossover events between the two loci should have been captured. As illustrated in Fig. 3, none of the 126 offspring of the recombinant mice showed separation of incubation time phenotype and *Prn-p* genotype.

Genes not linked to *Prn-p* also influence prion disease phenotype

The most plausible explanation for the mice with discordant time phenotype and *Prn-p* genotype in earlier crosses invokes minor genes affecting scrapie incubation time. The effects of such genes are clearly indicated by the differences in incubation time among inbred strains with identical *Prn-p* genotype (Carlson *et al.*, 1988a; Dickinson and MacKay, 1964; Kingsbury *et al.*, 1983); similarly, genes unlinked to *Prn-p* determine the ~170 day incubation times seen in mice carrying the *d* and *e* haplotypes of *Prn-p* (Carlson *et al.*, 1988a; Race *et al.*, 1990). Simultaneous inheritance of several 'minor' loci, each with a tendency to shorten incubation period, might result in an extremely 'short' phenotypic variant mouse, in spite of a *Prn-p^a/Prn-p^b* genotype; a similar argument can be invoked for long incubation periods in *Prn-p^a/Prn-p^a* mice. Supporting the likelihood that the cumulative effect of genes unlinked to *Prn-p* was responsible for the putative recombinants is the fact that no deviant mice were observed among the 126 offspring of *we–un* recombinant mice that were inoculated with scrapie prions. C57BL/6 (B6) and C57BL/10 (B10) mice are closely related and differ for only 6 of the approximately 235 loci that have been typed in both strains (Roderick and Guidi, 1989). I/LnJ and NZW/LacJ were independently derived; both have been typed for only 54 loci, but 12 of these differ between the strains. The lack of deviant mice in the 'recombinant capture' experiments might reflect the similarity of the B6 and B10 background of the mice used in the cross, while the deviants in the I/Ln × NZW crosses may reflect reproducible cosegregation of non-*Prn-p* genes.

Identification of non-*Prn-p* genes influencing scrapie disease phenotype could prove extremely valuable in understanding PrP metabolism, the mechanism of PrP^Sc formation, and the disease process. PrP primary structure may not account for several features of inherited prion disease including age of onset and duration of illness. For example, a thirty year span in age of onset has been seen in a single family with leucine-mutant PrP GSS (Hsiao and Prusiner, 1989). Because differences in PrP primary structure are not found in affected individuals within a family, the disparity in age of onset must be due to other factors. A similar variation in the time of disease onset (from 125 to >275 days) has been seen in hemizygous Tg (GSS PrP) mice (Hsiao *et al.*, 1990); the founding parents of this line were (C57BL/6J × SJL/J)F1 and numerous alleles, in addition to the transgene, are segregating in this Tg line.

Compatibility of PrP primary structure in the prion and in the host influences initiation of prion replication

In spite of the lack of evidence for recombination between *Prn-p* and the scrapie incubation time gene, definitive evidence linking control of the rate of prion replication with PrP primary structure is lacking. However, PrP sequence may determine the efficiency of interaction between PrP^Sc in the inoculum and host PrP^C in initiating prion replication. It is clear that prions from short incubation time *Prn-p^a* mice and long incubation time *Prn-p^b* mice contain distinct PrP^Sc allotypes, PrP^Sc-A and PrP^Sc-B, that differ by two amino acids. Passage of the Chandler scrapie isolate through various inbred mouse strains and F1 hybrids revealed a mouse strain barrier to prion passage and suggested that the presence of allogeneic PrP^Sc in the inoculum prolonged incubation time and increased the variance of incubation times among recipients (Carlson *et al.*, 1989). For example, the incubation time of I/LnJ mice inoculated with PrP-A prions from Swiss mice was 314 ± 13 days; PrP-B prions produced by a single passage in I/LnJ mice gave an incubation time of 193 ± 6 days. Conversely, the Swiss PrP-A isolate produced incubation times of 113 ± 2 days in *Prn-p^a* NZW mice compared with a 129 ± 4 day incubation time for PrP-B prions. One explanation for these results is that host PrP^C interacts with PrP^Sc in the inoculum, with interaction between homologous molecules being more efficient in initiating conversion of host PrP^C to PrP^{Sc}. For example, the conversion of PrP^C-B to PrP^{Sc}-B molecules by PrP^{Sc}-A prions may occur at a much lower frequency than occurs with homologous PrP^{Sc}-B prions; after PrP^{Sc}-B prions are produced, the rate of replication is controlled by the scrapie incubation time gene (which may be *Prn-p* itself).

The species barrier to scrapie transmission is analogous to the PrP allotype barrier, and results from Tg mice reinforce the concept that interactions between PrP^C and PrP^{Sc} feature in prion replication (Scott *et al.*, 1989; Prusiner *et al.*, 1990). Only a minority of mice inoculated with prions from Syrian hamsters develop scrapie, and then only after more than 600 days. The prions produced contain mouse PrP^{Sc} and cause scrapie efficiently in mice but not in hamsters, indicating that establishment of infection across this species barrier is a stochastic process. In Tg mice expressing hamster PrP (HaPrP) transgenes, in addition to their endogenous *Prn-p* gene, initiation of hamster prion replication was a non-stochastic process. All individuals in four Tg lines expressing HaPrP became ill following inoculation with hamster prions and produced HaPrP^{Sc}-containing prions that were readily transmissible to hamsters but not to mice. Each Tg line had its own distinct hamster prion incubation time for hamster prions that was inversely proportional to steady-state levels of HaPrP^C (Prusiner *et al.*, 1990). Therefore, two effects of HaPrP^C are evident in these Tg mice: first, interaction with HaPrP^{Sc} in inoculated prions to initiate infection; second, quantitative control of the rate of prion replication.

Prion protein genes and scrapie susceptibility in mice and humans

Missense and insertional mutations in the human PRNP gene have been linked to GSS and familial CJD. One of these mutations, a leucine for proline substitution at codon 102, in the analogous position (codon 101) of a mouse *Prn-p* transgene caused neurologic dysfunction, spongiform degeneration and gliosis in high copy number transgenic mice (Hsiao *et al.*, 1990); the spontaneously arising disease has been recently transmitted to other animals with brain extracts from affected Tg mice (see Chapters 13 and 44). These experiments provide the strongest evidence that prions are devoid of functional nucleic acid, as suggested by many previous studies that have failed to identify a scrapie-specific polynucleotide. It is likely that mutant PrP^C spontaneously converts into a malignant isoform; once spontaneous conversion occurs, the process may become autocatalytic. While disease-associated PRNP mutations clearly influence disease susceptibility, they are not analogous to the *Prn-p* linked incubation time gene in mice. Neither *Prn-p^a* nor *Prn-p^b* mice develop spontaneous neurologic disease; the effect of *Sinc/Prn-i* is to control the rate of prion replication.

Recent evidence in humans raises the possibility that the nonpathogenic Met/Val polymorphism at PRNP codon 129 may be analogous to the mouse prion incubation time gene (Palmer *et al.*, 1991). Individuals affected with sporadic CJD were significantly more likely to be homozygous for one or the other polymorphism; only one individual among 22 cases was PRNP heterozygous while 46% of nonaffected individuals were heterozygous. One interpretation of these findings is that conversion PrP^C encoded by either allele to the disease-specific isoform is an extremely rare event, but that efficient prion replication depends on interaction between homologous

PrP^C and PrP^{Sc}. With some scrapie isolates *Prn-p* heterozygous mice have a much longer incubation time than either homozygous parent (Dickinson and Meikle, 1969). Similarly, the PrP allotype barrier in mice (Carlson *et al.*, 1989) and the PrP-dependent species barrier to scrapie transmission provide evidence that interaction between homologous isoforms features in the initiation of prion replication (Prusiner *et al.*, 1990).

Though data are limited, an influence of the non-pathogenic PRNP allele on the age of death has been described for a pedigree carrying a 144 bp insert on the Met-129 allele (Palmer *et al.*, 1991). Individuals with the non-insert carrying Met-129 allele in addition to the pathogenic PRNP mutant allele died at a significantly earlier age than those that possess the Val-129 allele. It is not known whether the codon 129 polymorphisms influence the rate of prion replication, as do the mouse *Sinc/Prn-i* alleles, or whether the PRNP polymorphisms act in analogy with the PrP allotype barrier by determining the efficiency with which prion replication is initiated.

CONCLUSION

Although genetic linkage between the prion incubation time gene and *Prn-p* in mice provided the first evidence indicating that PrP might be involved in genetic susceptibility to prion diseases, formal proof that *Prn-p* is equivalent to *Sinc/Prn-i* is lacking. However, it appears likely that control of scrapie incubation time is a pleiotropic effect of *Prn-p*. The strongest evidence for a distinct incubation time locus came from segregating crosses which produced rare offspring whose scrapie incubation time phenotype and *Prn-p* genotype were discordant. As discussed here, however, it is unlikely that these discordant mice were meiotic recombinants between *Prn-p* and *Sinc/Prn-i*. Results from experiments designed to capture *Prn-p–Sinc/Prn-i* recombinants suggested cumulative effects of genes unlinked to *Prn-p* could override *Prn-p* genotype and provided a more plausible explanation than *Prn-p–Sinc/Prn-i* crossovers for the discordant mice observed in earlier crosses. Identification of these modifier genes that are not linked to *Prn-p* could provide important new information on scrapie pathogenesis and the mechanisms of prion replication. Such genes also might account for rare individuals carrying pathogenic PRNP mutations who fail to develop disease.

'Scrapie incubation time' is the cumulative result of a variety of processes including initiation of prion replication, the rate of prion replication, and accumulation of neurological damage in 'clinical target areas'. It seems likely that interaction between PrP^C and PrP^{Sc} is involved in prion replication as indicated by the PrP allotype barrier and the species barrier to scrapie transmission; interaction between PrP^C and PrP^{Sc} that share the same primary structure appears to be most efficient in producing new PrP^{Sc} molecules. The role of PrP primary structure in controlling the rate of prion replication when PrP^{Sc} in the agent is homologous with host PrP^C is less clear. Transgenic mice expressing the *Prn-p^b* allele from long incubation time mice were constructed to address this question, but had shorter incubation times than non-Tg mice. This result does not provide evidence for a distinct incubation time locus; other factors, including shortening of incubation time due to PrP overexpression, may override an effect of PrP primary structure on the rate of prion replication. Definitive evidence that PrP primary structure determines the rate of prion replication might be provided through targeting the $Prn-p^b$ mutations to an endogenous $Prn-p^a$ allele. Alternatively, although gross differences in $Prn-p$ expression are not found between long and short incubation time mice (Westaway et al., 1987), subtle quantitative or regional differences in PrP expression might account for the different incubation time alleles. Elucidation of the mechanisms responsible for the control of scrapie incubation time in mice will provide new clues to the nature of prion replication, and shed new light on the most fascinating problem in prion biology—the biochemical basis for prion-specified information (Weissmann, 1991).

ACKNOWLEDGEMENTS

This work was supported by grants NS14069 and NS22786 from the National Institutes of Health, USPHS. We also thank M. Scott, C. Mirenda, M. Torchia, C. Ebeling, P. Hoppe, D. Foster and S.J. DeArmond for their vital contributions to these studies.

REFERENCES

Balling, R., Deutsch, U. and Gruss, P. (1988) Undulated, a mutation affecting the development of the mouse skeleton, has a point mutation in the paired box of Pax-1. *Cell* **55** 531–535.

Basler, K., Oesch, B., Scott, M., Westaway, D., Wächli, M., Groth, D.F., McKinley, M.P., Prusiner, S.B. and Weissmann, C. (1986) Scrapie and cellular PrP isoforms are encoded by the same chromosomal gene. *Cell* **46** 417–428.

Bruce, M.E., McConnell, I., Fraser, H. and Dickinson, A.G. (1991) The disease characteristics of different strains of scrapie in *Sinc* congeneic mouse lines: implications for the nature of the agent and host control of pathogenesis. *J. Gen. Virol.* **72** 595–603.

Carlson, G.A., Kingsbury, D.T., Goodman, P.A., Coleman, S., Marshall, S.T., DeArmond, S., Westaway, D. and Prusiner, S.B. (1986) Linkage of prion protein and scrapie incubation time genes. *Cell* **46** 503–511.

Carlson, G.A., Goodman, P.A., Lovett, M., Taylor, B.A., Marshall, S.T., Peterson-Torchia, S.T., Westaway, D. and Prusiner, S.B. (1988a) Genetics and polymorphism of the mouse prion gene complex: control of scrapie incubation time. *Mol. Cell Biol.* **8** 5528–5540.

Carlson, G.A., Westaway, D., Goodman, P.A., Peterson, M., Marshall, S.T. and Prusiner, S.B. (1988b) Genetic control of prion incubation period in mice. *Ciba Found. Symp.* **135** 84–100.

Carlson, G.A., Westaway, D., DeArmond, S.J., Peterson-Torchia, M. and Prusiner, S.B. (1989) Primary structure of prion protein may modify scrapie isolate properties. *Proc. Natl. Acad. Sci. USA* **86** 7475–7479.

Chandler, R.L. (1961) Encephalopathy in mice produced by inoculation with scrapie brain material. *Lancet* **i** 1378–1379.

Dickinson, A.G. and MacKay, J.M.K. (1964) Genetical control of the incubation period in mice of the neurological disease, scrapie. *Heredity* **19** 279–288.

Dickinson, A.G. and Meikle, V.M. (1969) A comparison of some biological characteristics of the mouse-passaged scrapie agents, 22A and ME7. *Genet. Res.* **13** 213–225.

Dickinson, A.G. and Meikle, V.M.H. (1971) Host-genotype and agent effects in scrapie incubation: change in allelic interaction with different strains of agent. *Mol. Gen. Genet.* **112** 73–79.

Dickinson, A.G. and Outram, G.W. (1979) The scrapie replication-site hypothesis and its implications for pathogenesis. In: Prusiner, S.B. and Hadlow, W.J. (eds) *Slow Transmissible Diseases of the Nervous System*, Vol. 2. Academic Press, New York, pp. 13–31.

Dickinson, A.G., Meikle, V.M.H. and Fraser, H.G. (1968) Identification of a gene which controls the incubation period of some strains of scrapie agent in mice. *J. Comp. Pathol.* **78** 293–299.

Eissenberg, J.C. and Elgin, S.C.R. (1991) Boundary functions in the control of gene expression. *TIG* **7** 335–340.

Goldgaber, D. (1991) Anticipating the anti-prion protein? *Nature (London)* **351** 106.

Hewinson, R.G., Lowings, J.P., Dawson, M.D. and Woodward, M.J. (1991) Anti-prions and other agents (letter). *Nature (London)* **352** 291.

Hsiao, K. and Prusiner, S. B. (1989) Inherited human prion diseases. *Ann. Neurol.* **26** 137.

Hsiao, K., Baker, H.F., Crow, T.J., Poulter, M., Owen, F., Terwilliger, J.D., Westaway, D., Ott, J. and Prusiner, S.B. (1989) Linkage of a prion protein missense variant to Gerstmann–Sträussler syndrome. *Nature (London)* **338** 342–345.

Hsiao, K., Scott, M., Foster, D., Groth, D.F., DeArmond, S.J. and Prusiner, S.B. (1990) Spontaneous neurodegeneration in transgenic mice with mutant prion protein. *Science* **250** 1587–1590.

Hunter, N., Hope, J., McConnell, I. and Dickinson, A.G. (1987) Linkage of the scrapie-associated fibril protein (PrP) gene and Sinc using congenic mice and restriction fragment length polymorphism analysis. *J. Gen. Virol.* **68** 2711–2716.

Kingsbury, D.T., Kasper, K.C., Stites, D.P., Watson, J.C., Hogan, R.N. and Prusiner, S.B. (1983) Genetic control of scrapie and Creutzfeldt–Jakob disease in mice. *J. Immunol.* **131** 491–496.

Manson, J. and Hope, J. (1991) Anti-prions and other agents (letter). *Nature (London)* **352** 291.

Oesch, B., Westaway, D., Wächli, M., McKinley, M.P., Kent, S.B.H., Aebersold, R., Barry, R.A., Tempst, P., Teplow, D.B., Hood, L., Prusiner, S.B. and Weissmann, C. (1985) A cellular gene encodes scrapie PrP 27-30 protein. *Cell* **40** 735–746.

Palmer, M.S., Dryden, A.J., Hughes, J.T. and Collinge, J. (1991) Homozygous prion protein genotype predisposes to sporadic Creutzfeldt–Jakob disease. *Nature (London)* **352** 340–342.

Prusiner, S.B., Scott, M., Foster, D., Pan, K., Groth, D., Mirenda, C., Torchia, M., Yang, S., Serban, D., Carlson, G.A., Hoppe, P.C., Westaway, D. and DeArmond, S. J. (1990) Transgenetic studies implicate interactions between homologous PrP isoforms in scrapie prion replication. *Cell* **63** 673–686.

Race, R.E., Graham, K., Ernst, D., Caughey, B. and Chesebro, B. (1990) Analysis of linkage between scrapie incubation period and the prion protein gene in mice. *J. Gen. Virol.* **71** 493–497.

Roderick, T.H. and Guidi, J.N. (1989) Strain distribution of polymorphic variants. In: Lyon, M.F. and Searle, A.G. (eds) *Genetic Variants and Strains of the Laboratory Mouse.* Oxford University Press, Oxford, pp. 663–772.

Scott, M., Foster, D., Mirenda, C., Serban, D., Coufal, F., Wächli, M., Torchia, M., Groth, D., Carlson, G., DeArmond, S.J., Westaway, D. and Prusiner, S.B. (1989) Transgenic mice expressing hamster prion protein produce species-specific scrapie infectivity and amyloid plaques. *Cell* **59** 847–857.

Weissmann, C. (1991) A 'unified theory' of prion propagation. *Nature (London)* **352** 679–683.

Westaway, D., Goodman, P.A., Mirenda, C.A., McKinley, M.P., Carlson, G.A. and Prusiner, S.B. (1987) Distinct prion proteins in short and long scrapie incubation period mice. *Cell* **51** 651–662.

Westaway, D., Mirenda, C.A., Foster, D., Zebarjadian, Y., Scott, M., Torchia, M., Yang, S.L., Serban, H., DeArmond, S.J., Ebeling, C., Prusiner, S.B. and Carlson, G.A. (1991) Paradoxical shortening of scrapie incubation times by expression of prion protein transgenes derived from long incubation time mice. *Neuron* **7** 59–68.

43

How can the 'protein only' hypothesis of prion propagation be reconciled with the existence of multiple prion strains?

Charles Weissmann

Prusiner has proposed that the prion, the agent that causes transmissible spongiform encephalopathies such as scrapie or bovine spongiform encephalopathy in animals, or kuru, Creutzfeldt–Jakob disease (CJD) and Gerstmann–Sträussler–Scheinker (GSS) disease in man, is devoid of nucleic acid and identical with PrP^{Sc}, a modified form of PrP^C (Prusiner, 1989; Prusiner and DeArmond, 1990; Prusiner, 1991). PrP^C is a normal host protein (Oesch *et al.*, 1985; Chesebro *et al.*, 1985; Hope *et al.*, 1986) encoded within a single exon of a single copy gene (Basler *et al.*, 1986) and is found predominantly on the surface of neurons, attached by a phospholidylinositol glycolipid anchor (Prusiner, 1989; Prusiner and DeArmond, 1990; Prusiner, 1991; Stahl *et al.*, 1987). PrP^{Sc}, in contrast to PrP^C, is relatively resistant to protease (Oesch *et al.*, 1985) and accumulates intracellularly, in cytoplasmic vesicles (Taraboulos *et al.*, 1990; McKinley *et al.*, 1990).

Prusiner suggested that PrP^{Sc}, when introduced into a normal cell, propagates by causing the conversion of PrP^{Sc} or its precursor into PrP^{Sc} (Oesch *et al.*, 1985, 1988; Prusiner *et al.*, 1990; Prusiner, 1991; Bolton and Bendheim, 1988). The nature of the conversion is unknown and could be due to a chemical or conformational modification, during or after its synthesis. However, the existence of many different strains of scrapie which can be propagated in one and the same inbred mouse line and the apparent mutability of the agent (Bruce and Dickinson, 1987) are better explained by the updated nucleoprotein or virino hypothesis, which holds that the infectious agent consists of a nucleic acid genome and the host-derived PrP, which is recruited

PrionModel910620.Fig(long)

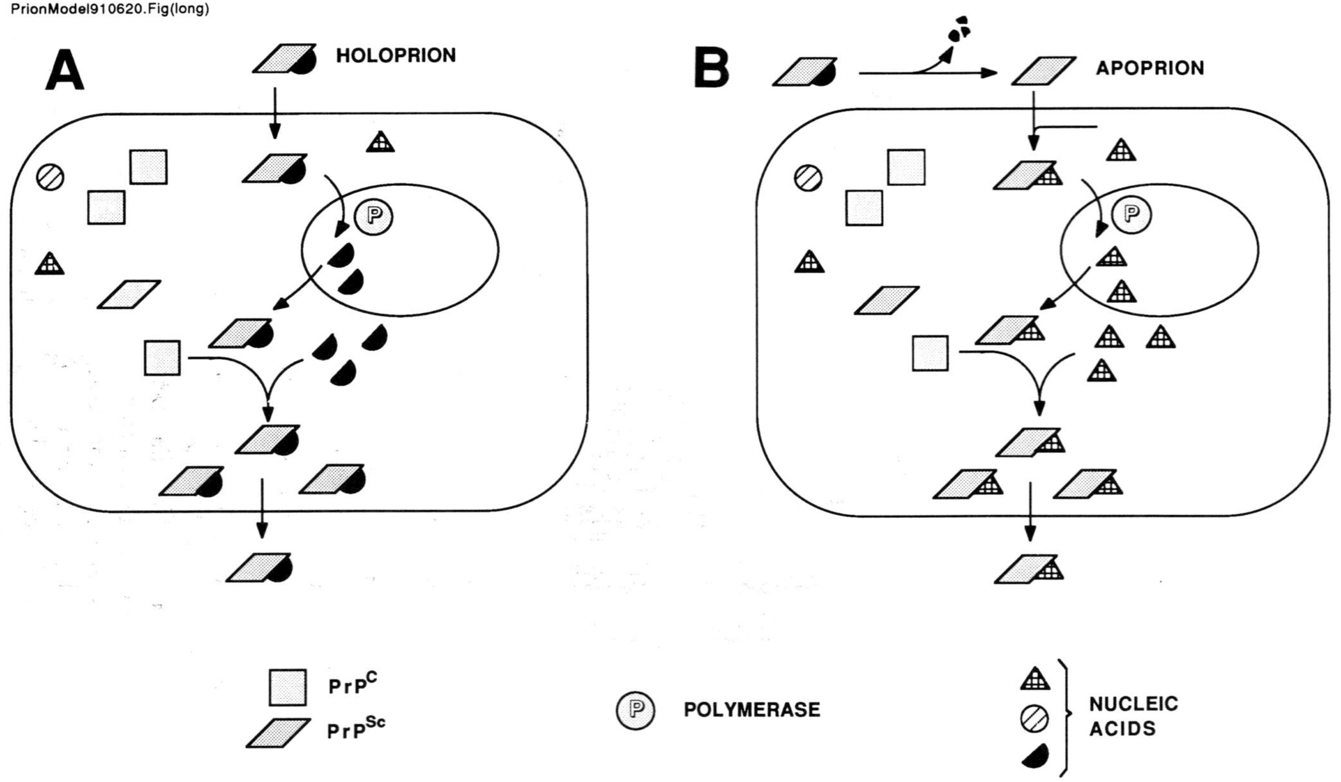

Fig. 1. A model for prion propagation. (A) the 'holoprion', consisting of PrP^{Sc} (the 'apoprion') and a nucleic acid (the 'coprion'), invades the cell. Conversion of PrP^C into PrP^{Sc} is mediated by PrP^{Sc}, the holoprion or the coprion, and involves either a chemical modification or a conformational change. The coprion nucleic acid is replicated by a cellular polymerase, a process which is promoted by or dependent on its association with PrP^{Sc}. PrP^{Sc} binds coprion nucleic acid and reforms holoprions. (B) The holoprion is subjected to nuclease treatment and/or purification, to yield PrP^{Sc}, the apoprion. After penetrating the cell, the apoprion may associate with a potential coprion, namely a cellular nucleic acid to which it has affinity and which can be replicated by a cellular polymerase in the presence of PrP^{Sc}. The resulting holoprion may have phenotypic properties which differ from those of the original holoprion.

as some sort of coat (Oesch *et al.*, 1985; Dickinson and Outram, 1988); however, no evidence for such a nucleic acid has yet been adduced (Aiken and Marsh, 1990; Meyer *et al.*, 1991; Oesch *et al.*, 1988; Prusiner, 1991).

I have proposed the 'unified theory' of prion propagation to reconcile the diametrically opposed views outlined above (Weissmann, 1991a). According to this theory, PrP^{Sc} by itself, the 'apoprion', can indeed initiate the pathogenic process, but the phenotypic properties which define strain differences are ascribed to a 'coprion', a nucleic acid which is usually associated with PrP^{Sc} but can also be recruited or exchanged within the host cell. The theory explains why scrapie agent preparations appear resistant to ionizing and non-ionizing radiation and nucleolytic agents: destruction or removal of nucleic acids should not affect infectivity as such, but only phenotypic features which were not scored in the inactivation experiments. Conversely, characterization of strains seems to have always been carried out with crude brain extracts, in which the conjectured phenotype-determining nucleic acid would naturally be present.

The 'unified theory' accepts the premise that the propagation of PrP^{Sc} occurs by the conversion of PrP^C or its (perhaps nascent) precursor into a replica of itself, be it by a covalent modification or by a conformational change. This conversion could be caused by PrP^{Sc} (with or without associated coprion), essentially as proposed by Prusiner and his colleagues (Prusiner *et al.*, 1990; Prusiner, 1991), or by association with the coprion. In this connection it is of particular interest to note that wild-type and mutant p53 have a different conformations, and that cotranslation of the two proteins forces the wild-type protein into the mutant conformation (Milner and Medcalf, 1991). Also, as pointed out recently (Hurst, 1991), the ATP-dependent assembly of hsp60 subunits into the 14-mer chaperonin complex, which is strictly dependent on pre-existing chaperonin complex (Cheng *et al.*, 1990), may serve as a paradigm for a catalysed conformational modification of PrP^C. Other examples of self-controlled self-assembly are set forth and analysed by Caspar (1991).

The conjectured coprion is most likely a nucleic acid, because it must be susceptible to replication and mutation. If it were essential for pathogenesis and propagation and not only involved in modulating the inherent properties of PrP^{Sc}, it would have to be present in all organisms that can be infected by highly purified PrP^{Sc} devoid of nucleic acid or which can develop prion disease without exogenous infection. Nucleic acid capable of serving as coprion may be either encoded by the 'normal' host genome and thus be ubiquitious, or acquired, subsisting as a widely distributed intracellular commensal. Whatever its origin, the coprion nucleic acid must be susceptible to direct replication in the cell, because otherwise different scrapie strains could not be maintained in the same inbred species. The nucleic acid could be an episomal DNA or an RNA; no estimate can be made regarding its size, because all target size determinations done so far have been scored on the basis of infectivity and not of phenotypic properties. Because coprion-related nucleic acid would be present in normal cells at some level, it may not be easily detected by differential hybridization or by screening a subtraction library.

How could a coprion introduced into a host cell be amplified? A DNA molecule might be replicated by a nuclear or mitochondrial DNA polymerase. If the coprion is an RNA, several possibilities may be envisaged. It has been claimed that normal mammalian cells contain an RNA replicase; however, the data are not compelling (Volloch, 1986; Volloch *et al.*, 1987). Retrotranscription and integration of the resulting DNA, followed by transcription, is a conceivable mechanism (Manuelidis *et al.*, 1988; Murdoch *et al.*, 1990). The replication of an RNA by DNA-dependent RNA polymerase is an intriguing possibility in view of the findings that viroid RNA in plant cells (Rackwitz *et al.*, 1981; Robertson and Branch, 1987; Semancik and Harper, 1984) and most likely hepatitis delta RNA in mammalian cells (Kuo *et al.*, 1989; Taylor, 1990) can be replicated by DNA-dependent RNA polymerase. Moreover, *Escherichi coli* DNA-dependent RNA polymerase (Biebricher and Orgel, 1973) and phage T7 DNA-dependent RNA polymerase (Konarska and Sharp, 1989, 1990) are able to replicate certain RNA species. A specific interaction between such RNA molecules and the polymerase must be postulated, because not any RNA can serve as template, and an RNA which can be replicated by one kind of DNA-dependent RNA polymerase is not necessarily replicated by another (Konarska and Sharp, 1989, 1990). This suggests that within the cell RNA variants may compete for the polymerase, leading to selection and amplification of particularly adapted RNA species, If a coprion RNA introduced into a cell is particularly suitable for intracellular replication, by virtue of selection in a preceding host, it will outcompete endogenous RNA species and provide a pool of coprions which can associate with the newly formed PrP^{Sc}. If PrP^{Sc} devoid of RNA is introduced into the cell, endogenous RNA may be recruited as coprion, and mutation and selection may render it more proficient for its role as coprion and template for replication. I postulate that the presence of PrP^{Sc} is required to promote the replication of the coprion RNA, just as the presence of the hepatitis delta antigen is required for the replication of the hepatitis delta RNA by a cellular enzyme (Glenn *et al.*, 1990).

Because the conjectured RNA that can assume the role of a coprion may be different in different animals, the passaging of holoprions through different hosts may occasionally lead to an exchange of the nucleic acid and thus to the appearance of different strains. Mutational events at the level of RNA synthesis followed by selection could of course also lead to the strain shifts reported in the literature. Within the framework of the 'protein only' hypothesis, the appearance of strains is explained by novel conformations resulting from the interaction of mismatched PrP^{C} and PrP^{Sc} (S. B. Prusiner, personal communication).

It is interesting to remember in this connection that the phenotypic consequences of many RNA virus infections are modulated by RNA molecules which are derived from the viral RNA or arise from unknown sources, so-called defective interfering (DI) RNAs (Perrault, 1981; Barrett and Dimmock, 1986; McLain *et al.*, 1988) and satellite RNAs (Francki, 1985; Gadani *et al.*, 1990), respectively. These are replicated in the infected host cells, where they may 'evolve' quite rapidly and adapt to and interfere with the viral RNA replication machinery (Giachetti and Holland, 1989).

It has been shown that mice inoculated first with a prion strain with a long incubation time, and then by one with a short incubation time, give rise to the strain injected first (Dickinson *et al.*, 1972, 1975; Dickinson and Outram, 1979; Kimberlin and Walker, 1985, 1988; Kimberlin, 1990). This phenomenon has been explained by the 'limited replication site' model, which proposes that a rate-limiting site or process is fully occupied by the first strain to invade the host, so that the second strain is deprived of a critical step in replication. In terms of the 'unified theory', the nucleic acid introduced with the first scrapie strain would be amplified extensively by the time the second strain is introduced, and the nucleic acid introduced with the second strain would not be able to compete with that of the first.

The 'unified theory' makes several testable predications, the first of which may serve to falsify it.

(1) *Strain-specific properties of prions are lost following removal of nucleic acids.* If preparations of two distinct scrapie strains are extensively purified and the nucleic acids effectively eliminated or destroyed, then the phenotypic properties of the two strains, such as incubation or lesion pattern, should revert to the same type in one and the same host strain.

(2) *The strain-specific properties of a prion can be changed by replacing the nucleic acid associated with it by that derived from another strain.* If PrP^{Sc} from a scrapie strain A, purified free of nucleic acid, is mixed with the deproteinized nucleic acid fraction of brain extract from an animal infected with a scrapie strain B, the resulting preparation should give rise to strain B when inoculated into animals. The reciprocal results should be obtained in the converse experiment.

(3) *PrP^{Sc} should specifically bind certain kinds of nucleic acids.* Purified PrP^{Sc} preparations devoid of nucleic acids should preferentially bind some species of radioactive nucleic acids from scrapie-infected brain or partially purified PrP^{Sc} preparations.

The 'unified theory' incorporates most features of the 'protein only' hypothesis and therefore adopts the explanations offered by the latter for the species barrier phenomenon and for the origin of sporadic and genetically determined prion diseases. The theory accepts the premise of the virino hypothesis that a nucleic acid is responsible for some phenotypic features and is replicated in the cell; however, it denies that this nucleic acid is required for infectivity and that it represents an independent, agent-specific genome.

I propose the 'unified theory' in order to reconcile apparently contradictory conclusions regarding on the one hand the requirement for a nucleic acid, invoked to explain the existence of scrapie strains, and on the other the apparent absence of nucleic acid in prions. However, the premises on which these conclusions are based are neither conclusive nor undisputed. Despite the fact that it seems unlikely, PrP^{Sc} derived from one and the same precursor might indeed assume many different conformations, making the 'unified theory' unnecessary. A small nucleic acid, required for infectivity, may yet be found. Less likely, the data interpreted as proof for the existence of strains may be explained otherwise. The suggestion that PrP^{Sc} is a conformational isomer of PrP^C is based on the failure to find a chemical difference between the two forms, but the infectious entity, even if it is derived from PrP^C, need not be PrP^{Sc}, as defined by its physical properties (Weissmann, 1991b), and could be chemically different, in regard either to its primary structure or to post-translational modifications. It has also been suggested that different 'strains' of PrP^{Sc} may come about if a particular modification, such as glycosylation, is specifically imparted by certain cells within the host, and this modification targets the resulting 'strain' specifically to the same type of cells, where PrP^{Sc} with the same modification is again generated. This too is a hypothesis which is currently not supported by experimental evidence, but would reconcile the 'protein only' hypothesis with the existence of multiple strains. Mechanisms which might generate variants of PrP differing in their amino acid sequence include editing at the mRNA level (Scott, 1989) or specific mistranslation (Wills, 1989); to explain stable scrapie strains, these variations would have to be steered specifically by the individual PrP species, for which a novel mechanism has to be invoked.

In closing, I would like to stress that the theory outlined in this article is based entirely on biochemical reactions for which precedents are known, even though the overall combination of these steps is unusual.

ACKNOWLEDGEMENT

I am grateful to Stan Prusiner for many productive discussions.

REFERENCES

Aiken, J.M. and Marsh, R.F. (1990) The search for scrapie agent nucleic acid. *Microbiol. Rev.* **54** 242–246.

Barrett, A.D. and Dimmock, N.J. (1986) Defective interfering viruses and infections of animals. *Curr. Top. Microbiol. Immunol.* **128** 55–84.

Basler, K., Oesch, B., Scott, M., Westaway, D., Wälchli, M., Groth, D.F., McKinley, M.P., Prusiner, S.B. and Weissmann, C. (1986) Scrapie and cellular PrP isoforms are encoded by the same chromosomal gene. *Cell* **46** 417–428.

Biebricher, C.K. and Orgel, L.E. (1973) An RNA that multiplies indefinitely with DNA dependent RNA polymerase selection from a random co polymer. *Proc. Natl. Acad. Sci. USA* **70** 934-938.

Bolton, D.C. and Bendheim, P.E. (1988) A modified host protein model of scrapie. In: Bock, G. and Marsh, J. (eds) *Novel Infectious Agents and the Central Nervous System.* Wiley, Chichester, pp. 164–177.

Bruce, M.E. and Dickinson, A.G. (1987) Biological evidence that scrapie agent has an independent genome. *J. Gen. Virol.* **68** 79–89.

Caspar, D.L.D. (1991) Self-control of self-assembly. *Curr. Biol.* **1** 30–32.

Cheng, M.Y., Hartl, F.-U. and Horwich, A.L. (1990) The mitchondrial chaperonin hsp60 is required for its own assembly. *Nature (London)* **348** 455–458.

Chesebro, B., Race, R., Wehrly, K., Nishio, J., Bloom, M., Lechner, D., Bergstrom, S., Robbins, K., Mayer, L., Keith, J.M., Garon, C. and Haase, A. (1985) Identification of scrapie prion protein-specific messenger RNA in scrapie-infected and uninfected brain. *Nature (London)* **315** 331–333.

Dickinson, A.G. and Outram, G.W. (1979) The scrapie replication-site hypothesis and its implications for pathogenesis. In: Prusiner, S.B. and Hadlow, W.J. (eds) *Slow Transmissible Diseases of the Nervous System,* Vol. 1. Academic Press, New York, pp. 13–31.

Dickinson, A.G. and Outram, G.W. (1988) Genetic aspects of unconventional virus infections: the basis of the virino hypothesis. *Ciba Found. Symp.* **135** 63–83.

Dickinson, A.G., Fraser, H., Meikle, V.M.H. and Outram, G.W. (1972) Competition between different scrapie agents in mice. *Nature (London)* **237** 244–245.

Dickinson, A.G., Fraser, H., McConnell, I., Outram, G.W., Sales, D.I. and Taylor, D.M. (1975) Extraneural competition between different scrapie agents leading to loss of infectivity. *Nature (London)* **253** 556.

Francki, R.I. (1985) Plant virus satellites. *Annu. Rev. Microbiol.* **39** 151–174.

Gadani, F., Mansky, L.M., Medici, R., Miller, W.A. and Hill, J.H. (1990) Genetic engineering of plants for virus resistance. *Arch. Virol.* **115** 1–22.

Giachetti, C. and Holland, J.J. (1989) Vesicular stomatitis virus and its defective interfering particles exhibit *in vitro* transcriptional and replicative competition for purified L-NS polymerase molecules. *Virology* **170** 264–267.

Glenn, J.S., Taylor, J.M. and White, J.M. (1990) *In-vitro*-synthesized hepatitis delta virus RNA initiates genome replication in cultured cells. *J. Virol.* **64** 3104-3107.

Hope, J., Morton, L.J., Farquhar, C.F., Multhaup, G., Beyreuther, K. and Kimberlin, R.H. (1986) The major polypeptide of scrapie-associated fibrils (SAF) has the same size, charge distribution and *N*-terminal protein sequence as predicted for the normal brain protein (PrP). *EMBO J* **5** 2591–2597.

Hurst, L.D. (1991) Prion infection. *Nature (London)* **351** 21.

Kimberlin, R.H. (1990) Scrapie and possible relationship with viroids *Semin. Virol.* **1** 153–162.

Kimberlin, R.H. and Walker, C.A. (1985) Competition between strains of scrapie depends on the blocking agent being infectious. *Intervirology* **23** 74–81.

Kimberlin, R.H. and Walker, C.A. (1988) Pathogenesis of experimental scrapie. In: Bock, G. and Marsh, J. (eds) *Novel Infectious Agents and the Nervous System.* pp. 37–62.

Konarska, M.M. and Sharp, P.A. (1989) Replication of RNA by the DNA-dependent RNA polymerase of phage T7. *Cell* **57** 423–431.

Konarska, M.M. and Sharp, P.A. (1990) Structure of RNAs replicated by the DNA-dependent T7 RNA polymerase. *Cell* **63** 609–618.

Kuo, M.Y.P., Chao, M. and Taylor, J. (1989) Initiation of replication of the human hepatitis delta virus genome from cloned DNA: role of delta antigen. *J. Virol.* **63** 1945–1950.

Manuelidis, L., Murdoch, G. and Manuelidis, E.E. (1988) Potential involvement of retroviral elements in human dementias. *Ciba Found. Symp.* **135** 117–134.

McKinley, M.P., Taraboulos, A., Kenaga, L., Serban, D., DeArmond, S.J., Stieber, A., Prusiner, S.B. and Gonatas, N. (1990) Ultrastructural localization of scrapie prion proteins in secondary lysosomes of infected cultured cells. *J. Cell Biol.* **111** (5, part 2) 316a.

McLain, L., Armstrong, S.J. and Dimmock, N.J. (1988) One defective interfering particle per cell prevents influenza virus-mediated cytopathology: an efficient assay system. *J. Gen. Virol.* **69** 1415–1419.

Meyer, N., Rosenbaum, V., Schmidt, B., Gilles, K., Mirenda, C., Groth, D., Prusiner, S.B. and Riesner, D. (1991) Search for a putative scrapie genome in purified prion fractions reveals a paucity of nucleic acids. *J. Gen. Virol.* **72** 37–49.

Milner, J. and Medcalf, E.A. (1991) Cotranslation of activated mutant p53 with wild type drives the wild type p53 protein into the mutant conformation. *Cell* **65** 766–774.

Murdoch, G.H., Sklaviadis, T., Manuelidis, E.E. and Manuelidis, L. (1990) Potential retroviral RNAs in Creutzfeldt–Jakob disease. *J. Virol.* **64** 1477–1486.

Oesch, B., Westaway, D., Walchli, M. *et al.* (1985) A cellular gene encodes scrapie PrP 27-30 protein. *Cell* **40** 735–746.

Oesch, B., Groth, D.F., Prusiner, S.B. and Weissmann, C. (1988) Search for a scrapie-specific nucleic acid: a progress report. *Ciba Found. Symp.* **135** 209–223.

Perrault, J. (1981) Origin and replication of defective interfering particles. *Curr. Top. Microbiol. Immunol.* **93** 151–207.

Prusiner, S.B. (1989) Scrapie prions. *Annu. Rev. Microbiol.* **43** 345–374.

Prusiner, S.B. (1991) Molecular biology of prion diseases. *Science* **252** 1515–1522.

Prusiner, S.B. and DeArmond, S.J. (1990) Prion diseases of the central nervous system. *Monogr. Pathol.* **32** 86–122.

Prusiner, S.B., Scott, M., Foster, D. *et al.* (1990) Transgenetic studies implicate interactions between homologous PrP isoforms in scrapie prion replication. *Cell* **63** 673–686.

Rackwitz, H.R., Rohde, W. and Saenger, H.L. (1981) DNA dependent RNA polymerase II of plant origin transcribes viroid RNA into full length copies. *Nature (London)* **291** 297–301.

Robertson, H.D. and Branch, A.D. (1987) The viroid replication process. In: Semancik, J. S. (ed.) *Viroids and Viroid-like Pathogens.* CRC Press, Boca Raton, FL, pp. 49–69.

Scott, J. (1989) Messenger RNA editing and modification. *Curr. Opinion Cell Biol.* **1** 1141–1147.

Semancik, J.S. and Harper, K.L. (1984) Optimal conditions for cell-free synthesis of citrus exocortis viroid and the question of specificity of RNA polymerase activity. *Proc. Natl. Acad. Sci. USA* **81** 4429–4433.

Stahl, N., Borchelt, D.R., Hsiao, K. and Prusiner, S.B. (1987) Scrapie prion protein contains a phosphatidylinositol glycolipid. *Cell* **51** 229–240.

Taraboulos, A., Serban, D. and Prusiner, S. B. (1990) Scrapie prion proteins accumulate in the cytoplasm of persistently infected cultured cells. *J. Cell Biol.* **110** 2117–2132.

Taylor, J.M. (1990) Hepatitis delta virus: cis and trans functions required for replication. *Cell* **61** 371–373.

Volloch, V. (1986) Cytoplasmic synthesis of globin RNA in differentiated murine erythroleukemia cells: possible involvement of RNA-dependent RNA polymerase. *Proc. Natl. Acad. Sci. USA* **83** 1208–1212.

Volloch, V., Schweitzer, B. and Rits, S. (1987) Synthesis of globin RNA in enucleated differentiating murine erythroleukemia cells. *J. Cell Biol.* **105** 137–143.

Weissmann, C. (1991a) A 'unified theory' of prion propagation. *Nature (London)* **352** 679–683.

Weissmann, C. (1991b) Spongiform encephalopathies. The prion's progress. *Nature (London)* **349** 569–571.

Wills, P.R. (1989) Induced frameshifting mechanism of replication for an information-carrying scrapie prion. *Microbial Pathogen.* **6** 235–249.

Part VIII   Overview of prion research

44

Prion biology

Stanley B. Prusiner

ABSTRACT

Many advances in our knowledge of the transmissible pathogens causing scrapie and other transmissible neurodegenerative diseases over the past decade support the hypothesis that these pathogens are novel and different from both viroids and viruses. After convincing evidence was obtained showing that scrapie infectivity depends upon a protein component (Prusiner *et al.*, 1981), the term 'prion' was introduced to distinguish these infectious pathogens from others including viroids and viruses (Prusiner, 1982). Enriching fractions from Syrian hamster (SHa) brain for scrapie prion infectivity led to the discovery of the prion protein (PrP) (Bolton *et al.*, 1982). Determination of the *N*-terminal sequence of the protease resistance core of PrP (Prusiner *et al.*, 1984) permitted retrieval of molecular clones encoding PrP from cDNA libraries (Oesch *et al.*, 1985; Chesebro *et al.*, 1985). The finding of PrP mRNA in uninfected tissues led to discovery of the normal PrP isoform denoted PrP^C (Oesch *et al.*, 1985). Deciphering the role of the abnormal PrP isoform designated PrP^{Sc} is a central focus of research on prion diseases which include scrapie of sheep, bovine spongiform encephalopathy of cattle, as well as Creutzfeldt–Jakob disease (CJD) and Gerstmann–Sträussler–Scheinker (GSS) syndrome of humans. Transgenic (Tg) mice expressing both SHa and mouse (Mo) PrP genes were used to probe the molecular basis of the species barrier and the mechanism of scrapie prion replication. Four Tg lines expressing SHaPrP exhibited distinct incubation times ranging from 48 to 277 days after SHa prion inoculation, which were inversely correlated with the steady-state levels of SHaPrP mRNA and SHaPrP^C (Prusiner *et al.*, 1990). Bioassays of brain extracts from two scrapie-infected Tg lines showed that the prion inoculum dictates which prions are synthesized *de novo*, even though the cells express both PrP genes. Tg mice inoculated with SHa prions had $\sim 10^9$ ID_{50} units of SHa prions per gram of brain while <10 units of Mo prions were found. Conversely, Tg mice inoculated with Mo prions had $\sim 10^6$ ID_{50} units of Mo prions and <10 units of SHa prions. Our results argue that the species barrier for scrapie prions resides in the primary structure of PrP (Scott *et al.*, 1989) and formation of infectious prions is initiated by a species-specific interaction between PrP[Sc] in the inoculum and homologous PrP[C]. Studies on Syrian, Armenian and Chinese hamsters suggest that the domain of the PrP molecule between codons 100 and 120 controls both the length of the incubation time and the deposition of PrP in amyloid plaques (Lowenstein *et al.*, 1990). Ataxic GSS in families was found to be linked genetically to a mutation in the PrP gene leading to the substitution of Leu for Pro at codon 102 (Hsiao *et al.*, 1989a). Discovery of point mutations in the PrP gene of humans with GSS syndrome or familial CJD established that prion diseases are unique among human illnesses—they are both genetic and infectious. Tg mice expressing MoPrP with the GSS point mutation spontaneously develop neurologic dysfunction, spongiform degeneration and astrocytic gliosis (Hsiao *et al.*, 1990). Inoculation of brain extracts prepared from these Tg(GSSMoPrP) mice into Syrian hamsters has produced neurodegeneration in recipient animals after prolonged incubation times (Hsiao *et al.*, 1991c). If convincing data on serial passage of prions from the inoculated recipients can be obtained, then these results will argue that prions are devoid of foreign nucleic acid. Studies of inherited prion diseases have revised thinking about sporadic CJD, suggesting it may arise from a somatic mutation. Pulse-chase radiolabelling experiments of scrapie-infected cultures of mouse neuroblastoma cells indicate that protease-resistant PrP[Sc] is synthesized during the chase period with $t_{1/2} \sim 1\text{--}3\,\text{h}$ from a protease-sensitive precursor (Borchelt *et al.*, 1990), consistent with the conclusion that PrP[C] and PrP[Sc] differ as a result of a post-translational event (Basler *et al.*, 1986). The acquisition of PrP protease resistance in scrapie-infected cultured cells was found to be independent of Asn-linked glycosylation (Taraboulos *et al.*, 1990a). Neither tunicamycin nor mutation of Asn-linked glycosylation sites prevented PrP[Sc] formation. PrP[C] is bound to external surface of cells by a glycoinositol phospholipid anchor (Stahl *et al.*, 1990a, b). In contrast, PrP[Sc] accumulates within cytoplasmic vesicles of cultured cells (Taraboulos *et al.*, 1990b; McKinley *et al.*, 1991b). Immunoaffinity chromatography with monoclonal antibodies to PrP 27-30 demonstrated copurification of PrP[Sc] and scrapie infectivity (Gabizon *et al.*, 1988b). The foregoing results assert that PrP[Sc] is a component of the transmissible particle, and the PrP amino acid sequence controls the neuropathology and species specificity of prion infectivity. Attempts to demonstrate a scrapie-specific nucleic acid within highly purified preparations of prions have been unrewarding to date (Meyer *et al.*, 1991). These results are in accord with the preliminary findings noted above that brain extracts prepared from Tg(GSSMoPrP) mice with spontaneous neurodegeneration transmit CNS disease to inoculated recipients and with many unsuccessful attempts to inactivate prion infectivity by procedures that specifically hydrolyse or modify nucleic acids (Prusiner, 1991). Although it seems likely that transmissible prions are composed of PrP[Sc] molecules alone, a hypothetical second component such as a small polynucleotide remains a formal possibility. Studies on the structure of PrP^{Sc} and PrP^C have been unsuccessful in defining a post-translational chemical modification that distinguishes one PrP isoform from the other (see Chapter 32). These findings suggest that the difference between PrP^{Sc} and PrP^C may be conformational. Whether distinct prion isolates or 'strains' with different properties result from multiple conformers of PrP^{Sc} remains to be established. The study of prion diseases seems to be emerging as a unique area of investigation at the interface of such disciplines as genetics, cell biology and virology.

INTRODUCTION

Prion diseases are uniquely both genetic and infectious. The transmissible prion particle is composed largely, if not entirely, of an abnormal isoform of the prion protein (PrP) designated PrP^{Sc} (Prusiner, 1991). These findings argue that prion diseases should be considered pseudoinfections since the particles transmitting disease appear to be devoid of a foreign nucleic acid and thus differ from all known microorganisms as well as viruses and viroids. Because much information especially about scrapie of rodents and Creutzfeldt–Jakob disease in humans has been derived using experimental techniques adapted from virology, we continue to use terms such as infection, incubation period, transmissibility and endpoint titration in studies of prion diseases. Indeed, discoveries in prion biology are creating a new area of investigation which lies at the intersection of cell biology, genetics and virology.

Basic studies on the molecular biology and chemical structure of prions promise to open new vistas into fundamental mechanisms of cellular regulation and homeostasis not previously appreciated (Prusiner, 1982, 1991). Medical investigations of prion diseases are beginning to elucidate mechanisms responsible for CNS degeneration. Individuals at risk for familial prion diseases can often be identified decades in advance of CNS dysfunction (Hsiao et al., 1989a), yet no effective theory exists to prevent these lethal disorders. Applied research directed at detecting prions in asymptomatic cattle and sheep is urgently needed. Bovine spongiform encephalopathy (BSE) threatens the beef industry of Great Britain (Bradley, 1990; Dealler and Lacey, 1990; Hope et al., 1988; Kirkwood et al., 1990; Pain, 1990; Scott et al., 1990; Wells et al., 1987; Wilesmith and Wells, 1991; Wilesmith et al., 1988; Winter et al., 1989) and possibly other countries. The production of pharmaceuticals involving cattle is also of concern. Control of sheep scrapie in many countries is a persistent and vexing problem (Parry, 1962, 1983).

Besides scrapie and BSE, five other disorders presently constitute the ensemble of prion diseases (Table 1). Like BSE, both transmissible mink encephalopathy (TME) and chronic wasting disease (CWD) of captive mule deer and elk are thought to result from the ingestion of prion-infected animal products. Kuru, CJD and GSS are all human neurodegenerative diseases that are frequently transmissible to laboratory animals (Gajdusek, 1977; Gajdusek et al., 1966; Gibbs et al., 1968; Masters et al., 1981a, b).

Since 1986, > 60 000 cattle have died of BSE in Great Britain (Dealler and Lacey, 1990; Wilesmith and Wells, 1991; Wilesmith et al., 1988). Neither the cause of BSE,

Table 1.   Prion diseases[a]

Disease	Natural host
Scrapie	Sheep and goats
Transmissible mink encephalopathy (TME)	Mink
Chronic wasting disease (CWD)	Mule deer and elk
Bovine spongiform encephalopathy (BSE)	Cattle
Kuru	Humans—Fore
Creutzfeldt–Jakob disease (CJD)	Humans
Gerstmann–Sträussler–Scheinker (GSS) syndrome	Humans
Fatal familial insomnia (FFI)	Humans

[a] Alternative terminologies include slow virus infections, subacute transmisssible spongiform encephalopathies, and unconventional slow virus diseases (Gajdusek, 1977).

often referred to as 'mad cow disease', nor methods of controlling the spread of this disorder are known. Many investigators contend that BSE resulted from the feeding of dietary protein supplements derived from rendered scrapie-infected sheep offal to cattle, a practice banned since 1988. Curiously, the majority of BSE cases have occurred in herds with a single affected animal within a herd; several cases of BSE in a single herd are infrequent (Dealler and Lacey, 1990; Wilesmith and Wells, 1991; Wilesmith et al., 1988). Whether the distribution of BSE cases within herds will change as the epidemic progresses and BSE will disappear with the cessation of feeding rendered meat and bone meal are uncertain.

Of particular importance to the BSE epidemic is kuru of humans, confined to the Fore region of New Guinea (Gajdusek, 1977; Gajdusek et al., 1966). Once the most common cause of death among women and children, kuru has almost disappeared with the cessation of ritualistic cannibalism (Alpers, 1987). These findings argue that kuru was transmitted orally as proposed for BSE. Of note are recent cases of kuru which have occurred in people exposed to prions more than three decades ago.

DEVELOPMENT OF THE PRION HYPOTHESIS

The unusual biological properties of the scrapie agent were first recognized in studies with sheep (Gordon, 1946). The experimental transmission of scrapie to mice (Chandler, 1961) gave investigators a more convenient laboratory model which yielded considerable information on the novelty of the infectious pathogen causing scrapie (Alper et al., 1966, 1967, 1978; Gibbons and Hunter, 1967; Griffith, 1967; Hunter, 1972; Latarjet et al., 1970; Millson et al., 1971; Pattison and Jones, 1967). However, progress was slow since quantitation of infectivity in a single sample required holding 60 mice for one year prior to scoring an endpoint titration (Chandler, 1961).

The development of a more rapid and economical bioassay for the scrapie agent in Syrian golden hamsters accelerated work aimed at purification of the infectious particles (Prusiner et al., 1980, 1982b). Partial purification led to the discovery that a protein is required for infectivity (Prusiner *et al.*, 1981) in agreement with some earlier studies which raised the possibility that protein might be necessary (Cho, 1980; Hunter *et al.*, 1969; Hunter and Millson, 1967). Procedures which modify nucleic acids were found not to alter scrapie infectivity (Prusiner, 1982). Other investigators had demonstrated the extreme resistance of scrapie infectivity to both ultraviolet and ionizing radiation (Alper *et al.*, 1966, 1967, 1978; Hunter, 1972; Latarjet *et al.*, 1970) prompting speculation that the scrapie pathogen might be devoid of nucleic acid—a postulate dismissed by most scientists. These early radiobiological results were extended using reagents specifically modifying or damaging nucleic acids: nucleases, psoralens, hydroxylamine and Zn^{2+} ions—none of which was found to alter scrapie infectivity in homogenates (Prusiner, 1982), microsomal fractions (Prusiner, 1982), purified prion rod preparations or detergent–lipid–protein complexes (DLPCs) (Bellinger-Kawahara *et al.*, 1987a, b; Diener *et al.*, 1982; Gabizon *et al.*, 1988a; McKinley *et al.*, 1983b).

Based upon some of the foregoing results, the term 'prion' was introduced to distinguish the proteinaceous infectious particles that cause scrapie, CJD, GSS syndrome and kuru from both viroids and viruses (Prusiner, 1982). Hypotheses for the structure of the infectious prion particle included (1) proteins surrounding a nucleic acid encoding them (a virus), (2) proteins associated with a small polynucleotide, and (3) proteins devoid of nucleic acid. Mechanisms postulated for the replication of infectious prion particles ranged from those used by viruses to the synthesis of polypeptides in the absence of nucleic acid template to post-translational modifications of cellular proteins. Subsequent discoveries have narrowed hypotheses for both prion structure and the mechanism of replication.

DISCOVERY OF THE PRION PROTEIN

Progress in the study of prions and the CNS degenerative diseases that they cause was dramatically accelerated by the discovery of a protein designated prion protein or PrP (Bolton *et al.*, 1982; McKinley *et al.*, 1983a; Prusiner *et al.*, 1982a). Enriching fractions from hamster brain for scrapie infectivity led to the identification of a protease-resistant protein of $M_r = 27–30$ kDa, designated PrP 27-30, that was present in scrapie fractions but absent from controls. Purification of PrP 27-30 to homogeneity allowed determination of its *N*-terminal amino acid sequence (Prusiner *et al.*, 1984) which, in turn, permitted the synthesis of isocoding mixtures of oligonucleotides that were used to identify cDNA clones encoding PrP (Basler *et al.*, 1986; Chesebro *et al.*, 1985; Locht *et al.*, 1986; Oesch *et al.*, 1985). Subsequent studies revealed that PrP is encoded by a chromosomal gene and not by a nucleic acid within the infectious scrapie prion particle (Basler *et al.*, 1986). Levels of PrP mRNA remain unchanged throughout the course of scrapie infection—an observation which led to the identification of the normal PrP gene product, a protein of 33–35 kDa, designated PrP^C (Oesch *et al.*, 1985). PrP^C is protease-sensitive while PrP 27-30 was found to be the protease-resistant core of a 33–35 kDa disease-specific protein, designated PrP^{Sc} (Table 2).

Table 2. Properties of cellular and scrapie PrP isoforms

Property	PrP^C	PrP^{Sc}	References
Concentration in normal SHa brain	~1–5 μg/g	—	Prusiner et al. (1990)
Concentration in scrapie–infected SHa brain	~1–5 μg/g	~5–10 μg/g	Prusiner et al. (1990)
Presence in purified prions	—	+[a]	Bolton et al. (1982), Prusiner et al. (1982a, 1983, 1984); McKinley et al. (1983a)
Protease resistance	—	+[b]	Bolton et al. (1982) Prusiner et al. (1982a, 1983, 1984), McKinley et al. (1983a), Oesch et al. (1985)
Presence in amyloid rods	—	+[c]	Prusiner et al. (1982b, 1983), Hunter and Millson (1967), Bendheim et al. (1984), DeArmond et al. (1985), Kitamoto et al. (1986), Roberts et al. (1988), Tagliavini et al. (1991), McKinley et al. (1991a)
Subcellular localization in cultured cells	Cell surface	Primarily cytoplasmic vesicles	Stahl et al. (1987), Taraboulos et al. (1990b), McKinley et al. (1991b)
PIPLC[d] release from membranes	+	—	Stahl et al. (1987, 1990b)
Synthesis ($t_{1/2}$)	<0.1 h	~1–3 h[e]	Borchelt et al. (1990), Caughey et al. (1989), Borchelt et al. (1992)
Degradation ($t_{1/2}$)	~5 h	≫24 h	Borchelt et al. (1990), Caughey et al. (1989), Borchelt et al. (1992)

[a] Copurification of PrP^{Sc} and prion infectivity demonstrated by two protocols: (1) detergent extraction followed by sedimentation protease digestion, and (2) PrP 27–30 monoclonal antibody affinity chromatography.
[b] Limited proteinase K digestion of SHaPrP^{Sc} produces PrP 27–30.
[c] After limited proteolysis of PrP^{Sc} (PrP 27–30 is produced) and detergent extraction, amyloid rods form: except for length, the rods are indistinghishable from amyloid filaments forming plaques.
[d] PIPLC, phosphatidylinositol-specific phospholipase C.
[e] PrP^{Sc} de novo synthesis is a post-translational process.

Sequencing of molecular clones recovered from cDNA libraries constructed from mRNA isolated from scrapie-infected Syrian hamster (SHa) and mouse (Mo) brains showed that the SHa and Mo PrP cDNAs encode proteins of 254 amino acids (Fig. 1) (Basler et al., 1986; Chesebro et al., 1985; Locht et al., 1986; Oesch et al., 1985). Identical sequences were deduced from genomic clones derived from DNA isolated from uninfected, control animals (Basler et al., 1986). Human (Hu) PrP contains 253 residues (Kretzschmar et al., 1986b). Signal peptides of 22 amino acids at the N-terminus are cleaved during the biosynthesis of SHa and Mo PrP in the rough endoplasmic reticulum (Hope et al., 1986; Safar et al., 1990b; Turk et al., 1988). A 23

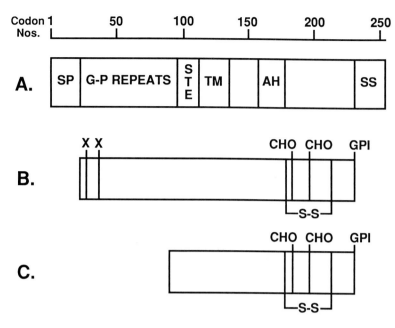

Fig. 1.   Structural features of the Syrian golden hamster prion protein. Codon numbers are indicated at the top of the figure. (A) NH$_2$-terminal signal peptide (SP) of 22 amino acids is removed during biosynthesis (Basler *et al.*, 1986; Hope *et al.*, 1986; Oesch *et al.*, 1985; Safar *et al.*, 1990b; Turk *et al.*, 1988). The NH$_2$-terminal region contains five Gly–Pro-rich (G–P) octarepeats and two hexarepeats; between codons 96 and 112 a domain controlling PrP topology is designated as the stop-transfer effector (STE) (Lopez *et al.*, 1990; Yost *et al.*, 1990); codons 113 to 135 encode a transmembrane (TM) α-helix; codons 157 to 177 encode an amphipathic helix (AH) (Bazan *et al.*, 1987; Hay *et al.*, 1987a,b; Lopez *et al.*, 1990; Yost *et al.*, 1990); and codons 232 to 254 encode a hydrophobic signal sequence (SS) which is removed when a GPI anchor is added (Stahl *et al.*, 1987, 1990a). (B) Unknown modifications (X) of the arginine residue at codons 25 and 37 in PrP^{Sc} and at least codon 25 in PrP^C result in a loss of the arginine signal in the Edman degradation, but these modifications are inconsistently reported (Turk *et al.*, 1988). Both PrP isoforms contain a disulphide (S–S) bond between Cys^{179} and Cys^{214} (Turk *et al.*, 1988); asparagine-linked glycosylation (CHO) occurs at residues 181 and 197 (Bolton *et al.*, 1985; Endo *et al.*, 1989; Haraguchi *et al.*, 1989; Manuelidis *et al.*, 1985; Rogers *et al.*, 1990), and a GPI anchor is attached to Ser^{231} (Stahl *et al.*, 1990a). (C) PrP 27–30. This molecule is derived from PrP^{Sc} by limited proteolysis that removes the NH$_2$-terminal 67 amino acids and leaves a protease-resistant core of 141 amino acids (Basler *et al.*, 1986; Oesch *et al.*, 1985).

residue peptide is removed from *C*-terminus of SHaPrP upon addition of a glycoinositol phospholipid (GPI) anchor (Baldwin *et al.*, 1990; Safar *et al.*, 1990a; Stahl *et al.*, 1987, 1990a). Two Asn-linked complex-type oligosaccharides are attached to sites within a loop formed by a disulphide bond (Bolton *et al.*, 1985; Endo *et al.*, 1989; Haraguchi *et al.*, 1989; Manuelidis *et al.*, 1985; Rogers *et al.*, 1990; Turk *et al.*, 1988). Limited proteolysis of PrP^{Sc} generates PrP 27-30 which lacks ∼ 67 amino acids from the *N*-terminus of PrP^{Sc} (Oesch *et al.*, 1985; Prusiner *et al.*, 1984). Neither gas-phase sequencing nor mass spectrometric analysis of PrP 27-30 has revealed any amino acid differences between the sequence determined by these methods and that deduced from the translated sequence of molecular clones (see Chapter 32). Con-

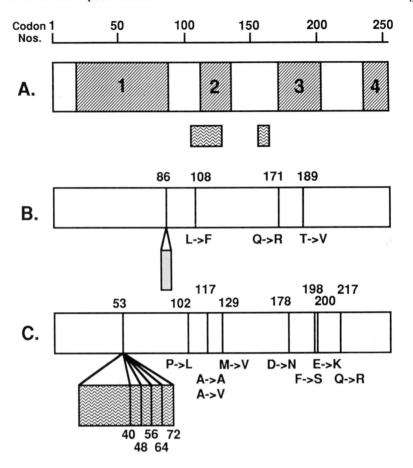

Fig. 2.   Genetic map of prion open reading frames. Codon numbers are indicated at the top of the figure. (A) Four regions among mammalian PrP molecules (hatched) (Basler *et al.*, 1986; Chesebro *et al.*, 1985; Goldmann *et al.*, 1990a,b; Kretzschmar *et al.*, 1986b; Locht *et al.*, 1986b; Lowenstein *et al.*, 1990; Oesch *et al.*, 1985; M. Scott *et al.*, unpublished data); regions of MoPrP homologous to a molecule found in fractions containing acetylcholine receptor-inducing activity in chickens (wave) (Falls *et al.*, 1991; M. Scott *et al.*, unpublished data). (B) Animal mutations and polymorphisms. Two alleles of bovine PrP identified, with one containing an additional octarepeat (stippled) at codon 86, and a polymorphism at codon 171 in sheep PrP resulting in the substitution of arginine for glutamine (Goldmann *et al.*, 1990b, 1991; M. Scott *et al.*, unpublished data). Mice with *Prn-p*[b] genes have long scrapie incubation times and amino acid substitutions at codon 108 (Leu→Phe) and 189 (Thr→Val) (Westaway *et al.*, 1987). (C) Human PrP mutations and polymorphisms. Octarepeat inserts of 32, 40, 48, 56, 64, and 72 amino acids have been found (Collinge *et al.*, 1989, 1990; Crow *et al.*, 1990; Goldfarb *et al.*, 1990c, 1991b; Owen *et al.*, 1989, 1990b). Inserts of 40, 48, 56, 64, and 72 amino acids are associated with familial CJD. Point mutations at codons 102 (Pro→Leu), 117 (Ala→Val), and 198 (Phe→Ser) are found in patients with GSS syndrome (Doh-ura *et al.*, 1989; Goldfarb *et al.*, 1990a,c,d; Goldgaber *et al.*, 1989; Hsiao and Prusiner, 1990; Hsiao *et al.*, 1989a,b, 1991b; Tateishi *et al.*, 1990). There are common polymorphisms at codons 117 (Ala→Ala) and 129 (Met→Val). Point mutations at codons 178 (Asp→Asn) and 200 (Glu→Lys) are found in patients with famial CJD (Gabizon *et al.*, 1991; Goldfarb *et al.*, 1990b, 1991a; Hsiao *et al.*, 1991a). Point mutations at codons 198 (Phe→Ser) and 217 (Gln→Arg) are found in patients with GSS syndrome who have PrP amyloid plaques and neurofibrillary tangles (Hsiao *et al.*, 1992). Single-letter code for amino acids is as follows: A, Ala; D, Asp; E, Glu; F, Phe; K, Lys; L, Leu; M, Met; Asn; P, Pro; Q, Gln; R, Arg; S, Ser; T, Thr; V, Val.

Table 3.   Evidence that PrP^Sc is a major and necessary component of the infectious prion

---

(1) Copurification of PrP 27–30 and scrapie infectivity by biochemical methods. Concentration of PrP 27–30 is proportional to prion titre (Bolton *et al.*, 1982; McKinley *et al.*, 1983a; Prusiner *et al.*, 1982a).

(2) Kinetics of proteolytic digestion of PrP 27–30 and infectivity are similar (McKinley *et al.*, 1983a).

(3) Copurification of PrP^Sc and infectivity by immunoaffinity chromatography, α-PrP antisera neutralization of infectivity (Gabizon *et al.*, 1988b; Gabizon and Prusiner, 1990).

(4) PrP^Sc detected only in clones of cultured cells producing prion infectivity (Butler *et al.*, 1988; Taraboulos *et al.*, 1990b).

(5) PrP amyloid plaques are specific for prion diseases of animals and humans (Bendheim *et al.*, 1984; DeArmond *et al.*, 1985; Kitamoto *et al.*, 1986; Roberts *et al.*, 1988). Deposition of PrP amyloid is controlled, at least in part, by the PrP sequence (Prusiner *et al.*, 1990).

(6) Correlation between PrP^Sc (or PrP^CJD) in brain tissue with prion diseases in animals and humans (Bockman *et al.*, 1985; Brown *et al.*, 1986; Serban *et al.*, 1990).

(7) Genetic linkage between MoPrP gene and scrapie incubation times (Carlson *et al.*, 1986, 1988; Hunter *et al.*, 1987; Race *et al.*, 1990). PrP gene of mice with long incubation times encodes amino acid substitutions at codons 108 and 189 as compared with mice with short or intermediate incubation times (Westaway *et al.*, 1987).

(8) SHaPrP transgene and scrapie PrP^Sc in the inoculum govern the 'species barrier', scrapie incubation times, neuropathology and prion synthesis in mice (Prusiner *et al.*, 1990; Scott *et al.*, 1989).

(9) Genetic linkage between human PrP gene mutation at codon 102 and development of GSS syndrome (Hsiao *et al.*, 1989a). Association between codon 200 point mutation or codon 53 insertion of six additional octarepeats and familial CJD (Collinge *et al.*, 1989, 1990; Crow *et al.*, 1990; Gabizon *et al.*, 1991; Goldfarb *et al.*, 1990b; Owen *et al.*, 1989, 1990b; Hsiao *et al.*, 1991a)..

(10) Mice expressing MoPrP transgenes with the point mutation of GSS spontaneously develop neurologic dysfunction, spongiform brain degeneration, and astrocytic gliosis (Hsiao *et al.*, 1990).

---

clusions about the covalent structure of PrP^Sc must be guarded since purified fractions contain ~10^5 PrP 27-30 molecules/ID_{50} unit (Prusiner *et al.*, 1982a). If <1% of the PrP 27-30 molecules contained an amino acid substitution or post-translational chemical modification which conferred scrapie infectivity, our methods would not detect such a change.

INFECTIOUS PRION PARTICLES

A remarkable convergence of information on PrP^Sc in prion diseases argues persuasively that prions are composed largely, if not entirely, of PrP^{Sc} molecules (Table 3). Although some investigators contend that PrP^{Sc} is merely a pathologic product of scrapie infection and that PrP^{Sc} coincidentally purifies with the 'scrapie virus' (Aiken *et al.*, 1989, 1990; Akowitz *et al.*, 1990; Braig and Diringer, 1985; Manuelidis and Manuelidis, 1989; Murdoch *et al.*, 1990; Sklaviadis *et al.*, 1989, 1990), no convincing data to support this view have been offered. No fractions containing <1 PrP^{Sc} molecule per ID$_{50}$ unit have been found; such fractions would argue that PrP^{Sc} is not required for infectivity. Some investigators have reported that PrP^{Sc} accumulation in hamsters occurs after the synthesis of infectivity (Czub *et al.*, 1986, 1988), but these results have been refuted by three groups working independently (Jendroska *et al.*, 1991; Pocchiari, 1990; R. E. Race *et al.*, unpublished data). The discrepancy appears to be due to comparisons of infectivity in crude homogenates with PrP^{Sc} concentrations measured in purified fractions.

The search for a second component within the prion particle has focused largely on a nucleic acid because it would most readily explain different isolates or 'strains' of infectivity (Bruce and Dickinson, 1987; Dickinson and Fraser, 1979; Dickinson and Outram, 1988; Kimberlin *et al.*, 1987). Multiple scrapie isolates, each with different incubation times, have been found to breed true in mice and hamsters (Bruce and Dickinson, 1987; Dickinson and Fraser, 1979; Dickinson and Outram, 1988; Kimberlin *et al.*, 1987). In addition, many other factors modulate scrapie incubation times, including PrP gene expression, murine genes linked to PrP designated *Prn-i* and *Sinc*, dose of inoculum, route of inoculation and the genetic origin of the prion inoculum, i.e. PrP^{Sc} sequence, as discussed below.

The search for a scrapie-specific nucleic acid molecule has been unrewarding using reagents that modify or hydrolyse polynucleotides, molecular cloning procedures and physicochemical techniques (Bellinger-Kawahara *et al.*, 1987a, b; Diedrich *et al.*, 1987; Diener *et al.*, 1982; Duguid *et al.*, 1988; Gabizon *et al.*, 1987, 1988a; McKinley *et al.*, 1983b; Meyer *et al.*, 1991; Oesch *et al.*, 1988; Weitgrefe *et al.*, 1985). The implications of an infectious pathogen that does not contain a nucleic acid are profound. Although available data do not permit its exclusion, finding a scrapie-specific polynucleotide seems unlikely. The possibility of a non-covalently bound cofactor such as a peptide, oligosaccharide, fatty acid, sterol or inorganic compound also deserves consideration (see Fig. 4).

PrP POLYMERS AND AMYLOID

The discovery of PrP 27-30 in fractions enriched for scrapie infectivity was accompanied by the electron microscopic identification of rod-shaped particles in rapidly sedimenting fractions in discontinuous sucrose gradients (Bolton *et al.*, 1982; McKinley *et al.*, 1983a; Prusiner *et al.*, 1982a, 1983). The rods are ultrastructurally indistinguishable from many purified amyloids and display the tinctorial properties of amyloids (Prusiner *et al.*, 1983). These findings were followed by the demonstration that amyloid plaques in prion diseases contain PrP as determined by immunoreactivity and amino acid sequencing (Bendheim *et al.*, 1984; DeArmond *et al.*, 1985; Kitamoto *et al.*, 1986; Roberts *et al.*, 1988; Tagliavini *et al.*, 1991). It has been claimed that scrapie-associated fibrils (SAFs) are synonymous with prion rods and are composed of ʻPrP although SAF was repeatedly distinguished from amyloids (Aiken and Marsh, 1990; Diener, 1987; Diringer et al., 1983; Kimberlin, 1990; Merz et al., 1981, 1983, 1984, 1987; Somerville et al., 1989).

Studies on the generation of prion rods have clearly shown that their formation requires limited proteolysis in the presence of detergent (McKinley et al., 1991a). Thus, the prion rods found in fractions enriched for scrapie infectivity are largely, if not entirely, artifacts of the purification protocol. While the prion rods gave the first clue that amyloid plaques in prion diseases might be composed of PrP (Prusiner et al., 1983), these insoluble structures have greatly retarded protein chemical analysis of PrP^{Sc}. Functional solubilization of PrP 27-30 in DLPCs with retention of infectivity (Gabizon et al., 1987) demonstrated that PrP polymers are not required for infectivity and permitted the immunoaffinity copurification of PrP^{Sc} and infectivity (Gabizon et al., 1988b; Gabizon and Prusiner, 1990).

PrP GENE STRUCTURE AND EXPRESSION

Mapping PrP genes to the short arm of human chromosome 20 and the homologous region of mouse chromosome 2 argues for the existence of PrP genes prior to the speciation of mammals (Liao et al., 1986; Robakis et al., 1986; Sparkes et al., 1986). Hybridization studies demonstrated <0.002 PrP gene sequences per ID$_{50}$ unit in purified prion fractions indicating that a gene encoding PrP^{Sc} is not a component of the infectious prion particle (Oesch et al., 1985). This is a major feature which distinguishes prions from viruses including those retroviruses that carry cellular oncogenes and from satellite viruses that derive their coat proteins from other viruses previously infecting plant cells.

The entire open reading frame (ORF) of all known PrP genes is contained within a single exon eliminating the possibility that variant forms of PrP arise from alternative RNA splicing (Basler et al., 1986; Westaway et al., 1987, 1991), but not excluding such mechanisms as RNA editing or protein splicing (Blum et al., 1990; Kane et al., 1990). The two exons of the SHaPrP gene are separated by a 10 kb intron: exon 1 encodes a portion of the 5′ untranslated leader sequence while exon 2 encodes the ORF and 3′ untranslated region (Basler et al., 1986). The MoPrP gene comprises three exons with exon 3 analogous to exon 2 of the hamster (Westaway et al., 1991). The promoters of both the SHa and MoPrP genes contain copies of G–C-rich repeats 3 and 2, respectively, but are devoid of TATA boxes. These G–C nonamers represent a motif which may function as a canonical binding site for the transcription factor Sp1 (McKnight and Tjian, 1986).

Although PrP mRNA is constitutively expressed in the brains of adult animals (Oesch et al., 1985), it is highly regulated during development. In the septum, levels of PrP mRNA and choline acetyltransferase were found to increase in parallel during development (Mobley et al., 1988). In other brain regions, PrP gene expression occurred at an earlier age. In situ hybridization studies show that the highest levels of PrP mRNA are found in neurons (Kretzschmar et al., 1986a).

Four regions of the mammalian PrP gene ORF are highly conserved when the translated amino acid sequences are compared (Fig. 2) (Basler *et al.*, 1986; Chesebro *et al.*, 1985; Goldmann *et al.*, 1990a, b; Kretzschmar *et al.*, 1986b; Locht *et al.*, 1986; Lowenstein *et al.*, 1990; Oesch *et al.*, 1985; M. Scott *et al.*, unpublished data). While the function of PrP^C is unknown, the MoPrP sequence is ~30% identical with a putative factor from chickens possessing acetylcholine receptor inducing activity (ARIA) (Falls *et al.*, 1990; Harris *et al.*, 1989). Twenty-three of 24 amino acids encoded by mouse *Prn-p^a* correspond to codons 104 to 127 and are identical to those found in ARIA. With the *N*-terminal conserved regions of mammalian PrP, five Gly:Pro-rich octarepeats and two hexarepeats (Fig. 1) have been found while chicken ARIA has eight hexarepeats.

SYNTHESIS OF PrP ISOFORMS

Metabolic labelling studies of scrapie-infected cultured cells have shown that PrP^C is synthesized and degraded rapidly while PrP^{Sc} accumulates slowly (Borchelt *et al.*, 1990; Caughey *et al.*, 1989; Caughey and Raymond, 1991; Borchelt *et al.*, 1992). These observations are consistent with earlier findings showing that PrP^{Sc} accumulates to high levels in the brains of scrapie-infected animals, yet PrP mRNA levels remain unchanged (Oesch *et al.*, 1985).

Both PrP isoforms appear to transit through the Golgi apparatus where their Asn-linked oligosaccharides are modified and sialylated (Bolton *et al.*, 1985; Endo *et al.*, 1989; Haraguchi *et al.*, 1989; Manuelidis *et al.*, 1985; Rogers *et al.*, 1990). PrP^C is presumably transported within secretory vesicles to the external cell surface where it is anchored by a GPI moiety (Baldwin *et al.*, 1990; Safar *et al.*, 1990a; Stahl *et al.*, 1987, 1990a, b). In contrast, PrP^{Sc} accumulates primarily within cells where it is deposited in cytoplasmic vesicles, many of which appear to be secondary lysosomes (Table 2) (Butler *et al.*, 1988; McKinley *et al.*, 1991b; Taraboulos *et al.*, 1990b). Although most of the difference in mass of PrP 27-30 predicted from the amino acid sequence and that observed after post-translational modification is due to complex-type oligosaccharides, these sugar chains are not required for the synthesis of protease-resistant PrP in scrapie-infected cultured cells based on experiments with the Asn-linked glycosylation inhibitor tunicamycin and on site-directed mutagenesis studies (Taraboulos *et al.*, 1990a). Whether unglycosylated PrP^{Sc} is associated with scrapie prion infectivity remains to be established, but experiments with transgenic mice may resolve this issue.

Cell-free translation studies have demonstrated two forms of PrP: a transmembrane form which spans the bilayer twice at the transmembrane (TM) and amphipathic helix (AH) domains and a secretory form (Fig. 1) (Bazan *et al.*, 1987; Hay *et al.*, 1987a, b; Lopez *et al.*, 1990; Yost *et al.*, 1990). The stop transfer effector (STE) domain controls the topogenesis of PrP. That PrP contains both a TM domain and a GPI anchor poses a topologic conundrum. It seems likely that membrane-dependent events feature in the synthesis of PrP^{Sc} especially since Brefeldin A, which selectively destroys the Golgi stacks (Doms *et al.*, 1989; Lippincott-Schwartz *et al.*, 1989), prevents PrP^{Sc} synthesis in scrapie-infected cultured cells (Taraboulos *et al.*, 1991). For many years, the association of scrapie infectivity with membrane fractions has been appreciated (Gibbons and Hunter, 1967; Griffith, 1967; Millson *et al.*, 1971); indeed, hydrophobic interactions are thought to account for many of the physical properties displayed by infectious prion particles (Gabizon *et al.*, 1987; Prusiner *et al.*, 1978, 1980).

GENETIC LINKAGE OF PrP WITH SCRAPIE INCUBATION TIMES

Studies of PrP genes (*Prn-p*) in mice with short and long incubation times demonstrated genetic linkage between the *Prn-p* restriction fragment length polymorphism (RFLP) and a gene modulating incubation times (*Prn-i*) (Carlson *et al.*, 1986). Other investigators have confirmed the genetic linkage and one group has shown that the incubation time gene *Sinc* is also linked to PrP (Carlson *et al.*, 1988; Hunter *et al.*, 1987; Race *et al.*, 1990). *Sinc* was first described by Dickinson and colleagues over 20 years ago (Dickinson *et al.*, 1968); whether the genes for PrP, *Prn-i* and *Sinc* are all congruent remains to be established. The PrP sequences of NZW (*Prn-p^a*) and I/Ln (*Prn-p^b*) mice with short and long scrapie incubation times, respectively, differ at codons 108 (L → F) and 189 (T → V) (Fig. 2) (Westaway *et al.*, 1987). While these amino acid substitutions argue for the congruency of *Prn-p* and *Prn-i*, experiments with *Prn-p^a* mice expressing *Prn-p^b* transgenes demonstrated a paradoxical shortening of incubation times (Westaway *et al.*, 1991) instead of a prolongation as predicted from (*Prn-p^a* × *Prn-p^b*) F1 mice which exhibit long incubation times that are dominant (Carlson *et al.*, 1986, 1988; Dickinson *et al.*, 1968; Hunter *et al.*, 1987; Race *et al.*, 1990). Whether this paradoxical shortening of scrapie incubation times in Tg(*Prn-p^b*) mice results from high levels of PrP^C-B expression remains to be established (Westaway *et al.*, 1991).

Host genes also influence the development of scrapie in sheep. Parry argued that natural scrapie is a genetic disease which could be eradicated by proper breeding protocols (Parry, 1962, 1983). He considered its transmission by inoculation of importance primarily for laboratory studies and communicable infection of little consequence in nature. Other investigators viewed natural scrapie as an infectious disease and argued that host genetics only modulates susceptibility to an endemic infectious agent (Dickinson *et al.*, 1965). The incubation time gene for experimental scrapie in Cheviot sheep called *Sip* is said to be linked to a PrP gene RFLP (Hunter *et al.*, 1989), a situation perhaps analogous to *Prn-i* and *Sinc* in mice. However, the null hypothesis of non-linkage has yet to be tested and this is important, especially in view of earlier studies which argue that susceptibility of sheep to scrapie is governed by a recessive gene (Parry, 1962, 1983). In a Suffolk sheep, a polymorphism in the PrP ORF was found at codon 171 (R → Q) (Fig. 2(B)) (Goldmann *et al.*, 1990a, b; D. Westaway *et al.*, unpublished data); whether it segregates with a *Sip* phenotype in Cheviot sheep is unknown.

HUMAN FAMILIAL PRION DISEASES

In humans, genetics were first thought to have a role in CJD with the recognition that ~10% of cases are familial (Gajdusek, 1977; Masters *et al.*, 1981b). Like sheep scrapie, the relative contributions of genetic and infectious aetiologies in the human prion diseases remained puzzling. The discovery of the PrP gene raised the possibility that mutation might feature in the hereditary human prion diseases. A point mutation at codon 102 (P → L) was shown to be linked to development of GSS syndrome (Fig. 2(C)) (Hsiao *et al.*, 1989a). This mutation may be due to the deamination of a methylated CpG in a germline PrP gene resulting in the substitution of a T for C.

An insert of 144 bp at codon 53 containing six octarepeats has been described in patients with CJD from four families all residing in southern England (Fig. 2(C)) (Collinge *et al.*, 1989, 1990; Crow *et al.*, 1990; Owen *et al.*, 1989, 1990b, 1991). This mutation must have arisen through a complex series of events since the HuPrP gene contains only five octarepeats indicating that a single recombination event could not have created the insert. Genealogic investigations have shown that all four families are related, arguing for a single founder born more than two centuries ago (Crow *et al.*, 1990). Five, six, seven, eight or nine octarepeats (in addition to the normal five) were found in individuals with CJD, whereas deletion of one octarepeat or four additional octarepeats have been identified without the neurologic disease (Collinge *et al.*, 1989, 1990; Goldfarb *et al.*, 1991b; Laplanche *et al.*, 1990; Owen *et al.*, 1989, 1990b).

For many years the unusually high incidence of CJD among Israeli Jews of Libyan origin was thought to be due to the consumption of lightly cooked sheep brain or eyeballs (Alter and Kahana, 1976; Herzberg *et al.*, 1974; Kahana *et al.*, 1974). Recent studies have shown that some Libyan and Tunisian Jews in families with CJD have a PrP gene point mutation at codon 200 resulting in an E → K substitution (Gabizon *et al.*, 1991; Goldfarb *et al.*, 1990d; Hsiao *et al.*, 1991a). One patient was homozygous for the mutation, but her clinical presentation was similar to that of heterozygotes (Hsiao *et al.*, 1991b) arguing that familial prion diseases are true autosomal dominant disorders like Huntington's disease (Wexler *et al.*, 1987). The codon 200 mutation has also been found in Slovaks originating from Orava in North Central Czechoslovakia (Goldfarb *et al.*, 1990d).

Other point mutations at codons 117, 178, 198 and 217 also segregate with inherited prion diseases (Doh-ura *et al.*, 1989; Goldfarb *et al.*, 1991a; Hsiao *et al.*, 1991c; Hsiao *et al.*, 1992). Some patients once thought to have familial Alzheimer's disease are now known to have prion diseases on the basis of PrP immunostaining of amyloid plaques and PrP gene mutations (Farlow *et al.*, 1989; Ghetti *et al.*, 1989; Giaccone *et al.*, 1990; Nochlin *et al.*, 1989). Patients with the codon 198 mutation have numerous neurofibrillary tangles that stain with antibodies to τ and have amyloid plaques (Farlow *et al.*, 1989; Ghetti *et al.*, 1989; Giaccone *et al.*, 1990; Nochlin *et al.*, 1989) that are composed largely of a PrP fragment extending from residues 58 to 150 (Tagliavini *et al.*, 1991).

At PrP codon 129, an amino acid (Met/Val) polymorphism (Fig. 2) has been identified (Owen *et al.*, 1990a). Patients with CJD following treatment with human pituitary growth hormone (Fradkin *et al.*, 1991; Buchanan *et al.*, 1991) or gonadotrophin have a significant preponderance of the Val allele (Collinge *et al.*, 1991) compared with the general population. Sporadic CJD patients were found to be homozygous for the Met or Val allele at codon 129 but were rarely heterozygous

(Palmer *et al.*, 1991). This finding was interpreted (Palmer *et al.*, 1991; Hardy, 1991) as consistent with the hypothesis that PrP^C/PrP^{Sc} heterodimers feature in the replication of prions (Prusiner *et al.*, 1990; Prusiner, 1991).

DE NOVO SYNTHESIS OF PRIONS IN TRANSGENIC MICE EXPRESSING GSS MUTANT MoPrP

When the codon 102 point mutation was introduced into MoPrP in transgenic mice, spontaneous CNS degeneration occurred, characterized by clinical signs indistinguishable from experimental murine scrapie and neuropathology consisting of widespread spongiform morphology and astrocytic gliosis (Hsiao *et al.*, 1990). By inference, these results suggest that PrP mutations cause GSS syndrome and familial CJD. It is unclear whether low levels of protease-resistant PrP in the brains of transgenic mice with the GSS mutation is PrP^{Sc} or residual PrP^C. Undetectable or low levels of PrP^{Sc} in the brains of these transgenic mice are consistent with the results of transmission experiments that suggest low titres of infectious prions. Brain extracts transmit CNS degeneration to inoculated recipients and the *de novo* synthesis of prions has been demonstrated by serial passage (Hsiao *et al.*, 1991c). If the possibility of contamination can be eliminated, then these observations indicate that prions are devoid of foreign nucleic acid, in accord with studies that use other experimental approaches (Bellinger-Kawahara *et al.*, 1987a, b; Diedrich *et al.*, 1987; Diener *et al.*, 1982; Duguid *et al.*, 1988; Gabizon *et al.*, 1988a; McKinley *et al.*, 1983b; Meyer *et al.*, 1991; Neary *et al.*, 1991; Oesch *et al.*, 1988; Weitgrefe *et al.*, 1985).

One view of the PrP gene mutations has been that they render individuals susceptible to a common 'virus' (Aiken and Marsh, 1990; Kimberlin, 1990). In this scenario, the putative scrapie virus is thought to persist with a worldwide reservoir of humans, animals or insects without causing detectable illness. However, $1/10^6$ individuals develop sporadic CJD and die from a lethal 'infection' while $\sim 100\%$ of people with PrP point mutations or inserts appear to develop eventually neurologic dysfunction. That germline mutations found in the PrP genes of patients and at-risk individuals are the cause of familial prion diseases is supported by experiments with Tg(GSS MoPrP) mice described above (Hsiao and Prusiner, 1990; Hsiao *et al.*, 1991c; Weissmann, 1991b). The transgenic mouse studies also argue that sporadic CJD might arise from the spontaneous conversion of PrP^C to PrP^{CJD} due to either a somatic mutation of the PrP gene or rare event involving modification of wild-type PrP^C (Prusiner, 1991).

TRANSGENETICS AND SPECIES BARRIERS

Passage of prions between species is a stochastic process characterized by prolonged incubation times (Pattison, 1965, 1966; Pattison and Jones, 1967). Prions synthesized *de novo* reflect the sequence of the host PrP gene and not that of the PrP^{Sc} molecules in the inoculum (Bockman *et al.*, 1987). On subsequent passage in an homologous host, the incubation time shortens to that recorded for all subsequent passages and it becomes a non-stochastic process. The species barrier concept is of practical

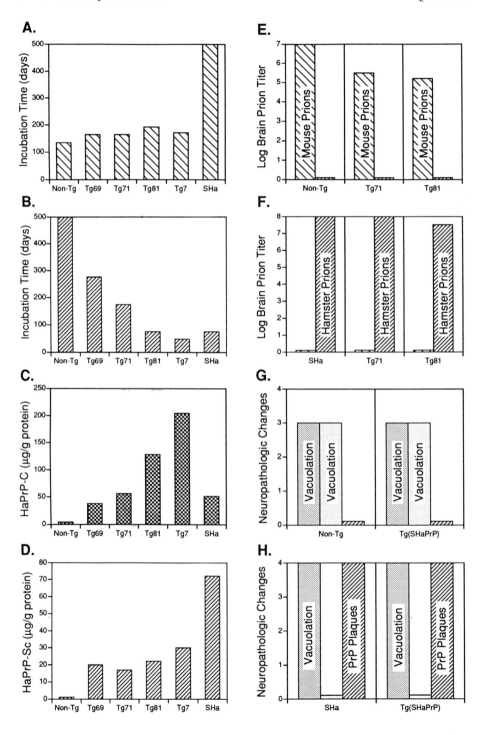

importance in assessing the risk for humans of developing CJD after consumption of scrapie-infected lamb or BSE beef.

To test the hypothesis that differences in PrP gene sequences might be responsible for the species barrier, transgenic mice expressing SHaPrP were constructed (Prusiner *et al.*, 1990; Scott *et al.*, 1989). The PrP genes of Syrian hamsters and mice encode proteins differing at 16 positions. Incubation times in four lines of Tg(SHaPrP) mice inoculated with Mo prions were prolonged compared with those observed from non-transgenic, control mice (Fig. 3(A)). Inoculation of Tg(SHaPrP) mice with SHa prions demonstrated abrogation of the species barrier resulting in abbreviated incubation times due to a non-stochastic process (Fig. 3(B)) (Prusiner *et al.*, 1990; Scott *et al.*, 1989). The length of the incubation time after inoculation with SHa prions was inversely proportional to the level of SHaPrP^C in the brains of Tg(SHaPrP) mice (Fig. 3(B) and 3(C)) (Prusiner *et al.*, 1990). SHaPrP^{Sc} levels in the brains of clinically ill mice were similar in all four Tg(SHaPrP) lines inoculated with SHa prions (Fig. 3(D)). Bioassays of brain extracts from clinically ill Tg(SHaPrP) mice inoculated with Mo prions revealed that only Mo prions but no SHa prions were produced (Fig. 3(E)). Conversely, inoculation of Tg(SHaPrP) mice with SHa prions led to only the synthesis of SHa prions (Fig. 3(F)). Thus, the *de novo* synthesis of prions is species specific and reflects the genetic origin of the inoculated prions. Similarly, the neuropathology of Tg(SHaPrP) mice is determined by the genetic origin of prion inoculum. Mo prions injected into Tg(SHaPrP) mice produced a neuropathology characteristic of mice with scrapie. A moderate degree of vacuolation in both the gray and white matter was found while amyloid plaques were rarely detected (Fig. 3(G)). Inoculation of Tg(SHaPrP) mice with SHa prions produced intense vacuolation of the gray matter, sparing of the white matter and numerous SHaPrP amyloid plaques characteristic of Syrian hamsters with scrapie (Fig. 3(H).

These studies with transgenic mice establish that the PrP gene influences virtually all phases of scrapie including: (1) species barrier, (2) replication of prions, (3) incubation times, (4) synthesis of PrP^{Sc} and (5) neuropathologic changes.

Fig. 3. Transgenic mice expressing SHa prion protein exhibit species-specific scrapie incubation times, infectious prion synthesis and neuropathology (Prusiner *et al.*, 1990). (A) Scrapie incubation times in non-transgenic mice (Non-Tg) and four lines of transgenic mice expressing SHaPrP and Syrian hamsters inoculated intracerebrally with ~10^6 ID$_{50}$ units of Chandler Mo prions serially passaged in Swiss mice. The four lines of transgenic mice have different numbers of transgene copies: Tg69 and 71 mice have two to four copies of the SHaPrP transgene, whereas Tg81 have 30 to 50 and Tg7 mice have >60. Incubation times are number of days from inoculation to onset of neurologic dysfunction. (B) Scrapie incubation times in mice and hamsters inoculated with ~10^7 ID$_{50}$ units of Sc237 prions serially passaged in Syrian hamsters and as described in (A). (C) Brain SHaPrP^C in transgenic mice and hamsters. SHaPrP^C levels were quantitated by an enzyme-linked immunoassay. (D) Brain SHaPrP^{Sc} in transgenic mice and hamsters. Animals were killed after exhibiting clinical signs of scrapie. SHaPrP^{Sc} levels were determined by immunoassay. (E) Prion titres in brains of clinically ill animals after inoculation with Mo prions. Brain extracts from Non-Tg, Tg71, and Tg71, and Tg81 mice were bioassayed for prions in mice (left) and hamsters (right). (F) Prion titres in brains of clinically ill animals after inoculation with SHa prions. Brain extracts from Syrian hamsters as well as Tg71 and Tg81 mice were bioassayed for prions in mice (left) and hamsters (right). (G) Neuropathology in Non-Tg mice and Tg(SHaPrP) mice with clinical signs of scrapie after inoculation with Mo prions. Vacuolation in gray (left) and white matter (centre); PrP amyloid plaques (right). Vacuolation score: 0, none; 1, rare; 2, modest; 3, moderate; 4, intense. (H) Neuropathology in Syrian hamsters and transgenic mice inoculated with SHa prions. Degree of vacuolation and frequency of PrP amyloid plaques as described in (G)

A.

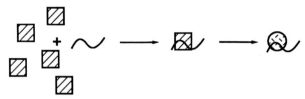

B.

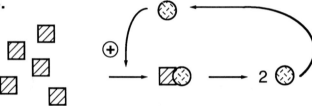

C.

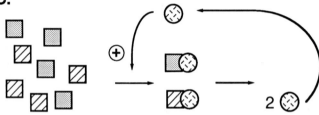

D.

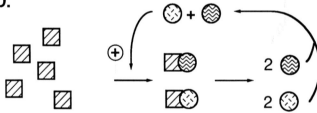

E.

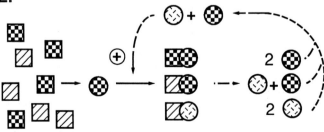

PRION REPLICATION

The mechanism by which prions multiply is unknown. Although the search for a scrapie-specific nucleic acid continues to be unrewarding, some investigators steadfastly cling to the notion that this putative polynucleotide drives prion replication. If prions are found to contain a scrapie-specific nucleic acid, then such a molecule would be expected to direct scrapie agent replication using a strategy similar to that employed by viruses (Fig. 4(A)). In the absence of any chemical or physical evidence for a scrapie-specific polynucleotide (Fig. 3(A)) (Aiken *et al.*, 1989, 1990; Akowitz *et al.*, 1990; Bellinger-Kawahara *et al.*, 1987a, b; Braig and Diringer, 1985; Diedrich *et al.*, 1987; Diener *et al.*, 1982; Duguid *et al.*, 1988; Gabizon *et al.*, 1988a; Manuelidis and Manuelidis, 1989; McKinley *et al.*, 1983b; Meyer *et al.*, 1991; Murdoch *et al.*, 1990; Neary *et al.*, 1991; Oesch *et al.*, 1988; Sklaviadis *et al.*, 1989, 1990; Weitgrefe *et al.*, 1985), it seems reasonable to consider some alternative mechanisms that might feature in prion biosynthesis. The multiplication of prion infectivity is an exponential process in which the post-translational conversion of PrP^C or a precursor to PrP^{Sc} appears to be obligatory (Borchelt *et al.*, 1990). As illustrated in Fig. 4(B), two PrP^{Sc} molecules combine with two heterodimers which are subsequently transformed into two homodimers. In the next cycle, four PrP^{Sc} molecules combine with four PrP^C molecules giving rise to four homodimers that dissociate to combine with eight PrP^C molecules creating an exponential process. Studies with Tg(SHaPrP) mice argue that prion synthesis involves 'replication', not merely 'amplification' (Prusiner *et al.*, 1990). Assuming prion biosynthesis simply involves amplification of post-translationally altered PrP molecules, we might expect Tg(SHaPrP) mice to produce both SHa and Mo prions after inoculation with either prion since these mice produce both SHa and $MoPrP^C$. However, Tg(SHaPrP) mice synthesize only those prions present in the inoculum (Figs 3(E) and 3(F)). These results argue that the incoming prion and PrP^{Sc} interact with the homologous PrP^C substrate to replicate more of the same prions (Fig. 4(C)).

In the absence of any candidate post-translational chemical modifications (N. Stahl *et al.*, unpublished data) that differentiate PrP^C from PrP^{Sc}, we are forced to consider

Fig. 4 Some possible mechanisms of prion replication. (A) Two-component prion model. Prions contain a putative, as yet unidentified, nucleic acid or another second component (solid, thick wavy line) that binds to PrP^C (squares) and stimulates conversion of PrP^C or a precursor to PrP^{Sc} (circles). (B) One-component prion model—prions devoid of nucleic acid. PrP^{Sc} binds to PrP^C forming heterodimers that function as replication intermediates in the synthesis of PrP^{Sc}. Repeated cycles of this process result in an exponential increase in PrP^{Sc}. (C) Prion synthesis in transgenic mice (Prusiner *et al.*, 1990). $SHaPrP^{Sc}$ (broken, cross-hatched circles) binds to $SHaPrP^C$ (diagonal squares), leading to the synthesis of PrP^{Sc}. Binding to $MoPrP^C$ (stippled squares) does not produce PrP^{Sc}. Species barrier for scrapie between mice and hamsters represented by $MoPrP^C$–$SHaPrP^{Sc}$ heterodimer. (D) Scrapie isolates (in circles) bind to PrP^C and constrain the conformational changes that PrP^C undergoes during its conversion into PrP^{Sc}. (E) Inherited prion diseases in humans and transgenic mice. Mutant PrP molecules (checkered pattern in squares) might initiate the conversion of PrP^C to PrP^{Sc} (or PrP^{CJD}). Infectious prions are produced (dashed lines), then they stimulate the synthesis of more PrP^{CJD} in humans and PrP^{Sc} in experimental animals. Alternatively, prion infectivity is not generated, but the host develops neurologic dysfunction, spongiform degeneration, astrocytic gliosis, and possibly PrP amyloid plaques (Doh-ura *et al.*, 1989; Gajdusek, 1977; Goldfarb *et al.*, 1990a–d; Goldgaber *et al.*, 1989; Hsiao *et al.*, 1989a,b, 1991a,c; Hsiao and Prusiner, 1990; Masters *et al.*, 1981b; Tateishi *et al.*, 1990).

the possibility that conformation distinguishes these isoforms. Various 'strains' or isolates of scrapie prions (Bruce and Dickinson, 1987; Dickinson and Fraser, 1979; Dickinson and Outram, 1988; Kimberlin et al., 1987) could be accommodated by multiple conformers that act as templates for the folding of de novo synthesized PrP^{Sc} molecules during prion 'replication' (Fig. 4(D)). Although this proposal is rather unorthodox, it is consistent with observations generated from Tg(SHaPrP)Mo studies contending that PrP^{Sc} in the inoculum binds to homologous PrP^C or a precursor to form a heterodimeric intermediate in the replication process (Prusiner et al., 1990). Whether foldases, chaperonins or other types of molecules feature in the conversion of the PrP^C/PrP^{Sc} heterodimer to a PrP^{Sc} homodimer is unknown. The molecular weight of a PrP^{Sc} homodimer is consistent with the ionizing radiation target size of $55\,000 \pm 9000$ Da as determined for infectious prion particles independent of their polymeric form (Bellinger-Kawahara et al., 1988).

In humans carrying point mutations or inserts in their PrP genes, mutant PrP^C molecules might spontaneously convert into PrP^{Sc} (Fig. 4(E)). While the initial stochastic event may be inefficient, once it happens then the process becomes autocatalytic. The proposed mechanism is consistent with individuals harbouring germline mutations who do not develop CNS dysfunction for decades and with studies on Tg(GSS MoPrP) mice that spontaneously develop CNS degeneration (Hsiao et al., 1990). Whether all GSS and familial CJD cases contain infectious prions or some represent inborn errors of PrP metabolism in which neither PrP^{Sc} nor prion infectivity accumulates is unknown. If the latter is found then, presumably, mutant PrP^C molecules alone can produce CNS degeneration.

Some investigators have suggested that scrapie agent multiplication proceeds through a crystallization process involving PrP amyloid formation (Gajdusek, 1988, 1990; Gajdusek and Gibbs, 1990). Against this hypothesis is the absence of rarity of amyloid plaques in many prion diseases as well as the inability to identify any amyloid-like polymers in cultured cells chronically synthesizing prions (McKinley et al., 1991a; Prusiner et al., 1990). Purified infectious preparations isolated from scrapie-infected hamster brains exist as amorphous aggregates; only if PrP^{Sc} is exposed to detergents and limited proteolysis does it then polymerize into prion rods exhibiting the ultrastructural and tinctorial features of amyloid (McKinley et al., 1991a). Furthermore, dispersion of prion rods into DLPCs results in a 10- to 100-fold increase in scrapie titre and no rods could be identified in these fractions by electron microscopy (Gabizon et al., 1987).

DISTINCT PRION ISOLATES OR 'STRAINS'

While a wealth of data argue that PrP^{Sc} alone can transmit scrapie prion infectivity, the molecular basis of prion diversity remains enigmatic (Weissmann, 1991a). Dickinson, Kimberlin and colleagues have provided convincing evidence for the existence of distinct prion isolates or 'strains' with different properties (Bruce and Dickinson, 1987; Kimberlin et al., 1987). Prion 'strains' have been characterized by their incubation times and neuropathologic lesion profiles in mice and hamsters.

Studies with Tg(SHaPrP) mice have shown that the PrP gene and PrP^{Sc} in the inoculum govern scrapie incubation times, neuropathology and prion synthesis (Fig. 3) (Prusiner *et al.*, 1990). These results suggest that transgenetic studies may provide a unique opportunity to dissect the molecular basis of scrapie 'strains'. Results with two isolates or 'strains' of SHa prions inoculated into Tg(SHaPrP) mice argue that the expression and metabolism of PrP may profoundly influence the isolate phenotype. Sc237 (similar to 263 K) isolate of SHa prions produces 77 ± 1 day ($n=48$) incubation times in Syrian hamsters while the 139H isolate yields 168 ± 7 day ($n=54$) incubation times (Kimberlin *et al.*, 1987). SHaPrP^C expression in Tg(SHaPrP)7 mice, about 4- to 8-fold higher than in Syrian hamsters, gives 48 ± 1 day ($n=26$) incubation times with Sc237 and 40 ± 3 day ($n=11$) incubation times with 139H (Hecker *et al.*, 1992). One interpretation of these observations is that Sc237 prions have a higher affinity for PrP^{Sc} than 139H prions. Increased expression of PrP^C substrate in Tg(SHaPrP) mice might saturate the PrP^{Sc} conversion process resulting in diminution of the incubation times for both prion isolates and obliteration of the differences between the two. In Chinese and Armenian hamsters whose PrP gene sequences differ from that of the Syrian at 7 and 8 codons, respectively (Lowenstein *et al.*, 1990), 139H produces incubation times either shorter than or similar to those observed with Sc237. In this case, the amino acid sequence of PrP may modulate the affinities of PrP^{Sc} in the two isolates for PrP^C molecules; indeed, the formation of PrP^C/PrP^{Sc} heterodimers may be the rate-limiting step in prion biosynthesis which determines scrapie incubation times.

Defining the molecular basis of 'prion strains' is one of the most challenging and perplexing problems in biology. A variety of hypotheses have been offered to explain distinct isolates of prions. Despite the lack of physical, chemical and biological evidence for a scrapie-specific nucleic acid, the remote scenario that such a molecule exists remains a formal possibility to explain the distinct prion isolates. A second hypothesis suggests that PrP^{Sc} alone can cause scrapie with the properties of Sc237 but that the 139H isolate contains an accessory RNA molecule which modulates the properties of this prion (Weissmann, 1991a). Such accessory RNAs are of cellular origin and thus would not be detected by subtractive hybridization and differential cloning studies. Synthesis of such RNA molecules would be triggered by PrP^{Sc} or the hypothetical accessory RNA in the prion inoculum. A third hypothesis considers the possibility that scrapie 'strains' may have their origin in a non-PrP molecule which purifies with PrP^{Sc} but is not a nucleic acid and as yet it is undetected. A fourth possibility is that different chemical and conformational modifications of PrP^{Sc} are responsible for the particular biological properties exhibited by scrapie prion 'strains' (Prusiner, 1991).

SIGNIFICANCE OF PRION RESEARCH

Defining the complete molecular structure of the infectious prion particle and learning how prions replicate lie at the centre of both basic and applied future advances in prion biology. Whether prions are composed entirely of PrP^{Sc} molecules or contain a second component needs to be resolved. Learning the molecular events which feature in prion replication should help to decipher the structural basis for scrapie isolates or 'strains' which give different incubation times in the same host. Elucidating the function of PrP^C might extend our understanding of the pathogenesis of prion diseases and provide clues to other macromolecules which may participate in a variety of human and animal diseases of unknown aetiology. It seems likely that lessons learned from prion diseases may give insights into the aetiologies as well as pathogenic mechanisms of such common CNS degenerative disorders afflicting older people as Alzheimer's disease, amyotropic lateral sclerosis and Parkinson's disease.

The advances in our knowledge of prions derived from mice expressing foreign and mutant PrP transgenes demand the production and analysis of many new lines of Tg(PrP) mice, but such research is necessarily limited by the economics of prolonged bioassays. Transgenetics offers new approaches to dissecting the mechanism of prion replication and to deciphering the enigma posed by 'strains' of prions. PrP gene targetting has produced mice with both PrP alleles ablated (Büeler *et al.*, 1992). These mice with ablated PrP genes will greatly enhance transgenetic studies since they provide a method for removal of the host MoPrP genes while leaving the PrP transgene of interest.

Although the results of many studies indicate that prions are a new class of pathogens distinct from both viroids and viruses, it is unknown whether different types of prions exist. Are there prions that contain modified proteins other than PrP^{Sc}? Assessing how widespread prions are in nature and defining their subclasses are subjects for future investigation. Elucidation of the mechanism by which brain cells cease to function and die in prion diseases after a long delay may offer approaches to understanding how neurons develop, mature, transmit signals for decades, and eventually grow senescent.

REFERENCES

Aiken, J.M. and Marsh, R.F. (1990) The search for scrapie agent nucleic acid. *Microbiol. Rev.* **54** 242–246.

Aiken, J.M., Williamson, J.L. and Marsh, R.F. (1989) Evidence of mitochondrial involvement in scrapie infection. *J. Virol.* **63** 1686–1694.

Aiken, J.M., Williamson, J.L., Borchardt, L.M. and Marsh, R.F. (1990) Presence of mitochondrial D-loop DNA in scrapie-infected brain preparations enriched for the prion protein. *J. Virol.* **64** 3265–3268.

Akowitz, A., Sklaviadis, T., Manuelidis, E.E. and Manuelidis, L. (1990) Nuclease-resistant polyadenylated RNAs of significant size are detected by PCR in highly purified Creutzfeldt–Jakob disease preparations. *Microb. Pathog.* **9** 33–45.

Alper, T., Haig, D.A. and Clarke, M.C. (1966) The exceptionally small size of the scrapie agent. *Biochem. Biophys. Res. Commun.* **22** 278–284.

Alper, T., Cramp, W.A., Haig, D.A. and Clarke, M.C. (1967) Does the agent of scrapie replicate without nucleic acid? *Nature (London)* **214** 764–766.

Alper, T., Haig, D.A. and Clarke, M.C. (1978) The scrapie agent: evidence against its dependence for replication on intrinsic nucleic acid. *J. Gen. Virol.* **41** 503–516.

Alpers, M. (1987) Epidemiology and clinical aspects of kuru. In: Prusiner, S. B. and McKinley, M.P. (eds) *Prions—Novel Infectious Pathogens Causing Scrapie and Creutzfeldt–Jakob Disease.* Academic Press, Orlando, FL, pp. 451–465.

Alter, M. and Kahana, E. (1976) Creutzfeldt–Jakob disease among Libyan Jews in Israel. *Science* **192** 428.

Baldwin, M.A., Stahl, N., Reinders, L.G., Gibson, B.W., Prusiner, S.B. and Burlingame, A. L. (1990). Permethylation and tandem mass spectrometry of oligosaccharides having free hexosamine: analysis of the glycoinositol phospholipid anchor glycan from the scrapie prion protein. *Anal. Biochem.* **191** 174–182.

Basler, K., Oesch, B., Scott, M., Westaway, D., Wälchli, M., Groth, D.F., McKinley, M.P., Prusiner, S.B. and Weissmann, C. (1986) Scrapie and cellular PrP isoforms are encoded by the same chromosomal gene. *Cell* **46** 417–428.

Bazan, J.F., Fletterick, R.J., McKinley, M.P. and Prusiner, S.B. (1987) Predicted secondary structure and membrane topology of the scrapie prion protein. *Protein Eng.* **1** 125–135.

Bellinger-Kawahara, C., Cleaver, J.E., Diener, T.O. and Prusiner, S.B. (1987a) Purified scrapie prions resist inactivation by UV irradiation. *J. Virol.* **61** 159–166.

Bellinger-Kawahara, C., Diener, T.O., McKinley, M.P., Groth, D.F., Smith, D.R. and Prusiner, S. B. (1987b) Purified scrapie prions resist inactivation by procedures that hydrolyze, modify, or shear nucleic acids. *Virology* **160** 271–274.

Bellinger-Kawahara, C.G., Kempner, E., Groth, D.F., Gabizon, R. and Prusiner, S. B. (1988) Scrapie prion liposomes and rods exhibit target sizes of 55,000 Da. *Virology* **164** 537–541.

Bendheim, P.E., Barry, R.A., DeArmond, S.J., Stites, D.P. and Prusiner, S.B. (1984) Antibodies to a scrapie prion protein. *Nature (London)* **310** 418–421.

Blum, B., Bakalara, N. and Simpson, L. (1990) A model for RNA editing in kinetoplastid mitochondria: "guide" RNA molecules transcribed from maxicircle DNA provide edited information. *Cell* **60** 189–198.

Bockman, J.M., Kingsbury, D.T., McKinley, M.P., Bendheim, P.E. and Prusiner, S.B. (1985) Creutzfeldt–Jakob disease prion proteins in human brains. *N. Engl. J. Med.* **312** 73–78.

Bockan, J.M., Prusiner, S.B., Tateishi, J. and Kingsbury, D.T. (1987) Immunoblotting of Creutzfeldt–Jakob disease prion proteins: host species-specific epitopes. *Ann. Neurol.* **21** 589–595.

Bolton, D.C., McKinley, M.P. and Prusiner, S.B. (1982) Identification of a protein that purifies with the scrapie prion. *Science* **218** 1309–1311.

Bolton, D.C., Meyer, R.K. and Prusiner, S.B. (1985) Scrapie PrP 27-30 is a sialoglycoprotein. *J. Virol.* **53** 596–606.

Borchelt, D.R., Scott, M., Taraboulos, A., Stahl, N. and Prusiner, S.B. (1990) Scrapie and cellular prion proteins differ in their kinetics of synthesis and topology in cultured cells. *J. Cell Biol.* **110** 743–752.

Borchelt, D.R., Taraboulos, A. and Prusiner, S.B. (1992) Evidence for synthesis of scrapie prion proteins in the endocytic pathway. *J. Biol. Chem.*, in press.

Bradley, R. (1990) Bovine spongiform encephalopathy: the need for knowledge, balance, patience and action. *J. Pathol.* **160** 283–285.

Braig, H. and Diringer, H. (1985) Scrapie: concept of a virus-induced amyloidosis of the brain. *EMBO J.* **4** 2309–2312.

Brown, P., Coker-Vann, M., Pomeroy, K., Franko, M., Asher, D.M., Gibbs, C.J., Jr, and Gajdusek, D.C. (1986) Diagnosis of Creutzfeldt–Jakob disease by Western blot identification of marker protein in human brain tissue. *N. Engl. J. Med.* **314** 547–551.

Bruce, M.E. and Dickinson, A.G. (1987) Biological evidence that the scrapie agent has an independent genome. *J. Gen. Virol.* **68** 79–89.

Buchanan, C.R., Preece, M.A. and Miller, R.D.G. (1991) Mortality, neoplasia, and Creutzfeldt–Jakob disease in patients treated with human pituitary growth hormone in the United Kingdom. *Br. Med. J.* **302** 824–828.

Büeler, H., Fischer, M., Lang, Y., Bluethmann, H., Lipp, H.-P., DeArmond, S.J., Prusiner, S.B., Aguet, M. and Weissmann, C. (1992) Normal development and behaviour of mice lacking the neuronal cell-surface PrP protein. *Nature* **356** 577–582.

Butler, D.A., Scott, M.R.D., Bockman, J.M., Borchelt, D.R., Taraboulos, A., Hsiao, K.K., Kingsbury, D.T. and Prusiner, S.B. (1988) Scrapie-infected murine neuroblastoma cells produce protease-resistant prion proteins. *J. Virol.* **62** 1558–1564.

Carlson, G.A., Kingsbury, D.T., Goodman, P.A., Coleman, S., Marshall, S.T., DeArmond, S.J., Westaway, D. and Prusiner, S.B. (1986) Linkage of prion protein and scrapie incubation time genes. *Cell* **46** 503–511.

Carlson, G.A., Goodman, P.A., Lovett, M., Taylor, B.A., Marshall, S.T., Peterson-Torchia, M., Westaway, D. and Prusiner, S.B. (1988) Genetics and polymorphism of the mouse prion gene complex: the control of scrapie incubation time. *Mol. Cell. Biol.* **8** 5528–5540.

Caughey, B., Race, R.E., Ernst, D., Buchmeier, M.J. and Chesebro, B. (1989) Prion protein biosynthesis in scrapie-infected and uninfected neuroblastoma cells. *J. Virol.* **63** 175–181.

Caughey, B. and Raymond, G.J. (1991) The scrapie-associated form of PrP is made from a cell surface precursor that is both protease- and phospholipase-sensitive. *J. Biol. Chem.* **226** 18217–18223.

Chandler, R.L. (1961) Encephalopathy in mice produced by inoculation with scrapie brain material. *Lancet* **1** 1378–1379.

Chesebro, B., Race, R., Wehrly, K., Nishio, J., Bloom, M., Lechner, D., Bergstrom, S., Robbins, K., Mayer, L., Keith, J.M., Garon, C. and Haase, A. (1985) Identification of scrapie prion protein-specific mRNA in scrapie-infected and uninfected brain. *Nature (London)* **315** 331–333.

Cho, H.J. (1980) Requirement of a protein component for scrapie infectivity. *Intervirology* **14** 213–216.

Collinge, J., Harding, A.E., Owen, F., Poulter, M., Lofthouse, R., Boughey, A.M., Shah, T. and Crow, T.J. (1989) Diagnosis of Gerstmann–Sträussler syndrome in familial dementia with prion protein gene analysis. *Lancet* **2** 15–17.

Collinge, J., Owen, F., Poulter, H., Leach, M., Crow, T., Rosser, M., Hardy, J., Mullan, H., Janota, I. and Lantos, P. (1990) Prion dementia without characteristic pathology. *Lancet* **336** 7–9.

Crow, T.J., Collinge, J., Ridley, R.M., Baker, H.F., Lofthouse, R., Owen, F. and

Harding, A.E. (1990) Mutations in the prion gene in human transmissible dementia. (abstract) *Seminar on Molecular Approaches to Research in Spongiform Encephalopathies in Man, Medical Research Council, London.*

Czub, M., Braig, H.R. and Diringer, H. (1986) Pathogenesis of scrapie: study of the temporal development of clinical symptoms of infectivity titres and scrapie-associated fibrils in brains of hamsters infected intraperitoneally. *J. Gen. Virol.* **67** 2005–2009.

Czub, M., Braig, H.R. and Diringer, H. (1988) Replication of the scrapie agent in hamsters infected intracerebrally confirms the pathogenesis of an amyloid-inducing virosis. *J. Gen. Virol.* **69** 1753–1756.

Dealler, S.F. and Lacey, R.W. (1990) Transmissible spongiform encephalopathies: the threat of BSE to man. *Food Microbiol.* **7** 253–279.

DeArmond, S.J., McKinley, M.P., Barry, R.A., Braunfeld, M.B., McColloch, J.R. and Prusiner, S.B. (1985) Identification of prion amyloid filaments in scrapie-infected brain. *Cell* **41** 221–235.

Dickinson, A.G. and Fraser, H. (1979) An assessment of the genetics of scrapie in sheep and mice. In: Prusiner, S.B. and Hadlow, W.J. (eds) *Slow Transmissible Diseases of the Nervous System*, Vol. 1. Academic Press, New York, pp. 367–386.

Dickinson, A.G. and Outram, G.W. (1988) Genetic aspects of unconventional virus infections: the basis of the virino hypothesis. In: Bock, G. and Marsh, J. (eds) *Novel Infectious Agents and the Central Nervous System*. Ciba Foundation Symposium 135, Wiley, Chichester, pp. 63–83.

Dickinson, A.G., Young, G.B., Stamp, J.T. and Renwick, C.C. (1965) An analysis of natural scrapie in Suffolk sheep. *Heredity* **20** 485–503.

Dickinson, A.G., Meikle, V.M.H. and Fraser, H. (1968) Identification of a gene which controls the incubation period of some strains of scrapie agent in mice. *J. Comp. Pathol.* **78** 293–299.

Diedrich, J., Weitgrefe, S., Zupancic, M., Staskus, K., Retzel, E., Haase, A.T. and Race, R. (1987) The molecular pathogenesis of astrogliosis in scrapie and Alzheimer's disease. *Microb. Pathog.* **2** 435–442.

Diener, T.O. (1987) PrP and the nature of the scrapie agent. *Cell* **49** 719–721.

Diener, T.O., McKinley, M.P. and Prusiner, S.B. (1982) Viroids and prions. *Proc. Natl. Acad. Sci. USA* **79** 5220–5224.

Diringer, H., Gelderblom, H., Hilmert, H., Ozel, M., Edelbluth, C. and Kimberlin, R.H. (1983) Scrapie infectivity, fibrils and low molecular weight protein. *Nature (London)* **306** 476–478.

Doh-ura, K., Tateishi, J., Sasaki, H., Kitamoto, T. and Sakaki, Y. (1989) Pro→Leu change at position 102 of prion protein is the most common but not the sole mutation related to Gerstmann–Sträussler syndrome. *Biochem. Biophys. Res. Commun.* **163** 974–979.

Doms, R.W., Russ, G. and Yewdell, J.W. (1989) Brefeldin A redistributes resident and itinerant Golgi proteins to the endoplasmic reticulum. *J. Cell. Biol.* **109** 61–72.

Duguid, J. R., Rohwer, R. G. and Seed, B. (1988) Isolation of cDNAs of scrapie-modulated RNAs by subtractive hybridization of a cDNA library. *Proc. Natl. Acad. Sci. USA* **85** 5738–5742.

Endo, T., Groth, D., Prusiner, S.B. and Kobata, A. (1989) Diversity of oligosaccharide structures linked to asparigines of the scrapie prion protein. *Biochemistry* **28** 8380–8388.

Falls, D.L., Harris, D.A., Johnson, F.A., Morgan, M.M., Corfas, G. and Fischbach, G.D. (1990) 42 kD ARIA: a protein that may regulate the accumulation of acetylcholine receptors at developing chick neuromuscular junctions. *Cold Spring Harbor Sympos. Quant. Biol.* **55** 397–406.

Farlow, M.R., Yee, R.D., Dlouhy, S.R., Conneally, P.M., Azzarelli, B. and Ghetti, B. (1989) Gerstmann–Sträussler–Scheinker disease. I. Extending the clinical spectrum. *Neurology* **39** 1446–1452.

Gabizon, R. and Prusiner, S.B. (1990) Prion liposomes. *Biochem. J.* **266** 1–14.

Gabizon, R., McKinley, M.P. and Prusiner, S.B. (1987) Purified prion proteins and scrapie infectivity copartition into liposomes. *Proc. Natl. Acad. Sci. USA* **84** 4017–4021.

Gabizon, R., McKinley, M.P., Groth, D.F., Kenaga, L. and Prusiner, S.B. (1988a) Properties of scrapie prion liposomes. *J. Biol. Chem.* **263** 4950–4955.

Gabizon, R., McKinley, M.P., Groth, D.F. and Prusiner, S.B. (1988b) Immunoaffinity purification and neutralization of scrapie prion infectivity. *Proc. Natl. Acad. Sci. USA* **85** 6617–6621.

Gabizon, R., Meiner, Z., Cass, C., Kahana, E., Kahana, I., Avrahami, D., Abramsky, O., Scarlato, G., Prusiner, S.B. and Hsiao, K.K. (1991) Prion protein gene mutation in Libyan Jews with Creutzfeldt–Jakob disease. *Neurology* **41** 160.

Gajdusek, D.C. (1977) Unconventional viruses and the origin and disappearance of kuru. *Science* **197** 943–960.

Gajdusek, D.C. (1988) Transmissible and non-transmissible amyloidoses: autocatalytic post-translational conversion of host precursor proteins to β-pleated sheet configurations. *J. Neuroimmunol.* **20** 95–110.

Gajdusek, D.C. (1990) Subacute spongiform encephalopathies: transmissible cerebral amyloidoses caused by unconventional viruses. In: Fields, B. N., Knipe, D. M., Chanock, R.M., Hirsch, M.S., Melnick, J.L., Monath, T.P. and Roizman, B. (eds) *Virology*, 2nd edn. Raven Press, New York, pp. 2289–2324.

Gajdusek, D.C. and Gibbs, C.J., Jr (1990) Brain amyloidoses—precursor protein and the amyloids of transmissible and nontransmissible dementias: scrapie–kuru–CJD viruses as infectious polypeptides or amyloid enhancing factor. In: Goldstein, A. (ed.) *Biomedical Advances in Aging*. Plenum, New York, pp. 3–24.

Gajdusek, D.C., Gibbs, C.J., Jr, and Alpers, M. (1966) Experimental transmission of a kuru-like syndrome to chimpanzees. *Nature (London)* **209** 794–796.

Ghetti, B., Tagliavini, F., Masters, C.L., Beyreuther, K., Giaccone, G., Verga, L., Farlo, M.R., Conneally, P.M., Dlouhy, S.R., Azzarelli, B. and Bugiani, O. (1989) Gerstmann–Sträussler–Scheinker disease. II. Neurofibrillary tangles and plaques with PrP-amyloid coexist in an affected family. *Neurology* **39** 1453–1461.

Giaccone, G., Tagliavini, F., Verga, L., Frangione, B., Farlow, M.R., Bugiani, O. and Ghetti, B. (1990) Neurofibrillary tangles of the Indiana kindred of Gerstmann–Sträussler–Scheinker disease share antigenic determinants with those of Alzheimer disease. *Brain Res.* **530** 325–329.

Gibbons, R.A. and Hunter, G.D. (1967) Nature of the scrapie agent. *Nature (London)* **215** 1041–1043.

Gibbs, C.J., Jr, Gajdusek, D.C., Asher, D.M., Alpers, M.P., Beck, E., Daniel, P.M. and Matthews, W. B. (1968) Creutzfeldt–Jakob disease (spongiform encephalopathy): transmission to the chimpanzee. *Science* **161** 388–389.

Goldfarb, L., Brown, P., Goldgaber, D., Garruto, R., Yanaghiara, R., Asher, D. and Gajdusek, D.C. (1990a) Identical mutation in unrelated patients with Creutzfeldt–Jakob disease. *Lancet* **336** 174–175.

Goldfarb, L., Korczyn, A., Brown, P., Chapman, J. and Gajdusek, D.C. (1990b) Mutation in codon 200 of scrapie amyloid precursor gene linked to Creutzfeldt–Jakob disease in Sephardic Jews of Libyan and non-Libyan origin. *Lancet* **336** 637–638.

Goldfarb, L.G., Brown, P., Goldgaber, D., Asher, D.M., Rubenstein, R., Brown, W.T., Piccardo, P., Kascsak, R.J., Boellaard, J.W. and Gajdusek, D.C. (1990c) Creutzfeldt–Jakob disease and kuru patients lack a mutation consistently found in the Gerstmann–Sträussler–Scheinker syndrome. *Exp. Neurol.* **108** 247–250.

Goldfarb, L.G., Mitrova, E., Brown, P., Toh, B.H. and Gajdusek, D.C. (1990d) Mutation in codon 200 of scrapie amyloid protein gene in two clusters of Creutzfeldt–Jakob disease in Slovakia. *Lancet* **336** 514–515.

Goldfarb, L.G., Haltia, M., Brown, P., Nieto, A., Kovanen, J., McCombie, W.R., Trapp, S. and Gajdusek, D.C. (1991a) New mutation in scrapie amyloid precursor gene (at codon 178) in Finnish Creutzfeldt–Jakob kindred. *Lancet* **337** 425.

Goldfarb, L.G., Brown, P., McCombie, W.R., Goldgaber, D., Swergold, G.D., Wills, P.R., Cervenakova, L., Baron, H., Gibbs, C.J., Jr, and Gajdusek, D.C. (1991b) Transmissible familial Creutzfeldt–Jakob disease associated with five, seven, and eight extra octapeptide coding repeats in the *PRNP* gene. *Proc. Natl. Acad. Sci. USA* **88** 10926–10930.

Goldgaber, D., Goldfarb, L.G., Brown, P., Asher, D.M., Brown, W.T., Lin, S., Teener, J.W., Feinstone, S.M., Rubenstein, R., Kascsak, R.J., Boellaard, J.W. and Gajdusek, D.C. (1989) Mutations in familial Creutzfeldt–Jakob disease and Gerstmann–Sträussler–Scheinker's syndrome. *Exp. Neurol.* **106** 204–206.

Goldmann, W., Hunter, N., Foster, J.D., Salbaum, J.M., Beyreuther, K.and Hope, J. (1990a) Two alleles of a neural protein gene linked to scrapie in sheep. *Proc. Natl. Acad. Sci. USA* **87** 2476–2480.

Goldmann, W., Hunter, N., Manson, J. and Hope, J. (1990b) The PrP gene of the sheep, a natural host of scrapie. In: *Proc. VIIIth Int. Congress of Virology, Berlin, August 26–31 (Abstracts)*. p. 284.

Goldmann, W., Hunter, N., Martin, T., Dawson, M. and Hope, J. (1991) Different forms of the bovine PrP gene have five or six copies of a short, G–C-rich element with the protein-coding exon. *J. Gen. Virol.* **72** 201–204.

Gordon, W.S. (1946) Advances in veterinary research. *Vet. Res.* **58** 516–520.

Griffith, J.S. (1967) Self-replication and scrapie. *Nature (London)* **215** 1043–1044.

Haraguchi, T., Fisher, S., Olofsson, S., Endo, T., Groth, D., Tarantino, A., Borchelt, D.R., Teplow, D., Hood, L., Burlingame, A., Lycke, E., Kobata, A. and Prusiner, S.B. (1989) Asparagine-linked glycosylation of the scrapie and cellular prion proteins. *Arch. Biochem. Biophys.* **274** 1–13.

Hardy, J. (1991) Prion dimers—a deadly duo. *Trends Neurosci.* **14** 423–424.

Harris, D.A., Falls, D.L., Walsh, W. and Fischbach, G.D. (19890 Molecular cloning of an acetylcholine receptor-inducing protein. *Soc. Neurosci.* **15** 70.7.

Hay, B., Barry, R.A., Lieberburg, I., Prusiner, S.B. and Lingappa, V.R. (1987a) Biogenesis and transmembrane orientation of the cellular isoform of the scrapie prion protein. *Mol. Cell. Biol.* **7** 914–920.

Hay, B., Prusiner, S.B. and Lingappa, V.R. (1987b) Evidence for a secretory form of the cellular prion protein. *Biochemistry* **26** 8110–8115.

Hecker, R., Taraboulos, A., Scott, M., Pan, K.-M., Torchia, M., Jendroska, K., DeArmond, S.J. and Prusiner, S.B. (1992) Replication of distinct prion isolates is region specific in brains of transgenic mice and hamsters. *Genes and Development* **6** 1213–1228.

Herzberg, L., Herzberg, B.N., Gibbs, C.J., Jr, Sullivan, W., Amyx, H. and Gajdusek, D.C. (1974) Creutzfeldt–Jakob disease: hypothesis for high incidence in Libyan Jews in Israel. *Science* **186** 848.

Hope, J., Morton, L.J.D., Farquhar, C.F., Multhaup, G., Beyreuther, K. and Kimberlin, R.H. (1986) The major polypeptide of scrapie-associated fibrils (SAF) has the same size, charge distribution and *N*-terminal protein sequence as predicted for the normal brain protein (PrP). *EMBO J.* **5** 2591–2597.

Hope, J., Reekie, L.J.D., Hunter, N., Multhaup, G., Beyreuther, K., White, H., Scott, A.C., Stack, M.J., Dawson, M. and Wells, G.A.H. (1988) Fibrils from brain of cows with new cattle disease contain scrapie-associated protein. *Nature (London)* **336** 390–392.

Hsiao, K. and Prusiner, S.B. (1990) Inherited human prion diseases. *Neurology* **40** 1820–1827.

Hsiao, K., Baker, H.F., Crow, T.J., Poulter, M., Owen, F., Terwilliger, J.D., Westaway, D., Ott, J. and Prusiner, S.B. (1989a) Linkage of a prion protein missense variant to Gerstmann-Sträussler syndrome. *Nature (London)* **338** 342-345.

Hsiao, K., Baker, H.F., Crow, T.J., Poulter, M., Owen, F., Terwilliger, J.D., Westaway, D., Ott, J. and Prusiner, S.B. (1989a) Linkage of a prion protein missense variant to Gerstmann-Sträussler syndrome. *Nature (London)* **338** 342-345.

Hsiao, K.K., Doh-ura, K., Kitamoto, T., Tateishi, J. and Prusiner, S.B. (1989b) A prion protein amino acid substitution in ataxic Gerstmann–Sträussler syndrome. *Ann. Neurol.* **26** 137.

Hsiao, K.K., Scott, M., Foster, D., Groth, D.F., DeArmond, S.J. and Prusiner, S.B. (1990) Spontaneous neurodegeneration in transgenic mice with mutant prion protein of Gerstmann–Sträussler syndrome. *Science* **250** 1587–1590.

Hsiao, K., Meiner, Z., Kahana, E., Cass, C., Kahana, I., Avrahami, D., Scarlatto, G., Abramsky, O., Prusiner, S.B. and Gabizon, R. (1991a) Mutation of the prion protein in Libyan Jews with Creutzfeldt–Jakob disease. *N. Engl. J. Med.* **324** 1091–1097.

Hsiao, K.K., Cass, C., Schellenberg, G.D., Bird, T., Devine-Gage, E. and Prusiner, S.B. (1991b) A prion protein variant in a family with the telencephalic form of Gerstmann–Sträussler–Scheinker syndrome. *Neurology* **41** 681–684.

Hsiao, K.K., Groth, D., Scott, M., Yang, S.-L., Serban, A., Rapp, D., Foster, D.,

Torchia, M., DeArmond, S.J. and Prusiner, S.B. (1991c) Neurologic disease of transgenic mice which express GSS mutant prion protein is transmissible to inoculated recipient animals. *Prion Diseases in Humans and Animals Symp., London, September 2–4, 1991 (Abstracts)*.

Hsiao, K., Dlouhy, S.R., Farlow, M.R., Cass, C., Da Costa, M., Conneally, P.M., Hodes, M.E., Ghetti, B. and Prusiner, S.B. (1992) Mutant prion proteins in Gerstmann–Sträussler–Scheinker disease with neurofibrillary tangles. *Nature Genetics* **1** 68–71.

Hunter, G.D. (1972) Scrapie: a prototype slow infection. *J. Infect. Dis.* **125** 427–440.

Hunter, G.D. and Millson, G.C. (1967) Attempts to release the scrapie agent from tissue debris. *J. Comp. Pathol.* **77** 301–307.

Hunter, G.D., Gibbons, R.A., Kimberlin, R.H. and Millson, G.C. (1969) Further studies of the infectivity and stability of extracts and homogenates derived from scrapie affected mouse brains. *J. Comp. Pathol.* **79** 101–108.

Hunter, N., Hope, J., McConnell, I. and Dickinson, A.G. (1987) Linkage of the scrapie-associated fibril protein (PrP) gene and Sinc using congenic mice and restriction fragment length polymorphism analysis. *J. Gen. Virol.* **68** 2711–2716.

Hunter, N., Foster, J.D., Dickinson, A.G. and Hope, J. (1989) Linkage of the gene for the scrapie-associated fibril protein (PrP) to the Sip gene in Cheviot sheep. *Vet. Rec.* **124** 364–366.

Jendroska, K., Heinzel, F.P., Torchia, M., Stowring, L., Kretzschmar, H.A., Kon, A., Stern, A., Prusiner, S.B. and DeArmond, S.J. (1991) Proteinase-resistant prion protein accumulation in Syrian hamster brain correlates with regional pathology and scrapie infectivity. *Neurology* **41** 1482-1490.

Kahana, E., Milton, A., Braham, J. and Sofer, D. (1974) Creutzfeldt–Jakob disease: focus among Libyan Jews in Israel. *Science* **183** 90–91.

Kane, P.M., Yamashiro, C.T., Wolczyk, D.F., Neff, N., Goebl, M. and Stevens, T.H. (1990) Protein splicing converts the yeast *TFP1* gene product to the 69-kD subunit of the vacuolar H^+-adenosine triphosphatase. *Science* **250** 651–657.

Kimberlin, R.H. (1990) Scrapie and possible relationships with viroids. *Semin. Virol.* **1** 153–162.

Kimberlin, R.H., Cole, S. and Walker, C.A. (1987) Temporary and permanent modifications to a single strain of mouse scrapie on transmission to rats and hamsters. *J. Gen. Virol.* **68** 1875–1881.

Kirkwood, J.K., Wells, G.A. H., Wilesmith, J.W., Cunningham, A.A. and Jackson, S.I. (1990) Spongiform encephalopathy in an arabian oryx (*Oryx leucoryx*) and a greater kudu (*Tragelaphus strepsiceros*). *Vet. Rec.* **127** 418–420.

Kitamoto, T., Tateishi, J., Tashima, I., Takeshita, I., Barry, R.A., DeArmond, S.J. and Prusiner, S.B. (1986) Amyloid plaques in Creutzfeldt–Jakob disease stain with prion protein antibodies. *Ann. Neurol.* **20** 204–208.

Kretzschmar, H.A., Prusiner, S.B., Stowring, L.E. and DeArmond, S.J. (1986a) Scrapie prion proteins are synthesized in neurons. *Am. J. Pathol.* **122** 1–5.

Kretzschmar, H.A., Stowring, L.E., Westaway, D., Stubblebine, W.H., Prusiner, S.B. and DeArmond, S.J. (1986b) Molecular cloning of a human prion protein cDNA. *DNA* **5** 315–324.

Laplanche, J.-L., Chatelain, J., Launay, J.-M., Gazengel, C. and Vidaud, M. (1990) Deletion in prion protein gene in a Moroccan family. *Nucleic Acids Res.* **18** 6745.

Latarjet, R., Muel, B., Haig, D.A., Clarke, M.C. and Alper, T. (1970) Inactivation of the scrapie agent by near monochromatic ultraviolet light. *Nature (London)* **227** 1341–1343.

Liao, Y.-C., Lebo, R.V., Clawson, G.A. and Smuckler, E.A. (1986) Human prion protein cDNA: molecular cloning, chromosomal mapping, and biological implication. *Science* **233** 364--367.

Lippincott-Schwartz, J., Yuan, L.C., Bonifacino, J.S. and Klausner, R.D. (1989) Rapid redistribution of Golgi proteins into the ER in cells treated with Brefeldin A: evidence for membrane cycling from the Golgi to ER. *Cell* **56** 801–813.

Locht, C., Chesebro, B., Race, R. and Keith, J.M. (1986) Molecular cloning and complete sequence of prion protein cDNA from mouse brain infected with the scrapie agent. *Proc. Natl. Acad. Sci. USA* **83** 6372–6376.

Lopez, C.D., Yost, C.S., Prusiner, S.B., Myers, R.M. and Lingappa, V.R. (1990) Unusual topogenic sequence directs prion protein biogenesis. *Science* **248** 226–229.

Lowenstein, D.H., Butler, D.A., Westaway, D., McKinley, M.P., DeArmond, S.J. and Prusiner, S.B. (1990) Three hamster species with different scrapie incubation times and neuropathological features encode distinct prion proteins. *Mol. Cell. Biol.* **10** 1153–1163.

Manuelidis, L. and Manuelidis, E.E. (1989) Creutzfeldt–Jakob disease and dementias. *Microb. Pathog.* **7** 157–164.

Manuelidis, L., Valley, S. and Manuelidis, E.E. (1985) Specific proteins associated with Creutzfeldt–Jakob disease and scrapie share antigenic and carbohydrate determinants. *Proc. Natl. Acad. Sci. USA* **82** 4263–4267.

Masters, C.L., Gajdusek, D.C. and Gibbs, C.J., Jr (1981a) Creutzfeldt–Jakob disease virus isolations from the Gerstmann–Sträussler syndrome. *Brain* **104** 559–588.

Masters, C.L., Gajdusek, D.C. and Gibbs, C.J., Jr (1981b) The familial occurrence of Creutzfeldt–Jakob disease and Alzheimer's disease. *Brain* **104** 535–558.

McKinley, M.P., Bolton, D.C. and Prusiner, S.B. (1983a) A protease-resistant protein is a structural component of the scrapie prion. *Cell* **35** 57–62.

McKinley, M.P., Masiarz, F.R., Isaacs, S.T., Hearst, J.E. and Prusiner, S.B. (1983b) Resistance of the scrapie agent to inactivation by psoralens. *Photochem. Photobiol.* **37** 539–545.

McKinley, M.P., Meyer, R., Kenaga, L., Rahbar, F., Cotter, R., Serban, A. and Prusiner, S.B. (1991a) Scrapie prion rod formation *in vitro* requires both detergent extraction and limited proteolysis. *J. Virol.* **65** 1440–1449.

McKinley, M.P., Taraboulos, A., Kenaga, L., Serban, D., Stieber, A., DeArmond, S.J., Prusiner, S.B. and Gonatas, N. (1991b) Ultrastructural localization of scrapie prion proteins in cytoplasmic vesicles of infected cultured cells. *Lab. Invest* **65** 622–630.

McKnight, S. and Tjian, R. (1986) Transcriptional selectivity of viral genes in mammalian cells. *Cell* **46** 795–805.

Merz, P.A., Somerville, R.A., Wisniewski, H.M. and Iqbal, K. (1981) Abnormal fibrils from scrapie-infected brain. *Acta Neuropathol. (Berlin)* **54** 63–74.

Merz, P.A., Wisniewski, H.M., Somerville, R.A., Bobin, S.A., Masters, C.L. and Iqbal,

K. (1983) Ultrastructural morphology of amyloid fibrils from neuritic and amyloid plaques. *Acta Neuropathol.* (*Berlin*) **60** 113–124.

Merz, P.A., Rohwer, R.G., Kascsak, R., Wisniewski, H.M., Somerville, R.A., Gibbs, C.J., Jr, and Gajdusek, D.C. (1984) Infection-specific particle from the unconventional slow virus diseases. *Science* **225** 437–440.

Merz, P.A., Kascsak, R.J., Rubenstein, R., Carp, R.I. and Wisniewski, H.M. (1987) Antisera to scrapie-associated fibril protein and prion protein decorate scrapie-associated fibrils. *J. Virol.* **61** 42–49.

Meyer, N., Rosenbaum, V., Schmidt, B., Gilles, K., Mirenda, C., Groth, D., Prusiner, S.B. and Riesner, D. (1991) Search for a putative scrapie genome in purified prion fractions reveals a paucity of nucleic acids. *J. Gen. Virol.* **72** 37–49.

Millson, G., Hunter, G.D. and Kimberlin, R.H. (1971) An experimental examination of the scrapie agent in cell membrane mixtures. II. The association of scrapie infectivity with membrane fractions. *J. Comp. Pathol.* **81** 255–265.

Mobley, W.C., Neve, R.L., Prusiner, S.B. and McKinley, M.P. (1988) Nerve growth factor increases mRNA levels for the prion protein and the beta-amyloid protein precursor in developing hamster brain. *Proc. Natl. Acad. Sci. USA* **85** 9811–9815.

Murdoch, G.H., Sklaviadis, T., Manuelidis, E.E. and Manuelidis, L. (1990) Potential retroviral RNAs in Creutzfeldt–Jakob disease. *J. Virol.* **64** 1477–1486.

Neary, K., Caughey, B., Ernst, D., Race, R.E. and Chesebro, B. (1991) Protease sensitivity and nuclease resistance of the scrapie agent propagated *in vitro* in neuroblastoma-cells. *J. Virol.* **65** 1031–1034.

Nochlin, D., Sumi, S.M., Bird, T.D., Snow, A.D., Leventhal, C.M., Beyreuther, K. and Masters, C.L. (1989) Familial dementia with PrP-positive amyloid plaques: a variant of Gerstmann–Sträussler syndrome. *Neurology* **39** 910–918.

Oesch, B., Westaway, D., Wälchli, M., McKinley, M.P., Kent, S.B.H., Aebersold, R., Barry, R.A., Tempst, P., Teplow, D.B., Hood, L.E., Prusiner, S.B. and Weissmann, C. (1985) A cellular gene encodes scrapie PrP 27-30 protein. *Cell* **40** 735–746.

Oesch, B., Groth, D.F., Prusiner, S.B. and Weissmann, C. (1988) Search for a scrapie-specific nucleic acid: a progress report. In: Bock, G. and Marsh, J. (eds) *Novel Infectious Agents and the Central Nervous System.* Ciba Foundation Symposium 135, Wiley, Chichester, pp. 209–223.

Owen, F., Poulter, M., Lofthouse, R., Collinge, J., Crow, T.J., Risby, D., Baker, H.F., Ridley, R.M., Hsiao, K. and Prusiner, S.B. (1989) Insertion in prion protein gene in familial Creutzfeldt–Jakob disease. *Lancet* **1** 51–52.

Owen, F., Poulter, M., Collinge, J. and Crow, T.J. (1990a) Codon 129 changes in the prion protein gene in Caucasians. *Am. J. Hum. Genet.* **46** 1215–1216.

Owen, F., Poulter, M., Shah, T., Collinge, J., Lofthouse, R., Baker, H., Ridley, R., McVey, J. and Crow, T. (1990b) An in-frame insertion in the prion protein gene in familial Creutzfeldt–Jakob disease. *Mol. Brain Res.* **7** 273–276.

Owen, F., Poulter, M., Collinge, J., Leach, M., Shah, T., Lofthouse, R., Chen, Y.F., Crow, T.J., Harding, A.E. and Hardy, J. (1991) Insertions in the prion protein gene in atypical dementias. *Exp. Neurol.* **112** 240–242.

Pain, S. (1990) BSE: what madness is this? *New Scientist* (June 9) 32–34.

Palmer, M.S., Dryden, A.J., Hughes, J.T. and Collinge, J. (1991) Homozygous prion protein genotype predisposes to sporadic Creutzfeldt–Jakob disease. *Nature (London)* **352** 340–342.

Parry, H.B. (1962) Scrapie: a transmissible and hereditary disease of sheep. *Heredity* **17** 75–105.

Parry, H.B. (1983) *Scrapie Disease in Sheep*. Academic Press, New York, 192 p.

Pattison, I. H. (1965) Experiments with scrapie with special reference to the nature of the agent and the pathology of the disease. In: Gajdusek, D.C., Gibbs, C.J., Jr, and Alpers, M.P. (eds) *Slow, Latent and Temperate Virus Infections*. NINDB Monograph 2, US Government Printing Office, Washington, DC, pp. 249–257.

Pattison, I. H. (1966) The relative susceptibility of sheep, goats and mice to two types of the goat scrapie agent. *Res. Vet. Sci.* **7** 207–212.

Pattison, I.H. and Jones, K.M. (1967) The possible nature of the transmissible agent of scrapie. *Vet. Rec.* **80** 1–8.

Pocchiari, M. (1990) Methodological aspects of the validation of purification procedures of human/animal-derived products to remove unconventional slow viruses. In: *Symp. on Virological Aspects of the Safety of Biological Products, London, November 8–9, 1990 (Abstracts)*. p. 13.

Prusiner, S.B. (1982) Novel proteinaceous infectious particles cause scrapie. *Science* **216** 136–144.

Prusiner, S.B. (1991) Molecular biology of prion diseases. *Science* **252** 1515–1522.

Prusiner, S.B., Hadlow, W.J., Garfin, D.E., Cochran, S.P., Baringer, J.R., Race, R.E. and Eklund, C. M. (1978) Partial purification and evidence for multiple molecular forms of the scrapie agent. *Biochemistry* **17** 4993–4997.

Prusiner, S.B., Groth, D.F., Cochran, S.P., Masiarz, F.R., McKinley, M.P. and Martinez, H.M. (1980) Molecular properties, partial purification, and assay by incubation period measurements of the hamster scrapie agent. *Biochemistry* **19** 4883–4891.

Prusiner, S.B., McKinley, M.P., Groth, D.F., Bowman, K.A., Mock, N.I., Cochran, S.P. and Masiarz, F.R. (1981) Scrapie agent contains a hydrophobic protein. *Proc. Natl. Acad. Sci. USA* **78** 6675–6679.

Prusiner, S.B., Bolton, D.C., Groth, D.F., Bowman, K.A., Cochran, S.P. and McKinley, M.P. (1982a) Further purification and characterization of scrapie prions. *Biochemistry* **21** 6942–6950.

Prusiner, S.B., Cochran, S.P., Groth, D.F., Downey, D.E., Bowman, K.A. and Martinez, H.M. (1982b) Measurement of the scrapie agent using an incubation time interval assay. *Ann. Neurol.* **11** 353–358.

Prusiner, S.B., McKinley, M.P., Bowman, K.A., Bolton, D.C., Bendheim, P.E., Groth, D.F. and Glenner, G.G. (1983) Scrapie prions aggregate to form amyloid-like birefringent rods. *Cell* **35** 349–358.

Prusiner, S.B., Groth, D.F., Bolton, D.C., Kent, S.B. and Hood, L.E. (1984) Purification and structural studies of a major scrapie prion protein. *Cell* **38** 127–134.

Prusiner, S.B., Scott, M., Foster, D., Pan, K.-M., Groth, D., Mirenda, C., Torchia, M., Yang, S.-L., Serban, D., Carlson, G.A., Hoppe, P.C., Westaway, D. and DeArmond, S.J. (1990) Transgenetic studies implicate interactions between homologous PrP isoforms in scrapie prion replication. *Cell* **63** 673–686.

Race, R.E., Graham, K., Ernst, D., Caughey, B. and Chesebro, B. (1990) Analysis of linkage between scrapie incubation period and the prion protein gene in mice. *J. Gen. Virol.* **71** 493–497.

Robakis, N.K., Devine-Gage, E.A., Kascsak, R.J., Brown, W.T., Krawczun, C. and Silverman, W.P. (1986) Localization of a human gene homologous to the PrP gene on the p arm of chromosome 20 and detection of PrP-related antigens in normal human brain. *Biochem. Biophys. Res. Commun.* **140** 758–765.

Roberts, G.W., Lofthouse, R., Allsop, D., Landon, M., Kidd, M., Prusiner, S.B. and Crow, T.J. (1988) CNS amyloid proteins in neurodegenerative diseases. *Neurology* **38** 1534–1540.

Rogers, M., Taraboulos, A., Scott, M., Groth, D. and Prusiner, S.B. (1990) Intracellular accumulation of the cellular prion protein after mutagenesis of its Asn-linked glycosylation sites. *Glycobiology* **1** 101–109.

Safar, J., Ceroni, M., Piccardo, P., Liberski, P.P., Miyazaki, M., Gajdusek, D.C. and Gibbs, C.J., Jr (1990a) Subcellular distribution and physicochemical properties of scrapie associated precursor protein and relationship with scrapie agent. *Neurology* **40** 503–508.

Safar, J., Wang, W., Padgett, M.P., Ceroni, M., Piccardo, P., Zopf, D., Gajdusek, D.C. and Gibbs, C.J., Jr (1990b) Molecular mass, biochemical composition, and physicochemical behavior of the infectious form of the scrapie precursor protein monomer. *Proc. Natl. Acad. Sci. USA* **87** 6373–6377.

Scott, M., Foster, D., Mirenda, C., Serban, D., Coufal, F., Wälchli, M., Torchia, M., Groth, D., Carlson, G., DeArmond, S.J., Westaway, D. and Prusiner, S.B. (1989) Transgenic mice expressing hamster prion protein produce species-specific scrapie infectivity and amyloid plaques. *Cell* **59** 847–857.

Scott, A.C., Wells, G.A.H., Stack, M.J., White, H. and Dawson, M. (1990) Bovine spongiform encephalopathy: detection and quantitation of fibrils, fibril protein (PrP) and vacuolation in brain. *Vet. Microbiol.* **23** 295–304.

Serban, D., Taraboulos, A., DeArmond, S.J. and Prusiner, S.B. (1990) Rapid detection of Creutzfeldt–Jakob disease and scrapie prion proteins. *Neurology* **40** 110–117.

Sklaviadis, T.K., Manuelidis, L. and Manuelidis, E.E. (1989) Physical properties of the Creutzfeldt–Jakob disease agent. *J. Virol.* **63** 1212–1222.

Sklaviadis, T., Akowitz, A., Manuelidis, E.E. and Manuelidis, L. (1990) Nuclease treatment results in high specific purification of Creutzfeldt–Jakob disease infectivity with a density characteristic of nucleic acid–protein complexes. *Arch. Virol.* **112** 215–229.

Somerville, R.A., Ritchie, L.A. and Gibson, P.H. (1989) Structural and biochemical evidence that scrapie-associated fibrils assemble *in vivo*. *J. Gen. Virol.* **70** 25–35.

Sparkes, R.S., Simon, M., Cohn, V.H., Fournier, R.E. K., Lem, J., Klisak, I., Heinzmann, C., Blatt, C., Lucero, M., Mohandas, T., DeArmond, S.J., Westaway, D., Prusiner, S.B. and Weiner, L.P. (1986) Assignment of the human and mouse prion protein genes to homologous chromosomes. *Proc. Natl. Acad. Sci. USA* **83** 7358–7362.

Stahl, N., Borchelt, D.R., Hsiao, K. and Prusiner, S.B. (1987) Scrapie prion protein contains a phosphatidylinositol glycolipid. *Cell* **51** 229–240.

Stahl, N., Baldwin, M.A., Burlingame, A.L. and Prusiner, S.B. (1990a) Identification of glycoinositol phospholipid-linked and truncated forms of the scrapie prion protein. *Biochemistry* **29** 8879–8884.

Stahl, N., Borchelt, D.R. and Prusiner, S.B. (1990b) Differential release of cellular and scrapie prion proteins from cellular membranes by phosphatidylinositol-specific phospholipase C. *Biochemistry* **29** 5405–5412.

Stahl, N., Baldwin, M.A., Teplow, D., Hood, L., Beavis, R., Chait, B., Gibson, B.W., Burlingame, A.L. and Prusiner, S.B. (1992) Cataloguing post-translational modifications of the scrapie prion protein by mass spectrometry. In: Prusiner, S. B., Collinge, J., Powell, J. and Anderton, B. (eds) *Prion Diseases of Humans and Animals.* Ellis Horwood, Chichester, Chap. 32.

Tagliavini, F., Prelli, F., Ghisto, J., Bugiani, O., Serban, S., Prusiner, S.B., Farlow, M.R., Ghetti, B. and Frangione, B. (1991) Amyloid protein of Gerstmann–Sträussler–Scheinker disease (Indiana kindred) is an 11-kd fragment of prion protein with an *N*-terminal glycine at codon 58. *EMBO J.* **10** 513–519.

Taraboulos, A., Rogers, M., Borchelt, D.R., McKinley, M.P., Scott, M., Serban, D. and Prusiner, S.B. (1990a) Acquisition of protease resistance by prion proteins in scrapie-infected cells does not require asparagine-linked glycosylation. *Proc. Natl. Acad. Sci. USA* **87** 8262–8266.

Taraboulos, A., Serban, D. and Prusiner, S.B. (1990b) Scrapie prion proteins accumulate in the cytoplasm of persistently-infected cultured cells. *J. Cell Biol.* **110** 2117–2132.

Taraboulos, A., Raeber, A., Borchelt, D., McKinley, M.P. and Prusiner, S.B. (1991) Brefeldin A inhibits protease resistant prion protein synthesis in scrapie-infected cultured cells. *FASEB J.* **5** A1177.

Tateishi, J., Kitamoto, T., Doh-ura, K., Sakaki, Y., Steinmetz, G., Tranchant, C., Warter, J.M. and Heldt, N. (1990) Immunochemical, molecular genetic, and transmission studies on a case of Gerstmann–Sträussler–Scheinker syndrome. *Neurology* **40** 1578–1581.

Turk, E., Teplow, D.B., Hood, L.E. and Prusiner, S.B. (1988) Purification and properties of the cellular and scrapie hamster prion proteins. *Eur. J. Biochem.* **176** 21–30.

Weissmann, C. (1991a) A "unified theory" of prion propagation. *Nature (London)* **352** 679–683.

Weissmann, C. (1991b) Spongiform encephalopathies—the prion's progress. *Nature (London)* **349** 569–571.

Weitgrefe, S., Zupancic, M., Haase, A., Chesebro, B., Race, R., Frey, W., II, Rustan, T. and Friedman, R.L. (1985) Cloning of a gene whose expression is increased in scrapie and in senile plaques. *Science* **230** 1177–1181.

Wells, G.A.H., Scott, A.C., Johnson, C.T., Gunning, R.F., Hancock, R.D., Jeffrey, M., Dawson, M. and Bradley, R. (1987) A novel progressive spongiform encephalopathy in cattle. *Vet. Rec.* **121** 419–420.

Westaway, D., Goodman, P.A., Mirenda, C.A., McKinley, M.P., Carlson, G.A. and Prusiner, S.B. (1987) Distinct prion proteins in short and long scrapie incubation period mice. *Cell* **51** 651–662.

Westaway, D., Mirenda, C.A., Foster, D., Zebarjadian, Y., Scott, M., Torchia, M., Yang, S.-L., Serban, H., DeArmond, S.J., Ebeling, C., Prusiner, S.B. and Carlson, G.A. (1991) Paradoxical shortening of scrapie incubation times by expression of prion protein transgenes derived from long incubation period mice. *Neuron* **7** 59–68.

Wexler, N.S., Young, A.B., Tanzi, R.E., Travers, H., Starosta-Rubinstein, S., Penney, J.B., Snodgrass, S.R., Shoulson, I., Gomez, F., Ramos Arroyo, M.A., Penchaszadeh, G.K., Moreno, H., Gibbons, K., Faryniarz, A., Hobbs, W., Anderson, M.A., Bonilla, E., Conneally, P.M. and Gusella, J.F. (1987) Homozygotes for Huntington's disease. *Nature (London)* **326** 194–197.

Wilesmith, J. and Wells, G.A. H. (1991) Bovine spongiform encephalopathy. *Curr. Top. Microbiol. Immunol.* **172** 21–38.

Wilesmith, J.W., Wells, G.A.H., Cranwell, M.P. and Ryan, J.B. M. (1988) Bovine spongiform encephalopathy: epidemiological studies. *Vet. Rec.* **123** 638–644.

Winter, M.H., Aldridge, B.M., Scott, P.R. and Clarke, M. (1989) Occurrence of 14 cases of bovine spongiform encephalopathy in a closed dairy herd. *Br. Vet. J.* **145** 191–194.

Yost, C.S., Lopez, C.D., Prusiner, S.B., Meyers, R.M. and Lingappa, V.R. (1990) A non-hydrophobic extracytoplasmic determinant of stop transfer in the prion protein. *Nature (London)* **343** 669–672.

45

Prions and neurodegenerative diseases

John Collinge and Stanley B. Prusiner

'Neurodegenerative diseases' is a broad term used to describe many chronic progressive neurologic disorders of the central nervous system (CNS) in which a degenerative process is found upon clinical and neuropathologic evaluation. The aetiologies of most neurodegenerative diseases are unknown. The most common of these neurodegenerative disorders is Alzheimer's disease, followed by Parkinson's disease and amyotrophic lateral sclerosis (ALS). Kuru, Creutzfeldt–Jakob disease and Gerstmann–Sträussler–Scheinker syndrome are all prion diseases and are also classified as neurodegenerative diseases. These three diseases illustrate the infectious, sporadic, and genetic manifestations of the prion disorders. Huntington's disease is also a neurodegenerative disease, but it is only manifest as a genetic disorder with an autosomal dominant pattern of inheritance. On the other hand, Alzheimer's disease, like the prion diseases of humans, is largely a sporadic disorder with only about 10% to 15% of individuals clearly identified as members of families with the disease.

The availability of DNA markers of inherited prion diseases has enabled a re-evaluation of the phenotypic range of these disorders. Previous diagnostic criteria, whether based on clinicopathological descriptions or transmissibility, are by definition restrictive; the clinical and pathological phenotypes can now be directly examined and defined, non-transmissible types could well be demonstrated. Such findings challenge the traditional classification system of neurodegenerative diseases. It is remarkable that the prion diseases have provided the first challenge to our traditional thinking. Molecular genetics may lead to a fundamental reclassification of all neurodegenerative diseases as further aetiological markers become available. As occurred with prion diseases, the identification of genetic mutations in the rarer familial forms may cast considerable light on the aetiology of the much commoner sporadic forms of these diseases.

In the closing pages of a book on the prion diseases, it is reasonable to ask where studies of these disorders will take us in the future, how fast they will proceed, and whether they will have significant implications for the more common neurodegenerative diseases. Our ability to see into the future of a scientific discipline is very poor. It is doubtful that anyone could have predicted where the study of prion diseases has moved over the past ten years. Indeed, a decade ago it is doubtful whether anyone could have predicted the course of events that has brought us to our present state of knowledge and understanding in research on prions.

Viewed in the most broad perspective, the study of prion diseases may open up new vistas in our understanding of a wide variety of degenerative processes, both those within the CNS and those afflicting other organs. Will we learn that prion diseases have some important parallels with such disorders as adult onset diabetes? Will events in the pathogenesis of the prion diseases tell us about how autoimmune disorders are initiated? Alternatively, will the prion diseases give us new insights into such perplexing diseases as multiple sclerosis and Guillain Barre syndrome? Only further research will answer these questions.

While the parallels between the prion diseases, Alzheimer's disease, Parkinson's disease and ALS are striking both clinically and neuropathologically, we cannot predict whether insights gained from the studies of prion diseases will significantly alter our understanding of these common neurodegenerative disorders. Only further research on the prion diseases and this trio of common neurodegenerative disorders will help us shed our ignorance and bring an understanding of the aetiologies of these diseases.

As the details of the molecular pathogenesis of the prion diseases are revealed by further research and we learn about the molecular structure of the prion particle including the transformation of PrP^C into PrP^{Sc}, it should be possible to design highly effective and specific pharmacological reagents. Because the prion diseases are unprecedented, it seems likely that the therapeutic reagents required to prevent, alleviate or ameliorate these disorders may be unique. If indeed that proves to be the case, then these drugs will certainly be candidates for empirical testing in the trio of more common neurodegenerative diseases. While it seems quite unlikely that the prion protein is involved in the pathogenesis of Alzheimer's disease, Parkinson's disease or ALS, there may be other proteins which undergo a similar post-translational transformation to that involved in the conversion of PrP^C to PrP^{Sc} which features in the prion diseases. Identifying those proteins which change may not be easy, but might be greatly facilitated once we understand the process that converts PrP^C to PrP^{Sc}.

The study of prions may lead us to uncover new mechanisms previously unknown in cell biology. Certainly, the process by which PrP^C is converted to PrP^{Sc} is likely not to be confined to prion proteins. Whether this conversion process will have more widespread implications for the normal cell or be featured only in disease processes remains to be determined.

Since prion diseases generally occur later in life, it may be that the study of these disorders will give us new insights into the aging process, in particular processes whereby neuronal cells develop, mature, transmit signals for decades, and then ultimately undergo senescence.

Index